THE COLOR LINE
AND THE QUALITY OF LIFE
IN AMERICA

THE COLOR LINE
AND THE QUALITY OF LIFE
IN AMERICA

Reynolds Farley
Walter R. Allen

for the
National Committee for Research
on the 1980 Census

Oxford University Press
New York   Oxford
1989

Oxford University Press

Oxford New York Toronto
Delhi Bombay Calcutta Madras Karachi
Petaling Jaya Singapore Hong Kong Tokyo
Nairobi Dar es Salaam Cape Town
Melbourne Auckland

and associated companies in
Berlin Ibadan

Copyright © 1987 by Russell Sage Foundation

First published in 1987 by Russel Sage Foundation, New York
First issued in paperback in 1989 by Oxford University Press, Inc.,
200 Madison Avenue, New York, New York 10016
Reprinted by arrangement with author

Oxford is a registered trademark of Oxford University Press

Library of Congress Cataloging-in-Publication Data

Farley, Reynolds, 1938–
    The color line and the quality of life in America / Reynolds
Farley, Walter R. Allen for the National Committee for Research on
the 1980 Census.
        p. cm.
    Bibliography: p.
    Includes index.
    ISBN 0-19-506029-6
    1. Afro-Americans—Economic conditions. 2. Afro-Americans—Social
conditions—1975– 3. Afro-Americans—Population. 4. Quality of
life—United States. I. Allen, Walter Recharde. II. National
Committee for Research on the 1980 Census. III. Title.
E185.8.F36 1989
305.8'96073—dc19   88-37161
                                                             CIP

Printing (last digit): 9 8 7 6 5 4 3 2 1

Printed in the United States of America
on acid-free paper

The National Committee for Research on the 1980 Census

The committee is sponsored by the Social Science Research Council, the Russell Sage Foundation, and the Alfred P. Sloan Foundation, in collaboration with the U.S. Bureau of the Census. The opinions, findings, and conclusions or recommendations expressed in the monographs supported by the committee are those of the author(s) and do not necessarily reflect the views of the committee or its sponsors.

Foreword

The Color Line and the Quality of Life in America is one of an ambitious series of volumes aimed at converting the vast statistical yield of the 1980 census into authoritative analyses of major changes and trends in American life. This series, "The Population of the United States in the 1980s," represents an important episode in social science research and revives a long tradition of independent census analysis. First in 1930, and then again in 1950 and 1960, teams of social scientists worked with the U.S. Bureau of the Census to investigate significant social, economic, and demographic developments revealed by the decennial censuses. These census projects produced three landmark series of studies, providing a firm foundation and setting a high standard for our present undertaking.

There is, in fact, more than a theoretical continuity between those earlier census projects and the present one. Like those previous efforts, this new census project has benefited from close cooperation between the Census Bureau and a distinguished, interdisciplinary group of scholars. Like the 1950 and 1960 research projects, research on the 1980 census was initiated by the Social Science Research Council and the Russell Sage Foundation. In deciding once again to promote a coordinated program of census analysis, Russell Sage and the Council were mindful not only of the severe budgetary restrictions imposed on the Census Bureau's own publishing and dissemination activities in the 1980s, but also of the extraordinary changes that have occurred in so many dimensions of American life over the past two decades.

The studies constituting "The Population of the United States in the 1980s" were planned, commissioned, and monitored by the National Committee for Research on the 1980 Census, a special committee appointed by the Social Science Research Council and sponsored by the Council, the Russell Sage Foundation, and the Alfred P. Sloan Foundation, with the collaboration of the U.S. Bureau of the Census. This committee includes leading social scientists from a broad range of fields—demography, economics, education, geography, history, political science, sociology, and statistics. It has been the committee's task to select the main topics for research, obtain highly qualified specialists to carry out that research, and provide the structure necessary to facilitate coordination among researchers and with the Census Bureau.

The topics treated in this series span virtually all the major features of American society—ethnic groups (blacks, Hispanics, foreign-born); spatial dimensions (migration, neighborhoods, housing, regional and metropolitan growth and decline); and status groups (income levels, families and households, women). Authors were encouraged to draw not only on the 1980 Census but also on previous censuses and on subsequent national data. Each individual research project was assigned a special advisory panel made up of one committee member, one member nominated by the Census Bureau, one nominated by the National Science Foundation, and one or two other experts. These advisory panels were responsible for project liaison and review and for recommendations to the National Committee regarding the readiness of each manuscript for publication. With the final approval of the chairman of the National Committee, each report was released to the Russell Sage Foundation for publication and distribution.

The debts of gratitude incurred by a project of such scope and organizational complexity are necessarily large and numerous. The committee must thank, first, its sponsors—the Social Science Research Council, headed until recently by Kenneth Prewitt; the Russell Sage Foundation, under the direction of president Marshall Robinson; and the Alfred P. Sloan Foundation, led by Albert Rees. The long-range vision and day-to-day persistence of these organizations and individuals sustained this research program over many years. The active and willing cooperation of the Bureau of the Census was clearly invaluable at all stages of this project, and the extra commitment of time and effort made by Bureau economist James R. Wetzel must be singled out for special recognition. A special tribute is also due to David L. Sills of the Social Science Research Council, staff member of the committee, whose organizational, administrative, and diplomatic skills kept this complicated project running smoothly.

The committee also wishes to thank those organizations that contributed additional funding to the 1980 Census project—the Ford Foundation and its deputy vice president, Louis Winnick, the National Science Foundation, the National Institute on Aging, and the National Institute of Child Health and Human Development. Their support of the research program in general and of several particular studies is gratefully acknowledged.

The ultimate goal of the National Committee and its sponsors has been to produce a definitive, accurate, and comprehensive picture of the U.S. population in the 1980s, a picture that would be primarily descriptive but also enriched by a historical perspective and a sense of the challenges for the future inherent in the trends of today. We hope our readers will agree that the present volume takes a significant step toward achieving that goal.

CHARLES F. WESTOFF

Chairman and Executive Director
National Committee for Research
on the 1980 Census

Acknowledgments

Financial support for this research was provided by the National Committee for Research on the 1980 Census and by the Cornerhouse Fund. We thank Charles Y. Glock for his assistance in arranging the Cornerhouse monies. David L. Sills of the Social Science Research Council was extremely helpful on both the scholarly and financial aspects of this project. Grants from the C. S. Mott Foundation and the Ford Foundation to support related research projects also contributed to the completion of this book.

We appreciate the reviewers whose criticisms helped shape the final version. In particular, Charles F. Westoff, Chairman and Executive Director of the National Committee for Research on the 1980 Census, read and commented on several drafts. William H. Sewell, who chaired the advisory panel for this study, made enlightening suggestions, particularly with regard to which topics should be included or deleted. Patricia Berman, the representative of the Bureau of the Census on this panel, provided numerous helpful suggestions. We also thank the other members of this panel, Karl Taeuber and Harold Rose. Herbert Morton provided editorial guidance for parts of this manuscript.

Many colleagues read, discussed, and commented on individual chapters. We thank Suzanne Bianchi, Nesha Hanniff, Bruce Hare, Robert Hill, Stanley Lieberson, Cora Marrett, Aldon Morris, John Reid, Howard Schuman, Margaret Spenser, Daphne Spain, William Wilson, and Robin Williams for their suggestions. Commentaries and ideas were also provided by Christopher Davis, Michael Sawson, Donald Deskins, Richard English, Darnell Hawkins, Masipula Sithole, A. Wade Smith, Gail Thomas, and Ernest Wilson.

Albert Hermalin, director of the University of Michigan's Population Studies Center, provided an extremely supportive research environment along with encouragement that facilitated completion of the book. The work could not have been executed without the extraordinarily competent assistance of the research staff of this Center. We thank Albert Anderson, Michael Coble, and Lisa Neidert for their help in preparing the extract files from the Census Bureau's public use samples and in analyzing these data. Amy Hsu aided in the preparation of graphs and figures. Judy Mullin entered the entire text and tables and made numerous revisions with the steadiness and good humor that assured completion of the project. We have special thanks for Kathleen Duke, the editor at the Population Studies Center. Her never-failing efforts dramatically improved the presentation of our findings.

Encouragement and assistance from Niara Sudarkasa and Thomas Holt, successive directors of the University of Michigan's Center for Afro-American and African Studies, were also essential. In addition, the University of Zimbabwe's Center for Applied Social Sciences, directed by Marshall Murphree, provided facilities to Walter Allen while he wrote several chapters.

Finally, Priscilla Lewis and Dianne Garda of the Russell Sage Foundation guided this manuscript into a book. Skillful copy editing was provided by Sylvia Newman, and Ian Tucker compiled an impressively complete index.

Primary responsibility for the first and last chapters of this monograph and for the chapters on family and education belong to Walter Allen; Chapter 12 was written collaboratively; and Reynolds Farley authored the remaining chapters. It is hoped that this extensive review of demographic and socioeconomic changes will provide a useful reference for readers interested in the status of blacks in the United States a century and a quarter after Emancipation.

REYNOLDS FARLEY
WALTER R. ALLEN

Contents

List of Tables   *xv*
List of Figures   *xix*
List of Appendix Tables   *xxiii*

1   RACE IN AMERICA: The Dilemma Continues   *1*

2   BLACK POPULATION GROWTH FROM COLONIAL TIMES TO
        WORLD WAR II   *7*

3   AN ANALYSIS OF MORTALITY: 1940 to the Present   *32*

4   FERTILITY TRENDS: 1940 to the Present   *58*

5   THE REDISTRIBUTION OF THE BLACK POPULATION AND
        RESIDENTIAL SEGREGATION   *103*

6   BLACK FAMILY, WHITE FAMILY: A Comparison of Family
        Organization   *160*

7   THE SCHOOLING OF AMERICA: Black-White Differences in
        Education   *188*

8   EMPLOYMENT   *209*

9   RACIAL DIFFERENCES IN OCCUPATIONAL ACHIEVEMENT   *256*

10 PERSONAL INCOME   283

11 THE EARNINGS OF EMPLOYED WORKERS   316

12 RACE, ANCESTRY, AND SOCIOECONOMIC STATUS: Are West
     Indian Blacks More Successful?   362

13 A WORLD WITH NO COLOR LINE: Race and Class in Twenty-
     First Century America   408

     Bibliography   439
     Name Index   469
     Subject Index   477

List of Tables

2.1	Death Rates, Standardized for Age, in the Death Registration States of 1900 and 1910	27
3.1	Death Rates by Race, Age, and Sex, 1940–1984	38
3.2	Death Rates by Cause for Black and White Men and Women, 1950, 1970, and 1983	42
3.3	Expectation of Life for Black and White Men and Women, 1940–1980	53
4.1	Birthrates for Married and Unmarried Black and White Women, 1940–1984	79
4.2	Differential Fertility Among White and Black Women Aged 18–34 and 35–49, 1980	90
4.3	Net Effects on Fertility of Social and Economic Factors for Ever-Married Black and White Women Aged 18–34 and 35–49, 1980	95
4.4	Net Effects on Fertility of Social and Economic Factors for Married-Spouse-Present Black and White Women Aged 18–34 and 35–49, 1980	100
5.1	Black Out-Migration from the South, 1870–1970	113
5.2	Black and White Interregional Migration, 1965–1980	118
5.3	Characteristics of Black and White Male Migrants and Nonmigrants Aged 25–64, 1980	124
5.4	Proportion of Black and White Men in a Region in 1975 Who Lived in the Other Region in 1980, Classified by Region of Birth, Educational Attainment, and Age	127
5.5	Estimated Proportion of Black Men Who Lived in One Region in 1975 and Moved to the Other Region by 1980, Classified by Education, Age, and Region of Birth	128

5.6   Indexes of Racial Residential Segregation for the 25
        Central Cities with Largest Black Populations   141
5.7   Indexes of Racial Residential Segregation for
        Metropolitan Areas with 250,000 or more Black
        Residents in 1980   143
5.8   Indexes of the Residential Segregation of Blacks,
        Hispanics, and Asians from Non-Hispanic Whites
        for Metropolitan Areas, 1980   145
5.9   Indexes of the Residential Segregation of Blacks and
        Selected Ethnic Groups from the English Ethnic
        Group for Metropolitan Areas, 1980   147
5.10   Indexes of Racial Residential Segregation,
        Controlling for Income and Education, for
        Metropolitan Areas, 1980   149

6.1   Household Type by Race and Household
        Income, 1980   174
6.2   Household Type by Race and Employment Status of
        Householder, 1980   175
6.3   Child's Age by Race and Family Income, 1980   176
6.4   Household Extendedness by Race and Family
        Income, 1980   178
6.5   Multiple Classification Analysis of Per Capita
        Income of Households in 1980, by Race   182

7.1   Educational Attainment by Race for the Population
        Aged 16 and Over, 1980   192
7.2   Educational Attainment by Race and Sex for the
        Population Aged 16 and Over, 1980   193
7.3   Educational Attainment by Race and Age for the
        Population Aged 16 and Over, 1980   194
7.4   Educational Attainment by Race and Region for the
        Population Aged 16 and Over, 1980   196
7.5   Educational Attainment for Household Heads Aged
        16 and Over, 1980   197
7.6   Educational Attainment by Race and Employment
        Status, 1980   199
7.7   Educational Attainment by Race and Occupation for
        Persons Aged 16 and Over, 1980   200
7.8   Educational Attainment of Household Head by Race
        and Annual Household Income, 1980   202
7.9   School Attendance by Race and Type of School for
        Persons Aged 4–15, 1980   203
7.10   Scholastic Aptitude Test Scores by Race, 1983   205

7.11   Scholastic Aptitude Test Scores by Race and Family
       Socioeconomic Status, 1983   *206*
7.12   Changes in Enrollment and Racial Segregation in
       School Districts of Largest U.S. Cities, by Region,
       1967–1978   *207*
8.1   Change in Unemployment Rate and Change in
       Proportion Out of the Labor Force Associated
       with a 1 Percent Change in the Gross National
       Product, 1950–1951 to 1984–1985   *216*
8.2   Reported Activities of Persons Aged 16–24 and 25–54
       Who Were Outside the Labor Force, by Sex and Race,
       1967 and 1985   *221*
8.3   Employment Information for the Black and White
       Population Aged 25–54, 1980   *226*
8.4   Employment Information for the Black and White
       Population Aged 16–24, 1980   *238*
9.1   Occupational Distributions by Race and Sex,
       1940–1980   *264*
9.2   Earnings by Occupation in 1979, Occupational
       Distributions, and Employment by Occupation by
       Race and Sex, 1970–1986   *272*
9.3   Percentage of Employed Workers in Executive,
       Administrative, Managerial, and Professional
       Specialty Occupations and Index of Occupational
       Dissimilarity for Black and White Men and Women
       Aged 25–64, 1980   *276*
10.1   Sources and Amounts of Personal Income for Blacks
       and Whites Aged 15 and Over by Sex, 1984   *287*
10.2   Households Owning Assets and Mean Value of
       Holdings by Race, 1984   *289*
10.3   Median Net Worth of Households by Race, Monthly
       Income, and Type, 1984   *290*
10.4   Median Income for Black and White Adults Who
       Received Income, 1939–1985   *298*
10.5   Per Capita Income for the Adult Population by Race
       and Sex, 1949–1985   *300*
10.6   Median Income by Region for Black and White Men
       and Women Who Received Income, 1939–1984   *302*
10.7   Median Income by Educational Attainment for Black
       and White Men and Women Who Received Income,
       1949–1984   *305*
10.8   Median Income by Age for Black and White Men and
       Women Who Received Income, 1949–1984   *307*

11.1   Hourly Wage Rates in 1979 for Full-Time, Year-
       Round Workers in Selected Occupations with
       Specified Educations, by Race, Sex, and Age   *318*
11.2   National Studies of Racial Differences in the
       Earnings or Income of Employed Blacks and
       Whites   *322*
11.3   Characteristics of Black and White Men and Women
       Aged 25–64 in 1980 by Hours of Employment
       in 1979   *330*
11.4   Average Hourly Wage Rate, Average Hours of
       Employment, and Average Annual Earnings for
       Blacks and Whites Aged 25–34 in 1980   *352*
12.1   Leading Places of Birth of the Foreign-Born
       Population Classified by Race, 1980   *374*
12.2   Distribution of Foreign-Born Population by Date of
       Arrival in the United States and Proportion
       Citizens, 1980   *376*
12.3   Distribution of the Population by Race, Nativity,
       and State or Division of Residence, 1980   *380*
12.4   Proportion of Population Living in Selected
       Metropolitan Areas for the Foreign-Born and Native,
       Classified by Race, 1980   *381*
12.5   Proportion Speaking a Language Other Than English
       in Their Home, by Nativity and Race, 1980   *383*
12.6   Age Distribution and Median Age of the Population
       by Race and Nativity, 1980   *385*
12.7   Males per 1,000 Females for the Population by Race
       and Nativity, 1980   *386*
12.8   Occupational Distribution of Employed Men and
       Women by Race, Sex, and Nativity, 1980   *398*
12.9   Estimated Annual Earnings in 1979 for Men Aged 40
       Who Worked Full Time for Entire Year, Specific for
       Occupation, Education, Place of Residence, and
       Race-Nativity   *402*
12.10 Average Income of Families in 1979 by Race, Type,
       and Nativity   *404*

List of Figures

2.1   Total and Black Populations and Proportion Black,
      1790–1986   10
2.2   Average Annual Growth Rates of White and Black
      Populations, 1790–1986   12
2.3   Children Under Age 5 per 1000 Women Aged 15–49,
      1850–1985   17
2.4   Total Fertility Rates of the Black and White
      Populations, 1860–1984   19
2.5   Expectations of Life at Birth by Race and Sex,
      1900–1984   26
3.1   Crude and Age-Adjusted Death Rates for the Black
      Population, 1940–1985   34
3.2   Age-Adjusted Death Rates by Race and Sex,
      1940–1985   36
3.3   Years Added to the Life Span with the Elimination of
      a Disease, 1982   46
3.4   Infant Mortality Rates by Race, 1940–1984   48
4.1   Trends in the Total Fertility Rate and Net
      Reproduction Rate by Race, 1940–1984   62
4.2   Completed Fertility for Birth Cohorts of Black and
      White Women, Aged 40–44   64
4.3   Trends in Fertility Rates by Age of Women, for
      Blacks and Whites, 1940–1984   66
4.4   Proportion of White and Nonwhite Women Childless
      at Ages 20–24 and 30–34, 1940–1982   69
4.5   Trends in Percent of Total Births Delivered to
      Unmarried Women by Race, 1940–1984   78
4.6   Components of Change in the Percentage of Births
      Out of Wedlock for Blacks and Whites, 1960–1984   81

4.7   Methods of Contraception Used by Ever-Married
      Black and White Women Aged 15–44 at Risk of
      Becoming Pregnant, 1965 and 1982   86
5.1   Distribution of the Black and White Population of
      the United States by Region, 1790–1985   105
5.2   Percent of Population Black for the United States
      and Regions, 1790–1985   108
5.3   Migration Rates for Blacks and Whites in the Four
      Regions by Age, 1960–1970 and 1970–1980   120
5.4   Estimated Earnings in 1979 for Black and White Men
      Aged 35, Classified by Region of Birth and Region
      of Residence in 1980   130
5.5   Proportion of Black and White Population Living in
      Urban Places for the United States and its
      Regions, 1870–1980   135
6.1   Types of Household by Race, 1980   169
8.1   Proportion of Labor Force Unemployed by Race, Sex,
      and Age, 1950–1985   214
8.2   Trends in Proportion of Population Out of the Labor
      Force by Race, Sex, and Age, 1950–1985   219
8.3   Proportion of Total Population at Work by Race, Sex,
      and Age, 1950–1985   223
8.4   Estimated Unemployment Rates in April 1980 for
      Blacks and Whites, Aged 25 to 34, by Education and
      Marital Status   230
8.5   Estimated Hours of Employment in 1979 for Black
      and White Men Living in the South by Educational
      Attainment, Age, and Marital Status   232
8.6   Estimated Unemployment Rates for Persons Aged 20
      and 21 by Race, Sex, Educational Attainment,
      and Enrollment, 1980   240
9.1   Proportion of Employed Workers Holding
      Professional, Technical, Managerial, and
      Administrative Jobs, 1958–1982   268
9.2   Measures of the Similarity of the Occupations of
      Whites and Nonwhites and of Men and Women,
      1958–1979   269
9.3   Proportion of Labor Force Participants Holding
      Executive, Managerial, or Professional Speciality
      Occupations by Race, Sex, Age, and Educational
      Attainment, 1980   279

10.1   Median Income for Black and White Men and
Women Aged 14 and Over Who Reported Income,
1948–1985   *296*

10.2   Black Median Income as a Percentage of White and
Racial Differences in Income by Age for Birth
Cohorts   *310*

10.3   Median Incomes for Income Recipients by Age for
Black and White Men and Women, 1980   *312*

11.1   Earnings of Black and White Men and Women
Aged 25–64, Reported in Censuses of 1960 and 1980   *332*

11.2   Rates of Return Associated with Labor Market
Characteristics   *333*

11.3   Change in Annual Earnings, 1960 to 1980, and
Components of Change   *336*

11.4   Actual and Expected Annual Earnings of Blacks as a
Percentage of the Actual Earnings of Whites,
1960 and 1980   *338*

11.5   Actual and Expected Annual Earnings of Black
Women as a Percentage of the Actual Earnings of
White Women in the Same Marital Status, 1980   *341*

11.6   Actual and Expected Annual Earnings of Black Men
and Women as a Percentage of the Actual
Earnings of Whites, by Region, 1960 and 1980   *344*

11.7   Actual and Expected Annual Earnings of Black Men
and Women as a Percentage of the Actual Earnings
of Whites, by Educational Attainment, 1960
and 1980   *347*

11.8   Actual and Expected Annual Earnings of Black Men
and Women as a Percentage of the Actual Annual
Earnings of Whites, by Age, 1960 and 1980   *349*

11.9   Actual and Expected Earnings of Black Men and
Women and Actual Earnings of Whites Aged 25–64;
1970, 1980, and 1985   *354*

12.1   Distribution of Immigrants by Country of Origin,
1820–1965 and 1966–1979   *369*

12.2   Composition of the Native- and Foreign-Born
Population, 1980   *373*

12.3   Distribution of the Foreign-Born Population by Race
and Date of Arrival, 1980   *377*

12.4   Marital Status of the Population Aged 15 and Over,
by Race, Sex, and Nativity, 1980   *389*

12.5   Educational Attainment for the Population Aged 25
          and Over, Standardized for Age and Classified by
          Race and Nativity, 1980   *392*
12.6   Labor Force Status of the Population Aged 16 and
          Over, Standardized for Age and Classified by Race
          and Nativity, 1980   *395*

List of Appendix Tables

5.1   Effects Parameters from a Logit Model Which
Treats the Log of the Odds of Interregional
Migration as the Dependent Variable for Black and
White Men Aged 25 and Over, Classified by
Region of Residence in 1975   *158*

5.2   Regression of Log of Earnings in 1979 on Age,
Age-Squared, Educational Attainment, and Hours of
Employment for Black and White Men Aged 25–64,
Classified by Region of Birth and Region of Residence   *159*

8.1   Ratio of a Group's Unemployment Rate to That of
White Men Aged 25–54 and Time Trend, 1950–1985   *251*

8.2   Average Annual Change in the Proportion of the
Population Out of the Labor Force, 1950–1985   *252*

8.3   Analysis of Determinants of Unemployment in
April 1980 for Black and White Men and Women
Aged 25–54   *252*

8.4   Regression of Hours of Employment in 1979 on
Years of Education, Age, Age-Squared, Region,
Marital Status, Work Limitation, and Presence of
Child Under Age 6 for Blacks and Whites Who Were
Employed in 1979   *254*

8.5   Analysis of Determinants of Unemployment Among
Men and Women Aged 16–24, 1980   *255*

9.1   Effects Parameters for Logit Model Which Treats the
Odds of Employment in Executive, Administrative,
Managerial, and Professional Specialty Occupations
for Labor Force Participants, by Race and Sex,
1980   *282*

10.1   Comparison of Income Data Gathered in the March
         1980 Current Population Survey and the April 1980
         Census for the Population Aged 15 and Over   *315*
11.1   Mean Value of Variables and Parameters of
         Regression Equations for Earnings of Black and White
         Men and Women, 1960 and 1980   *359*
11.2   Sample Size and Deletions   *360*
11.3   Actual Average Annual Earnings of Whites, and
         Actual and Expected Average Annual Earnings of
         Blacks, Classified by Region, Educational Attainment,
         and Age, 1960 and 1980   *361*
12.1   Regression of Log of Hourly Wage Rate on
         Explanatory Variables for Men Aged 25–64 Employed
         in 1979, Classified by Race and Nativity   *406*
A.1   Net Census Undercount by Age, Sex, and Race,
         1950–1980   *423*
A.2   Racial, Age, and Marital Status Characteristics of the
         Enumerated, Substituted, Allocated, and Total
         Population in 1980   *427*
A.3   Labor Force, Educational, Occupational, and Income
         Characteristics of the Population Before and After
         Allocations, 1980   *432*
A.4   Comparison of the Regional, Age, and Marital Status
         Distributions of the Black and White
         Populations, 1980, from Complete Count Data, the
         Census Bureau's Sample Data, and the Sample
         Drawn for This Monograph   *435*
A.5   Comparison of Years of Schooling, Occupation, and
         Household Income by Race, 1980, from the Census
         Bureau's Sample Drawn for this Monograph   *436*

THE COLOR LINE
AND THE QUALITY OF LIFE
IN AMERICA

RACE IN AMERICA:
THE DILEMMA CONTINUES

The problem of the twentieth century is the problem of the color-line,—the relation of the darker to the lighter races of men in Asia and Africa, in America and the islands of the sea.[1]

TWENTY years ago the United States was a country in flames. Fires flickered from Paterson, New Jersey, to Detroit, Michigan, to Harlem in New York City. Racial conflict was at the heart of this national conflagration; over 50 cities burned. The powder keg that touched off a year of "Burn, Baby, Burn" was a police incident in a black community—Watts, Los Angeles—on a hot, muggy night, August 11, 1965. However, the fuse leading to the eventual explosion had been lit and burning for generations.

The 1960s provided a frightening glimpse of a dismal American future. That future was characterized by war between the races; an aroused, largely urban black population was in armed conflict with the largely white forces of social control—the police, the National Guard, and the Army. The potential scenario seemed certain to parallel the conflicts between Catholics and Protestants in Northern Ireland, between Hindus and Muslims in India, and between Christians and Muslims in Lebanon. The United States seemed destined for the kind of black–white conflict characteristic of South Africa—a level of conflict capable of destroying our society. Yet somehow the United States managed to avoid the ravages of such internecine struggle; the flames were doused, or so it seemed.

[1]Dubois (1961), p. 23.

In the summer of 1980, fifteen years after Watts, Liberty City in Miami, Florida, exploded. Once again, the specter of irreconcilable differences confronted the country. Television news commentaries, newspaper editorials, articles in scholarly journals, and everyday conversation over coffee or beer all asked the same questions: How could this happen in the United States? Hadn't a successful "war on poverty" been waged? Wasn't the United States a model for the rest of the world in its founding ethic of equality among men? Didn't the United States provide avenues for upward mobility to all its citizens without prejudice? The reality of an aggrieved, militant minority— urban blacks—presented a palpable contradiction to the ideal of an egalitarian society.

This book is about race and the difference race makes in the lives of Americans. Our question is a straightforward one: Is the United States indeed a "nation divided by color"?—as the Report of the National Advisory Commission on Civil Disorders concluded after an exhaustive study of the conditions and circumstances leading up to the urban warfare of the 1960s.[2] We wish to establish the extent to which racial identity influences opportunities and outcomes for Americans. Looking at census data from 1980 and earlier, we compare the relative statuses of blacks and whites in this country. Do blacks and whites have comparable educational attainment? Have blacks achieved economic parity with whites? Is the racial difference in mortality decreasing or increasing? To what extent do the two races live in the same neighborhoods? Has the legacy of black deprivation growing out of legal slavery and perpetuated by continuing racial discrimination in the society been overcome?

Race and Economic Status in America

Racial and ethnic competition has been a hallmark of American society from the moment the first European settlers arrived to discover an established Native American presence. Periodically, this competition has spilled over into violent conflict. Whether the struggle occurred in political patronage systems, in labor-organizing halls, in university classrooms, or in the street, there has been a long history of struggle between racial and ethnic groups. The Irish, West Indians, Germans, Chinese, Cubans, Italians, Mexicans, Native Americans, and Afro-Americans have all struggled among themselves and with each other over how societal wealth and privileges were to be distributed. So ax-

[2]National Advisory Commission on Civil Disorders (1968).

iomatic was strife along racial lines that the noted scholar W. E. B. Du-Bois was prompted to conclude that "the color line is the problem of the twentieth century."[3]

Despite the long tradition of ethnic and racial conflict in our society, we are without general models which effectively summarize race relations.[4] The absence of such a general model of race relations should not be construed, however, as a failure to study the problem, for race and race relations has been one of the most exhaustively analyzed problems in American scholarship. Rather, the complexity of the subject has impeded the development of generally applicable models. Majority and minority group relations, specifically relations between blacks and whites, represent the complex intertwining of historical, economic, political, social, and cultural factors. Thus, any serious study of these relationships must attend to the broad range of factors that affect relations between blacks and whites.

This book began as a descriptive study of whites and blacks; its task was to compare the relative statuses of the two groups. Over time, the book evolved into a more broadly defined examination of the nexus between economic status and racial identity in American society. Time and time again, as we compared the status of whites and blacks in education, fertility, employment, earnings, and family structure, the power of economic factors was observed. Economic status was found to make an important difference, at points exceeding racial identity as the primary explanatory factor.

Economic status is critical in the lives of Americans generally and in the lives of black Americans in particular. This should come as no surprise given the nature of our society. In the United States the quality of a person's life is often closely related to available economic resources. Modifying the aphorism that "you are what you eat," we can conclude in this instance that "you are what you can afford." Our society is founded upon a market economy; as such, the "good life" is a commodity to be bought and sold. The basic essentials, such as food, clothing, and shelter, and the elaborate enhancements, such as sporty cars, ocean cruises, and private school educations, are both allocated in relation to purchasing power. Even physical health and well-being is influenced by a person's economic standing.

Against this backdrop of the importance of economics in the lives of Americans, it is useful to consider briefly the economic histories of most Americans. Long viewed as the land of opportunity, the United States has been a place for dreamers who aspire to great heights. The dream has generally been put in the form of an Horatio Alger "rags to

[3]DuBois (1944), p. 23.
[4]Frazier (1966) chap. 4; Blalock (1967).

riches" saga where a person made a fortune or achieved success by the sweat of his brow. It matters little that more people failed (or succeeded modestly) than duplicated Alger's remarkable feats. The point was the dream itself and the possibilities held out for the future.

If the dream of success has not always availed itself to individuals, it has been real for different racial and ethnic groups in the United States.[5] Thus, people fleeing persecution, famine, or grinding poverty in their home countries have managed to find a bright alternative in the United States. They have risen to successful positions in the health fields, industry, education, politics, and commerce. In fact, most immigrant groups in this country have claimed great forward progress after several generations.[6]

Common perceptions to the contrary, black Americans have also made tremendous strides in this country. The fact that their current position is one of considerable disadvantage relative to whites is sometimes viewed as reason for dismissing the progress that black Americans have made. In the span of three generations, beginning with Emancipation, black Americans have profoundly altered their social, economic, political, and psychological characteristics. Despite the experiences of chattel slavery, sharecropping in the feudal South and *lumpen* labor in factories of the industrializing North, black Americans managed to progress economically. The counterpoints to this remarkably accelerated development of the black population are provided by the disproportionate numbers of blacks who continue to be mired in poverty and by the examples of immigrant groups who arrived in this country generations after blacks had arrived, only to catch up with and pass black Americans on the ladder to success. This book hopes to shed some light on this apparent contradiction.

Data and Methods of Study

The data for this study are from the 1980 Census of Population and more recent Census Bureau publications. The purpose of the decennial census is to gather information about the population's characteristics and housing patterns for use by governmental units in their planning and operation. Census data are also routinely used by researchers to study questions of scholarly interest.

This study of black and white differences relies primarily on census data. In many instances, however, the 1980 census data are supplemented with information from earlier censuses or from other sources,

[5]Sowell (1981).
[6]Lieberson (1980).

4

such as national surveys, local studies, and other census surveys. This study's reliance on census data is both an asset and a liability. A major strength of census data is its scale and representativeness, since these data provide information on many thousands of Americans. A major liability of census data is the general nature of the information which they provide. Since census data are collected primarily for governmental purposes, these data are usually in the form of general population statistics. Moreover, the sheer size of the population on which census statistics are collected limits both the number and detail of questions that can be asked.

Census data provide information about income, family organization and size, labor force participation, and educational attainment. Using demographic techniques to view this information from a variety of perspectives, we are able to illustrate the comparative status of blacks and whites in 1980.

Many of the census data are from the 1980 Public Use Microdata Samples (PUMS) computer tapes, prepared for public release by the Bureau of the Census and containing records for a sample of households. For each housing unit sample, information on the characteristics of that unit and its occupants is provided.[7]

The sample used in analyses for this book is drawn from a file containing 250 variables for 12 million people. Whites, Hispanics, and others were sampled from the A file at the rate of 1 in 1,000; blacks were sampled at the rate of 1 in 100. The rationale for oversampling blacks was to provide a sample of sufficient size for comparative analyses with the proportionately larger white sample. Our sample consisted of 70 economic or social variables on approximately 450,000 people.

Overview of the Volume

Each chapter in this book examines the comparative status of black and white Americans. We look at the internal diversity of blacks and whites; we also look at the groups in relation to one another and over time. Chapters 2–5 examine the dynamic population processes of fertility, mortality, growth, and migration. Over time, these processes have changed not only the size but also the age structure and geographic location of the black and white populations in the United States.

Chapter 6 describes marital status and family structure; Chapter 7 describes educational attainment and school enrollment; Chapters 8–11

---

[7]U.S. Bureau of the Census (1983), *Census of Population and Housing, 1980,* Public Use Microdata Samples: Technical Documentation.

focus on the economic statuses of blacks and whites as revealed in the 1980 census. Of special importance in this connection is the relationship of economic status to other characteristics of the individual. In Chapter 12, we compare foreign-born and native-born blacks.

While this book relies heavily on the "race-differences paradigm" that has characterized so much of the research on blacks and whites in our society, we continue to be mindful of the characteristics shared by the two races. Indeed, we are struck by the fact that race itself is in so many ways more an attributed quality than a real one. As many scholars have noted, race in a heterogenous society like the United States is often no more than a socially constructed and defined characteristic. It is not at all uncommon in the United States, for example, to find considerable overlap between blacks and whites in terms of their values, personal histories, and blood lines. The idea of race is a much purer concept than the reality of race in American society. How else is one to understand the possibility of blacks "passing" as whites or of white jazz pianists qualifying for the title of "blue-eyed soul brother"?

In this book we consciously examine the impact of race on the lives of black and white Americans and find that the sociological reality of race is more important than its biological reality. Race exerts profound influence over the lives of people in this society. W. I. Thomas reminded us that what men perceive to be real becomes real in its consequences.[8] This powerful assessment still rings true in American racial issues.

[8]Cited in Janowitz (1966) p. xl.

BLACK POPULATION GROWTH FROM COLONIAL TIMES TO WORLD WAR II

I N THE mid-1980s the black population of the United States approached 30 million.[1] Approximately one resident in eight was black. This chapter describes the history of black population growth to 1940. From the founding of the colonies in the 1600s through the mid-nineteenth century, the black population grew rapidly. Indeed, its growth rate exceeded that of whites for much of the pre-Revolutionary era, and in 1790 about one American in five was either an enslaved or free black.[2] During the late nineteenth century and into the twentieth century, the growth rate of the black population slackened, while the white population was augmented by migration from Europe. By 1940 black representation fell to less than 10 percent of the total.

Population Increases Before the Revolution

In 1619 the first blacks arrived in the English colonies which later formed the United States.[3] Twenty Africans were brought to Jamestown and treated as indentured servants, that is, they were servants who

[1]U.S. Bureau of the Census, *Current Population Reports*, series P-25, no. 990 (July 1986), table 2.
[2]Rossiter (1909), table 7.
[3]Franklin (1966), p. 71.

owed a five- to seven-year labor obligation in exchange for food and lodging.[4]

In 1790 the first census counted 757,000 blacks. Not only had the black population grown rapidly—an average of 6 percent each year—but the colonies had established a system of slavery such that fewer than 10 percent of the nation's blacks were free at the time of the Revolution.[5]

It is difficult to disentangle the demographic, economic, and social processes that led to the rapid growth of black population in the colonial era, but some aspects are well known. The earliest settlements developed at or near the points where major rivers entered the Atlantic: Charleston in the South, New Amsterdam and Philadelphia in the mid-Atlantic colonies, and Boston and Newport in New England. In these thriving ports, merchants exchanged furs and agricultural goods from the hinterlands for products from Europe, the West Indies, or other colonies.[6]

A need arose for laborers. Indians were one source, and the British made sporadic attempts to enslave or employ them, but Indians were not inclined to work in cities, and those held in bondage could easily escape to a wilderness which they knew better than did their captors.[7] Indentured servants offered another source of labor, but the mercantilist policies of the European nations restricted emigration. Furthermore, the colonist who sought indentured servants had to pay for their transport and would have use of their labor for only a few years.

Blacks were a third source of labor, and they quickly became the most popular one. By the end of the seventeenth century, colonial courts established that blacks could be held for the duration of their lives, that they could be exchanged in the same manner as other property, that they lacked standing before courts, and that the status of a child followed that of his or her mother.[8] Slaves became the preferred source of labor since they were held in perpetuity, had no legal rights, and reproduced themselves.

In the cities of the North slaves worked at skilled and unskilled jobs, while in the South most of them worked in agriculture. Since rice and indigo grew as complementary crops, South Carolina planters were among the first to efficiently use gang labor. The expanding world market for tobacco in the 1700s led to large increases in the slave population in the Carolinas, Virginia, and Maryland.

The colonies, especially those in the North, took many censuses, supplying some demographic information about blacks prior to the Rev-

[4]Jordan (1968), pp. 71–82.
[5]Rossiter (1909), p. 80.
[6]Bridenbaugh (1964), chaps. 1 and 2.
[7]Green (1968), p. 20; Jordan (1968), p. 68.
[8]Jordan (1968), chap. 2; Harding (1983), chap. 2.

olution. Two conclusions may be drawn from these enumerations. First, blacks made up a considerable proportion of the population in the seaports during the early 1700s. In New York just under one quarter of the residents were slaves, in Boston one sixth were slaves; in Newport almost one fifth were slaves, and in Charleston the black and white populations were about equal.[9] The proportion black in northern cities was higher in the eighteenth century than in the nineteenth and early twentieth centuries.

Second, the colonial enumerations suggest that the black population grew rapidly. In Connecticut, for example, the black population increased 3 percent per year between 1756 and 1774; in Rhode Island, about 4 percent per year between 1708 and 1774; and at a similarly high rate in New York and New Jersey. The only southern colony that took several censuses was Maryland, and the demographic evidence is consistent with the view that slavery was increasing rapidly in the tobacco region since that colony's black population rose by more than 4 percent per year between 1704 and 1762.[10] An estimate of a similar annual growth rate in South Carolina was made for the eighteenth century.[11] These rates are very high compared with the present; the nation's total population grew by only 1.1 percent and the black population by 1.6 percent per year during the 1970s.[12]

These high growth rates resulted from the importation of slaves and the excess of births over deaths. Given the absence of import taxes and accurate censuses, it is impossible to ascertain what role the slave trade played in the population growth. The most widely cited figures were developed more than a century ago by Carey, who contended that about 260,000 slaves entered prior to the Revolution. Several analysts believe that this figure is an underestimate, while others argue that it is too large.[13] Presumably, almost all of the present black residents of the United States are descendants of 200,000 to 300,000 slaves brought to the colonies several centuries ago.

It is equally challenging to determine the vital rates of blacks during this period. A few colonial censuses asked about age, permitting the calculation of child-to-woman ratios and other indexes of fertility for blacks.[14] In addition, the New Jersey enumeration of 1772 asked about births or deaths that occurred in the previous year. This source suggests that the birthrate of blacks was 36 per 1,000 and the death rate was 16,

[9]Bridenbaugh (1964), p. 249; Wells (1975), p. 100.
[10]Wells (1975), pp. 90, 98, 112, 135, and 147.
[11]Rossiter (1909), tables 76–103.
[12]Passell, Siegel, and Robinson (1982), table 1.
[13]Rossiter (1909), p. 36; Potter (1965), p. 641; Curtin (1969); Fogel and Engerman (1974a), pp. 30–31.
[14]Wells (1975), p. 137; Potter (1965), p. 658.

implying a rate of natural increase of 2 percent. Among whites, the corresponding rates were 31 and 10.[15] Fragmentary evidence of this nature is consistent with the view that the black population increased rapidly because of both natural increase and the slave trade.

Population Growth Since 1790

The census of 1790 found that just under 20 percent of the 3.9 million residents were black. Fewer than 2 percent in New England were

FIGURE 2.1

Total and Black Populations and Proportion Black, 1790–1986

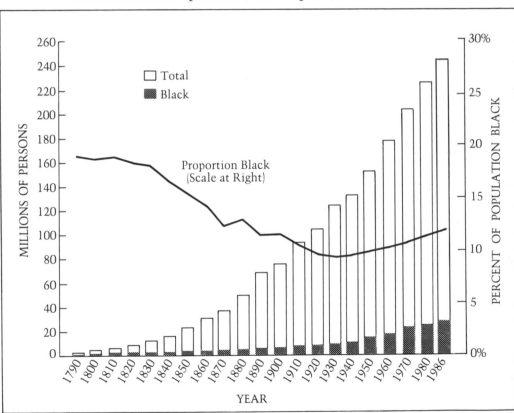

SOURCES: U.S. Bureau of the Census, *Historical Statistics of the United States: Colonial Times to 19£* series A-45, A-46, A-59, and A-65; *Census of Population: 1980*, PC80-1-B1, table 40; *Current Populati* *Reports*, series P-25, No. 990 (July 1986), table 2.

[15]Wells (1975), p. 141.

black, about 6 percent in the Mid-Atlantic states, and more than 33 percent in the South.[16]

The black population reached 1 million when the second census was taken in 1800, grew to 5 million immediately following the Civil War, and reached 10 million after World War I. Figure 2.1 shows total and black populations at each census date.

During the first third of the nineteenth century, the black and white populations increased at similar rates, but in the 1840s the potato crops in Ireland and northern Europe were destroyed. The resulting famine hastened the first large-scale migration to the United States, and primarily because of this movement the white population grew more rapidly than the black for the next 100 years.[17] Figure 2.2 shows the decennial rates of population growth for whites and blacks. (See Appendix, page 421, for discussion of fluctuating growth rates in the nineteenth century.)

These changing growth rates altered the composition of the population. Between 1790 and 1840 about 20 percent of the population was black. Because of the higher growth rates of whites, black representation declined and reached a minimum in 1930 when fewer than 10 percent were black. Since that time black fertility rates have exceeded white, and by 1986, 12 percent of the population was black.[18]

The Changing Black Population: Growth from the Revolutionary War to the Civil War

The need for agricultural labor greatly increased following the invention of the cotton gin in 1793. Southern planters had a valuable but labor-intensive crop which brought them great profits, and after the War of 1812 they found that they could sell cotton almost as fast as they could grow it.[19] A major demographic change occurred as the Gulf Coast states were settled, largely because of the need for more lands for cotton cultivation. In addition to the Louisiana Purchase, the United States added Florida in 1819 and Texas in the late 1840s. The development of the steamboat for river transport in the 1820s and 1830s, the removal of Indians from the Southeast by President Jackson, the construction of railroads, and the use of steam-powered ocean vessels allowed

[16]Rossiter (1909), table 108.

[17]Thompson and Whelpton (1933), pp. 7–10; Taeuber and Taeuber (1971), pp. 595–596.

[18]U.S. Bureau of the Census, *Negroes in the United States 1920–32* (1935) table 12; *Current Population Reports*, series P-25, no. 985 (April 1986), table 1; *Current Population Reports*, series P-25, no. 990 (July 1986), table 2.

[19]Fogel and Engerman (1974b), pp. 89–93.

FIGURE 2.2

Average Annual Growth Rates of White and Black Populations, 1790–1986

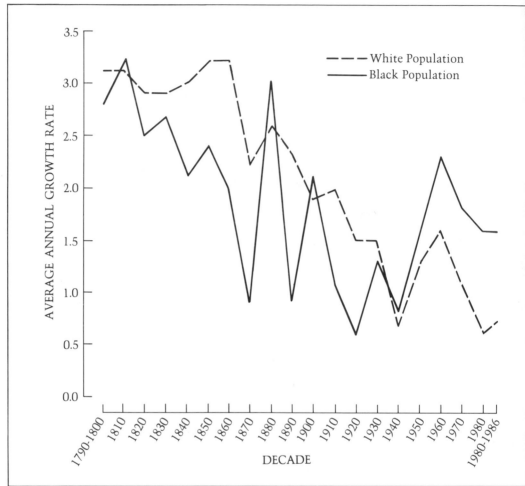

SOURCES: U.S. Bureau of the Census, *Historical Statistics of the United States: Colonial Times to 197?* *(1975)*, pt. 1, series A 91–104; *Census of Population: 1980*, PC80-1-B1, table 40; Current Population Reports, series P-25, no. 990 (July 1986), table 2.

cotton to be grown on southern farms and sold throughout Western Europe.

Between 1790 and the Civil War the black population increased from 757,000 to 4.4 million, implying a growth rate of 2.5 percent annually. There was a major geographic shift, but it was not a movement from the slave states of the South to the free states of the North. In fact, the proportion of blacks outside the South held constant at 9 percent for the first century of the nation's history.[20] Rather, the movement was from the tobacco plantations of the eastern tidewater to the cotton plantations of the Gulf Coast states. The census of 1860 found that more than one half of the nation's blacks lived in states added to the Union after the Revolution.[21] Despite the pleas, the political pressures, and the occasional militant efforts of abolitionists, the proportion of blacks who were free changed only a little in this period, growing from 8 percent of the total black population in 1790 to a peak of 14 percent in 1830 and falling back to 11 percent in 1860.[22] On the eve of the Civil War, approximately one third of the white families in the South held slaves, with an average of nine chattels per owner.[23]

The black population was doubling every 30 years prior to the Civil War. Was this rapid growth due to the importing of slaves or high rates of natural increase? The Constitutional Convention agreed that there would be no federal law banning the international slave trade until 1808. As the cotton economy expanded, the southern states considered amending their eighteenth-century laws which discouraged the slave trade, but South Carolina was the only state to do so; and between 1804 and 1807, 39,000 slaves were imported to Charleston.[24]

Undoubtedly there was some illegal slave trade, but demographic indicators suggest that migration did not play a large role in the high growth rate of the black population. Male slaves had market value far in excess of female slaves, and the typical slave ship carried more men than women. If the importing of slaves continued at high rates, we would expect to find a predominance of black men. The census of 1820, the first to tabulate blacks by sex, found a ratio of 103 men to 100 women.[25] This is a low sex ratio, much lower, for example, than the sex ratio observed among whites in 1910 (about 107 men to 100 women) when the population was greatly influenced by migration from Europe.[26] Between 1820 and 1860 the female black population grew more rapidly than the male and the sex ratio fell, further suggesting the absence of an international slave trade.

[20]U.S. Bureau of the Census, *Negro Population of the United States: 1790–1915* (1918), p. 33.
[21]U.S. Bureau of the Census, *Negro Population of the United States: 1790–1915* (1918), pp. 43–45.
[22]U.S. Bureau of the Census, *Negro Population of the United States: 1790–1915* (1918), p. 53.
[23]U.S. Bureau of the Census, *Negro Population of the United States: 1790–1915* (1918), p. 56. Rossiter (1909), chap. 14.
[24]Mannix (1962), pp.187–88.
[25]U.S. Bureau of the Census, *Negro Population of the United States: 1790–1915* (1918), p. 147.
[26]U.S. Bureau of the Census, *Census of Population: 1970*, PC(1)-B1, table 53.

In the decade prior to the Civil War the demand for cotton rose, leading to higher market prices for slaves.[27] However, the census of 1860 did not show any evidence of the recent arrival of young blacks. In fact, the growth rate for blacks in the 1850s was lower than it had been in any previous decade.[28]

Prior to the Civil War the black population grew rapidly because of natural increase, not because numerous slaves were being imported from the West Indies or Africa. If a population is closed to migration, we can determine a great deal about its levels of fertility and mortality. People enumerated in the age group 30 to 39 at one census date are the survivors of people enumerated in the age group 20 to 29 ten years before. Survival rates may be estimated in this fashion to develop estimates of mortality conditions. If the death rate and the growth rate are known, the level of fertility may be estimated. Several investigators have described the vital rates of the black population prior to the Civil War using these procedures.[29] They describe a population that had high mortality rates but even higher fertility rates.

Crude death rates must have been on the order of 30 to 35 per 1,000 persons each year, with an infant mortality rate in the range of 250 to 300 deaths per 1,000 births. These are extremely high mortality rates compared with those of the black population in the late twentieth century. In 1984, for example, the crude death rate for blacks was 8 per 1,000 and the infant mortality rate 19 per 1,000.[30] In the mid-nineteenth century, a black woman at birth could expect to live approximately 35 years. If she survived the great mortality risks in childhood and reached age 10, she could look forward to another 40 to 45 years.

Mortality conditions among blacks prior to the Civil War were worse than those of some western European nations and probably worse than those of American whites. During the early decades of the nineteenth century in Sweden, a woman could expect to live 45 years at birth or 10 more years than American blacks, and at the same time, life expectation for women in France was about 40 years.[31] It is challenging to determine the life span of American whites in the ante-bellum period because there was no registration of vital events and migration from Europe added to the population. One analyst used data from Massachusetts and Maryland and estimated a life span for all women in the

[27]Conrad and Meyer (1964), table 17.
[28]U.S. Bureau of the Census, *Negro Population of the United States: 1790–1915* (1918), p. 29.
[29]Eblan(1972, 1974); Farley (1965; 1970, chap. 2); McClelland and Zeckhauser (1982); Zelnick (1966).
[30]National Center for Health Statistics, *Monthly Vital Statistics Report*, vol. 33, no. 9, supplement (December 20, 1984), figure 5 and table 1; vol. 33, no. 13 (September 26, 1985), table 5.
[31]Keyfitz and Flieger (1968), pp. 312 and 476.

United States of 43 years in 1850, or approximately 8 years longer than that of black women.[32]

Historical accounts provide information about the living conditions of blacks before the Civil War, and three conclusions may be drawn regarding mortality. First, there was never a period in American history in which the black population was decimated by contagious disease or by crop failures similar to those that occurred in parts of Europe. Yellow fever killed blacks in some local areas at specific times, and in the early 1830s the cholera pandemic raised death rates. However, the black population—as well as the white population—was largely spared the tragic consequences of epidemics and famines.[33]

Second, there is consensus that slaves' diet was sufficient in quantity, consisting largely of corn, wheat, and pork. Fresh fruits and vegetables were eaten when in season, and some slaves apparently supplemented their diet with fish and game. However, there is disagreement about whether the diet was sufficiently varied. One view is that it was so restricted to grains and pork that slaves often suffered from dietary deficiencies.[34] Fogel and Engerman express the opposite view; their analysis of plantation records leads them to conclude that the diet of slaves was so varied that it exceeded the nutritional standards called for today.[35]

Third, there is agreement that modern medicine and public health techniques had few beneficial consequences for blacks. Although some masters may have called doctors for their slaves, the training of medical students in the nineteenth century was rudimentary and their cures may have been little better than folk medicine. In some southern cities modern hospitals were slowly developing, but blacks made little use of them.[36]

Using information about growth and mortality rates, we estimate that the birthrate for blacks in the first half of the nineteenth century must have been 55 to 60 per 1,000. This is a very high rate compared with the birthrate in 1984 of 21 for blacks and 16 for the total population.[37] The demographic investigations suggest that there was a slight fall in fertility during the ante-bellum period, but little can be said about the timing or magnitude since the census of 1850 was the first one to provide detailed information about blacks by age.[38] A black

[32]Jacobson (1957).

[33]McClelland and Zeckhauser (1982), p. 77; Stampp (1956), pp. 299–303; Curry (1981), p. 141; Savitt (1978), chap. 7.

[34]Stampp (1956), p. 304; Genovese (1965), p. 44.

[35]Fogel and Engerman (1974b), p. 115.

[36]Curry (1981), p. 140; Savitt (1978), chap. 6.

[37]National Center for Health Statistics, *Monthly Vital Statistics Report*, vol. 35, no. 4, supplement (July 18, 1986), table 1.

[38]Zelnick (1966); Farley (1970), p. 33; McClelland and Zeckhauser (1982), p. 14.

woman who bore children in the 1850s and survived through her 40s typically bore in excess of seven children.[39]

Many writers assert that slave breeding was commonly practiced, but others who have investigated this find little evidence.[40] Some owners who needed cash sold their young slaves when they reached working age, and young women whose fertility was proven may have brought greater prices than childless women, although men consistently sold for much higher prices than women.[41] An important question to ask is what could have been done to raise or lower the fertility of blacks. Little was known about the biology of reproduction, and modern cures for sterility were unavailable. Perhaps slave owners punished their slaves who refused to cohabit or rewarded female slaves if they bore many offspring, but accounts of slave life fail to mention such practices. Work loads were apparently reduced for pregnant slaves, but they were still expected to work in the fields or in the kitchen up to delivery and resume work shortly thereafter. Most blacks lived near many other blacks, matings were frequent, and the demographic evidence suggests that women became pregnant early in life and frequently thereafter.[42]

Population Growth Following the Civil War
Trends in Fertility

Presumably, populations in developed countries have gone through a common pattern of demographic change.[43] At a first stage, rates of birth and death are high and the population grows slowly. At a second stage, improvements in the standard of living and, perhaps, the introduction of public health measures lead to a fall in death rates, but the birthrates remain high so the population grows rapidly. At a third stage, couples control their fertility, birthrates decline, and the population moves toward a lower growth rate. When did the demographic transition occur among blacks?

The growth rate of the black population decreased throughout a 75-year period following 1860. After increasing at rates that caused it to double every 35 years, the black population growth slowed so that by the 1930s, the doubling time was 70 years. This shift toward a lower

[39]The census of 1910 asked women the number of children they had. Black women born 1835 to 1840 who survived to 1910 averaged 6.7 live births; U.S. Bureau of the Census, *Sixteenth Census of the United States: 1940, Population; Differential Fertility: 1940 and 1910, Women by Number of Children Ever Born*, table 12.

[40]Bennett (1984); pp.82–86; Fogel and Engerman (1974b); pp.78–86; Phillips (1966); p. 361; Franklin (1967); pp. 177–78; Stampp (1956); pp. 245–51.

[41]Fogel and Engerman (1974b), p.76; Gutman (1976), p.76.

[42]Gutman (1976), p. 50.

[43]Coale (1974).

FIGURE 2.3
Children Under Age 5 per 1000 Women Aged 15–49, 1850–1985

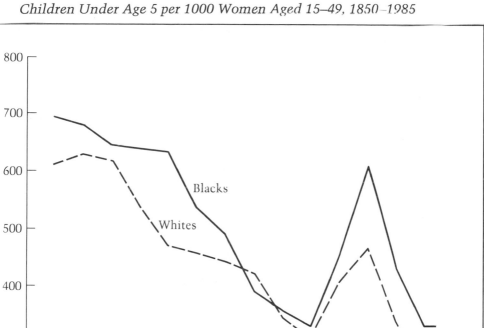

URCES: U.S. Bureau of the Census, *Historical Statistics of the United States: Colonial Times to 1970* 75), pt. 1, series A 119–234; *Census of Population: 1980*, PC80-1-B1, table 44; *Current Population orts*, series P-25, no. 985 (April 1986), table 1.

growth rate occurred primarily because of reductions in childbearing. Figure 2.3 shows the ratio of children under age 5 to women aged 15 to 49 between 1850 and 1985. The sharp drop in fertility can be seen in the change from 700 children per 1,000 women among blacks in 1850 to about 300 per 1,000 in 1940. After World War II women resumed bearing children rapidly, and the ratio of children to women among blacks was as high in 1960 as it had been in the late nineteenth century. This rise was followed by a sharp drop, and by 1980 the ratio of children to women was as low as it had been at any previous date.

Fertility trends among whites paralleled those of blacks: a decline throughout the late nineteenth century and first four decades of the twentieth followed by, first, a sharp rise and then an equally precipitous decline. Traditionally, birthrates have been higher among blacks, but the racial difference contracted during the period of decline and was small in the 1920s and 1930s. After 1940 the racial difference grew larger.

The ratio of children to women may be readily calculated from census data, but it is influenced by changes in mortality, especially in infant mortality. A more frequently used index of childbearing for specific time periods is the total fertility rate, which reports how many babies a woman would bear if she lived to age 45 and had children at the rates observed in that period.

Figure 2.4 shows total fertility rates for whites and blacks between 1860 and 1984. Since the mid-1930s there has been a national system for registering vital events; for earlier years, the rates were estimated on the basis of the age composition of the population reported in different censuses.[44]

The fertility rates of the Civil War era imply that black women who survived to menopause averaged 7 to 8 children, or about 2.5 children more than white women. Total fertility rates of the 1930s suggest fewer than 3 children per black woman, or just about .5 child more than whites.

There is no agreement as to why fertility rates among blacks fell between the Civil War and the Depression and rose during the following decades. Three changes may reduce birthrates: (1) patterns of marriage or cohabitation may change such that women are less exposed to the risk of becoming pregnant; (2) couples might adopt effective methods of birth control or use abortion to terminate unwanted pregnancies; (3) fertility-inhibiting diseases, such as venereal diseases and tuberculosis, may become more common.

We cannot determine which factor was most important in reducing black childbearing before 1940.[45] Apparently there was not a shift in marriage or cohabitation patterns which was sufficient to reduce the total fertility rate from 7.7 to 2.6. Since 1890 the censuses have provided information about the marital status of adults. In that year, for example, 62 percent of the black women aged 20 to 24 had been married.[46] There

---

[44]Coale and Zelnick (1963); Coale and Rives (1973).
[45]Lantz and Hendrix (1978); McFalls and Masnick (1981); Tolnay (1981); Engerman (1978); Meeker (1977).
[46]U.S. Census Office, *Eleventh Census of the United States: 1890*, pt. 1.

FIGURE 2.4
Total Fertility Rates of the Black and White Populations, 1860–1984

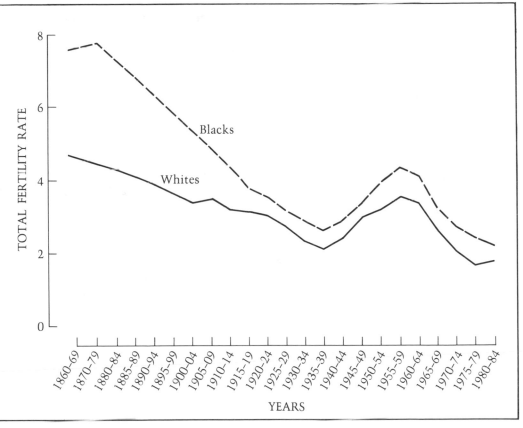

SOURCES: A. J. Coale and M. Zelnik, *New Estimates of Fertility and Population in the United States* (Princeton, N.J.: Princeton University Press, 1963), table 2; A. J. Coale and N. W. Rives, Jr., "A Statistical Reconstruction of the Black Population of the United States: Estimates of True Numbers by Age and Sex," *Population Index* 39 (January 1973): table 7; National Center for Health Statistics, *Vital Statistics of the United States: 1970*, vol. 1, table 1–6; *Monthly Vital Statistics Report*, vol. 35, no. 4, supplement (July 18, 1986), table 4.

was a slight shift toward earlier marriage in the next four decades; the census of 1930 found that 67 percent of the black women aged 20 to 24 had been married.[47] We cannot attribute the decline in fertility among blacks to a shift to much older ages at marriage since the data suggest earlier marriage.

[47]U.S. Bureau of the Census, *Census of Population: 1960*, PC(1)-1D, table 177.

19

Since 1910 the censuses have provided information about the proportion of adult women who live with a husband. In 1910, 51 percent of the women aged 15 to 44 were married and lived with a husband; in 1940, 49 percent.[48] This suggests that changes in marriage and cohabitation cannot explain the sharp decrease in fertility rates among blacks.

It is difficult to assess the use of birth control or abortion by black women prior to 1960 since there were no national surveys of this topic. One might assume that the sharp declines in childbearing occurred because men were using condoms or women were being fitted with diaphragms. However, studies of contraceptive use by blacks in the 1920s and 1930s suggest that few women knew about contraception or had access to the medical assistance which would allow them to control their fertility.

The largest early study of contraception in the United States was conducted by Raymond Pearl in the 1930s.[49] He interviewed some 5,600 urban black women and found that only one sixth of them had ever used any technique to avoid conception, and many reported relatively ineffective means such as douching and periodic abstinence. He compared the fertility rates of black women who reported use with those of women who said they had never practiced birth control. Among whites, the birthrate of users was about 30 percent lower than that of the control group, but among blacks there was no difference, leading Pearl to conclude that contraception played no role in reducing black fertility.[50]

If birth control had little effect on the childbearing of the urban black women, it is difficult to believe that it reduced fertility among the majority of blacks who lived in the small towns and farms of the South. Other investigators corroborate this view. A study of married black women in Spartanburg, South Carolina, in the 1930s found that only 20 percent had ever used a contraceptive technique and in rural West Virginia only 7 percent.[51] In both studies, douching was reported as the most common method of birth control. Not surprisingly, estimated fertility rates during intervals when contraception was used were just about as high as the fertility rates of control groups of nonusers.

McFalls convincingly argues that tuberculosis has had a major impact upon the fertility of blacks since Emancipation.[52] If untreated, this disease leads to the development of nodules that close the Fallopian tubes or the epididymis. Since TB ordinarily develops about the time of puberty, sterility is common in a population suffering from this disease.

[48]U.S. Bureau of the Census, *Sixteenth Census of the United States: 1940, Population; Differential Fertility: 1940 and 1910, Fertility by Duration of Marriage,* tables 42 and 44.

[49]Pearl (1933, 1934, 1936a, 1936b, 1937, 1939).
[50]Pearl (1936b), p. 505.
[51]Beebe (1942); Stix (1941).
[52]McFalls and McFalls (1984): pp. 482–500.

There was a sharp increase in TB among blacks in the decades following the Civil War. Kiple and King claim that the disease exploded in epidemic form.[53] The diet of blacks, which may have been adequate during slavery, probably deteriorated as they became sharecroppers and entered the cash economy to buy food. Fogel and Engerman claim that by the 1890s tenant farmers' diets included few proteins or vitamins,[54] a diet that would encourage TB. Those blacks who moved to cities probably increased their risk because of their poverty, the absence of health programs, and their greater exposure to the infected white population.[55]

Since about 1905 tuberculosis has declined in the United States, with the rate of decrease among nonwhites being much greater after 1935 than before.[56] Trends in fertility among blacks—especially childlessness—match the trends in TB. Among married black women who were born prior to the Civil War, fewer than 10 percent remained childless for their entire lives; among those born in the first decade of this century, an exceptionally high 28 percent had no children. In contrast, among black married women who were born in the early 1940s, fewer than 6 percent will be childless at menopause.[57] The elimination of tuberculosis occurred concurrently with a drastic decrease in apparent sterility.

It is possible that the spread of venereal diseases also accounts for a significant proportion of the decline in childbearing.[58] Gonorrhea produces pelvic infections in women, including salpingitis, which may eventually close the Fallopian tubes.[59] Sterility is not a direct consequence of syphilis, but a pregnant woman with this disease is at high risk of experiencing a stillbirth or spontaneous abortion. Investigations suggest that 30 to 50 percent of the pregnancies occurring in syphilitic women who receive no care end in fetal deaths; upwards of 20 percent of the infants born to such women die of congenital syphilis.[60]

Although there are no national studies on this topic, a variety of investigations suggest that many blacks had untreated veneral diseases during the era of declining fertility.[61] An investigation in Philadelphia

[53]Kiple and King (1981), p. 146.
[54]Fogel and Engerman (1974a), p. 261.
[55]McFalls and McFalls (1984), p. 489; Rabinowitz (1978), pp. 120–22, 128–52.
[56]Lowell, Edwards, and Palmer (1969), figure 4-1.
[57]U.S. Bureau of the Census, *Sixteenth Census of the United States; 1940, Population; Differential Fertility: 1940 and 1910, Women by Number of Children Ever Born,* table 12; *Current Population Reports,* series P-20, no. 108 (July 12, 1961), table 1; series P-20, no. 395 (November 1983), table 10, pt. C.
[58]Increases in rickets and postpartum infections also apparently reduced black fertility rates between the Civil War and the Depression decade. See Cutright and Shorter (1979).
[59]Nelson and Crain (1938); Wilcox (1964).
[60]Weiss and Joseph (1951); Kampmeier (1943); Moore (1941).
[61]Jones (1981), chaps. 2 and 3.

attributed the very low fertility of urban black women to venereal diseases. Employment opportunities were so restricted that many black girls and women became prostitutes. Lane estimates that about 15 percent of females of childbearing age worked at this occupation at the turn of the century, thereby exposing themselves to the fertility-inhibiting venereal diseases.[62] Early in this century hospitalized black patients were given Wasserman tests that indicated that one quarter to one third suffered from venereal diseases.[63] In the 1920s the Julius Rosenwald Fund sponsored numerous investigations of the health of rural blacks. One of their first efforts was to develop syphilis control programs since it was believed that this disease accounted for the high infant and maternal death rates. Later in the 1920s the U.S. Public Health Service conducted tests with 33,000 blacks in four southern counties and found that 20 percent of the adults had positive reactions to a syphilis test.[64]

Undoubtedly these diseases reduced birthrates in the 1920s and 1930s, but were they responsible for the longer-run fertility decline? If so, we should demonstrate that these diseases became more common during the period following 1860 and that their elimination was marked by a rise in childbearing.

The evidence about trends in venereal disease is indirect, but supports this hypothesis. Those few writers who comment about the health conditions of slaves seldom mention these diseases. Instead, they focus their attention on yellow fever, cholera, and sometimes malaria. Venereal diseases are absent from the typical list of ailments that affected slaves.[65]

Among blacks, the death rates from syphilis increased between 1910 and 1930. To be sure, we are limited to evidence from those states included in the Death Registration Area, and we know little about the accuracy with which this cause was diagnosed. Nevertheless, the overall death rates from syphilis tripled from 30 per 100,000 in 1910 to 90 in 1935.[66] At this time, syphilis was a common cause of infant deaths. In the mid-1930s, there were about 3 infant deaths from syphilis per 1,000 black births, which is a rise from the rates reported for previous decades.[67]

The serologic tests given by the military indicate increasing venereal disease. During World War I about 7 percent of the black men were found to have syphilis; at the start of World War II, the rate was 27 per-

[62]Lane (1986), p. 158.
[63]McNeil (1916); Hazen (1936).
[64]Clark (1932); Hazen (1936).
[65]Fogel and Engerman (1974b), p. 121; Phillips (1966), pp. 58 and 300; Savitt (1978).
[66]Linder and Grove (1947), tables 16 and 18.
[67]Linder and Grove (1947), table 32.

cent.[68] Wright and Pirie analyzed data from the tests administered by the Selective Service system at the start of World War II. Overall, 25 percent of the black and 2 percent of the white recruits in six southern states with large black populations had positive reactions. Syphilis rates at the county level were estimated and then related to the childbearing of blacks in the 1935–40 period. A very strong relationship was evident since counties with the highest prevalence of syphilis had the lowest fertility,[69] suggesting that venereal diseases help account for the low birthrates of blacks in the 1930s.

While the evidence linking venereal disease to declining fertility is no more than suggestive, we can be quite certain that the effective fight against these diseases occurred at the same time black fertility rates rose. One provision of the Social Security Act of 1935 provided federal support for public health programs in areas where they were poorly developed. The larger cities and prosperous states had extensive public health programs before the Depression decade, but most southern counties provided no public health service to blacks until there was a flow of federal monies in the late 1930s.[70] Because so many draftees for World War II had venereal diseases, Congress appropriated special funds to eradicate what was defined as the nation's leading health problem.[71] The Public Health Service established rapid treatment centers which became especially effective after the discovery in 1943 that penicillin could cure both syphilis and gonorrhea, and venereal disease rates dropped dramatically during the World War II decade. During the span in which venereal diseases were effectively controlled, the fertility rates of blacks went from historic lows to levels approaching those of the nineteenth century.

Trends in Mortality

Age distributions from censuses clearly outline the shift toward smaller families that began after the Civil War. It is more difficult to describe mortality trends before 1940 since there were no national systems for registering deaths. Indeed, it was not until 1933 that all states registered at least 90 percent of their deaths and thereby qualified for inclusion in the Death Registration Area (DRA).[72]

[68]Davie (1949), p. 240; Vonderlehr and Usilton (1942), p. 1370.
[69]Wright and Pirie (1984), pp. 29–61.
[70]Woodson's study (1930) of rural blacks in the 1920s reported that treatments were occasionally available for smallpox and yellow fever but that there were no public health facilities treating pneumonia, tuberculosis, or venereal diseases.
[71]Vonderlehr and Usilton (1942), p. 1938.
[72]Linder and Grove (1947), p. 95.

Because there was no registration of deaths, several investigators used census data to estimate death rates among blacks for the decades following the Civil War. Although there are ambiguities in using these census survival procedures, there is consensus that health conditions improved very little among blacks in the late nineteenth century. Eblan found that the life span of black women increased from about 34 years in the 1850s to 35 years in 1900.[73] Meeker, using different assumptions about census errors, concluded that the late nineteenth century should be divided into two spans so far as black mortality rates are concerned: Between 1860 and 1880 economic conditions worsened, death rates increased, and the life span for black women dipped from about 30 to 28 years. After the end of Reconstruction, economic conditions stabilized or even improved and, according to Meeker's estimates, the life span for black women once again reached 30 years by 1900.[74] The most pessimistic view is presented by Farley, who reports a gradual increase in mortality rates throughout the late nineteenth century: The life span of black women, which had been about 30 years in 1860, fell to 27 in 1900.[75] This is consistent with an analysis of mortality which specifically focused on racial differences at the turn of this century and placed the life span of black women in 1900 in the 25-to 30-year range.[76] The life span among blacks in 1900 could not have been much different than it was during the Civil War.[77]

The life expectation of blacks was quite short by historical standards and much briefer than the expectation of life in developing countries today. Sweden is the nation with the longest history of accurate demographic statistics, and by the late 1700s the Swedish life span exceeded that of American blacks in 1900.[78] Among countries reporting mortality rates in 1980, the shortest life spans were found in impoverished African countries such as Zambia, Guinea, and Mali, but even in these places the life span was seven to ten years longer than it was for American blacks at the turn of the century.[79]

[73]Eblan (1974), table 2.
[74]Meeker (1976), table 2.
[75]Farley (1970), tables 2-2 and 3-9.
[76]Demeny and Gingrich (1967), p. 825.
[77]Data from the census of 1900, showing women by number of children ever born and number of children surviving to 1900, have recently been released. These data may be used to estimate survival rates for the childhood ages, and these mortality rates can be used with model life tables to estimate the life span. A use of these techniques suggests a life expectation of 31 years for black women living in the Death Registration Area but 42 years for the total black female population of the country. Preston and Haines (1984), tables 1 and 2.
[78]Keyfitz and Flieger (1968), p. 37.
[79]United Nations (1982) *Demographic Yearbook 1980*, table 34.

Mortality Trends: The Early Twentieth Century

Data from the DRA imply that the life span of blacks increased by 20 years between 1900 and 1940. These data, shown in Figure 2.5, suggest a substantial reduction in infant mortality and in infectious diseases since these types of death are common when life expectation is low. Can we trust these mortality trends, or are they confounded by the expansion of the DRA from 10 states in 1900 to the entire nation in 1933? This is a particularly important question since only two of the areas in the 1900 Registration Area—the District of Columbia and New York State—had numerous black residents.

There are several reasons to believe that the mortality trends shown in Figure 2.5 accurately describe the direction of change. If the analysis is restricted to that small group of states included in the DRA of 1900, we still find large declines in mortality. Table 2.1 presents death rates for 1900–40 for the DRA of 1900 and rates for 1910–40 for the DRA of 1910. These are age-standardized data so the rate for any year may be compared with that for any other, knowing that a difference represents a change in mortality. Similarly, age-standardized rates for blacks may be compared with those for whites knowing that they reflect mortality differences.

Table 2.1 shows that little progress was made in reducing black death rates between 1900 and 1910 in the original DRA. In the following two decades there were modest declines in mortality, followed by a much larger drop during the 1930s. Among whites, the reductions in mortality were spread more evenly across the four decades, and at all dates the age-standardized death rate for blacks was about 60 percent greater than that for whites. Turning to the DRA of 1910, we again find a small decline in black mortality between 1910 and 1920, some change in the 1920s, and much greater reductions in the 1930s.

The demographic analysis that used data from the census enumerations of 1890–1910 estimated a life span for all blacks in 1900 close to that estimated for those in the DRA of 1900: approximately 30 to 35 years. If these census survival estimates for 1900 are correct, and if the 1940 figure that is based on the registration of deaths in all states is accurate, then the life span of blacks must have increased between 1900 and 1940 by just about 20 years.

Was There a Turning Point in Black Mortality?

We can be certain that there was no sharp downturn in death rates among blacks in the decades immediately after the Civil War but that

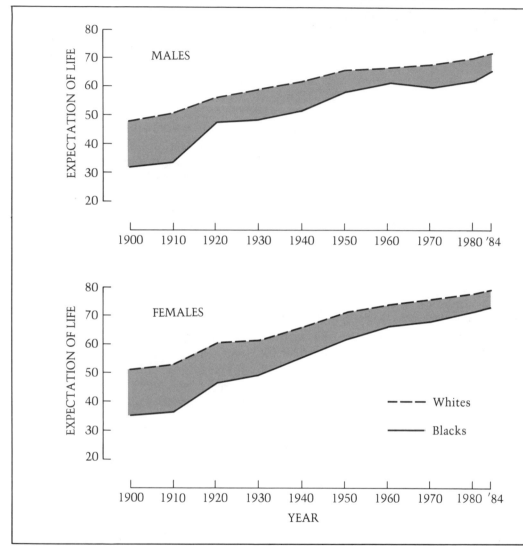

FIGURE 2.5

Expectations of Life at Birth by Sex and Race, 1900–1984

NOTE: Data for 1900 to 1930 refer to the Death Registration Area. Estimates for 1950 and 1960 refer to nonwhites.

SOURCES: L. I. Dublin, A. J. Lotka, and M. Spiegelman, *Length of Life* (New York: Ronald Press, 1949), tables 12 and 14; National Center for Health Statistics, *United States Life Tables: 1959–61*, vol. 1, no. 1, tables 5, 6, 8, and 9; *United States Life Tables: 1969–71*, tables 5, 6, 10, and 11; *Monthly Vital Statistics Report*, vol. 35, no. 6, supplement (2) (September 26, 1986): table 4.

TABLE 2.1

Death Rates, Standardized for Age,
in the Death Registration States of 1900 and 1910

Year	Nonwhites of Both Sexes	Whites of Both Sexes
DEATH REGISTRATION AREA, 1900		
1900	27.8	17.6
1905	27.5	16.4
1910	27.4	16.1
1915	25.0	15.2
1920	24.5	14.6
1925	22.1	13.1
1930	21.8	12.1
1935	18.9	11.2
1940	16.6	10.2
DEATH REGISTRATION AREA, 1910		
1910	24.0	15.0
1915	23.2	14.4
1920	23.7	14.1
1925	22.4	12.8
1930	21.0	12.1
1935	19.0	10.8
1940	16.7	9.3
Total U.S.A., 1985	6.9	5.2

NOTE: DRA states of 1900 are Connecticut, District of Columbia, Indiana, Maine, Massachusetts, Michigan, New Hampshire, New Jersey, New York, Rhode Island, and Vermont.

The following states were added to the DRA of 1910: California, Colorado, Maryland, Minnesota, Montana, Ohio, Pennsylvania, Utah, Washington, and Wisconsin.

SOURCES: Forrest E. Linder and Robert D. Grove, *Vital Statistics Rates in the United States: 1900–1940* (Washington, DC: U.S. Government Printing Office, 1947), tables 7 and 8; U.S. National Center for Health Statistics, *Monthly Vital Statistics Report,* vol. 34, no. 13 (September 19, 1986), table 5.

considerable progress had been made by 1940. Can we identify a turning point, or was progress gradual over a long span?

If death rates fall from a high level, one or more of four changes must occur: (1) the virulence of disease decreases; (2) medical advances eliminate some diseases or reduce mortality from others; (3) public health measures lessen the prevalence or impact of contagious diseases; (4) increases in the standard of living improve the health of the population, enabling them to withstand diseases.

Although we lack detailed data, there is no reason to think that a change in the virulence of diseases added many years to the life span of blacks.[80] Medical advances probably had limited effects on the health of blacks until well into the twentieth century since most blacks lived in rural areas where they had little access to physicians or hospitals. Public health activities may have had some effect on black death rates in urban areas, but the development of an extensive public health system serving the majority of blacks who lived in the poor states of the South awaited passage of the Social Security Act in 1935.[81]

Were there improvements in the standard of living which lengthened the life span of blacks during the late nineteenth and early twentieth centuries? Historians and commentators provide contradictory answers to this question. Those who examine the status of blacks during the first two decades of this century conclude that their welfare had improved greatly in the half-century following the Civil War. Higher standards of living meant improvements in the diets of blacks, improvements in the quality of housing, and better clothing. The result may have been lower infant mortality rates and less tuberculosis, diphtheria, and dietary deficiency diseases.

These writers stress several indicators of progress, such as literacy. When the census of 1910 was taken, 70 percent of the blacks aged 10 and over reported that they could read and write compared with only 43 percent in 1890. By 1910, 59 percent of the black children aged 6 to 14 were enrolled in school compared with a figure of close to zero 50 years earlier.[82] Presumably, the spread of literacy and the high rates of school enrollment reflect higher standards of living.

Another indication of improvement is property ownership. In 1860 no more than a handful of blacks owned property, but five decades later many rural blacks owned their farms while some in cities owned their houses. In 1910 just under 900,000 blacks told the census takers that they were farmers, and about one quarter of them either owned or were buying their farms; the remainder were tenant farmers.[83] The proportion of blacks living off-farm who owned their homes rose from 16 percent in 1890—when the tenure question was first asked—to 22 percent in 1910.[84] W.E.B. DuBois estimated that in 1908 the aggregate wealth of the 10.5 million blacks in the United States was $500,000,000, or about $50 per capita.[85] Although a very small figure, it represented a substantial improvement from 50 years earlier.

[80]Higgs (1977), p. 22.
[81]Jones (1981), pp. 36–44.
[82]U.S. Bureau of the Census, *Negro Population of the United States: 1790–1915* (1918), pp. 391 and 404.
[83]U.S. Bureau of the Census, *Negro Population of the United States: 1790–1915* (1918), p. 572.
[84]U.S. Bureau of the Census, *Negro Population of the United States: 1790–1915* (1918), p. 461.
[85]DuBois (1970), p. 111.

The third indication of social and economic gain is level of occupation. There is some evidence of occupational upgrading among blacks and the emergence of an urban middle class. The first census to gather data about employment or occupation was that of 1890. The census question asked about industrial activity and found that the overwhelming majority of blacks worked in agriculture or in domestic service. However, the proportion of black men working in manufacturing, trade, transportation, or the professions rose from 17 percent in 1890 to 27 percent in 1910.[86] In most of the larger cities, a few blacks were employed as teachers while others ran small retail businesses or served as preachers, again reflecting a change from 1860.

Although data about the actual standards of living of blacks in 1860 are unavailable, some analysts look at the status of blacks in 1910 and conclude that there were improvements, especially after 1890. Higgs argues that there must have been significant reductions in black mortality because of these changes in living standards.[87]

A contrary view is much more pessimistic, placing the decline in mortality at a later date. Utilizing observations and data from the 1920s and 1930s, several writers suggest that the living conditions of blacks in these decades were no better than they had been during the era of slavery. Two of the most influential studies were Charles Johnson's *Shadow of the Plantation* and Carter Woodson's *The Rural Negro*.[88] Neither of these investigations involved a systematic sample of blacks, but both writers claim that their findings are applicable to a large proportion of the nation's rural blacks. In 1920 two thirds of the black population lived on farms or in places with populations of less than 2,500.[89] With regard to health issues, these authors stress several points. First, rural diets were extremely deficient; many rural blacks lived on corn bread, salt meat, sorghum, and beans and seldom ate vegetables; and only the more well-to-do rural blacks owned milk cows.[90] The authors suggest that rural diets deteriorated after Emancipation because the steady increase in population and declines in cotton prices encouraged farmers to devote all of their acreage to a cash crop—cotton. Tenant farmers were not allowed to cultivate food crops or keep animals of their own since this would reduce cotton acreage. In the mid-1930s the Department of Agriculture surveyed 750 black families who did not qualify for relief because of their high incomes and found that 80 percent of them had poor diets that failed to meet minimum nutritional

[86]U.S. Bureau of the Census, *Negro Population of the United States: 1790–1915*, (1918), p. 526.
[87]Higgs (1977), pp. 20–24.
[88]Johnson (1934); Woodson (1930).
[89]U.S. Bureau of the Census, *Negroes in the United States: 1920–1932* (1935), p. 48.
[90]Johnson (1934), pp. 100–102.

standards. Myrdal, analyzing these data as well as Bureau of Labor Statistics surveys of food consumption, reported that he was startled to find that in the 1920s and 1930s many black families with young children never consumed milk, eggs, fruit, vegetables, or potatoes.[91]

Johnson and Woodson also contend that medical and public health facilities provided no assistance for rural blacks. In many areas no doctors or nurses were willing to treat black patients, and blacks who needed hospitalization had to find an urban medical facility that had a Jim Crow ward. In addition, the ignorance of the population contributed to the health problem. Woodson observed that if they were ill, most rural blacks hoped the ailment would cure itself and, if it did not, they relied almost exclusively on folk cures.[92] Johnson reported a great misunderstanding of biological processes; the 600 black respondents in his survey of Macon County, Alabama, confused syphilis with many other diseases, and none of them realized it was transmitted by sexual contact.[93]

In their demographic monograph for President Hoover's Committee on Social Trends, Thompson and Whelpton moved beyond observations for specific areas and calculated age-standardized death rates for the DRA of 1920, an area that included most southern states. They observed that not only was the age-standardized death rate for blacks about 80 percent higher than that for whites, but it also increased by about 7 percent during the 1920s, while that for whites fell by 10 percent, implying that black mortality rates were actually increasing for some part of this century.[94]

One of the clearest arguments suggesting that living standards for blacks did not improve after the Civil War is that of Gunnar Myrdal:

> The agricultural South is over-populated, and this over-population affects Negroes much more than whites. This applies particularly to the Old South, including the Delta district, which contains the main concentration of Negroes. In this Black Belt the over-population has—on the whole—been steadily increasing. Since 1860 the amount of land in southeastern farms has remained stationary, new lands being cleared about as rapidly as old land was exhausted, while the number of male agricultural workers in the same area rose from around 1,132,000 in 1860 to 2,102,000 in 1930. A cultural heritage from times of pioneering, colonization, and slavery makes the conditions even worse than can be visualized by the ratio of population to land alone.[95]

[91]Myrdal (1944), pp. 373–75.
[92]Woodson (1930), pp. 16–21.
[93]Johnson (1934), p. 201.
[94]Thompson and Whelpton (1933), pp. 245–47.
[95]Myrdal (1944), p. 231.

It is impossible to know whether the optimistic or pessimistic view is more accurate. No one can tell us what living conditions or mortality rates were for blacks in 1860, so it is difficult to judge progress. It is also possible that living conditions for rural blacks were better around the turn of this century than in the 1920s or 1930s, further complicating the search for a turning point in black mortality.

This analysis of mortality trends leads to four conclusions. First, there were no overall reductions in death rates in the latter half of the nineteenth century; the life span of blacks in 1900 was approximately equal to that of the final decade of slavery.

Second, mortality rates declined in the twentieth century in the DRAs of 1900 or 1910, with the greatest gains recorded during the 1930s. The black population of these states primarily lived in northern cities, so we can be certain that health conditions improved in these locations.

Third, early this century there apparently was not a large difference between the life span of blacks in the urban North and in the rural South. Ethnographic accounts report that urban blacks lived in overcrowded housing, but the sanitary facilities, water systems, public health services, and access to a varied diet had beneficial effects.[96]

Fourth, the most significant turning point toward lower death rates occurred fairly recently among blacks. The health programs of foundations such as the Julius Rosenwald Fund began to reduce the death rates of rural blacks in the 1920s, a process that accelerated after the Social Security Act provided federal monies for local health services.

We can be certain that the black population of the United States did not go through the demographic transition that characterized many Western nations. The fall in fertility preceded rather than followed the decline in mortality and was not attributable so much to the intentional control of fertility as it was to an increase in health problems: more tuberculosis and venereal diseases. During the era of fertility decline, death rates among blacks remained quite high, and the transition to low rates of growth occurred without the elimination of infectious disease. The transition to low rates of birth and mortality took place after 1940. In the next chapter, we examine more recent trends in mortality.

[96]Borchert (1980), pp. 182–85; Pleck (1979), pp. 35–36; Zunz (1982), p. 376.

AN ANALYSIS OF MORTALITY:
1940 TO THE PRESENT

THROUGHOUT the twentieth century, death rates declined in developed nations. Four reasons are cited for these reductions.[1] First, there were improvements in standards of living. As incomes went up, people could afford better food, clothing, and shelter, and, if ill, they were able to secure medical assistance. Second, there were improvements in the diagnosis and treatment of diseases. The use of antibiotics and chemotherapy drastically reduced death rates from infections and contagious diseases. In recent years, changes in the treatment of high blood pressure and improved therapy for the victims of strokes have lowered cerebrovascular death rates.[2] Third, in many nations there has been a sharp rise in government spending for public health and medical services. Finally, there have been changes in the lifestyles and health practices of individuals, some of which have led to lower death rates. The recent trends toward more exercise and less smoking may be reducing heart disease and stroke death rates.

Since 1940, death rates have fallen among both black and white populations,[3] and each of the four factors outlined above played a role in extending the life span. The standard of living of blacks—as measured by income and education—has risen. In 1939 nonwhite men who

[1]Mechanic (1968), chap. 7.
[2]Kleinman, Fingerhut, and Feldman (1980).
[3]Klebba, Maurer, and Glass (1974); Klebba (1966).

worked full time for the entire year earned an average of $4,760; in 1984 the income for black men who worked full time was $16,900.[4] Since these amounts are expressed in constant 1984 dollars, we can be certain that the purchasing power of these black men rose by 355 percent. For similar white men, the rise in income was 225 percent. In 1940 the typical adult black (aged 25 and over) reported six years of educational attainment, but by 1981 this had risen to 11 years. The gain among blacks—5 years—was much greater than that among whites—2.5 years.[5] A variety of investigators have shown that death rates are strongly linked to social and economic status; those at the top of the educational or income distribution have much lower death rates than those at the bottom.[6] The rise in educational attainment among blacks and their improved economic standing led to lower death rates.

The government's growing support for health services should have been especially beneficial to low-income groups such as blacks. In 1940, 4 percent of the nation's Gross National Product was devoted to health care; in 1984, 10.6 percent.[7] To a considerable degree, governmental agencies assumed responsibility for providing health care. In 1940 about 80 percent of all medical expenses were paid by private individuals and only 20 percent by governmental agencies. Forty-four years later the share paid by individuals fell to 58 percent while the government's share rose to 42 percent. On a per capita basis, governmental spending for health— in constant 1984 dollars—rose from $43 in 1940 to $654 in 1984.[8]

We would expect that improvements in the social and economic status of blacks and the government's assumption of a large share of the financial responsibility for health care would not only reduce the mortality rates of blacks but would lessen the racial differences. In this chapter we describe trends in death rates to determine what changes have occurred and whether the black–white difference in mortality was smaller in the 1980s than it was on the eve of World War II.

Crude and Age-Adjusted Death Rates

The most widely used mortality index is the crude death rate—the ratio of deaths during a year to the mid-year population. Figure 3.1

[4]U.S. Bureau of the Census, *Current Population Reports*, series P-60, no. 80 (October 4, 1971), table 59; series P-60, no. 149 (August 1985), table 8.

[5]U.S. Bureau of the Census, *Sixteenth Census of the United States: 1940, Population; Characteristics by Age*, table 18; *Current Population Reports*, series P-20, no. 390 (August 1984), table 1.

[6]Moriyama and Guralnik (1956); Kitagawa and Hauser (1973).

[7]National Center for Health Statistics, *Health, United States, 1985*, table 80.

[8]National Center for Health Statistics, *Health, United States, 1985*, table 83.

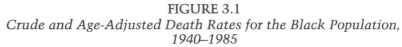

FIGURE 3.1

Crude and Age-Adjusted Death Rates for the Black Population,
1940–1985

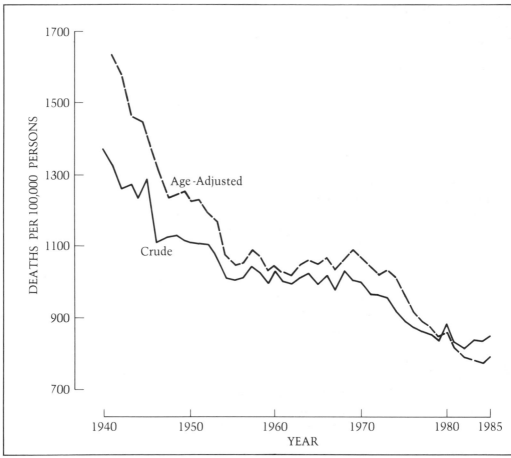

NOTE: Rates for 1940–59 and for 1961 refer to whites and nonwhites. Rates for other years refer to blacks and whites.

SOURCES: Robert D. Grove and Alice M. Hetzel, *Vital Statistics Rates in the United States: 1940–1960* (Washington, DC: U.S. Government Printing Office, 1968), tables 53 and 54; National Center for Health Statistics, *Monthly Vital Statistics Report*, vol. 31, no. 6, supplement (September 30, 1982), table 2; vol. 34, no. 13 (September 19, 1986), table 5.

shows trends in the crude death rate for 1940–85.[9] In 1940 there were 1,383 deaths per 100,000 blacks; by 1950 this declined to 1,119; and in 1985 the crude death rate had fallen to 841 per 100,000.

The crude death rate of a population is strongly influenced by its

[9]For much of this century, federal agencies tabulated health statistics for whites and nonwhites only. In Figure 3.1, the rates for 1940–59 and for 1961–62 refer to whites and nonwhites. Rates for other years refer to blacks and whites.

age structure. In the late 1970s Austria had one of the highest death rates in the world—1,220 deaths per 100,000—and Costa Rica one of the lowest—640 deaths per 100,000.[10] This does not come about because diseases are more virulent and doctors less competent in Austria than in Costa Rica. Rather, Austria has an elderly population because of its history of low fertility while Costa Rica, with its recent history of high fertility, has a very young population.

To overcome this age-structure difficulty, demographers calculate age-adjusted death rates. The death rates for the population in a given age range, such as 1 to 4, are weighted by a standard age distribution to determine the age-adjusted death rate for the total population. In the United States the age distribution of the total population in 1940 is the standard set of weights.[11] Age-adjusted death rates computed for the same population for different years or for different populations can be compared knowing that differences come about because of mortality differences, not because of differences in age structure.

Between 1940 and 1985 changes in the age structure of blacks disguised some of the decline in mortality. The age-adjusted death rate fell from 1,635 deaths per 100,000 in 1940 to 779 in 1985, or a decline of 52 percent. The crude death rate—or the unadjusted death rate—fell by only 39 percent. In other words, the trend in the crude death rate underestimates the gains made by blacks. To compare mortality trends among blacks and whites, Figure 3.2 shows age-adjusted death rates. Since death rates are very much lower among women than among men, separate panels are shown for each sex.

There is an unambiguous trend toward lower mortality rates, but there are persistent and substantial differences between the races. Among men, this span may be divided into three distinct periods.[12] During the 1940s death rates declined rapidly, but from 1950 to 1970 there was very little improvement. Indeed, the age-adjusted death rates of black men in the late 1960s were at the same level as they had been 20 years earlier. The years since 1970 form a third period since death rates among men have fallen rapidly just as they did in the post–World War II decade, and by 1985 mortality rates for both blacks and whites were close to their all-time lows.

Trends among women are quite different from those among men because there was no two-decade hiatus in the pattern of decline. In all periods, the death rates of women decreased, although there was more rapid improvement in the 1940s and 1970s than in the intervening decades.

[10]United Nations, *Statistical Yearbook: 1981*, table 20 (1983).
[11]Grove and Hetzel (1968), table F.
[12]Crimmins (1981).

FIGURE 3.2
Age-Adjusted Death Rates by Race and Sex, 1940–1985
(deaths per 100,000)

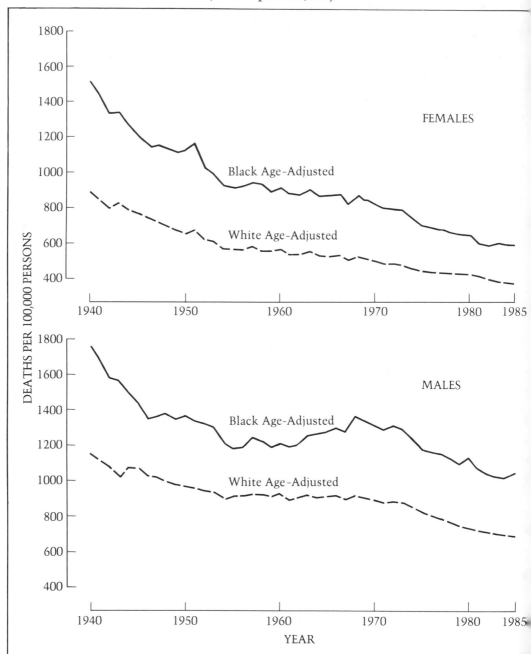

SOURCES: Robert D. Grove and Alice M. Hetzel, *Vital Statistics in the United States: 1940–1960)* (Washington, DC: U.S. Government Printing Office, 1968), tables 53 and 54; National Center for Health Statistics, *Monthly Vital Statistics Report*, vol. 31, no. 6, supplement (September 30, 1982), table 2; vol. 34, no. 13 (September 19, 1986), table 5.

Changes in Death Rates by Age

Age-adjusted death rates summarize mortality trends, but they do not tell us why or how changes occur. We can understand more about this if we examine death rates for specific age groups and from various causes. Since 1940 there has been steady progress in reducing mortality among children while the risk of dying increased for young adult men for much of this period. In the 1970s a new pattern emerged: rapidly falling death rates among the elderly.[13]

Table 3.1 presents information for ten specific age groups for 1940–84. Mortality rates are shown for 1940, 1950, 1970, and 1984. These dates were selected because 1940 is the first year in the interval; 1950 and 1970 roughly mark the beginning and end of a span in which male death rates remained pretty much constant; and 1984 is the most recent year. The first rows of data in the table show the age-adjusted death rates for the total population. Infant death rates are discussed in a subsequent section of this chapter.

During the 1940s death rates declined rapidly for men of both races and at all ages. Decreases were generally greater among blacks, leading to a slight convergence of mortality rates. An annual decrease of 2 percent will cut a death rate in half in 35 years; a 3 percent decline will do it in 24 years.

For both races death rates at ages under 15 continued to fall between 1950 and 1970 because of reductions in the infectious diseases which once killed many children.[14] At older ages, there was little change in male death rates between 1950 and 1970. The rise in black male mortality during this span—shown in Figure 3.2—came about solely because of higher death rates among men aged 15 to 44.

Changes after 1970 are quite like those of the 1940s; death rates declined at a rapid rate for black and white men of almost all ages. A feature of the recent mortality trends is the sharp decrease in death rates among men aged 55 to 74. Throughout this century death rates among the older population have been resistant to change, but this is no longer the case, perhaps because Medicare now provides assistance to these people.

At the end of the Depression, the age-adjusted mortality rate for nonwhite men was 53 percent greater than that for white men, as indicated in Table 3.1. Forty-four years later, it was 46 percent greater. Despite urbanization, improvements in economic status, and an expansion of governmental spending for health, the ratio of black to white

[13]Fingerhut (1982); Fingerhut and Rosenberg (1981); Crimmins (1981).
[14]Dauer, Korns, and Schuman (1968).

TABLE 3.1
Death Rates by Race, Age, and Sex, 1940–1984

		Deaths Per 100,000				Average Annual Percent Change in Death Rates			Black Rate as Percent of White	
		1940	1950	1970	1984	1940-1950	1950-1970	1970-1984	1940	1984
MEN										
All Ages	Black	1,764	1,373	1,319	1,016	-2.5%	-0.2%	-1.9%	153%	146%
	White	1,155	963	893	695	-1.8	-0.4	-1.8		
1–4	Black	527	271	151	84	-6.7	-2.9	-4.2	188	162
	White	281	136	84	52	-7.3	-2.4	-3.4		
5–14	Black	164	97	67	44	-5.3	-1.9	-3.0	146	154
	White	112	67	48	28	-5.1	-1.7	-3.8		
15–24	Black	499	290	321	163	-5.4	+0.5	-4.8	253	115
	White	197	152	172	142	-2.6	+0.6	-1.4		
25–34	Black	853	496	560	329	-5.4	+0.6	-3.8	304	205
	White	281	185	178	161	-4.2	-0.2	-0.7		
35–44	Black	1,318	861	957	609	-4.3	+0.5	-3.2	260	260
	White	507	381	344	234	-2.9	-0.5	-2.8		
45–54	Black	2,453	1,857	1,778	1,291	-2.8	-0.2	-2.2	215	204
	White	1,139	985	883	633	-1.5	-0.5	-2.4		
55–64	Black	3,711	3,480	3,257	2,657	-0.6	-0.3	-1.5	147	164
	White	2,522	2,304	2,203	1,622	-0.9	-0.2	-2.2		
65–74	Black	6,283	5,795	5,803	4,992	-0.8	0.0	-1.1	116	132
	White	5,400	4,865	4,810	3,783	-1.0	-0.1	-1.7		
75–84	Black	10,877	9,030	9,454	8,869	-1.9	+0.2	-0.4	89	104
	White	12,202	10,526	10,099	8,511	-1.5	-0.2	-1.2		
85+	Black	19,972	16,022	12,222	14,708	-2.2	-1.3	+1.3	79	79
	White							0.0		

WOMEN

Age	Race									
All Ages	Black	1,505	1,107	814	586	-3.1%	-1.5%	-2.3%	171%	150%
	White	879	645	502	391	-3.1	-1.3	-1.8		
1–4	Black	443	230	129	68	-6.6	-2.9	-4.6	184	176
	White	241	112	66	39	-7.7	-2.6	-3.8		
5–14	Black	144	75	44	25	-6.5	-2.7	-4.0	178	132
	White	81	45	30	19	-5.9	-2.0	-3.3		
15–24	Black	503	216	112	69	-8.5	-3.2	-3.5	359	138
	White	140	72	62	50	-6.6	-0.7	-1.5		
25–34	Black	740	390	231	130	-6.4	-2.6	-4.1	339	222
	White	218	113	84	58	-6.6	-1.5	-2.6		
35–44	Black	1,175	754	533	309	-4.4	-1.7	-3.9	320	251
	White	367	236	193	123	-4.4	-1.0	-3.2		
45–54	Black	2,109	1,555	1,044	641	-3.0	-2.0	-3.5	282	182
	White	747	546	463	352	-3.1	-0.8	-2.0		
55–64	Black	3,318	2,763	1,986	1,526	-1.8	-1.7	-1.9	197	176
	White	1,684	1,294	1,015	867	-2.6	-1.2	-1.2		
65–74	Black	5,228	4,611	3,861	2,882	-1.3	-0.9	-2.1	126	142
	White	4,154	3,243	2,471	2,031	-2.5	-1.4	-1.4		
75–84	Black	8,414	7,065	6,692	6,095	-1.7	-0.3	-0.7	80	118
	White	10,483	8,482	6,699	5,162	-2.1	-1.2	-1.9		
85+	Black	15,971	13,367	10,707	10,730	-1.8	-1.1	0.0	68	77
	White	23,495	19,680	15,980	13,910	-1.8	-1.0	-1.0		

NOTE: Age-adjustment rates for 1940 and age-specific rates for 1940 and 1950 refer to nonwhites.

SOURCES: National Center for Health Statistics, *Health, United States, 1981*, table 14; *Monthly Vital Statistics Report*, vol. 33, no. 13 (September 26, 1985), table 5.

death rates for men changed only a little in the last four decades. In many age groups there was a similarly small change in the excess mortality of black men, but the largest racial gaps were at ages 25 to 44 where the mortality rates of black men in both 1940 and 1984 were 2 to 2.5 times those of white men.

At the very oldest ages death rates of blacks fall below those of whites. It is possible that there is a racial selection such that blacks who survive to older ages are healthier than similar whites. A more plausible explanation is that errors in age reporting produce the favorable picture of black mortality at the oldest ages. There is a tendency to exaggerate age, apparently much more common among blacks than among whites,[15] which has the effect of increasing the denominators, thereby reducing death rates at the oldest ages.[16]

The data for women, shown in Table 3.1, are much easier to summarize. During the 1940s there were substantial decreases in the death rates of black and white women at all ages, with especially large declines occurring among black women under age 35. Among black children, this can be explained by reduced mortality from contagious diseases such as influenza, pneumonia, and tuberculosis, diseases which had been eliminated among whites before 1940.[17] Among women, the hospitalization of births also cut the maternal mortality rate. In 1940 there were about 800 maternal deaths per 100,000 nonwhite births, which was reduced to about 200 deaths by 1950. For whites, the change was from 320 to 60.[18]

During the 1950s and 1960s there was a slowdown in the shift toward lower mortality rates for women, but they continued to fall. In the 1970s a fast-paced decline in mortality resumed, with unusually large improvements among older women where death rates from heart diseases and cerebrovascular causes have been falling rapidly.[19]

There has been a gradual racial convergence of death rates among women. In 1940 the age-adjusted rate for black women was 71 percent greater than that for white women; in 1984 only 50 percent. Quite

[15]Coale (1955), pp. 41–43; Siegel (1974); Land, Hough, and McMillan (1984), tables 9 and 10.

[16]In recent years, several scholars have suggested that a crossover of black-white mortality rates is reasonable given the differential selectivity of death rates at younger ages. That is, blacks who survive to the oldest ages may be more highly selected than whites and therefore more resistant to disease. A longitudinal study covering 20 years of experience and accurate data in Evans County, Georgia, found that the death rates of elderly blacks were lower than those of whites. Wing et al. (1985); Manton, Poss, and Wing (1979); Manton and Stollard (1981).

[17]Shapiro, Schlesinger, and Nesbitt (1968), chap. 9.

[18]Grove and Hetzel (1968), table 44.

[19]Fingerhut and Rosenberg (1981).

likely, the death rates of black and white women will converge long before those of men.

Changes in Death Rates by Cause

When mortality rates by cause are examined, we find a complicated picture of racial similarities and differences. For most causes, the rates of blacks now exceed those of whites by a wide margin, but this has not always been the case, and for two leading causes of death—suicide and motor vehicle accidents—rates are now much lower among blacks.

Table 3.2 presents information about mortality by cause. Rates are presented for 12 of the most common causes. In 1983 these causes accounted for 76 percent of black deaths and 81 percent of white deaths. [20] Rates are not presented for 1940 since there was a major change in the classification of cardiovascular deaths during the 1940s. Indeed, because of revisions to the International List of Causes of Death, there is lack of comparability to a varying degree from one revision to the next. Data are shown for 1950, 1970, and 1983. The cause-specific death rates in Table 3.2 are age-adjusted so they may be compared across race, sex, time, and cause.

Looking at data for men, we see that in 1950 black men had lower death rates than white men from malignant neoplasms of the respiratory system, cirrhosis, and suicide. They had much higher death rates than white men from homicide, tuberculosis, influenza, and stroke (cerebrovascular diseases). In the two decades following 1950, mortality rates among black men rose for some causes—cancer, diabetes, motor vehicle accidents, and suicide—but declined for others, including heart disease and cerebrovascular causes.

Death rates which had been slowly declining in the 1950s and 1960s, such as those from heart diseases, stroke, and influenza, fell much more rapidly after 1970. Causes which had been increasing in the earlier span, such as cirrhosis, diabetes, and auto accidents, began to decline and among black men homicide rates also fell. Suicide and cancer are the major exceptions since death rates from these causes have continued their increase among black men. A closer inspection of cancer mortality rates among men in the 1970s reveals that for both races the increase is primarily attributable to respiratory cancer. Death rates for cancer of other sites have been rising among blacks, but constant or declining among whites.

[20]National Center for Health Statistics, *Monthly Vital Statistics Report*, vol. 34, no. 6, supplement (2) (September 26, 1985), table 6.

TABLE 3.2

Death Rates by Cause for Black and White Men and Women

	Age-Adjusted Deaths per 100,000			Actual Percent Change in Death Rates		Black Rate as Percent of Whit	
	1950	1970	1983	1950-1970	1970-1983	1950	1983
MEN							
Deaths from All Causes							
Black	1,373	1,319	1,020	−0.2%	−2.0%	143%	146%
White	963	893	698	−0.4	−1.9		
Diseases of the Heart							
Black	416	376	308	−0.5	−1.5	109	120
White	381	348	258	−0.5	−2.3		
Cerebrovascular Diseases							
Black	146	124	64	−0.8	−5.1	168	182
White	87	69	35	−1.2	−5.2		
Malignant Neoplasms of the Respiratory System							
Black	17	55	83	+5.9	+3.2	77	143
White	22	49	58	+4.0	+1.3		
Malignant Neoplasms of All Other Sites							
Black	109	130	149	+0.8	+1.0	100	147
White	109	105	101	−0.2	−0.3		
Influenza and Pneumonia							
Black	64	54	24	−0.8	−6.2	237	159
White	27	26	15	−0.2	−4.2		
Tuberculosis							
Black	81	11	4	−10.0	−9.1	352	700
White	23	3	<1	−10.2	−13.4		
Cirrhosis							
Black	9	33	23	+6.5	−2.8	75	170
White	12	19	13	+2.3	−2.9		
Diabetes							
Black	12	21	18	+2.8	−1.2	109	192
White	11	13	9	+0.8	−2.8		
Motor Vehicle Accidents							
Black	40	50	26	+1.1	−5.0	111	95
White	36	40	28	+0.3	−2.7		
All Other Accidents							
Black	66	69	40	+0.2	−4.2	147	166
White	45	36	24	−1.1	−3.1		
Suicide							
Black	7	10	11	+1.8	+0.7	39	54
White	18	18	19	0.0	+0.4		
Homicide							
Black	51	82	54	+2.4	−3.4	1,275	640
White	4	7	8	+5.6	+1.0		

TABLE 3.2 (*continued*)

	Age-Adjusted Deaths per 100,000			Actual Percent Change in Death Rates		Black Rate as Percent of White	
	1950	1970	1983	1950–1970	1970–1983	1950	1983
WOMEN							
Deaths from All Causes							
Black	1,107	814	590	−1.5%	−2.5%	172%	150%
White	645	502	393	−1.3	−1.9		
Diseases of the Heart							
Black	350	252	192	−1.6	−2.1	156	151
White	224	168	127	−1.4	−2.2		
Cerebrovascular Diseases							
Black	156	108	54	−1.8	−5.3	195	182
White	80	56	30	−1.8	−4.8		
Malignant Neoplasms of the Respiratory System							
Black	4	10	22	+4.6	+6.1	80	105
White	5	10	21	+3.5	+5.7		
Malignant Neoplasms of All Other Sites							
Black	127	108	109	−0.8	+0.1	110	124
White	115	98	88	−0.8	−0.8		
Influenza and Pneumonia							
Black	50	29	10	−2.7	−8.2	263	119
White	19	15	9	−1.2	−3.9		
Tuberculosis							
Black	51	4	1	−12.7	−11.6	510	500
White	10	1	<1	−11.5	−13.4		
Cirrhosis							
Black	6	18	11	+5.5	−3.7	100	180
White	6	9	6	+2.1	−3.1		
Diabetes							
Black	23	31	21	+1.5	−3.0	144	245
White	16	13	9	−1.0	−2.8		
Motor Vehicle Accidents							
Black	10	14	8	−1.7	−4.3	91	73
White	11	14	10	+1.2	−2.6		
All Other Accidents							
Black	28	22	14	−1.2	−3.5	140	179
White	20	13	8	−2.2	−3.7		
Suicide							
Black	2	3	2	+2.0	−3.1	40	38
White	5	7	6	+1.7	−1.2		
Homicide							
Black	12	15	11	+1.1	−2.4	1,200	400
White	1	2	3	+3.5	+3.1		

SOURCES: National Center for Health Statistics, *Health, United States, 1981*, table 14; *Monthly Vital Statistics Report*, vol. 34, no. 6, supplement (September 26, 1985), table 10.

Among men, there is very little evidence of a racial convergence of mortality rates by causes. Table 3.2 shows the death rate of blacks as a percentage of that of whites in 1950 and 1983. The advantages black men experienced in 1950 with regard to respiratory cancer and cirrhosis mortality disappeared, and death rates from those causes are now considerably higher among blacks. There has been some movement toward convergence of the suicide rate, since the death rate of blacks increased faster than that of whites, and of the homicide rate, where the white rate has been rising. Despite recent changes, homicide is the leading cause of death among black men aged 15 to 34 and is second only to heart disease among men aged 35 to 44.[21] Homicide is now the fourth leading killer of black men, and the age-adjusted rate for blacks is almost seven times that of whites. If the trend toward the improved diagnosis and treatment of hypertension continues, homicide will soon become the third most common mode of death for black men.[22]

Death rates for women are also reported in Table 3.2. Throughout the 1950–83 period, the mortality rates of women from most causes of death declined, with the rates of improvement greater after 1970. However, there has been a rise in cancer of the respiratory system, but not in other types of cancer. Traditionally, breast cancer has been a more frequent cause of death among women than respiratory cancer, but deaths from lung cancer will soon outnumber deaths from breast cancer, a change which may be attributed to increased smoking by women.[23] That is, because of cohort patterns in smoking, in the future an increasing proportion of women will reach their 50s with the history of tobacco use which is associated with respiratory cancer. Interpersonal violence, as indexed by the homicide rate, increased among white women in the 1970s and 1980s, but declined among blacks.

The most drastic changes in mortality rates among women have been the rapid decreases in deaths from heart disease and cerebrovascular causes. Since 1950 death rates from these causes have fallen much more rapidly among women than among men, leading to a much larger gender disparity. The increasing gender difference in the life span of men and women is largely attributable to the rapid decline in cardiovascular and stroke deaths among women.

For most causes of death, rates declined more rapidly among black women than white women. Nevertheless, black women are much more at risk of dying from 9 of the 12 causes listed in Table 3.2. The exceptions are motor vehicle accidents, where black women are experiencing

[21]National Center for Health Statistics, *Vital Statistics of the United States: 1979*, vol. 2, table 1–9.
[22]Danchik and Kleinman (1980).
[23]Kleinman, Fingerhut, and Feldman (1980).

an increasing advantage over white women, and suicide, which is quite low among black women. As a result of the changes, there has been some convergence of mortality rates among women. In 1950 the age-adjusted death rate for black women exceeded that of white women by 72 percent; in 1983 by 50 percent. Improvement occurred primarily because cardiovascular death rates became more alike. If this trend continues, there will be a further convergence of death rates among women.

Estimates of Life Span
with the Elimination of Causes of Death

Racial and gender differences in mortality may be illustrated clearly by asking how much longer the life span would be if a given cause of death were eliminated. Using the most recently available (1982) mortality rates specific for age, sex, cause, and race, we calculated estimates of life expectation, assuming that each of nine causes was eliminated. Among the four race-sex groups, white women had the longest life span, 79 years, or 14 more years than black men, the group with the shortest life expectation. The years which would be added if there were no deaths from specific causes are shown in Figure 3.3.

Among both races, the greatest additions to the life span will be made if death rates from heart disease, cancer, and stroke are reduced or eliminated. At present, abolishing heart disease would add five years to the life span of three groups, and four years to that of white women. Eliminating malignant neoplasm deaths would add three years to the life span of three groups, and four years to that of black men.

Homicide is now more costly for black men, in terms of life span, than are cerebrovascular deaths. At present, the actual number of black men dying from strokes exceeds the number who are homicide victims, but there is a great difference in their ages, since homicide victims tend to be young, while stroke victims are elderly. As a result, an effective program to prevent homicide would add more to the life span of black men than would the elimination of stroke deaths.

Eliminating the racial difference in the life span of men depends primarily on reducing black deaths from heart disease, cancer, and homicide. If the mortality rates of black men from these causes were brought down to those of white men, the racial gap in life expectation would be cut in half. Reducing the excess black mortality from cirrhosis, diabetes, and accidents will do little to eliminate the racial gap.

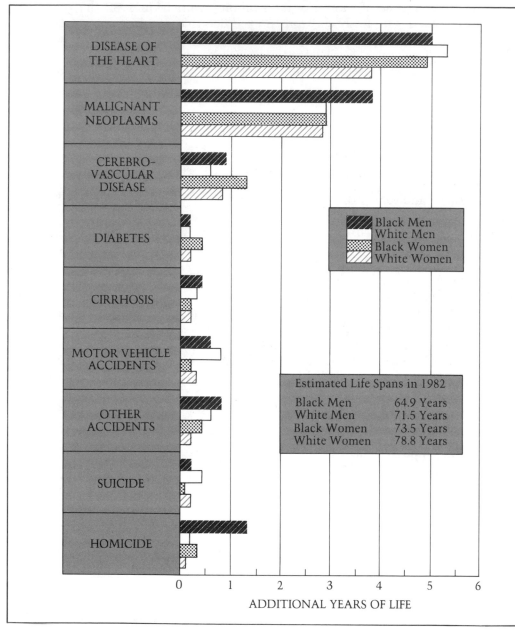

FIGURE 3.3
Years Added to the Life Span with the Elimination of a Disease, 1982

SOURCE: National Center for Health Statistics, *Monthly Vital Statistics Report*, vol. 33, no. supplement (December 20, 1984), table 3.

Among women, the racial difference is attributable to the much higher cerebrovascular and heart disease death rates of blacks. At ages 25 to 54 death rates from heart disease are currently three times as high for blacks as whites, while at ages 45 to 64, black women have similarly elevated death rates from cerebrovascular disease.[24] The racial difference in the life span of women is just under five years, which would be reduced to two years if racial differences in death rates from heart diseases and stroke were eliminated.

Infant Mortality

The death rate which receives the most attention is that for the first year of life: the infant mortality rate. Press reports often stress that children are much more likely to die shortly after birth in the United States than in prosperous foreign countries. In 1984, for example, when the infant mortality rate was about 11 deaths per 1,000 births in the United States, it was only 7 or 8 deaths per 1,000 births in Japan, the Scandinavian countries, and the Netherlands.[25] The high death rate of infants in this country is often taken as a sign that public health policies are not effective or that governmental spending for the care of pregnant women and their babies is inadequate. Presumably, the development of new programs or the allocation of additional funds could bring the infant mortality rate of the United States in line with that of other developed countries.

There is a major racial difference in infant mortality: black children are about twice as likely as white children to die before attaining their first birthday. White infant death rates in the United States are somewhat above those of several Western European countries and Japan, while the death rates of black children exceed those of these nations by a large margin. In such nations as Jamaica, Cuba, and Greece, the reported infant mortality rates are lower than they are for blacks in the United States.[26]

Substantial progress has been made in reducing infant mortality, but there is no indication that the racial gap will soon disappear. Figure 3.4 shows infant mortality rates by race for 1940–1984.[27] Death rates fell rapidly during the 1940s and at a more modest pace in the next

[24]National Center for Health Statistics, *Vital Statistics of the United States: 1979*, vol. 2, pt. A, table 1–9.
[25]National Center for Health Statistics, *Health, United States, 1984*, tables 12 and 14.
[26]National Center for Health Statistics, *Health, United States, 1985*, table 14.
[27]Infant mortality rates for 1940–59 refer to whites and nonwhites.

FIGURE 3.4
Infant Mortality Rates by Race, 1940–1984

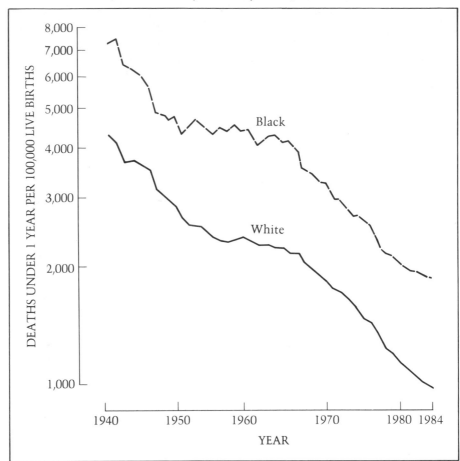

NOTE: Data for 1950, 1955, 1960, 1965, and 1970–84 refer to blacks and whites. For other years, data refer to nonwhites and whites.

SOURCES: Robert D. Grove and Alice M. Hetzel, *Vital Statistics Rates in the United States: 1940–1960* (Washington, DC: U.S. Government Printing Office, 1968), table 38; National Center for Health Statistics, *Health, United States, 1983,* table 11; *Vital Statistics of the United States: 1981,* vol. II, Part A, table 2-1; *Monthly Vital Statistics Report,* vol. 34, no. 13 (September 19, 1986), table 10.

decade. Since the mid-1960s the infant mortality rates have declined rapidly.

Many factors account for the steady decline in infant deaths which occurred in the 1940s and early 1950s. These include urbanization of the population, the hospitalization of births, and drastic reductions in infant deaths from infectious and parasitic diseases.[28] There were particularly sharp decreases in deaths during the postneonatal period—ages 2 to 11 months—reflecting rises in the standard of living and, presumably, the greater educational attainment of mothers. These changes benefited both blacks and whites.

After a decade of relative stability, infant mortality rates resumed their decline in the mid-1960s. While we have yet to ascertain the specific contribution of each cause, many factors are cited, including greater use of prenatal care; the federal nutritional programs, especially the supplemental food program for women, infants, and children; and major advances in neonatology.[29] Undoubtedly, higher standards of living and greater educational attainment of women played a role in this period, just as it did previously. Considerable evidence suggests that the increasing use of birth control and abortion accounts for a substantial fraction of the post-1960 decline in black infant mortality. In the early 1960s black women adopted the oral contraceptive and intrauterine device in large numbers, and after 1970, when federal programs made family planning services available to low-income couples, there was a dramatic rise in the use of abortion by blacks.[30] By 1981 there were 366 abortions per 1,000 births for blacks in an 11-state reporting area.[31] These changes apparently had the effect of reducing the number of births at high risk of neonatal mortality.

Figure 3.4 shows trends in infant mortality rates using a logarithmic scale. If the distance between the trend line for blacks and that for whites remains the same, the ratio of death rates remains constant; thus, the figure suggests that there has been little racial convergence. During the 1940s the rates moved a bit closer together, but since 1950 the rate for blacks has been double that for whites. Should we conclude that no progress has been made in reducing the racial disparity in infant mortality? We must answer this question by focusing on the difference between absolute change and relative change.

In terms of absolute change, there has been much more progress among blacks than among whites. Between 1940 and 1984 the infant mortality rate fell from 74 deaths per 1,000 to 18, a decline of 56 points. Blacks who bear children now are much less likely to lose an infant than were their parents and grandparents who bore children 40 years ago. This important improvement in health helps account for the lengthening life span of blacks. Among whites, the change was from 43

[28]Shapiro, Schlesinger, and Nesbitt (1968).
[29]Godley and Wilson (1980), p. 11.
[30]Grossman and Jacobowitz (1981).
[31]National Center for Health Statistics, "Induced Terminations of Pregnancy: Reporting States, 1981," *Monthly Vital Statistics Report*, vol. 34, no. 4, supplement (2) (July 30, 1985), table B.

deaths per 1,000 births in 1940 to 9 in 1984, a reduction of 34 points, a much smaller decline than among blacks. In 1940 the racial difference was 31 deaths per 1,000 births; in 1984 it was only 9.

Blacks are catching up with whites in terms of the absolute level of infant mortality, but the relative gap is not closing. Since the end of World War II, black children have been twice as likely as white children to die during their first year of life. While in the past the infant mortality rate for blacks was at a much higher level than that for whites and fell by a larger amount, the rate of decline was just about identical for both races. In relative terms, there has been no progress for blacks.

The answer an analyst gives to the question of progress depends on whether the focus is on absolute or relative rates, and often there are compelling scientific reasons for focusing on either one or the other. This is not the case with infant mortality. If the infant mortality rates of both races were cut in half, observers would hail this as progress, but they would not overlook the higher death rate of blacks. The excess mortality of blacks—even if the death rates themselves are low—will be seen as a penalty associated with race since the white rates are much lower.

Why are the infant mortality rates of blacks persistently higher than those of whites? Three factors are important in explaining this racial difference: a higher proportion of black births occurring to women at greater risk of infant loss, the lower birth weights of black infants, and timing and extent of prenatal care of black mothers.

Infant mortality rates are associated with age of mother since they are lowest for women aged 20 to 34 and higher for younger and older women.[32] In 1983 about 30 percent of the black births were to women under age 20 or over age 35 compared with less than 20 percent of white births. Higher infant mortality rates are also associated with order of birth,[33] and in 1983 14 percent of the black births were fourth-order or higher, in contrast to 9 percent of the white. Infant mortality rates are inversely related to the mother's educational attainment;[34] 34 percent of black women giving birth in 1983 had less than a complete high school education compared with 19 percent of white women. Finally, the death rate of infants is higher for those born to unmarried women[35] and 58 percent of the 1983 black births occurred to such women— almost five times the proportion for whites.[36]

[32]Vavra and Querec (1973), table A; Chase (1972), figure 6.
[33]MacMahon, Kovar, and Feldman (1973), figure 5; Vavra and Querec (1973), figure 2.
[34]Armstrong (1972); MacMahon, Kovar and Feldman (1972), table 9.
[35]Vavra and Querec (1973), table E.
[36]National Center for Health Statistics, "Advance Report of Final Natality Statistics: 1983," Monthly Vital Statistics Report, vol. 34, no. 6, supplement (September 20, 1985), tables 2, 17, and 21.

The second factor that accounts for racial differences is birth weight.[37] In the United States, about 12 percent of the babies whose weight is less than 5.5 pounds at delivery die in their first year compared with about 0.5 percent of those whose weight exceeds 5.5 pounds.[38]

Women in the United States bear lightweight infants compared with women in Japan or in the prosperous Western European nations.[39] However, black women are about twice as likely as white women to deliver children weighing less than 5.5 pounds. Throughout the last decade, about 13 percent of the black infants compared with 6 percent of the white infants were under this critical minimum. The most recent national study of infant deaths by weight and race involved births occurring in 1960.[40] The investigators found that the major reason for the higher infant mortality rate of blacks was the large proportion of lightweight births. If the weight-specific mortality rates were unchanged but black infants had the distribution by weight of white infants, the racial difference in infant mortality would have been cut by 60 percent.

There appears to be a real racial difference in birth weight. A recent investigation of birth weights in South Carolina found that after controlling for educational attainment of the mother, her use of prenatal care, the parity of the infant, and the interval since the last birth, there remained a distinct racial effect, with black women bearing lighter-weight children.[41]

In recent years there has been little change in the distribution of births by weight, and thus most of the reduction in infant mortality has come about because of the increased survival of lightweight infants through the neonatal period.[42] This results from improvements in medical technology and the development of more clinics specializing in the treatment of premature births. Programs that will reduce the incidence of lightweight births among blacks offer the best hope for producing a racial convergence of infant mortality rates.

A third explanation for the racial difference in the survival of infants involves prenatal care. Although almost all women in the United States currently obtain care early in their pregnancy, there is still a racial difference in favor of whites. Among those women who became mothers in 1983, 79 percent of the whites compared with 62 percent of the blacks began their prenatal care within three months of conception.

[37]Taffel (1980).
[38]Kleinman (1981), table A.
[39]Kleinman (1981), table A.
[40]Armstrong (1972), table 2; Chase (1972).
[41]Carlson (1984).
[42]Goldenberg et al. (1983).

Only 1 percent of the white women and 3 percent of the black women failed to get any care before delivery.[43]

Racial Differences in the Life Span

To describe racial trends and differences, it is useful to consider life expectation since this provides one number that describes the current mortality experience of a population. Analysts quite often look at the life span of blacks and whites shown in the official life tables and conclude that between 1900 and 1984 the racial difference fell from 16 years to 6 years among men and from 16 years to 5 years among women.[44]

These figures may be misleading. Life tables for the early decades of this century refer to that small fraction of the black population living in the Death Registration Area, so that we cannot be precise about racial differences in the life span before 1940. Also, life tables for some recent years pertain to nonwhites and the life span of this group exceeds that of blacks.[45] Finally, these life tables are not corrected for census undercount. Since 10 to 15 percent of the black male population has been missed in recent censuses, the published death rates of black men are artificially high.[46]

To overcome these difficulties, we calculated new life tables for blacks and whites for the 1940–80 span, the only years for which requisite data were available. Recorded deaths of blacks were used along with population figures corrected for census undercount. This adjustment, which assumes that only a trivial number of deaths are unregistered, adds about one year to the life span of blacks while the similar correction adds little to life expectation of whites because their undercount rates are low. Table 3.3 presents these findings.

In 1940 the white advantage among men was about 7 years, decreasing to 5 years in 1950. The following 20 years were not good in terms of male death rates, especially for young adult blacks. As a result,

[43]National Center for Health Statistics, "Advance Report of Final Natality Statistics: 1983," Monthly Vital Statistics Report, vol. 34, no. 6, supplement (September 20, 1985), table 24.

[44]National Center for Health Statistics, Vital Statistics of the United States: 1979, vol. 2, pt. A, table 6–5; Monthly Vital Statistics Report, vol. 33, no. 13 (September 26, 1985), table 7.

[45]In 1984 the life span for nonwhites was 71.3 years; for blacks, 69.9 years; National Center for Health Statistics, Monthly Vital Statistics Report, vol. 33, no. 13 (September 26, 1985), table 7.

[46]Siegel (1974), table 5; U.S. Bureau of the Census, Current Population Reports, series P-23, no. 115 (February 1982), table 4.

TABLE 3.3

Expectation of Life for Black and White Men and Women, 1940–1980
(corrected for net census undercount)

	Published Life Span from Vital Statistics	Men				Published Life Span from Vital Statistics	Women			
		Birth	Age 20	Age 40	Age 60		Birth	Age 20	Age 40	Age 60
BLACKS										
1940	52.3	55.6	42.6	27.4	15.5	55.6	57.4	43.4	28.3	15.9
1950	58.9*	61.4	46.3	29.4	16.8	62.7*	65.2	49.4	32.4	18.9
1960	61.5*	61.7	46.1	28.6	14.7	66.5*	66.5	50.1	32.1	17.5
1970	60.0	61.8	45.3	28.9	15.6	68.3	69.9	52.8	34.8	20.1
1980	63.7	64.9	47.3	30.1	15.8	72.3	73.1	55.1	36.5	20.7
WHITES										
1940	62.8	62.8	47.9	30.0	15.0	67.3	67.4	51.6	33.4	17.1
1950	66.3	66.7	49.9	31.5	16.0	72.0	72.4	55.0	36.1	19.0
1960	67.6	67.6	50.3	31.7	16.0	74.2	74.5	56.6	37.5	20.0
1970	67.9	68.3	50.6	32.2	16.4	75.5	75.9	57.7	38.6	21.3
1980	70.7	70.7	52.4	34.0	17.5	78.1	78.1	59.4	40.1	22.4
WHITE ADVANTAGE										
1940	10.5	7.2	5.3	2.6	−0.5	11.7	10.0	8.2	5.1	1.2
1950	7.4	5.3	3.6	2.1	−0.8	9.3	7.2	5.6	3.7	0.1
1960	6.1	5.9	4.2	3.1	1.3	7.7	8.0	6.5	5.4	2.5
1970	7.9	6.5	5.3	3.3	0.8	7.2	6.0	4.9	3.8	1.2
1980	7.0	5.8	5.1	3.9	1.7	5.8	5.0	4.3	3.6	1.7

NOTE: Asterisked data are for nonwhites, not for blacks.

SOURCES: National Center for Health Statistics, *Health, United States, 1981*, table 8; *Monthly Vital Statistics Report*, vol. 32, no. 4, supplement (August 11, 1983), table 1; Ansley J. Coale, "The Population of the United States in 1950 Classified by Age, Sex and Color—A Revision of Census Figures," *Journal of the American Statistical Association* 50 (1955): 16–54; Jacob S. Siegel, "Estimates of Coverage of the Population by Sex, Race and Age in the 1970 Census," *Demography* 11 (February 1974): table 6; U.S. Bureau of the Census, *Current Population Reports*, series P-23, no. 108.

the white advantage in life expectation at birth was even greater in 1970 than it had been in 1950. The previous sections of this chapter show that during the 1970s death rates at most ages and for many causes declined more rapidly among blacks than among whites, and thus the white advantage in life span decreased. Nevertheless, the racial difference in life expectation at birth for men in 1980, 5.8 years, was larger than it had been in 1950. When we consider men aged 20 and over, we find that blacks were also further behind whites in 1980 than they had been in 1950, reflecting the very slow declines in the death rates of adult black men.

Among women, the racial gap grew larger during the 1950s because death rates fell more rapidly among whites. Since the 1960s there has been a clear pattern of racial convergence because the additions to the life span have been greater for blacks. In 1980 the white advantage in life expectation at birth—5 years—was only half what it had been four decades earlier.

The progress of black women can also be gauged by comparing their life span with that of white men. In 1960 life expectation at birth for black women fell about one year short of that of white men. During the 1960s white men lost this advantage, and by 1980 black women could expect to live 2.4 years longer than white men, and thus the sex difference in mortality is now much larger than the racial difference.

The life span calculations, based on the corrected data shown in Table 3.3, imply that in 1980 black men could expect to live 65 years and black women 73 years. These are 4 to 6 years shorter than the life spans reported for most western or eastern European nations and for the prosperous Asian countries. The life span of blacks in the United States is also shorter than that of the principally black population of the French Caribbean,[47] but the life span of blacks is about 2 years longer than that reported for most South American nations and at least 15 years longer than the estimated life spans in developing African countries.[48]

Has there been a long-term trend toward elimination of the racial gap in life expectation?[49] On the basis of evidence presented in these chapters, it seems likely that there was some convergence between 1930 and 1950, perhaps because public health measures effectively controlled those infections and parasitic diseases that were very costly to blacks. Since 1960 there has been a convergence in life expectation for women but not for men (see Table 3.3). As a result, racial differences remain substantial. According to the corrected life table for 1980, if 1,000 black men and 1,000 white men begin their occupational careers at age 20, 400 of the black men but only 220 of the white men will die before they reach age 65. Suppose a couple marry when the man and woman are both 25. If they are white, the chances are 11 in 100 that at least one of them will die before age 50; for blacks, the chances are 21 in 100.

---

[47]*Population*, vol. 39, no. 4–5 (July–October 1984), p. 710.

[48]United Nations, *Demographic Yearbook: 1983*, table 4.

[49]The census of 1900 asked women how many children they had and how many were still alive. The release of microdata files from this census permits a new analysis of mortality levels in 1900. A preliminary investigation implies that the racial difference in life span at birth in 1900 was 10 years. Apparently this declined to about 5 years in 1980. Preston and Haines (1984), table 1.

Why Do Death Rates of Blacks Remain Higher Than Those of Whites?

At the start of this chapter, we observed that death rates were declining because of rising standards of living, improvements in diagnosis and treatment, the expansion of federal spending for health, and changes in diet and style of life. Why have these trends not eliminated racial differences in mortality rates in the United States?

First, there may be biological differences which account for the white advantage, a popular explanation in the nineteenth century and emphasized by the Social Darwinists, but such ideas disappeared as medical research demonstrated the basic similarity of the races.[50] Because there is some racial isolation of breeding pools, certain genetically linked diseases are more common among one race, such as sickle cell anemia among blacks, but this is a rare cause of death—responsible for only 342 of the 220,000 deaths of blacks in 1979.[51] Genetic factors may play a role in other causes of death, but it seems unlikely that biological differences account for a large fraction of the five- or six-year life span advantage currently enjoyed by whites.

Second, racial differences in socioeconomic status may help account for persistent racial differences in mortality. Given the strong link between educational attainment and mortality,[52] some of the excess mortality of blacks may be associated with this factor. The persistence of high levels of poverty among blacks—in 1984, 34 percent of the blacks and 12 percent of the whites fell below the poverty line[53]— accounts for some of the racial difference in death rates. Poverty and the more limited educational attainments of blacks may lead to higher death rates in a variety of ways. As we reported, infant mortality rates are linked to a mother's educational attainment, and in 1983 black mothers averaged one less year of schooling than whites.[54] The fact that the death rate among blacks from fires is three times that of whites may reflect the concentration of blacks in older central city housing.[55]

Third, black-white differences in health practices such as smoking,

[50]Hoffman (1896); Stanton (1966); Jones (1981), chap. 3.
[51]National Center for Health Statistics, *Vital Statistics of the United States: 1979*, vol. 2, pt. A, table 1–24.
[52]Kitagawa and Hauser (1973), chap. 5.
[53]U.S. Bureau of the Census, *Current Population Reports*, series P-60, no. 149 (August 1985), table 15.
[54]National Center for Health Statistics, "Advance Report of Final Natality Statistics: 1983," *Monthly Vital Statistics Report*, vol. 34, no. 6, supplement (September 20, 1985), table 21.
[55]National Center for Health Statistics, *Vital Statistics of the United States: 1979*, vol. 2, pt. A, table 1–22.

drinking, dietary intake, and exercise may help explain racial differ-ences in mortality. Racial differences in solving personal or interper-sonal problems may account for the high death rates from suicide among whites and from homicide among blacks. Although there is no racial difference in the consumption of cigarettes by women—30 per-cent of both black and white women said they smoked in 1980—black men are more likely to be smokers than white men. Forty-five percent of the adult black men smoked compared with 37 percent of the white men, a factor which may be related to the higher heart disease and malignant neoplasm death rates of black men.[56]

Between 1971 and 1974 the National Center for Health Statistics interviewed 28,000 persons and asked about their health status, food consumption, and use of medical services. The dietary studies report that the overwhelming majority of Americans, regardless of age, sex, race, or poverty status, had intakes of protein, calcium, and vitamins equal to or above the minimum daily requirement,[57] but blacks were considerably more likely than whites to have a deficient intake of cal-cium and vitamins A and C, and, perhaps because of differences in iron consumption, were more likely to be anemic than whites.[58]

If dietary deficiencies limited the growth of blacks, one might ex-pect that black children and adults would, on average, be shorter and weigh less than whites. This is not the case. Studies in the 1960s showed that black children were at least as tall and as heavy as white children.[59] More recent studies demonstrate that among adults there are no significant racial differences in height.[60] A consistent finding with regard to size is that adult black women tend to weigh much more than their white peers—about 20 pounds on average.[61] Both above and below the poverty line, obesity was more prevalent among black women than among white women.[62] National surveys in the late 1970s reported that one half of adult black women were overweight compared with one quarter of white women. In the 1960s there was no difference in the average weights of adult black and white men, but by 1980 the propor-tion overweight was greater among black men.[63] This racial difference in obesity is linked to the racial difference in diabetes mortality.

[56]National Center for Health Statistics, *Health, United States, 1984*, table 33. These are age-adjusted data. Apparently, there is a substantial racial difference in the amount of smoking, with whites more likely to be heavy smokers than blacks. In 1980, 14% of adult white men consumed 25 or more cigarettes per day compared with 6% of black men.
[57]Abraham et al. (1979), figures 19 and 20.
[58]Singer et al. (1982).
[59]Hamill, Johnston, and Lemeshow (1972).
[60]Abraham, Johnson, and Najjar (1979), figure 7.
[61]Abraham, Johnson, and Najjar (1979), table 11.
[62]Fulwood, Abraham, and Johnson (1981), table 15.
[63]National Center for Health Statistics, *Health, United States, 1984*, table 39.

Fourth, there is the possibility of racial differences in access to health care such that blacks are less likely to get an early diagnosis of an ailment or receive treatment while they can be easily cured. National surveys conducted in the early 1980s found that there was almost no variation by age, sex, race, or family income in the proportion of the population who had seen a physician within the previous year. For all groups, including low-income blacks, about three quarters reported a visit to a doctor within the last twelve months. This is a substantial change from the 1960s when surveys reported that a higher proportion of whites than blacks had seen a physician in the last year: 65 percent for whites and 57 percent for blacks.[64] There is a difference in the type of visit, however, since blacks are more likely than whites to see physicians in hospital clinics or in emergency rooms.[65]

There has also been a change in the use of hospitalization. In 1964 there were 134 discharges per 1,000 whites but only 106 per 1,000 blacks, perhaps reflecting the difficulties blacks had in securing and paying for health care. By 1982 the racial difference was reversed and blacks were more likely to be hospitalized than whites; in that year there were 147 discharges per 1,000 blacks and 125 per 1,000 whites. The average length of stay was also about two days longer for blacks—9 days versus 7 days.[66] In summary, blacks now utilize health care services somewhat more than whites.

By the late 1980s there were no substantial racial differences in the use of many preventive health services for adults, such as electrocardiograms, breast examinations, or glaucoma tests.[67] Although hypertension is much more common among blacks than among whites, sample surveys find that the proportion of adults with undiagnosed hypertension is actually lower among blacks than among whites.[68] Among children, there are persistent racial differences; white parents are much more likely than black parents to have their preschool children vaccinated for measles, rubella, polio, and mumps.[69]

The current racial difference in the life span—a six-year advantage for white men and a five-year advantage for white women—results from differences in socioeconomic status, lifestyle, and health practices, but it is impossible to specify the exact contributions of each factor. Similarly, there is no way to predict whether racial differences will gradually disappear in the future or will grow larger as gender differences have.

[64]National Center for Health Statistics, *Health, United States, 1984*, table 43; Godley and Wilson (1980), table 2.
[65]Drury (1978).
[66]National Center for Health Statistics, *Health, United States, 1984*, table 51.
[67]Moss and Wilder (1977).
[68]Roberts and Rowland (1981), tables 21, 24, and 26.
[69]National Center for Health Statistics, *Health, United States, 1984*, table 29.

4

FERTILITY TRENDS:
1940 TO THE PRESENT

T HE BIRTHRATES of blacks have traditionally exceeded those of whites. In 1940, for example, the total fertility rate—which is the number of children a woman would bear in her lifetime if the rates of a given year remained fixed—was 2.9 for blacks and 2.2 for whites.[1] Since 1940 there have been four major changes that would be expected to reduce the birthrates of blacks and bring them close to those of whites. First, the population became more urbanized and more extensively educated, which, presumably, leads to smaller family size. The proportion of black women of childbearing age who lived in the rural South declined from 42 percent in 1940 to 14 percent in 1980.[2] The median educational attainment of young black women increased from less than 7 years in 1940 to more than 12 years in 1980.[3]

Second, the approval and use of contraception or abortion increased. At one time, birth control was not widely accepted and printed information about it was considered pornography. The Comstock Law, enacted in 1873 and not repealed until 1971, made it a federal crime to

[1]Grove and Hetzel (1968), table 13.
[2]U.S. Bureau of the Census, *Sixteenth Census of the United States: 1940, Population; Characteristics by Age*, table 3; *Census of Population: 1980*, PC80-1-B1, table 55.
[3]U.S. Bureau of the Census, *Sixteenth Census of the United States: 1940, Population; Characteristics of the Nonwhite Population by Race*, table 143; *Current Population Reports*, series P-20, no. 390 (August 1984), table 1.

take any article designed to prevent conception across a state line or send it through the mail. Many states passed similar laws, and until 1965, for example, doctors in Connecticut were prohibited from providing married couples with contraceptives.[4]

Undoubtedly, there was a great deal of attitude change prior to 1940, but some of the traditional ideas about birth control remained long after that. A national study in 1960 found that 20 percent of the black married women and 10 percent of the white women disapproved of contraception. By the late 1960s, however, endorsement of contraceptives became almost universal and differences by race, educational attainment, income, and religion—which had once been substantial—pretty much disappeared.[5]

Third, the development of better contraceptives should have reduced black fertility. In 1940 the pill was unknown; few women relied on intrauterine devices and sterilization was seldom used as a contraceptive. Couples may now select among any array of effective methods which avoid the clumsiness, mess, or time pressure associated with traditional birth control. If they fail in preventing unwanted conceptions, they increasingly use abortion. In 1981 there were about 43 induced abortions per 100 live births, and increase from about 22 abortions per 100 live births in 1973.[6]

Fourth, the federal government has assumed responsibility for providing family planning services to low-income couples. In the 1950s President Eisenhower stated that no federal funds would be used for supplying birth control.[7] Studies in the 1960s showed that many women—especially low-income women—were bearing unwanted children because of difficulties in securing contraceptives from private practitioners. During the Johnson Administration, the Department of Health, Education, and Welfare began supporting family planning clinics, particularly those serving the poor.[8] By 1983 the federal and state governments were spending $340 million annually to support family planning clinics, or an average of $8 per woman aged 15 to 39.[9]

Those who investigated childbearing patterns predicted that black birthrates would fall and that the racial difference in fertility would

[4]Pilpel and Ames (1972); *Griswold* v. *Connecticut*, 381 U.S. 479 (1965).
[5]Forrest and Henshaw (1983), p. 157; Ryder and Westoff (1971), table V-6.
[6]National Center for Health Statistics, *Health, United States, 1985*, tables 1 and 6.
[7]At a news conference on December 2, 1959, President Eisenhower was asked about his administration's policy with regard to family planning. He replied: "I cannot imagine anything more emphatically a subject that is not a proper political or governmental activity or function or responsibility. This government has not, and will not as long as I am here, have a positive political doctrine in its program that has to do with this problem on birth control." Nam and Gustavus (1976), p. 135.
[8]Dryfoos (1976), p. 81.
[9]Levitan (1980), pp. 96–100; Gold and Nextor (1985), pp. 25–30.

disappear because of the changes described. The first national survey to examine family formation among both blacks and whites was conducted in 1960.[10] Black women were classified into three demographic groups and compared with similar white women: (1) those raised on southern farms and still living there, (2) those raised on southern farms but living in cities by 1960, and (3) those who were raised in and lived in cities. Blacks in the first group had very much higher fertility than comparable whites; in the second group the excess of fertility was much lower; those in the third group differed little from whites in their fertility. The investigators concluded that the high birthrates of black couples were largely attributable to their rural background. Such women, they observed, spent few years in school, seldom used contraception effectively, and accepted both their poverty and their childbearing fatalistically.[11] Because fewer black women would come from rural backgrounds, they predicted that the fertility rates of the two races would soon converge.

Five years later, in 1965, Norman Ryder and Charles Westoff conducted another national survey of the determinants of fertility, similarly restricted to the childbearing of married women. They also predicted a disappearance of racial differences, since they found that black women who had limited educations or who came from rural backgrounds bore many more children than similar white women. Among those raised in cities or who had extensive educations, there were only small differences; and in some cases—for example, among women with college degrees—the blacks, not the whites, had smaller families.[12] The urbanization of blacks and their greater educational attainment would, they also predicted, erase racial differences. This study also reported that there was no racial difference in the number of children women desired. Noting a trend toward more effective use of contraceptives by black couples, Ryder and Westoff foresaw declining racial differences.[13] In 1970 another national survey was conducted by the same investigators. They reported sharp increases in the effective use of contraception by women who had previously not used birth control effectively, that is, among blacks, among women low in education, and among Catholics.[14]

Has the predicted racial convergence occurred? Conclusions drawn about convergence also depend on whether one focuses on relative or absolute differences. In 1960, when rates were close to their baby boom

[10]Whelpton, Campbell, and Patterson (1966), p. 335.
[11]Whelpton, Campbell, and Patterson (1966), p. 347.
[12]Ryder and Westoff (1971), pp. 114–30.
[13]Ryder and Westoff (1971), pp. 114–30; Westoff and Ryder (1970).
[14]Westoff and Ryder (1977), p. 335.

peak, the total fertility rate was 4.5 for blacks and 3.5 for whites, or a difference of 1 child. In 1984 the total fertility rates were 2.2 for blacks and 1.7 for whites, or an average difference of only 0.5 child.[15] This indicates convergence since the difference in fertility rates was cut in half. However, the *rates* of decline in childbearing were identical for both races, and in both 1960 and 1984 the total fertility rate of blacks was about 30 percent greater than that of whites. In absolute amounts, black fertility rates have declined sharply, but relative to the rates of whites, there has been no change.

Figure 4.1 shows trends in the total fertility rates of blacks and whites between 1940 and 1984. For two decades after 1940 the birthrates of both races rose rapidly, but they fell just as precipitously between 1959 and 1974. The shift to low fertility and slow population growth among blacks is also illustrated by the trend line for the net reproduction rates. The measure of population replacement takes mortality into account and determines how many daughters will be born to 1,000 women who survive to reproductive ages, assuming that the fertility and mortality rates of a given year remain constant. The figure has a horizontal line at 1,000. If the net reproduction rate equals 1,000, the next generation will be just as large as the present generation, implying zero population growth in the absence of immigration. In 1940 the net reproduction rate for whites was 1,000, and that for nonwhites was 1,200. When these rates reached their peaks in 1959, the rate for whites was 1,700, and that for nonwhites was 2,200, implying rapid population expansion. The era of declining fertility altered this in a short span, and in 1984 the net reproduction rate for blacks implied an increase of only 3 percent from one generation to the next, while that for whites implied a decrease of 18 percent.[16]

As Figure 4.1 shows, the total fertility and net reproduction rates of blacks remain above those of whites. With regard to some aspects of fertility—such as childlessness and the birthrate of married women— there has been a racial convergence. On the other hand, the proportion of births occurring to unmarried women has always been higher for blacks than for whites, and this racial discrepancy has grown larger. In addition, there appears to be a growing racial difference in the timing of family formation.

[15]National Center for Health Statistics, *Vital Statistics of the United States: 1979,* vol. 1, table 1–6; *Monthly Vital Statistics Report,* vol. 35, no. 4, supplement (July 18, 1986), table 4.
[16]National Center for Health Statistics, *Vital Statistics of the United States: 1979,* vol. 34, no. 6, supplement 2 (September 26, 1985), table 3; vol. 35, no. 4, supplement (July 18, 1986), table 4.

FIGURE 4.1

Trends in the Total Fertility Rate and Net Reproduction Rate by Race, 1940–1984

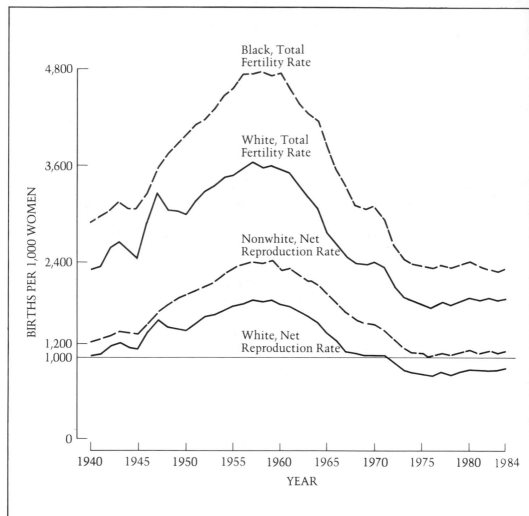

SOURCES: Robert D. Grove and Alice M. Hetzel, *Vital Statistics Rates in the United States: 1940–196*
(Washington, DC: U.S. Government Printing Office, 1968), table 12; National Center for Health Statistic
Vital Statistics of the United States: 1981. vol. 1, table 1-4; *Monthly Vital Statistics Report*, vol. 35, no.
supplement (July 18, 1986), table 4; vol. 35, no. 6 supplement (2) (September 26, 1986), table 2.

Family Size: Convergence or Divergence?

In the nineteenth and early twentieth centuries, average family size declined as more women remained childless and fewer had five or more children. This changed abruptly because women who began their fertility during World War II completed their families with many children and few remained childless. Figure 4.2 shows the number of children borne by women during their lives. These are *cohort fertility rates* since they report the total childbearing of birth cohorts of women.

Looking at this figure leads to several conclusions. First, there are substantial cohort differences in family size, but the patterns of change over time are identical for blacks and whites. Women whose childbearing years overlapped the Depression had unusually small families, much smaller than did women born earlier or later. In comparison, women who attained childbearing ages in the 1940s had unusually large families, although their completed families were considerably smaller than those of women born during the nineteenth century. This can be seen by looking at data in this figure showing family size of women born between 1865 and 1869.

Second, there has been no clear trend toward racial convergence in average family size. Among women born in the years immediately following the Civil War, black families averaged 1.2 children more than whites. This racial difference decreased to 0.5 child for the cohorts affected by the Depression but then increased to 0.7 child for more recent birth cohorts.

Third, with regard to childlessness, there has been a racial convergence. Among nineteenth-century women, about one tenth of both black and white women remained childless. A racial discrepancy later appeared, and among women born at the turn of this century one third of black women and one fifth of white women reached menopause without becoming mothers. Some of this increase is attributable to the greater use of birth control, but the spread of fertility-limiting ailments—especially tuberculosis and venereal disease among blacks—helped to account for high rates of childlessness.[17] The racial difference disappeared after World War II, and among the most recent cohorts to complete their fertility one tenth of both black and white women remained childless.

Finally, there is an increasing racial difference with regard to large families. Among whites, there has been a long-term trend away from families with five or more children, a trend which was hardly altered

[17]McFalls (1984), pp. 470, 480–81, 490; Whelpton, Campbell, and Patterson (1966), pp. 351–53; Grabill, Kiser, and Whelpton (1958), pp. 137–38; Cutright and Shorter (1979).

FIGURE 4.2
Completed Fertility for Birth Cohorts of Black and White Women Aged 40–44

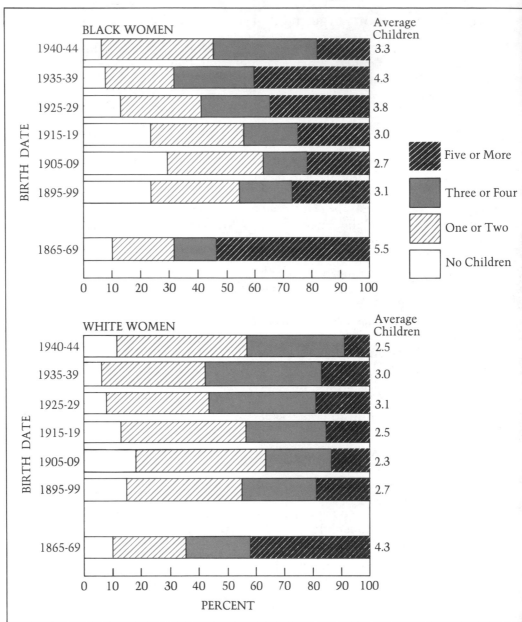

NOTE: Data refer to women who had married by the time of the census or survey. Women were 40 to at these dates.

SOURCES: U.S. Bureau of the Census, *Sixteenth Census of the United States: 1940, Population Differential Fertility, 1940 and 1910, Women by Number of Children Ever Born*, tables 1–6; *Census Population: 1950*, P-E no. 5C, tables 1 and 2; *Census of Population: 1960*, PC(2)-3A, tables 1–3; *Curr Population Reports*, series P-20, no. 375 (October 1982), table 12; no. 406 (June 1986), table 1.

during the post–World War II baby boom. The white women completing their childbearing most recently had record low proportions of women with five or more offspring, but this trend is not as pronounced among blacks, and thus there is a widening racial gap in the proportion of women who bear many children.

Fertility Rates by Age: Convergence or Divergence?

When analyzing racial differences, it is useful to distinguish completed family size from the timing of childbearing. If we compare two societies, we might find that upon reaching age 40 women in both societies bear the same average number of children. However, in one society they may have relatively few children while they are young and many when they are in their 30s, while in the other the timing of childbearing is just the opposite.

We now look at the rate at which women of different ages bear children. These age-specific rates are *period rates* since they describe the frequency of childbearing in a given interval: a one-year span. Figure 4.3 shows births per 1,000 women for eight age groups for 1940–1984.

When these rates are examined, we find three patterns of change in this four-decade interval. First, among black and white women aged 35 and over, there was no baby boom following World War II, and since 1960 the rates for this age group have fallen continuously. This reflects a long-term trend away from childbearing at older ages. Although there are some signs of convergence, the birthrates of black women remain above those of white women.

A second pattern is illustrated by the fertility rates of women aged 20 to 35. The frequency with which they had children increased sharply during the 1940s and 1950s and then declined after 1960, but this decline came to its end in the mid-1970s and since that time there has been little change in fertility. The racial difference in birthrates has pretty much disappeared at ages 25 to 34.

Finally, there is the distinct pattern of teenagers. There was a dip in the fertility rates of this group in the mid-1940s attributable to the mobilization of prospective husbands, but birthrates among teenagers increased in the baby boom era and dropped relatively little after 1960. To be sure, there has been a decline in the last two decades, but it is small compared with the changes in fertility of older women. Births to black women aged 15 to 19 fell from a peak of 170 per 1,000 in 1957 to 96 in 1984, while among whites the change was from 85 births per 1,000 to 43.[18] There has been little racial convergence of fertility rates

[18]National Center for Health Statistics, *Monthly Vital Statistics Report*, vol. 35, no. 4, supplement (July 18, 1986), table 4.

FIGURE 4.3
Trends in Fertility Rates by Age of Women, for Blacks and Whites,
1940–1984

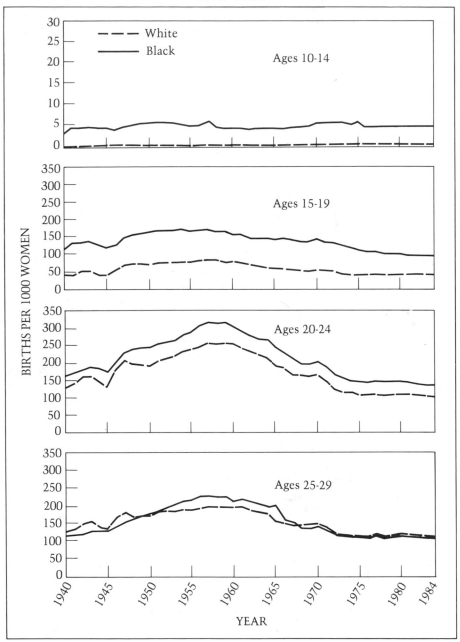

NOTE: Rates for 1940–59 and 1961–63 refer to whites and nonwhites.

FIGURE 4.3 (*continued*)

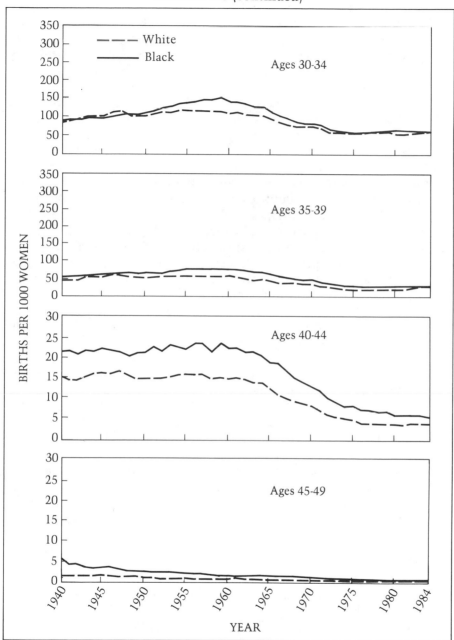

SOURCES: Robert D. Grove and Alice M. Hetzel, *Vital Statistics Rates in the United States: 1940–1960*, Public Health Service Publication no. 1677 (Washington, DC: U.S. Government Printing Office, 1968) table 13; National Center for Health Statistics, *Vital Statistics of the United States: 1981*, vol. 1, table 1–6; *Monthly Vital Statistics Report*, vol. 35, no. 4, supplement (July 18, 1986), table 4.

among teenagers, and in the mid-1980s, as in 1940, young black women were more than twice as likely to become mothers than were young white women.

One can focus attention on either the recent fall in the fertility rates of teenage blacks or the very elevated level of these rates. Between 1970 and 1984 the rate at which teenage black women have children fell by 35 percent compared with a decline of 30 percent for black women aged 30 and over and a drop of 33 percent for white teenage women.[19] On the other hand, the birthrates of teenage blacks are very high compared with most other populations. According to the rates of 1984, when 1,000 black women reach age 20, they will have borne 498 children. A group of 1,000 white women will bear 216 children by age 20. In western European countries, the birthrates for teenage women are very much lower; 46 births per 1,000 women attaining age 20 in the Netherlands, 84 in Denmark, 113 in Ireland, 125 in France, and 157 in England. Eastern European countries have high teenage birthrates, but they are still much lower than those of black women in the United States. For example, the number of births per 1,000 women reaching age 20 was 225 in Poland, 265 in Czechoslovakia, 342 in Hungary, and 352 in Romania.[20] The only nations with teenage birthrates approximating those of blacks in the United States are a cluster of Central American lands: Guatemala, Honduras, and Panama.[21]

The frequent use of abortion has played an important role in reducing teenage fertility. Although statistics are not available for the years before 1975, by 1980 there were about 65 abortions for every 100 live births to black teenagers. Among white teenagers, the ratio was about 85 abortions per 100 births.[22] Were there no use of abortion, the birthrates of teenage women in the 1980s would be as high as they were in the 1950s.

Finally, changes in fertility rates in the two most recent decades imply a widening racial difference in the timing of births; white women have delayed starting their families more than black women, so there has been a sharp increase in the proportion of white—but not black—women who are childless in their 30s.

Figure 4.4 shows the proportion of nonwhite and white women childless at ages 20 to 24 and 30 to 34.[23] Between 1940 and 1960 there

[19]National Center for Health Statistics, *Monthly Vital Statistics Report*, vol. 35, no. 4, supplement (July 18, 1986), table 4.
[20]Westoff, Calot, and Foster (1983), table 1.
[21]United Nations, *Demographic Yearbook: 1983*, table 11.
[22]Henshaw and O'Reilly (1983), p. 8; U.S. National Center for Health Statistics, *Monthly Vital Statistics Report*, vol. 33, no. 6, supplement (September 28, 1984), table 4.
[23]Tabulations of women by numbers of children ever born from the vital statistics system are available only for whites and nonwhites.

FIGURE 4.4
*Proportion of White and Nonwhite Women Childless at Ages 20–24
and 30–34, 1940–1982*

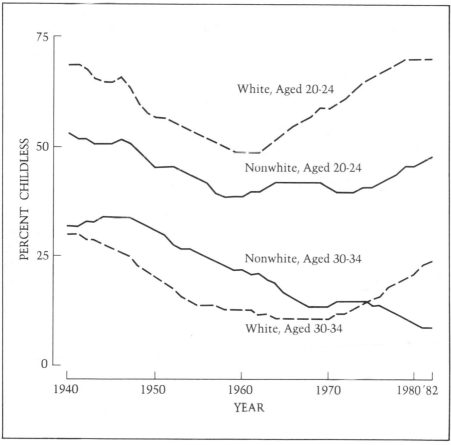

SOURCES: National Center for Health Statistics, *Fertility Tables for Birth Cohorts by Color: United States, 1917–73*, tables 7B and 7C; *Vital Statistics of the United States: 1981*, vol. 1, table 1–13.

was a shift toward earlier motherhood among both races, but after that date the proportion childless for young whites rose sharply and by 1980 reached a level even higher than it was in 1940. Quite clearly, we are witnessing a dramatic shift in the social roles of young white women, a shift which increasingly distinguishes them from their nonwhite peers, among whom the proportion childless has increased only a bit.

Racial differences in the delay of fertility are easily seen in child-lessness at ages 30 to 34. In 1982 a record low proportion of black women were childless—just 9 percent—but for whites there was a sharp rise throughout the 1970s, and a projection of recent trends implies that more than one quarter of all white women will enter their 30s childless.

The Baby Boom and the Birth Dearth: Why?

The most frequently cited explanation for the dramatic increase in fertility followed by a decline was developed by Richard Easterlin.[24] His argument focuses on the relative size of birth cohorts and the economic opportunities men face as they reach their 20s. If their own economic prospects are good compared with those of their fathers, they will have many children, prompted in large part by their feelings of economic security. If their own prospects are poor compared with those of their fathers, they will put off marriage and have few children.

During the 1920s and 1930s there were relatively few births, but when men born in those decades reached their 20s, they entered an economy which was bustling in the post–World War II economic boom. They could easily find high-paying and secure jobs even without college educations. They realized that their own standard of living was going to be much more prosperous than that of their parents who suffered through the Depression. This encouraged them to marry while they were young, purchase cars, and buy new homes. Since their wages were high, they could afford to have numerous children and so many of them decided to father a third, fourth, or fifth child. According to Easterlin, the baby boom came about because of the fortuitous economic circumstances of those men who reached childbearing age shortly after World War II.

Men born during the baby boom who entered the labor force in the 1970s or 1980s faced a much bleaker situation. These birth cohorts are the largest in the nation's history, meaning that there was great competition for jobs. In addition, the national economy had grown lethargically since 1974 so that these young men often experienced unemployment or limited opportunities as they started their careers. These men knew that their own parents were financially secure and that at young ages could afford cars, homes, vacations, and appliances. They judged that their own economic prospects were not as bright and thus delayed marriage and, if married, restricted family size to preserve their precarious status. This explanation accounts for both the rise and fall of fertility in terms of the relative economic opportunities available to young men, opportunities which are strongly influenced by the size of birth cohorts. Presumably, age at marriage will decline and fertility rates will

[24]Easterlin (1962, 1980).

rise in the 1990s if men in the small birth cohorts of the 1970s reach childbearing age during a period of economic expansion and believe that their own prospects are better than those of their parents.

An alternative explanation developed by William Butz and Michael Ward, takes into account the economic situation of men but also emphasizes the economic opportunities and wage rates of women.[25] For much of the baby boom period, few women were employed and the real wages of men were rising sharply. Beginning in the 1960s, the employment of wives increased sharply and their earnings rose, largely because they spent more hours on the job. As this trend continued into the 1970s, the economic welfare of a family increasingly depended on the wife's earnings. If her earnings were substantial, it would be costly for a wife to become a mother because she would probably drop out of the labor force for childbearing and childrearing. She might lose her seniority or the work experience which leads to better jobs and greater earnings, thereby undermining her family's economic status.

Butz and Ward account for the fluctuation of fertility between 1947 and 1975 by analyzing trends in the earnings of men and the employment of wives. In their view, the increasing labor force participation of wives was not a response to lower fertility rates; rather, more labor force participation encouraged women to put off having babies. If women continue to be employed at a high rate, the earnings of women will continue to be important for families, and fertility may remain low.

The empirical tests of these overarching explanations for childbearing trends seldom, if ever, deal specifically with blacks.[26] They describe either the total population or whites. On the one hand, if macro social and economic trends influence the fertility rates of whites, there is good reason to believe they similarly influence the childbearing of blacks because the rates of both races have moved in a similar manner. On the other hand, these explanations may need qualifications if they are to account for the childbearing of blacks. These theories deal with the fertility rates of married couples and use such indicators as the wages of husbands and the employment of wives as explanatory variables. In particular, these theories do not explain childbearing by unmarried women, although such women now account for about three fifths of black births and one eighth of white births.[27] Explanations dealing with the employment of women may apply to blacks quite differently than to whites since traditionally black women have had higher labor force participation rates than white women but also have had higher birthrates.

[25]Butz and Ward (1979).
[26]Devaney (1983); Lee (1976); Sanderson (1976); Smith (1981).
[27]National Center for Health Statistics, *Monthly Vital Statistics Report,* vol. 34, no. 6, supplement (September 20, 1985), table 17.

Very different theories seeking to explain the baby boom and subsequent birth dearth focus on psychological rather than economic factors. Glen Elder studied the experiences of people who grew up during the Depression and the consequences this had for their social values.[28] Because of the economic crisis, many husbands were unable to support their families and left permanently to avoid humiliation while others temporarily criss-crossed the country in search of jobs. Many wives and older children were forced to seek employment or devise ways to support themselves. As a result, teenagers during the Depression decade often assumed adult responsibilities by caring for their younger sibs, by getting part-time jobs, or by helping to hold the family together psychologically and emotionally.

Elder speculates that these experiences had lasting consequences for those people who started their own families after World War II. They placed a high value on strong families; indeed, they may have esteemed family life more highly than occupational achievement or economic success. They believed that families provided succor in a crisis and that teenagers were, in some cases, a form of financial security. Since they had favorable attitudes toward marriage and family life, they married at an early age, bore numerous children, and divorced at a low rate.[29] The economic boom of the post–World War II era helped them to achieve their family and fertility goals because they could afford to marry at young ages and raise numerous children.

A different set of psychological factors, some speculate, may explain the more recent decline in fertility rates. Children raised during the post–World War II baby boom were treated, it is argued, quite unlike earlier generations. Not only were they permitted more freedom by their parents, but the economic conditions were exceptionally favorable so that these children focused on their own interests and ambitions. Their parents were able to buy them clothes and cars and send them on trips or to college, and they were not forced to start working at age 16 or 17. The outcome, according to Christopher Lasch, is a narcissistic personality style.[30] Compared with earlier generations, people who became adults in the later 1960s or 1970s paid most attention to their personal achievements, often in the occupational or economic sphere. Because of their narcissism, they were reluctant to make the commitments necessary for marriage and, because they were raised in an era of permissiveness, they were willing to experiment with living arrangements that were once forbidden. If the narcissist marries, he or she may be unwilling to make the sacrifices necessary to remain married. Children, of

[28]Elder (1974).
[29]Cherlin (1981), p. 90.
[30]Lasch (1979).

course, are a problem to the narcissist since they are demanding and interfere with personal goals and economic achievements. As a result, in an age of narcissism adults delay marriage, many marriages end in divorce, and fertility rates remain low.

Undoubtedly, many people will find these psychological explanations more compelling than the economic ones. They appear to succinctly explain both the rise in fertility and its more recent decline, but they are difficult to prove or refute since they are based on either interviews with small numbers of people or the impressions of therapists who believe that there has been a sharp increase in narcissism and its concomitant ailment depression.

Many demographers contend that it is impossible to attribute the fluctuations in fertility to a single economic or psychological factor. As Charles Westoff points out, in the United States as in most Western European countries, the social roles of women and values about marriage and the family underwent profound changes in the post–World War II era, especially after 1960.[31] In these countries—just as among blacks and whites in the United States—there has been a rise in the age at marriage, cohabitation before marriage has become the norm, the proportion of marriages ending in divorce has increased sharply, and a record proportion of children have been born to unmarried women. The delay in marriage not only reduces the birthrate but provides many women with the opportunity to obtain education or job skills. The economic dependence of women on men is reduced.[32] With greater economic resources women are less likely to marry, and if their marriage is unsatisfactory separation or divorce is a realistic alternative, but these have the consequences of further reducing birthrates.

Black Fertility Rates: Why Are They So Much Higher Than Those of Whites?

Although the birthrates of blacks parallel those of whites and have now declined to a low level, they remain about 30 percent greater than those of whites. In relative terms, the racial difference has hardly changed.

Two hypotheses seek to explain this persistent difference: the *social characteristics hypothesis* and the *minority group status hypothesis*. Many investigators assume that as blacks become assimilated into the majority culture, they will take on the social and economic characteristics of whites and their fertility rates will eventually be in-

[31]Westoff (1978a, 1978b).
[32]Westoff (1978a), p. 81.

distinguishable from those of the majority. The high fertility of blacks, it is thought, is not racial in origin, but rather reflects their social and economic situations.

The minority group status hypothesis argues that even if a minority group is economically assimilated, it may have cultural values and preferences which lead to fertility rates unlike those of the majority population.[33] In the case of Hispanics and Catholics, it might be higher fertility; in the case of Jews and Japanese in the United States, it might be lower fertility. With regard to blacks, the unusually low fertility of those with college educations may represent an adjustment mechanism. Blacks entering the middle or upper rungs of society may face great economic challenges and their precarious status may be threatened if they have many children.

Numerous studies have compared the social characteristics and minority group status hypotheses to see which is the better explanation for current racial differences[34] There is no consensus about findings, although most studies find a net difference which cannot be attributed to racial differences in social or economic status. We will keep the distinction between the social characteristics and minority group status explanations in mind as we look at three major ways in which black and white fertility differ. We will consider teenage childbearing, racial differences in the martial status of mothers, and, finally, the use of contraception and abortion.

The Racial Difference in Teenage Fertility

The racial differences in rates of childbearing are greatest among teenagers. In recent years, about 37 percent of black women became mothers by age 20 compared with 18 percent of white women.[35] There has been a decline in the rate at which teenagers bear children, although teenage birthrates have fallen less rapidly than those of older women. As a result, teenage births constitute a growing share of the nation's total births. In 1960 teenage mothers bore 20 percent of the total black births; by 1983 this had increased to 24 percent.[36]

Why do black teenagers bear children at a much higher rate than white teenagers? There have been extensive investigations of the sexual

[33]Lee and Lee (1952); Smith (1960); Thomlinson (1965).

[34]Bean and Wood (1974); Johnson (1979); Johnson and Nishida (1980); Pohlmann and Walsh (1975); Roberts and Lee (1974); Rindfuss (1980); St. John (1982).

[35]National Center for Health Statistics, *Vital Statistics of the United States: 1981,* vol. 1, table 1–17.

[36]National Center for Health Statistics, *Vital Statistics Report,* vol. 35, no. 4, supplement (July 18, 1986), table 2.

activities of teenage women and their use of birth control, and they provide an explanation in terms of three factors.[37]

First, a higher proportion of black teenage women are sexually active and therefore at risk of conceiving. A 1979 survey which sampled residents of metropolitan areas found that 66 percent of the black teenage women compared with 47 percent of the white women reported having had intercourse. According to current rates, by the time they reach age 20 nine tenths of black single women and two thirds of white women have been sexually active. In addition, black women start at younger ages since the average age at first intercourse for women who are sexually active before age 20 is 15.5 for blacks and 16.5 for whites.[38]

Second, there are racial differences in the effective use of birth control. To be sure, teenagers are not faithful users of contraceptives, since only a small fraction do anything to prevent conception when they begin their sexual activity. Girls who first engage in sex at age 15 or younger are particularly unlikely to protect themselves against pregnancy; and among those who wait to the older teen years, a majority still begin their sexual activity without benefit of contraception.[39]

Sexually active white teenage women are more likely than black women to use birth control, which contributes to the higher birthrates of young blacks. In the 1979 study 31 percent of the blacks compared with 35 percent of the whites said they used birth control always, while 36 percent of the blacks compared with 24 percent of the whites said they never used contraceptives.[40]

These surveys explored why teenage women did not use birth control.[41] Only a minority reported that they were intentionally trying to become mothers. Many expressed the opinion that they did not need contraceptives because they were too young to conceive, because they had intercourse too infrequently, or because they were not ovulating at the time of intercourse. Others reported that intercourse was unplanned so they did not have time to use birth control. Many teenagers also reported that they did not go to family planning clinics because their parents might find out.[42]

Third, white teenagers are more likely than black teenagers to obtain abortions. Between 1973 and 1981 the number of legal abortions obtained by teenagers rose 80 percent, from about 200,000 to 365,000, while the number of teenage births fell 13 percent, from 620,000 to

[37]Zelnik, Kantner, and Ford (1981); Zelnik and Kantner (1980); Hogan and Kitagawa (1985).
[38]Zelnik and Kantner (1980), table 1.
[39]Zelnik, Kantner, and Ford (1981), chap. 4.
[40]Zelnik and Kantner (1980), table 7.
[41]Zelnik and Kantner (1979), table 3.
[42]Zabin and Clark (1981).

540,000.[43] This implies that an increasing proportion of teen pregnancies result in abortions, but teenage blacks are less likely than teenage whites to choose abortion. There are numerous problems in assembling accurate measures of induced abortion, but thirteen states now keep systematic records. In those states in 1983, if we consider those pregnancies to teenagers that resulted in either birth or abortion, we find that 42 percent of the white pregnancies compared with 40 percent of the black pregnancies were terminated by induced abortion.[44]

Racial Differences in the Marital Status of Mothers

One of the major racial differences in fertility—and one which is growing larger—concerns marital status of mothers. By the early 1980s almost six black babies out of ten were borne by women who were not married. Among whites, one birth in eight occurred to an unmarried woman.[45] The proportion of black births occurring to unmarried women is very high compared with the rates of Asian nations and considerably higher than those reported in Europe. The European nations with the greatest proportions out of wedlock—Denmark, Iceland, and Sweden—reported just under 40 percent, a rate that is three times the proportion for American whites but well below that for blacks. Several Caribbean islands and Central American nations, including Barbados, El Salvador, Panama, and St. Lucia, have proportions out of wedlock which exceed that of American blacks.[46]

Many observations are made about the negative consequences of out-of-wedlock births, but few studies have thoroughly analyzed the long-term effects of illegitimacy. These are difficult topics to investigate not only because people may conceal out-of-wedlock births, but also because of their close relationship to age of mother. A high proportion of nonmarital births are born to teenage women: 36 percent of both black and white out-of-wedlock births in 1984.[47] Bearing a child at a young age may truncate the mother's education; thus, it is difficult to disentangle the separate effects of illegitimacy, mother's age at birth, and mother's education attainment on the welfare of the mother or her baby.

[43]National Center for Health Statistics, *Health, United States, 1984*, table 5; *Vital Statistics of the United States: 1973*, vol. 1, table 1–53; *Monthly Vital Statistics Report*, vol. 32, no. 8, supplement (December 29, 1983), table 2.

[44]Burnhan (1983, 1981); Ezzard, Cates, Kramer, and Tietze (1982); Prager (1985), table A; Powell-Griner (1986), tables A and 1.

[45]National Center for Health Statistics, *Monthly Vital Statistics Report*, vol. 35, no. 4, supplement (July 18, 1986), table 18.

[46]Hartley (1975), chap. 2; United Nations, *Demographic Yearbook: 1981*, table 32.

[47]National Center for Health Statistics, *Monthly Vital Statistics Report*, vol. 35, no. 4, supplement (July 18, 1986), table 18.

Two of the deleterious correlates of nonmarital births are higher rates of infant mortality and poverty. The legitimacy status of a child has an effect on his or her chances of survival to age 1 apart from other important factors such as the mother's age or her educational attainment.[48] An out-of-wedlock child is not necessarily raised by only one parent, but the child often begins life in a family which does not include a father, a factor associated with poverty. Women who bear children before marriage are also more likely to separate or divorce than are other women, thereby increasing the probability that they will be raising children by themselves at a later point in their lives.[49] In 1985, for example, 67 percent of the black children living in families headed by mothers lived below the poverty line compared with 19 percent of the black children in husband-wife families. Among whites, the comparable proportions were 45 percent impoverished for children living with their mother only and 10 percent for children in husband-wife families.[50] Other studies report that out-of-wedlock children who grow up apart from a father experience more social problems than other children and perform less well in school.[51]

To the extent that there are problems associated with fertility outside marriage, they are affecting a rapidly increasing proportion of the nation's children. Figure 4.5 shows the reported proportion of births occurring to unmarried women between 1940 and 1984. Most of the change in the marital status of mothers occurred in the last two decades. That is, between 1940 and 1960 the proportion out of wedlock rose from 17 to 22 percent among blacks and rose to 59 percent in 1984. Among whites, there was almost no change during the 1940s or 1950s, but between 1960 and 1984 the proportion of births occurring to unmarried women increased from 2 to 13 percent.

These data must be interpreted cautiously. At present, only 41 states and the District of Columbia ascertain the marital status of a mother. On the basis of data from reporting states, national estimates are developed of births to married and unmarried women, a procedure which provides conservative estimates of illegitimate births.[52] Few studies have investigated the accuracy of reporting, whether there are racial differences in misreporting, or whether women are more willing now to report themselves as unmarried at the time of delivery than they were earlier.[53]

[48]Taffel (1980); Vavra and Querec (1973); Moore, Simms, and Betsey (1984), chap. 1.
[49]O'Connell and Rogers (1984), table 5.
[50]U.S. Bureau of the Census, *Current Population Reports*, series P-60, 154 (1986), table 16.
[51]Sklar and Berkov (1981), p. 42; Furstenberg (1976), pp. 214–15.
[52]Sklar and Berkov (1981), p. 24.
[53]Clague and Ventura (1968); Vincent (1961), p 53.

FIGURE 4.5

*Trends in Percent of Total Births Delivered to Unmarried Women
by Race, 1940–1984*

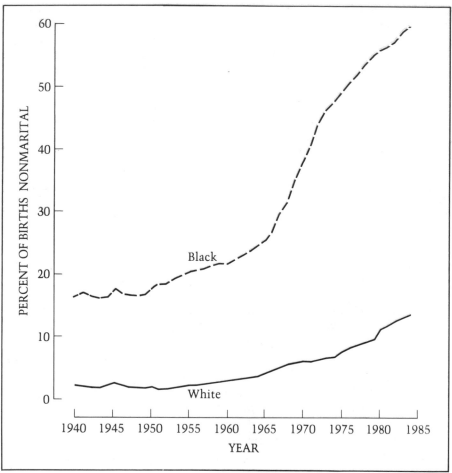

SOURCES: Robert D. Grove and Alice M. Hetzel, *Vital Statistics Rates in the United
States: 1940–1960*, National Center for Health Statistics, table 28; National Center for
Health Statistics, *Vital Statistics of the United States: 1981*, vol. 1, Natality, table 1–32;
Monthly Vital Statistics Reports, vol. 33, no. 6, supplement (September 28, 1984), table 17;
vol. 34, no. 6, supplement (September 20, 1985), table 17; vol. 35, no. 4 supplement (July
18, 1986), table 18. Data for years prior to 1969 refer to nonwhites.

The increase in the proportion of births to unmarried women
should not be interpreted as an increase in the birthrate of unmarried
women. In fact, the frequency with which unmarried black women bore
children actually *declined* during the 1970s. Attention is focused on the
proportion of births to umarried women when trends in legitimacy

status are discussed, but this measure is strongly influenced by the rate at which married women bear children. If married women reduce their childbearing but unmarried women continue having children at their previous rate, the proportion of births out of wedlock will necessarily rise.

To describe trends in fertility for married and unmarried women, rates were calculated for 1940–84. Table 4.1 shows births per 1,000 women classified by age and marital status. The birth rates for married women aged 15 to 19 are extremely high because marriages for these young women are often hastened by conception.

Table 4.1 illustrates that the marital fertility rates of black women peaked in 1960 and then fell rapidly; indeed, the age-specific fertility rates were just about cut in half. Among unmarried black teenagers, birth rates peaked in 1970, but among older unmarried black women they peaked in 1960. There has been a decline recently, and by 1984 unmarried black women were bearing children less frequently than they had in the past.

Trends among married white women resemble those among married black women: a high point in 1960 and a steady decline thereafter. Unmarried white women, however, differ from black women in that

TABLE 4.1

Birthrates for Married and Unmarried Black and White Women, 1940–1984

es	Births per 1,000 Married Women					Births per 1,000 Unmarried Women				
	1940	1950	1960	1970	1984	1940	1950	1960	1970	1984
ACK WOMEN										
5–19	340	475	713	533	322	43	69	77	97	87
0–24	264	292	364	263	236	46	105	167	131	111
5–29	150	180	224	148	132	33	94	172	101	80
0–34	105	114	142	81	72	23	64	104	72	45
5–44	56	47	54	29	18	9	20	36	22	13
IITE WOMEN										
5–19	401	399	532	432	409	3	5	7	11	19
0–24	263	281	355	244	201	6	10	18	23	28
5–29	168	193	219	165	145	4	9	18	21	25
0–34	103	116	121	78	81	3	6	11	14	16
5–44	34	39	39	20	16	1	2	4	4	5

TE: Data for 1940 and 1960 refer to nonwhites.

JRCES: National Center for Health Statistics, *Monthly Vital Statistics Report*, vol. 35, no. 4, plement (July 18, 1986) tables 3 and 18; *Vital and Health Statistics*, series 21, no. 36 (May 1980) table Robert D. Grove and Alice M. Hetzel, *Vital Statistics Rates in the United States: 1940–1960.* shington, DC: U.S. Government Printing Office, 1968) tables 14 and 15.

their fertility rates have risen continually since 1960—perhaps a result of the increasing tendency for whites to cohabit before marriage. As a result, racial differences in the fertility of unmarried women are slowly decreasing; however, unmarried black women are still three to four times as likely to become mothers as white women.

The proportion of births out of wedlock is also influenced by changes in marital status. If women delay marriage, they have more years in which to bear out-of-wedlock children and fewer years for marital births. We can ascertain how much of the total increase in the proportion of births out of wedlock is attributable to demographic components such as the rate of childbearing by married women, the rate of childbearing by unmarried women, the proportion of women married, and the age distribution of women. Figure 4.6 shows components of change in the proportion of births out of wedlock.

Between 1960 and 1984 the proportion of black births out of wedlock more than doubled from 21 to 59 percent. Of that 38 point increase, 16 points are due to the reduced fertility rates of married women. That is, if the only change in this span had been the observed decrease in the birthrate of married women, the proportion out of wedlock would have risen from 21 to 37 percent. Since the fertility rates of unmarried black women declined, this factor tended to reduce the proportion of births out of wedlock. If this had been the only change in this period, the proportion illegitimate for blacks in 1984 would have been 17 percent, that is, a proportion similar to what it was in 1940. The change in marital status was the biggest single factor since it contributed 31 points to the rise in the proportion of births out of wedlock. Changes in the age distribution of women also had the small net effect of increasing the proportion illegitimate.[54]

Among whites, the declining fertility rates of married women, the decrease in the proportion married, and the changing age distribution of women contributed to the higher proportion out of wedlock. However, unlike the situation among blacks, the change in the birthrates of unmarried women led to a greater proportion out of wedlock.

What can we conclude from this investigation? First, for both races a rapidly increasing proportion of children are borne by unmarried women. Second, the rise in the proportion of births out of wedlock is largely explained by declining marital fertility and by delays in marriage. Third, the rate at which unmarried women become mothers decreased among blacks while it rose modestly among whites.

We can begin to explain the high rates of childbearing by unmarried

[54]If we add the effects of these demographic components, we get a sum that is different from the total change indicated in figure 4.6. An interaction effect results from these different factors changing simultaneously, but it is not shown separately.

FIGURE 4.6
*Components of Change in the Percentage of Births Out of Wedlock
for Blacks and Whites, 1960–1984*

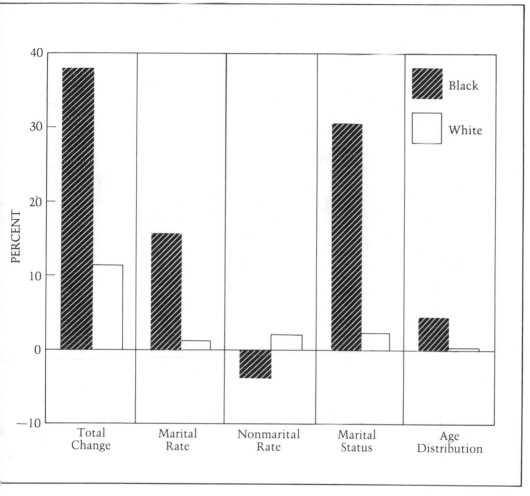

)TE: Data for 1960 refer to nonwhites.

URCES: National Center for Health Statistics, *Vital Statistics of the United States: 1960*, vol. 1, tables
, 2-13, and 5-2; *Monthly Vital Statistics Report*, vol. 35, no. 4, supplement (July 18, 1986), tables 2, 4,
18.

women and the persistent racial differences by observing that single
black women who become pregnant are much less likely to legitimize
the birth through marriage than are white women. In the late 1970s, for
example, about 34 percent of the black women aged 20 to 24 who con-

ceived premaritally and then bore children, married prior to delivery; among whites, 58 percent married following such a pregnancy.[55] Also, unmarried white women are more likely than black women to choose abortion. The National Center for Health Statistics (NCHS) provides information about the outcome of pregnancies which resulted in either births or induced abortions in the twelve-state reporting area.[56] Among unmarried black women of all ages, 34 percent of the pregnancies that resulted in either births or abortions were terminated by induced abortion; for unmarried white women, 62 percent ended in abortions, suggesting that unmarried white women are about twice as likely as black women to select abortion.

These findings raised important questions: Why do black women differ from white with regard to marriage and abortion? One view contends that the high illegitimacy rate is a cultural practice surviving from the time of slavery when few blacks could marry or maintain stable families.[57] Although this is frequently cited, it conflicts with historical evidence which indicates that blacks have always made an important distinction between legitimate and illegitimate births and that the majority of black families were husband-wife families for decades after Emancipation.[58] Furthermore, both the proportion of births out of wedlock and the rate at which unmarried black women bore children were much lower in 1940 than in 1970 (see Table 4.1), which casts doubt on a historical pattern surviving since slavery as an explanation for current fertility patterns.

Urban ethnographers who have lived in low-income neighborhoods provide insights about family formation for one component of the black population.[59] Apparently, many boys and girls initiate their sexual activity in their early teens and do not use birth control. To conceive a child is a mark of womanhood for a young girl and confirms masculinity for the father. Quite often, girls believe that it will be fulfilling to become mothers and fail to appreciate the responsibilities which ensue. The couple may love each other and hope to marry, but they recognize that the male has a limited earnings potential. Because he cannot support a family, the woman has little incentive to marry him. Indeed, she may have a mother, a grandmother, and other relatives who will assist her in caring for a baby. This extended family may provide more financial and emotional security than would marriage.

If impoverished black teenage women believe that their chances for

[55]O'Connell and Moore (1981), table 5.
[56]Powell-Griner (1986), table B.
[57]Vincent (1961), p. 94.
[58]Johnson (1934), pp. 66–70; Litwak (1979), pp. 71–75; Engerman (1978); Lantz and Hendrix (1978).
[59]Rainwater (1965); Liebow (1967); Hannerz (1969); Stack (1974).

a stable marriage and a successful career are small, they may not think it is costly to become mothers while young. Women from middle class backgrounds, on the other hand, may believe that early childbearing will seriously restrict their later opportunities and thus avoid becoming teenage mothers.[60] This may help account for the racial difference in the use of abortion and birth control by unmarried young women.

Those who comment about social policy argue that the expansion of federal welfare programs such as the Aid to Families with Dependent Children encourage unmarried women, especially blacks, to bear children.[61] But while benefit levels for these programs increased during the interval from the mid-1960s to the mid-1970s, the rate at which unmarried black women had babies actually decreased (see Table 4.1).

Many investigators have analyzed the direct impact of welfare programs on fertility,[62] and there is consensus that few women deliberately bear children to qualify for welfare payments. Indeed, these studies suggest that almost all pregnancies occurring to young unmarried women are unplanned.

On the other hand, the availability of welfare may influence decisions about marriage, abortion, and adoption. In the United States there has been a trend away from hasty marriage following premarital conceptions, a trend which may be partly attributable to the expansion of transfer payments. Among single black women aged 15 to 19 in the early 1980s, about 9 percent of those who became pregnant and had a child married between the time of conception and birth. In contrast, in the early 1950s about 28 percent of similar black women married before delivering their baby.[63] The availability of welfare may reduce the likelihood that a premaritally pregnant young woman will marry. Although the evidence is more disputable, welfare payments apparently also reduce the frequency with which premaritally pregnant women choose abortion and increase the proportion of women who keep out-of-wedlock children rather than placing them for adoption.[64]

Contraceptive Use: Persistent Racial Differences

Fertility studies conducted since World War II report that black women have desired family sizes no larger than those of white women.

[60]Hogan and Kitagawa (1985), p. 852.

[61]Murray (1984), pp. 125–33 and 156–62.

[62]Bernstein and Meezan (1975); Leibowitz, Eisen, and Chow (1980); Cutright (1970). These are evaluated and summarized in Moore and Burt (1982), pp. 108–13. See also Ellwood and Bane (1984); Wilson and Neckerman (1986).

[63]O'Connell and Rogers (1984), table 1; O'Connell and Moore (1981).

[64]O'Connell and Rogers (1984), table 1; O'Connell and Moore (1981); Moore and Burt (1982).

The 1960 study reported that black women, on average, actually wanted about 0.5 child less than white women.[65]

It is argued that the discrepancy between these findings and the actual birth rates of blacks comes about because black women do not effectively control their fertility. The studies of contraceptive use conducted before World War II found that married black women seldom tried to avoid pregnancy.[66] Apparently the use of birth control increased in the black community; the study conducted in 1960 reported that 59 percent of the black married women had used contraceptives, which was substantially below the 80 percent figure for white women.[67]

National studies conducted in 1965 and 1970 discovered widespread increases in the use of birth control, especially the effective oral contraceptives.[68] The authors of the 1970 study generalized:

> Group differences in the control of fertility have shown a distinct tendency toward convergence. The most effective methods are being used in about the same proportion by women of different educational achievement, by Catholics and non-Catholics, and by whites and blacks . . . .[69]

These investigators concluded that the excess fertility of black women was unplanned and because of the increasing use of birth control by blacks, racial differences in fertility would disappear, at least among married women:

> Nearly all the excess of black over white marital fertility is due to the considerably higher unwanted fertility among blacks. Given the substantial reduction in the unwanted fertility rate among blacks, a rapidly increasing convergence of black and white marital fertility can be expected during the current decade.[70]

Is there still a major racial difference in contraceptive use which explains why black birthrates remain above those of whites? Is the black excess entirely unplanned childbearing? These questions are difficult to answer because most studies of birth control have been limited to currently or formerly married women. However, the majority of black children are now born to unmarried women.

On the basis of recent studies, some generalizations may be made about differences in contraceptive use. First, there is a persistent racial

[65]Whelpton, Campbell, and Patterson (1966), table 20.
[66]Beebe (1942); Stix (1941); Pearl (1936).
[67]Whelpton, Campbell, and Patterson (1966), table 192.
[68]Westoff (1976); Westoff and Jones (1977).
[69]Westoff and Ryder (1977), p. 335.
[70]Westoff and Ryder (1977), p. 338.

difference in the proportion of married women who use birth control.[71] There are numerous measurement issues since it is not easy to define the population "at risk" of an unwanted pregnancy, but it is commonly accepted that women who are biologically capable of having children, who are not currently pregnant or in the postpartum period, and who do not wish to conceive are "at risk" of an unplanned pregnancy. In 1965 23 percent of the black married women "at risk" of unplanned pregnancy compared with 12 percent of the white women were not using birth control. In 1982 14 percent of the black married women and 6 percent of the white women "at risk" were not using birth control. These proportions are indicated in Figure 4.7, which shows the type of contraceptives used by national samples of black and white *married* women who were at risk of unplanned pregnancy in either 1965 and 1982.[72]

Contraceptives may be divided into modern methods, which are highly effective, and traditional methods, which have greater failure rates. Sterilizing operations, oral contraceptives, and the IUD—each of which requires assistance from medical personnel—are the modern methods, while all other forms of birth control are classified as traditional. Figure 4.7 shows that not only did the use of contraceptives become more common after 1965, but that women of both races adopted modern methods over the traditional ones. Considering only those married women who used birth control, the change among blacks was from 40 percent using modern methods in 1965 to 74 percent in 1982. Among white married women, the corresponding change was from 37 to 69 percent. For both races, female sterilization became the most relied-on contraceptive procedure by the 1980s. Although some differences remain among those who use birth control, blacks are now as likely as whites to use the effective modern methods. One large difference is that married black couples are much more likely than white couples to rely on female sterilization but are less likely to control fertility through use of vasectomy.

What do we know about the use of birth control by women who have never married? Single women were traditionally excluded from contraceptive studies, but the 1982 National Survey of Family Growth included such women for the first time. Single black women are less likely to use birth control than single white women, but those who do select the more effective methods. The data shown below refer to single women in 1982 who were not pregnant, were not trying to become pregnant, but had been sexually active within the three-month interval before the survey. For these sexually active single women, we show the

[71]Mosher (1981); Bachrach and Moser (1984), table 2.
[72]Mosher and Westoff (1982), tables 5 and 6.

FIGURE 4.7
Methods of Contraception Used by Ever-Married Black and White Women Aged 15–44 at Risk of Becoming Pregnant, 1965 and 1982

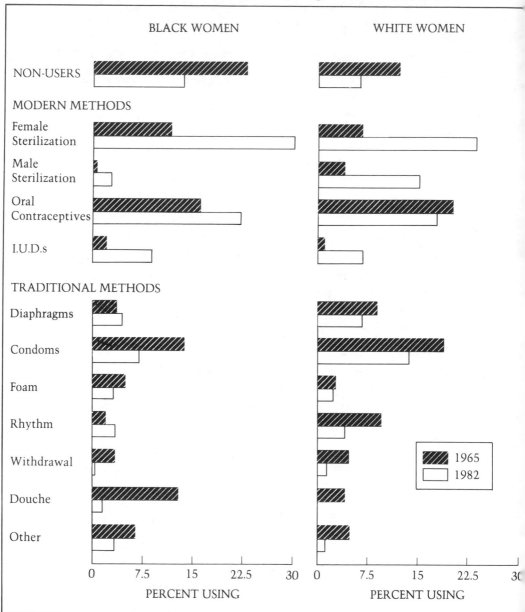

SOURCES: W. D. Mosher and C. F. Westoff, *Trends in Contraceptive Practice, 1965–76* (Washington, DC National Center for Health Statistics, 1982), tables 4 and 6; C. A. Bachrach and W. D. Mosher, "Use Contraception in the United States, 1982," National Center for Health Statistics, *Advancedata*, no. 1 (December 4, 1984), table 2.

	15 to 19 Years		20 to 44 Years	
	Black	White	Black	White
No Birth Control	36%	30%	23%	16%
Modern Methods	49	42	62	48
Traditional Methods	15	28	15	36

proportion who used no birth control, the proportion who used modern methods, and the proportion who relied on traditional methods.[73]

The use of birth control by single women increased with age, but among both teenagers and older women, black women were more likely than white women to be at risk of an unplanned conception. The differences in methods are largely attributable to oral contraceptives, which are much more popular with single black women than white. On the other hand, single teenage white women are more likely than black women to prevent contraception by male use of condoms, while older white women use diaphragms more often than black women. The proportion of single women at risk of becoming pregnant is considerably greater among blacks than whites, reflecting the racial difference in sexual activity.[74]

There are racial differences in the induced termination of pregnancy. Two agencies develop national estimates of the number and characteristics of women obtaining abortions, but their estimates differ by as much as 30 percent.[75] In addition, the National Center for Health Statistics compiles data for thirteen states in which the registration of abortion is thought to be accurate.[76] How frequently is abortion used?

The available data suggest that black women use abortion more frequently than white women. In 1981, according to the more liberal estimate of abortions, about 5 percent of nonwhite women of childbearing age and 2 percent of white women obtained abortions.[77] Considering pregnancies which ended in *either* live birth *or* abortion, 35 percent of the nonwhite and 24 percent of the white pregnancies were terminated by induced abortion in 1980.[78]

[73]Bachrach and Mosher (1984), table 3.

[74]In 1982, 55 percent of the single teenage black women said they had not been sexually active in the last three months compared with 67 percent of the white women. For single women aged 20 to 44, 84 percent of the black compared with 56 percent of the white reported sexual activity in the last three months. Bachrach and Mosher (1984), table 3.

[75]National Center for Health Statistics, *Health, United States, 1985*, table 6.

[76]Powell-Griner (1986).

[77]National Center for Health Statistics, *Health, United States, 1984*, tables 2, 3, and 5.

[78]For trends over time, see Henshaw and O'Reilly (1983), table 2.

Current Marital Status of Women	Black		White	
	1982	1972	1982	1972
Married-Spouse-Present	17%	28%	7%	11%
Divorced, Separated, or Widowed	28	37	12	19
Never-Married	41	n.a.	42	n.a.
Total	22%	n.a.	8%	n.a.

There is a further complication. Unmarried black women are less likely than unmarried white women to use abortion, but married black women are much more likely than married white women to choose abortion. Data from the NCHS abortion registration area report that in 1983, 62 percent of the pregnancies to *unmarried* white women ended in induced abortions, but among *unmarried* black women only 34 percent did. About 5 percent of the pregnancies to married white women but 13 percent of those to black women ended in abortion.[79]

Has the greater use of birth control and abortion eliminated the bearing of unwanted children? There has been a decline in such childbearing among both races, but unwanted children are still much more common among blacks. There are difficulties in measuring the "wantedness" of a child, but several national surveys have asked women about each of their conceptions which resulted in birth and sought to ascertain whether the woman wanted to have any additional children when she conceived. The table below reports the proportion of total births *unwanted* by women included in the 1973 and 1982 surveys.[80]

Among both blacks and whites, the proportion of births unwanted declined substantially in the 1970s, but the racial differences remain large. In the early 1980s black women reported that more than 20 percent of their children were unwanted at the time of their conception compared with less than 10 percent of the white children. Racial differences are evident at all marital statuses except among single women. These studies suggest that the higher fertility of blacks is largely unwanted childbearing and imply that the increasing use of effective family planning by blacks may further reduce the birthrate.

[79]Powell-Griner (1986), table B. These rates indicate abortions as a percentage of all pregnancies which ended in either a live birth or an abortion. The national estimates of abortion developed by the Alan Guttmacher Institute suggest that 72 percent of the pregnancies to unmarried white women and 52 percent of those to nonwhite women are terminated by abortion. For married women, the similar abortion rates are 8 percent for whites and 19 percent for nonwhites. Henshaw et al. (1985), table 4.

[80]Pratt and Horn (1985), tables 2 and 4.

Racial Differences Among Women of Childbearing Age

To further analyze racial differences in fertility, we divided women of childbearing age into two groups: those aged 35 to 49 and those aged 18 to 34 in 1980.[81] Women in the older group were born between 1930 and 1944 and reached childbearing ages during the peak years of the baby boom. They will bear few additional children after 1980 and will complete their fertility with a larger average family size than women born earlier or later in this century. Younger women, aged 18 to 34 in 1980, were born during the baby boom (1945 to 1962) and reached childbearing ages after fertility rates peaked. They will complete their families with fewer children than the older women, but, because many of them were still young in 1980, they could add numerous children later.

Table 4.2 classifies these black and white women by eight variables related to childbearing. Women have been sorted into categories of each variable describing such features as their age at first marriage and region of birth. For each category of every variable we show the average number of children ever born, how much this deviates from the overall average for that age group of women, and the racial difference. For example, black women aged 18 to 34 in 1980 who were married and lived with a husband had an average of 1.8 children. This is 0.5 more children than is typical for all black women aged 18 to 34. These black women were more fertile than similar white women, since whites had an average of 1.4 children, implying a racial difference of 0.4 child.

Marital status is the first variable presented in Table 4.2. Among both races and for both the younger and the older women, there were small differences in fertility among those who were living with a spouse, separated, or divorced. Widowed women reported larger numbers of children because they are older than other women in the same age group. Among whites, single women have few children. But single black women aged 35 to 49 reported an average of 2 children while younger black single women averaged 0.7 child, reflecting the higher rates of out-of-wedlock fertility in the black community.

Those who conducted the 1960 and 1965 national studies argued that black fertility rates remain high because many black women are reared in the rural South where the culture encourages large families but discourages the use of birth control. The decennial census does not tell us where a person was raised, but it provides information about state of birth. Black women born in the South have higher fertility rates

---

[81]The selection procedure obtained a 1 percent sample of persons living in a household headed by a black, and a 1-in 1,000 sample of all other persons. U.S. Bureau of the Census, *Census of Population and Housing: 1980*, Public Use Microdata Samples, Technical Documentation. See appendix for greater detail.

TABLE 4.2

Differential Fertility Among White and Black Women Aged 18–34 and 35–49, 1980

	18–34 Years					35–49 Years				
	Black		White		Racial Difference (Black Minus White)	Black		White		Racial Difference (Black Minus White)
	Average Children Ever Born	Deviation from Average	Average Children Ever Born	Deviation From Average		Average Children Ever Born	Deviation from Average	Average Children Ever Born	Deviation from Average	
Overall Average	1.3	0.0	1.0	0.0	+0.3	3.4	0.0	2.7	0.0	+ .7
MARITAL STATUS										
Married-Spouse-Present	1.8	+ .5	1.4	+ .4	+ .4	3.5	+ .1	2.8	+ .1	+ .7
Separated	2.2	+ .9	1.4	+ .4	+ .8	3.9	+ .5	3.0	+ .3	+ .9
Divorced	2.0	+ .7	1.3	+ .3	+ .7	3.2	– .2	2.5	– .2	+ .7
Widowed	2.5	+1.2	1.8	+ .8	+ .7	4.1	+ .7	2.9	+ .2	+1.2
Never Married	.7	– .6	.1	– .9	+ .6	2.0	–1.4	.2	–2.5	+1.8
AGE AT FIRST MARRIAGE										
<18 Years	2.6	+1.3	2.0	+1.0	+ .6	4.5	+1.1	3.5	+ .8	+1.0
18–19 Years	2.0	+ .7	1.6	+ .6	+ .4	3.8	+ .4	3.1	+ .4	+ .7
20–22 Years	1.7	+ .4	1.3	+ .3	+ .4	3.3	– .1	2.7	—	+ .6
≥ 23 Years	1.5	+ .2	.9	– .1	+ .6	2.8	– .6	2.1	– .6	+ .7
REGION OF BIRTH										
South	1.5	+ .5	1.1	+ .1	+ .4	3.5	+ .1	2.6	– .1	+ .9
Northeast	1.1	+ .2	.8	– .2	+ .3	2.8	– .6	2.6	– .1	+ .2
Midwest	1.2	– .1	1.0	—	+ .2	3.1	– .3	2.8	+ .1	+ .3
West	1.1	– .2	1.0	—	+ .1	2.8	– .6	2.8	+ .1	—
Foreign-Born	1.1	– .2	1.2	+ .2	– .1	2.7	– .7	2.7	—	—
REGION OF RESIDENCE										
South	1.4	+ .1	1.0	—	+ .4	3.7	+ .3	2.6	– .1	+1.1
Northeast	1.2	– .1	.8	– .2	+ .4	2.9	– .5	2.6	– .1	+ .3
Midwest	1.4	+ .1	1.0	—	+ .4	3.4	—	2.8	+ .1	+ .6
West	1.2	– .1	1.0	—	+ .2	3.0	– .4	2.7	—	+ .3

EDUCATIONAL ATTAINMENT

Elementary	2.2	+ .9	1.9	+ .9	+ .3	4.3	+ .9	3.3	+ .6	+1.0
High School, 1–3 Years	1.7	+ .4	1.3	+ .3	+ .4	4.1	+ .7	3.1	+ .4	+1.0
4 Years	1.3	—	1.0	—	+ .3	3.1	+ .3	2.7	—	+ .4
College, 1–3 Years	1.0	− .3	.7	− .3	+ .3	2.8	− .6	2.5	− .2	+ .3
4 Years	.7	− .6	.6	− .4	+ .1	2.0	−1.4	2.3	− .4	− .3
5+ Years	.8	− .5	.5	− .5	+ .3	1.9	−1.5	1.7	−1.0	+ .2
YEAR LAST WORKED										
Never Worked	1.2	− .1	1.2	+ .2	—	4.1	+ .7	3.2	+ .5	+ .9
Before 1975	2.3	+1.0	2.2	+1.2	+ .1	4.0	+ .6	2.9	+ .2	+1.1
1975 to 1978	1.7	+ .4	1.6	+ .6	+ .1	4.0	+ .6	2.8	+ .1	+1.2
1979	1.3	—	1.0	—	+ .3	3.7	+ .4	2.9	+ .2	+1.2
Worked in 1980	1.2	− .1	.7	− .3	+ .5	3.1	+ .3	2.5	− .2	+ .6
HOURS WORKED IN 1979										
None	1.5	+ .2	1.7	+ .7	− .2	4.0	+ .6	2.9	+ .2	+1.1
<1,000 Hours	1.2	− .1	.9	− .1	+ .3	3.6	+ .2	2.9	+ .2	+ .7
1,000–1,499	1.3	—	.8	− .2	+ .5	3.2	− .2	2.8	+ .1	+ .4
1,500–1,999	1.3	—	.6	− .4	+ .7	2.9	− .5	2.4	− .3	+ .5
≥2,000 Hours	1.3	—	.6	− .4	+ .7	3.0	− .4	2.4	− .3	+ .6
OCCUPATION OF EMPLOYED WOMEN										
Executive	1.1	− .2	.7	− .3	+ .4	2.6	− .8	2.2	− .5	+ .4
Professional Specialty	1.0	− .3	.7	− .3	+ .3	2.3	−1.1	2.2	− .5	+ .1
Technical	1.3	—	.7	− .3	+ .6	3.0	− .4	2.2	− .5	+ .7
Sales	1.0	− .3	.8	− .2	+ .2	3.0	− .4	2.8	+ .1	+ .2
Administrative Support	1.1	− .2	.8	− .2	+ .3	2.7	− .7	2.5	− .2	− .2
Domestic Service	1.6	+ .3	1.0	—	+ .6	3.9	+ .5	3.0	+ .3	+ .9
Other Service Occupations	1.5	+ .2	.9	− .1	+ .6	3.7	+ .3	3.0	+ .3	+ .7
Farming	1.7	+ .4	1.0	—	+ .7	4.6	+1.2	3.2	+ .5	+1.4
Precision Production	1.6	+ .3	1.1	+ .1	+ .5	3.2	− .2	2.6	− .1	+ .6
Operatives	1.7	+ .4	1.2	+ .2	+ .5	3.6	+ .2	3.0	+ .3	+ .6
Sample Size	43,090		28,835			21,089		16,317		

NOTE: "Deviation from Average" indicates how the average number of children ever born for women in a category differs from the overall average for her race and age group.

SOURCE: U.S. Bureau of the Census, *Census of Population and Housing: 1980*, Public Use Microdata Samples.

than black women born in other regions, a difference of at least 0.5 child for the younger women. Among whites, the region of birth differential is almost nonexistent.

A look at the racial difference columns in Table 4.2 indicates that black fertility almost always exceeds white. The place of birth variable is an exception since foreign-born black women bear no more children than foreign-born white women. Similarly, there is no racial difference in the fertility of older women born in the West. These findings suggest that region of birth still influences the fertility of blacks and that being born in the South is associated with higher fertility for blacks but not for whites.

Many studies of black childbearing have described large differences by place of current residence as well as by place of birth.[82] Table 4.2 classifies women by their region of residence in 1980. These differences are smaller than those associated with region of birth, but fertility clearly varies by place of residence. Black women living in the West or the Northeast bear fewer children than those in the South. Presumably, the continued existence of a rural black population in the South helps to explain the higher birthrates of that region. Among whites, differences in fertility by region of current residence are almost nonexistent.

For many women, marriage and schooling may be competing activities when they are young, and thus there are strong relationships between each of these variables and fertility. Those black women aged 35 to 49 in 1980 who married before age 18 averaged about 1.5 more children than those who waited until age 23 to marry. The same pattern is found among younger women, although the differences are smaller. There is also a clear educational difference: Women who complete five or more years of college bear about half the number of children of those women who do not finish high school. We find that the size of the black-white differences was much greater among those who married early and those who dropped out of high school than among those who married late or completed college. Black women who delay marriage or spend many years in school bear just about the same number of children as comparable white women.

For many women, decisions about careers and employment must be articulated with decisions about childbearing. A cross-sectional study such as the census provides information about a woman's current employment and tells us how many children she has borne. However, it does not provide the longitudinal data about work history and the timing of childbearing which is needed for a thorough study of how women balance employment and fertility. To learn something about this, fertil-

[82]Whelpton, Campbell, and Patterson (1966), pp. 340–42; Farley (1970), chap. 7.

ity is related to three variables in Table 4.2: the year the woman last worked; the number of hours she worked during 1979; and, if she was employed in 1980, her occupation.

In the older cohort, women who never worked bore considerably more children than women employed during the year before the census, and among both blacks and whites the more recently a woman was employed, the lower her fertility. The same relationship is evident among younger women, although those who have never worked report few children. Of course, many of these women are full-time students or young wives who have yet to begin either childbearing or careers.

Women who worked full time in 1979 reported fewer children than those who worked part time or not at all. Among young women, the relationship of fertility to employment was much *weaker* for blacks than for whites; that is, black women who did not work in 1979 had just about as many children as those who worked full time, but among white women there was a large fertility difference by amount of employment. This is consistent with the findings of other studies which suggest that white women who become mothers are much more likely than black women to drop out of the labor force or reduce their hours of work.[83]

We expect that women who invest in education or job training or who amass the seniority needed to obtain executive, professional, or technical jobs would have fewer children than women who work as domestic servants or machine operators. Employed women have been classified by their jobs using the new occupational categories introduced in the census of 1980. Among the older women, those working in executive positions or in the professional specialties such as engineering, medicine, or the law have the fewest children, while women working on farms or in the service occupations report the largest families. Among younger women, the inverse relationship between prestige of a woman's occupation and her fertility is also evident, but the fertility differences between those at the bottom and those at the top of the occupational ladder are much smaller.

Three observations may be drawn from this analysis of differential fertility. First, social, demographic, and economic factors influence the childbearing of black and white women in much the same manner. Being born in the South, however, is associated with higher fertility for blacks but not for whites.

Second, the fertility of black women exceeds that of white women in almost all categories of every variable. The higher childbearing of

[83]Hofferth and Moore (1979); Grabill, Kiser, and Whelpton (1958), table 95; Corcoran (1978), table A2.1b.

blacks is certainly not limited to those who married early, those who dropped out of high school, those who live in the South, or those who work at blue collar jobs.

Third, racial differences in fertility were much smaller among women aged 18 to 34 than among those aged 35 to 49, which may indicate that the racial gap is declining. But caution is needed when stating this conclusion because these young women have many years in which to add to their families, and if black women continue to bear more unplanned children than white, racial differences will become larger as these birth cohorts grow older.

The Net Effect of Factors Related to the Fertility of Married Women

The analysis thus far examined how factors such as place of birth, educational attainment, or employment relate to childbearing. Educational attainment, for example, is inversely related to the number of children a woman bears, but her education is also linked to age at marriage since women who spend many years in school marry at older ages. Perhaps educational attainment has no net effect on fertility; rather, its apparent effect may be entirely due to age at marriage. In this section we examine the net effects of variables associated with fertility, but the investigation is limited because we deal with cross-sectional data from the 1980 census rather than longitudinal information.

The technique used is multiple classification analysis, a procedure which is analogous to dummy variable regression analysis.[84] Table 4.3 presents findings for women aged 18 to 49 classified by age and race. The dependent variable is the number of children a woman has borne and ranges from 0 to 12. The independent variables which may influence fertility include age at first marriage, region of birth and region of current residence, educational attainment, current marital status, and the year last worked. The categories for each of these variables are listed in Table 4.3. For the region variable, the term "North" refers to all of the United States outside the South.

Multiple classification analysis treats every category of each independent variable as a dichotomous variable: A women either is or is not a member of the category. A statistical procedure that estimates a coefficient for every category of each variable measures the net effect of membership in that category and is similar to a partial regression coefficient. These coefficients are expressed as net deviations from the

[84]Blau and Duncan (1967), pp. 128–40; Sweet (1973), pp. 52–54; Morgan et al. (1962), app. E.

TABLE 4.3

Net Effects on Fertility of Social and Economic Factors for Ever-Married Black and White Women Aged 18–34 and 35–49, 1980

	18–34 Years				35–49 Years			
	Black		White		Black		White	
	Gross Deviation	Net Deviation	Gross Deviation	Net Deviation	Gross Deviation	Net Deviation	Gross Deviation	Net Deviation
Overall Average Children Ever Born	1.9		1.4		3.6		2.8	
AGE AT FIRST MARRIAGE								
<18 Years	+.6	+.4	+.6	+.5	+1.0	+.7	+.7	+.6
18–19 Years	+.1	+.1	+.2	+.1	+.3	+.3	+.3	+.2
20–22 Years	−.3	−.2	−.2	−.1	−.3	−.1	−.1	−.1
≥23 Years	−.4	−.3	−.5	−.4	−.8	−.7	−.7	−.6
REGION OF BIRTH BY REGION OF RESIDENCE								
Birth Residence								
South South	+.1	0	0	−.1	+.3	+.2	−.1	−.3
North South	−.4	−.3	.1	−.1	−.4	0	−.1	0
Foreign Either	−.3	−.2	+.2	+.1	−.7	−.3	−.1	−.1
South North	+.2	+.2	+.1	+.1	−.1	−.1	0	−.1
North North	−.1	−.1	0	0	−.4	−.3	+.1	+.1
EDUCATIONAL ATTAINMENT								
Elementary	+.9	+.7	+.9	+.6	+1.1	+.8	+.7	+.6
High School, 1–3 Years	+.6	+.4	+.4	+.2	+.7	+.5	+.4	+.2
High School, 4 Years	−.1	−.1	+.1	0	−.3	−.3	0	0
College, 1–3 Years	−.4	−.2	−.2	−.1	−.6	−.4	−.2	−.1
College, 4 Years	−.7	−.5	−.4	−.2	−1.4	−1.0	−.4	−.1
College, 5+ Years	−.8	−.5	−.6	−.2	−1.5	−1.1	−.8	−.5
CURRENT MARITAL STATUS								
Married-Spouse-Present	−.1	0	0	0	0	+.1	0	0
Separated	+.6	+.2	+.4	+.2	+.5	+.1	+.2	+.2
Divorced	+.1	+.1	−.1	−.1	−.4	−.3	−.3	−.2
Widowed	+.2	+.1	0	−.1	+.3	+.1	+.1	+.1
YEAR LAST WORKED								
Never Worked	+.3	+.2	+.6	+.3	+1.0	+.7	+.7	+.5
Before 1975	+.8	+.6	+.9	+.8	+.6	+.4	+.1	+.1
1975 to 1978	+.3	+.3	+.4	+.3	+.5	+.3	0	0
1979	+.1	0	+.1	0	+.3	+.2	+.2	+.1
Worked in 1980	−.2	−.1	−.3	−.2	−.3	−.2	−.1	−.1
SAMPLE SIZE	21,445		18,960		18,688		15,540	

ACCOUNTING FOR VARIANCE

SEPARATE VARIABLES				
Age at First Marriage	6.9%	7.9%	6.8%	7.5%
Region of Birth by Region of Residence	.8	.3	1.5	.2
Educational Attainment	8.7	7.3	8.7	4.4
Current Marital Status	4.0	.2	.9	.3
Year Last Worked	1.0	11.8	3.0	1.4
All Variables Together	14.3%	20.1%	14.7%	11.2%

NOTE: The net deviation indicates the net effect of being in a given category of a variable independent of the effects of all other variables. The multiple classification model assumes that the number of children born by ever-married women is an additive effect of age at first marriage, region of birth by region of current residence, educational attainment, current marital status, and the year the woman was last employed.

SOURCE: U.S. Bureau of the Census, *Census of Population and Housing: 1980,* Public Use Microdata Samples.

overall average number of children ever born, and the sum of their effects for all categories of one independent variable is zero when they are
weighted by the number of cases.[85]

This procedure also assesses the strength of a relationship by determining the proportion of variance of the dependent variable—children
ever born—attributable to any one or any combination of independent
variables.[86] As an example, consider educational attainment. Suppose
that fertility varied greatly from one level of attainment to another but
within levels of education there were little variation. In such a circumstance, the multiple classification analysis would reveal, first, that
there were substantial net differences in fertility by educational attainment and, second, that a large proportion of the total variance in fertility is attributable to educational attainment.

Table 4.3 shows how five variables are related to the fertility of
ever-married women. The gross effects are the difference between average fertility level within a category and the overall mean, while the net
effects measure the unique consequences of membership in a given
category of an independent variable. In discussing the relationship of a
variable to fertility, we can look at the gross effects, the net effects, or
the proportion of variance accounted for by that variable.

Several conclusions may be drawn from this analysis. First,
although the net effects are usually smaller than the gross ones, each of
the variables has net effects on childbearing. The direction of the effects
is quite similar for the women who participated in the baby boom and
for the younger cohorts. Second, educational attainment, employment,
and region affect the childbearing of whites and blacks in somewhat different ways. For older black women, the net effect of being born and
remaining in the South was an increase in family size. However, those
black women who were born in the South but moved away by 1980 had

[85]Similar to regression models, multiple classification analysis assumes that the effects of the independent variables are additive. The model we are using does not take into
account interactions except with regard to region of birth and region of residence. It assumes, for example, that the effect of current employment on fertility is the same for
women who completed either high school or college. It is not, however, a linear model
since it measures the net effect of being in a specific category of an independent variable.
This facilitates examining the effect of independent variables which are not linearly related to fertility, such as place of birth. Multiple classification analysis was chosen since it
provides easily understood estimates of the net effects of those variables which are related
to fertility.

[86]This is done by using the coefficients from multiple classification analysis to determine an expected mean value for children ever born for each category of every variable.
This expected mean is then compared with the observed values in that category and estimates are derived of explained and unexplained variance. If only one independent variable
is used, the proportion of variance explained is known as the correlation ratio. This is
similar to a one-way analysis of variance.

low fertility. The migration to the North offset the fertility-enhancing effects of southern birth.

The effects of education differ by race since a college education had a much more depressing effect on the fertility of black women than white women. Black women either make a choice between education and family building or, after completing college, they limit their families more than comparable white women. Finally, there is a racial difference in the way employment is related to fertility among young women. It appears that white women are more likely than black women to divide their time between periods of childrearing and periods of employment.

Because the multiple classification model is additive, we can specify the characteristics of a group and estimate its fertility. For instance, if we wish to know the number of children ever born by younger spouse-present women who married before age 18, who were born and still live in the South, who completed high school, and who were at work in 1980, we add the net effects for each variable to the overall average number of children.

The table below indicates the fertility of women with different characteristics. This procedure is useful for testing the hypotheses discussed earlier about why there are racial differences in fertility. These

Characteristics of Married Women	18 to 34 Years		35 to 49 Years	
	Black	White	Black	White
Spouse present, never worked, born and lived in South, married at 17 with 2 years of high school	2.9	2.3	5.8	3.8
Spouse present, worked in 1980, born in South but lived in North, married at 19 with a high school education	2.0	1.4	3.4	2.8
Separated, worked in 1980, born in South but lived in North, married at 22 with two years of college	1.8	1.3	2.9	2.6
Spouse present, worked in 1980, born and lived in North, married at 24 with a master's degree	.9	.6	1.4	1.7

calculations demonstrate that the social characteristics hypothesis is incorrect as applied to fertility; even when racial differences in social and economic status are taken into account, racial differences in child-bearing are real, since most groups of black women have higher fertility than comparable white women.

Characteristics of Husbands: Do They Influence Fertility?

What about a woman's husband? Does his educational attainment or economic status have an independent effect on fertility, or is child-bearing pretty much influenced by the woman's characteristics alone?

The census gathers information about a woman's *current* spouse, not her former husbands, so this analysis is restricted to women who were married and lived with husbands in 1980. Once again, additive multiple classification models are used since they allow us to describe the net effects of the husband's characteristics. In addition to the characteristics of wives used in the previous section, we added two characteristics of the husband: his educational attainment and his earnings in 1979. Since marital disruption may have an influence on the size of a woman's family, we included a variable which describes the couple's marital history; that is, whether the husband or the wife was married more than once. Results are presented in Table 4.4

The husband's educational attainment had a small net effect on childbearing apart from the wife's education or other variables. Husbands who spent many years in school had smaller families—among blacks but not among whites—than other men. For both races, men without high school educations fathered unusually large families. One might assume that a husband's earnings in 1979 indexed the couple's financial status and that those at the top end of the economic distribution had much smaller families than those at the bottom. However, the net influence of the husband's earnings on fertility was very small among either blacks or whites.

If a woman's first marriage ends and she remarries, one might expect that she would bear fewer children than comparable women who did not experience marital disruption since she was without a spouse for some time. On the other hand, once she remarries her second husband may wish to start his own family with her and, if this is the case, the net effect of remarriage could be an increase in fertility. Which effect is stronger? Among young women—both black and white—the new-family effect outweighs the interruption effect since remarried women have higher fertility. The effects are less clear among the older women, but for them the interruption effect seems to outweigh the new-family effect. The remarriage effect also varies by sex because marital disruption on the part of the husband seems to depress fertility among both blacks and whites. Women who marry for the first time but become a man's second or third wife bear fewer children, perhaps because he is responsible for supporting children from a previous marriage.

This analysis of differential fertility from the census of 1980 reveals that age at marriage and social and economic factors influence the fertility of blacks and whites in a similar fashion, but racial differences in social and economic characteristics cannot account for all of the racial differences in childbearing. Rather, there is a net effect of race which, in most cases, raises the fertility rate of blacks above that of similar whites.

Conclusion

According to demographic transition theory, the decline in fertility comes decades after the decline in mortality. There was, however, an unambiguous decrease in black birth rates between 1880 and 1940, that paralleled the decline in black mortality. It was followed by the sharp two-decade rise that characterized white fertility as well. Was this original decline in fertility the result of black couples choosing to have fewer children, or did it come about for other reasons? We may never be able to anser this question, but there is convincing evidence that the lower birth rates did not result from the widespread use of contraceptives but were attributable to tuberculosis, veneral disease, and other ailments that inhibited fertility. Wright and Pirie call this a false fertility transition, implying that the real one was that which has been occurring since 1960, a time when black couples became frequent users of contraception.[87]

At present, the birthrates of blacks are low, just above the replacement level. Have they converged with those of whites? We might expect so because of the increasing educational attainment of blacks, their urbanization, and the widespread use of effective birth control and abortion. The evidence reported in this chapter shows that racial differences in family size are getting much smaller; black women who complete their families in the late 1980s or 1990s will bear just over two children compared with an average of just under two for white women. However, there has been an increasing racial divergence in the timing of family formation.[88] Black women are increasingly differentiated from

[87]Wright and Pirie (1984).
[88]Evans (1986), pp. 269–271.

TABLE 4.4

Net Effects on Fertility of Social and Economic Factors for Married-Spouse-Present Black and White Women Aged 18–34 and 35–49, 1980

	Black		White		Racial Difference	Black		White		Racial Difference
	Adjusted Mean	Net Deviation	Adjusted Mean	Net Deviation		Adjusted Mean	Net Deviation	Adjusted Mean	Net Deviation	
Overall Average Children Ever Born	1.9	—	1.4	—	+ .5	3.5	—	2.8	—	+ .7
WIFE'S AGE AT FIRST MARRIAGE										
<18 Years	2.3	+ .4	1.9	+ .5	+ .4	4.3	+ .8	3.4	+ .6	+ .9
18–19 Years	2.0	+ .2	1.6	+ .2	+ .4	3.9	+ .4	3.0	+ .2	+ .9
20–22 Years	1.7	– .2	1.3	– .1	+ .4	3.5	—	2.8	—	+ .7
≥23 Years	1.6	– .3	1.1	– .3	+ .5	2.9	– .6	2.2	– .6	+ .7
WIFE'S REGION OF BIRTH BY REGION OF RESIDENCE										
Birth / Residence										
South / South	1.9	—	1.3	– .1	+ .6	3.7	+ .2	2.5	– .3	+1.2
North / South	1.6	– .3	1.4	—	+ .2	3.4	– .1	2.9	+ .1	+ .5
Foreign / Either	1.7	– .2	1.5	+ .1	+ .2	3.2	– .3	2.8	—	+ .4
South / North	2.0	+ .1	1.5	+ .1	+ .5	3.4	– .1	2.7	– .1	+ .7
North / North	1.8	– .1	1.5	+ .1	+ .3	3.3	– .2	3.0	+ .2	+ .3
WIFE'S EDUCATIONAL ATTAINMENT										
Elementary	2.6	+ .7	2.0	+ .6	+ .6	4.3	+ .8	3.3	+ .5	+1.0
High School, 1–3 Years	2.3	+ .4	1.6	+ .2	+ .7	3.9	+ .4	3.0	+ .2	+ .9
4 Years	1.9	—	1.4	—	+ .5	3.3	– .2	2.8	—	+ .5
College, 1–3 Years	1.7	– .2	1.4	– .1	+ .3	3.2	– .3	2.7	– .1	+ .5
4 Years	1.5	– .4	1.3	– .1	+ .2	2.8	– .7	2.7	+ .1	– .1
5+ Years	1.4	– .5	1.2	– .2	+ .2	2.7	– .8	2.3	– .5	+ .4
YEAR WIFE LAST WORKED										
Never Worked	2.0	+ .1	1.7	+ .3	+ .3	4.0	+ .5	3.3	+ .5	+ .7
Before 1975	1.3	– .6	2.2	+ .8	– .9	3.8	+ .3	2.9	+ .1	+ .9
1975 to 1978	2.1	+ .2	1.7	+ .3	+ .4	3.7	+ .2	2.8	—	+ .9
1979	1.9	—	1.4	—	+ .5	3.7	+ .2	2.9	+ .1	+ .8
Worked in 1980	1.8	– .1	1.2	– .2	+ .6	3.4	– .1	2.7	– .1	+ .7

		(1)		(2)		(3)		(4)	
		Adj. Mean	Net Dev.	Adj. Mean	Net Dev.	Adj. Mean	Net Dev.	Adj. Mean	Net Dev.
High School, 1–3 Years		2.0	+.1	1.3	-.1	3.6	+.1	2.8	+.8
4 Years		1.8	-.1	1.4	—	3.4	-.1	2.8	+.6
College, 1–3 Years		1.8	-.1	1.4	—	3.3	-.2	2.8	+.5
4 Years		1.7	-.2	1.3	-.1	3.2	-.3	2.8	+.4
5+ Years		1.8	-.1	1.5	+.1	3.2	-.3	2.8	+.4
HUSBAND'S EARNINGS IN 1979									
None		1.9	—	1.5	+.1	3.7	+.2	2.8	+.9
$1 to 4,999		1.8	-.1	1.2	-.2	3.6	+.1	3.0	+.6
$5,000 to 9,999		1.7	-.2	1.2	-.2	3.6	+.1	2.9	+.7
$10,000 to 14,999		1.8	-.1	1.3	-.1	3.5	—	2.8	+.7
$15,000 to 19,999		2.0	+.1	1.5	+.1	3.4	-.1	2.7	+.7
$20,000 to 24,999		2.1	+.2	1.6	+.2	3.5	—	2.8	+.7
$25,000 to 49,999		2.1	+.2	1.6	+.2	3.4	-.1	2.9	+.5
$50,000 or More		1.5	-.4	1.5	+.1	3.7	+.2	2.9	+.8
MARITAL HISTORY OF COUPLE (number of times married)									
Wife	Husband								
Once	Once	1.9	—	1.4	—	3.6	+.1	2.9	+.7
>Once	Once	2.2	+.3	1.6	+.2	3.3	-.2	2.8	+.5
Once	>Once	1.7	-.2	1.2	-.2	3.4	-.1	2.4	+1.0
>Once	>Once	2.1	+.2	1.7	+.3	3.2	-.3	2.6	+.6

ACCOUNTING FOR VARIANCE

	(1)	(2)	(3)	(4)
Wife's Age at First Marriage	7.6%	8.4%	7.3%	7.2%
Wife's Region of Birth by Region of Residence	.7	.3	1.8	.3
Wife's Educational Attainment	9.2	7.2	9.6	4.0
Year Wife Last Worked	3.6	11.8	2.5	1.4
Husband's Educational Attainment	4.8	3.4	6.5	2.1
Husband's Earnings in 1979	.4	1.7	2.1	.3
Marital History of Couple	1.3	1.9	.5	1.1
Sample Size	13,257	15,840	10,480	12,889

NOTES: The adjusted mean is the estimated number of children ever born for women in a given category of an independent variable, taking into account the effects of the other variables included in this model. The net deviation indicates the net effect of being in a given category of a variable independent of the effects of all other variables. The multiple classification model assumes that the number of children borne by a married-spouse-present woman is an additive effect of age at first marriage, wife's region of birth by region of residence, wife's educational attainment, the year the wife last worked, husband's educational attainment, husband's earnings in 1979, and the marital history of the couple.

SOURCE: U.S. Bureau of the Census, *Census of Population and Housing: 1980*, Public Use Microdata Samples.

white women because they initiate their childbearing at earlier ages; thus, they reach their desired family size while they are quite young. White women, on the other hand, are choosing to delay their fertility.

We used data from the 1980 census to investigate two conflicting explanations for the persistently higher fertility of blacks: the social characteristics hypothesis and the minority group hypothesis. Since racial differences in fertility appear at all economic levels, the minority group hypothesis seems to be the more plausible of the two explanations. However, we cannot confidently endorse this theory because blacks have been in the fertility-control stage of their demographic transition for only one quarter of a century, and it is impossible to tell if the higher fertility of blacks is the result of cultural values and preferences.

5

THE REDISTRIBUTION
OF THE BLACK POPULATION
AND RESIDENTIAL SEGREGATION

W HEN this nation conducted its first census two hundred years ago, nine out of ten blacks lived in southern states along the Atlantic Coast. Forty percent of the country's 750,000 blacks lived in Virginia.[1] During the era of slavery, blacks migrated into the Gulf Coast states but not toward the North, and between the Civil War and World War I there was a continuing shift of blacks within the South but only a small movement to the North.

The most substantial demographic change to affect blacks occurred in the fifty years following the outbreak of World War I. The out-migration of blacks from the South freed them from their three-century bond with agriculture and fundamentally altered the status of blacks by making racial issues national in character rather than regional.

Since 1900 another major demographic change has occurred—urbanization. At the turn of this century, the proportion living in urban places was about half as great for blacks as for whites: 23 percent compared with 43 percent.[2] During World War II, for the first time a major-

---

[1] U.S. Bureau of the Census, *Negro Population in the United States: 1790–1915* (1918), table 13.

[2] U.S. Bureau of the Census, *Census of Population: 1980*, PC80-1-B1, table 52. Traditionally, the Census Bureau has defined as urban those people who lived in incorporated places of 2,500 or more. Since 1950 those living in densely settled unincorporated suburban areas have been included among the urban population.

ity of blacks lived in cities rather than in the rural South, and by 1960 blacks were more urbanized than whites. This urbanization has been so complete that the concentration of blacks in large metropolitan areas has become a major issue in our society.

This massive redistribution of the black population may be coming to an end. Since 1970, and apparently for the first time in history, the South has gained in the exchange of black migrants with other regions. This chapter describes the shifting geographic distribution of blacks and then turns to an analysis of residential segregation.

Migration During the Period of Slavery

In both 1790 and 1860, 90 percent of blacks lived in the South, but this generalization conceals a dramatic population shift within that region. Figure 5.1 shows the regional distribution of the black and white populations between the presidencies of Washington and Reagan. To describe trends among blacks, the South was divided into its three divisions: the South Atlantic states, the East South Central states, and the West South Central states. The South Atlantic states extend from Delaware through Florida; the East South Central states comprise Alabama, Kentucky, Mississippi, and Tennessee; and Arkansas, Louisiana, Oklahoma, and Texas make up the West South Central division. The figure shows that increases took place in black population first in the East South Central states and after the war with Mexico in the West South Central division.[3]

Migration trends prior to the Civil War cannot be analyzed with precision, because the first census to present information about place of birth or age for blacks—the data required for a thorough migration analysis—was that of 1850.[4] But those who investigated the characteristics of black migration in this era agree that much of the young slave population shifted from Maryland, Virginia, and the Carolinas to the Deep South.[5] The most detailed study of the characteristics of these migrants shows that approximately equal numbers of male and female slaves were moved into the Gulf Coast states, many of them under age 25.[6]

There is consensus that economic factors account for this migration. The purchase of the Louisiana Territory and the perfecting of the

[3]U.S. Bureau of the Census, *Negro Population in the United States: 1790–1915* (1918), table 13; Taeuber and Taeuber (1966), pp. 98–100.
[4]U. S. Bureau of the Census, *Population and Housing Inquiries in U.S. Decennial Censuses: 1790–1970*, Working Paper no. 39.
[5]McClelland and Zeckhauser (1982), app. D.
[6]McClelland and Zeckhauser (1982), p. 8.

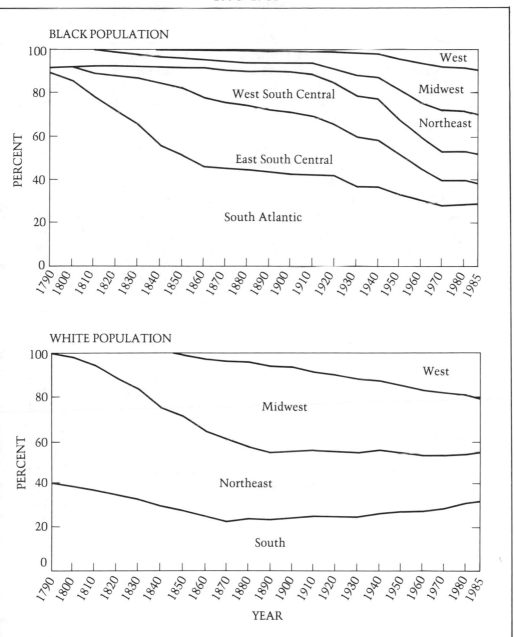

FIGURE 5.1
Distribution of the Black and White Population of the United States by Region,
1790–1985

NOTE: Estimates for 1985 refer to the noninstitutional population.

SOURCES: U.S. Bureau of the Census, *Historical Statistics of the United States: Colonial Times to 1970,*
1, series A 172-194. *Census of Population 1980,* PC80-1-B1, table 50; *Current Population Survey*
(March 1985), Public Use File.

105

cotton gin in the first decade of the 1800s were followed by the development of steamboats and later railroads. Because of these changes, a cotton economy flourished in the Gulf Coast states, creating favorable opportunities for settlers and a tremendous need for labor.

There are two conflicting views of the nature of the migration of blacks from the Atlantic to the Gulf Coast states. One perspective stresses that this was largely a domestic slave trade. With surplus labor, owners along the east coast knew they could realize a healthy profit by selling their chattels to a speculator who would ship them to New Orleans or march them there in a coffle. Since the nineteenth century, abolitionists, commentators, and some historians have argued that slaveholders in the upper South engaged in slave breeding to generate even greater profits through the sale of slaves which they exported to the Gulf Coast states.[7] Those who believe that slavery destroyed the black family contend that this migration played an important role because, they argue, children were frequently sold apart from their parents and spouses apart from their mates.

Using a different approach which involved the detailed analysis of economic records, Fogel and Engerman drew different conclusions about the migration of blacks before the Civil War.[8] They believe that expansion of the cotton economy in the Gulf Coast area depended on white settlers who knew how to grow the crop but found no opportunity to do so along the densely settled east coast. These whites, they contend, moved to Mississippi or Louisiana and brought their own slaves with them, often slaves held by their own families. They conclude that speculators and slave traders played no more than a small role in the redistribution of the black population and that more than 80 percent of the slaves who migrated to the new South did so along with their owners. The internal slave trade is not seen as being harmful to the stability of slave families since, they believe, black families migrated as units to the Gulf Coast states.

There was only a small movement of blacks to the northeastern states prior to Emancipation. During the era of slavery, 5 to 8 percent of the black population of the South was free,[9] and one might have expected that they would move to Philadelphia, New York, or the New England cities, where blacks had been living for two centuries. Perhaps three factors explain the absence of this migration from slave to free states. First, southern states passed many laws to rigorously control free blacks, especially after the Denmark Vesey affair in Charleston in 1822

[7]Franklin (1967), pp. 175–82; Phillips (1966), p. 50; Sutch (1975).

[8]Fogel and Engerman (1974), p. 48.

[9]U.S. Bureau of the Census, *Negro Population of the United States: 1790–1915* (1918), p. 55.

and the Nat Turner uprising in Virginia in 1831. State and local laws often insisted that free blacks could not travel without permission of a court.[10] In some cases, blacks who violated these strict laws could be sold into slavery.[11]

Second, some communities of free blacks were able to carve out economic niches for themselves in the South. Apparently this was the case in Baltimore, where free blacks monopolized the ship-caulking trade, and in Washington and New Orleans, where many were service workers.[12] Free blacks apparently were successful in some of the crafts and teamstering in cities of the lower South.[13] Their economic success, limited though it was, discouraged migration to the North.

Finally, blacks were not welcome in the North. Throughout the pre–Civil War span, there was increasing competition between free blacks and foreign migrants for manual jobs. Racial prejudice became more common, as Irish and German immigrants feared that former slaves would come north in great numbers and take their jobs.[14] During the 1830s and 1840s there were antiblack riots in Philadelphia, New York, and other northeastern cities.[15] Several of these were unintentionally triggered by the activities of abolitionists; that is, following a parade or meeting organized by antislavery crusaders, white immigrants who feared black competitors marched into black areas and burned homes and churches. Symbolic of the attitudes of this age were the efforts in many northern states to draw up codes which restricted the civil rights of blacks. Three states—Connecticut, New Jersey, and Pennsylvania—which once allowed blacks to vote, amended their constitutions to restrict suffrage to white males.[16]

There was some migration of free blacks and fugitive slaves into midwestern states prior to the Civil War. Figure 5.2 shows the proportion of the population which was black in the nation's regions since 1790. Although few blacks lived in the Midwest, the proportion black rose from 2 to 3 percent between 1800 and 1850, implying that the black population grew faster than the white. Perhaps it would have increased even more rapidly had whites not directed so much hostility toward black newcomers. Slavery was banned in the midwestern states by the Northwest Ordinance, but the white settlers feared an invasion by fugitive slaves, so these states enacted laws and constitutional provisions to keep their populations white. Illinois and Indiana adopted constitutions which forbade blacks—either free or enslaved—from entering

[10]Russell (1913), pp. 106–7.
[11]Franklin (1967), p. 218; Wade (1964), pp. 248–52.
[12]Fields (1985), p. 37; Frazier (1932), chap. 1.
[13]Curry (1981), chap. 2.
[14]Woodson (1918), pp. 41–47.
[15]Curry (1981), chap. 6; DuBois (1899), pp. 33–37.
[16]Litwak (1961), p. 77.

FIGURE 5.2
Percent of Population Black for the United States and Regions, 1790–1985

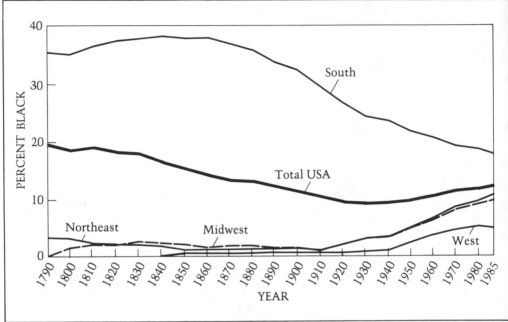

NOTE: Estimates for 1985 refer to the noninstitutional population.

SOURCES: U.S. Bureau of the Census, *Historical Statistics of the United States: Colonial Times to 19*
(1975), series A 91-93 and 172-176; *Census of Population: 1980*, PC80-1-B1, table 50; *Current Populati*
Survey (March 1985), Public Use File.

the state.[17] Ohio and Michigan adopted codes that allowed blacks to remain in the state only if they had court-issued proof of freedom, posted a cash bond, and registered themselves with local officials.[18]

These laws were not effective and black populations grew, discomforting whites. By the mid-1820s, about one tenth of the population of Cincinnati was black. White immigrants feared losing jobs and demanded that city officials enforce Ohio's black codes. Attempts were made to do so, but most blacks refused to either leave or post the required cash bonds. After several summers of sporadic violence, a riot occurred in 1829 in which whites tried to burn the black neighborhood and drive blacks from the city.[19] The next year a pogrom-like attack removed blacks from Portsmouth, Ohio.[20] In 1833 federal troops were

[17]Voegeli (1967), chap. 1; Litwak (1961), p. 70.
[18]Woodson (1918), p. 52; Katzman (1973), pp. 6–7.
[19]Wade (1959), pp. 224–29; Curry (1981), pp. 104–5.
[20]Woodson (1918), p. 57.

mustered to quell bloody racial riots in Detroit,[21] and eight years later in Cincinnati. Clearly free blacks and fugitive slaves faced immense hostility and had few job opportunities if they migrated from the South to the North.

Black Migration Between the Civil War and World War I

In 1860, 92 percent of the black population lived in the South and fifty years later 89 percent lived there, suggesting that there was no more than a small out-migration before World War I.[22] These figures conceal regional shifts within the South since new lands were opened to cotton cultivation. In the 1880s the black population of Arkansas grew rapidly because of immigration; in the 1890s this occurred in Texas, and during the first decade of this century rural Oklahoma attracted many black migrants.[23] Labor recruiters encouraged black farm workers to leave Atlantic Seaboard and Gulf Coast states for these opportunities in western states.[24] As a result, between 1860 and 1910 the proportion of the nation's blacks living in South Atlantic states fell from 46 to 42 percent while the proportion in the West South Central states grew from 15 to 20 percent.[25]

The Kansas Exodus of the 1870s was the first attempt by many blacks to escape the oppression and poverty of the rural South. Plantation owners who wished to reestablish white supremacy and guarantee themselves a source of cheap labor used many strategies to keep blacks in a condition resembling bondage. Just after the Civil War, blacks were often charged with vagrancy if they lacked jobs or had not entered into labor contracts. If arrested for vagrancy, nonpayment of debts, or other misdemeanors, they could be rented out to plantation owners in lieu of their fine. In many rural areas, the Ku Klux Klan and similar groups terrorized blacks who were deemed insufficiently subservient.

Henry Adams, a prosperous black civil rights advocate in Shreveport, concluded that blacks had no future in the South because he foresaw the reinstitution of slavery. Originally, he advocated a return to Africa, but later realized that it would be more feasible to go to Kansas where, he thought, land was readily available. Pappy Singleton was

[21]Katzman (1973), pp. 10–12.
[22]U.S. Bureau of the Census, *Negroes in the United States: 1920–1932* (1935), table 12.
[23]Eldridge and Thomas (1964), p. 92.
[24]Johnson and Campbell (1981), p. 62; Scott (1920), p. 3.
[25]U.S. Bureau of the Census, *Negro Population of the United States: 1790–1915* (1918), table 13.

another leader of the "Exoduster Movement." As a fugitive slave, he had lived in Canada and Michigan, and after the Civil War he returned to Nashville expecting to find improved conditions. By the mid-1870s he too concluded that southern whites would never respect the civil and economic rights of blacks, and urged blacks to leave the South. After visiting Kansas, he recommended that blacks populate the rural areas of that state.[26]

In the late 1870s "Kansas Fever" spread among blacks in the Gulf Coast states. White plantation owners opposed the migration and sometimes intimidated those who tried to leave. There are no accurate estimates of how many blacks moved to Kansas, although it could not have been more than a tiny fraction of the South's black population. The census, conducted in the summer of 1880, counted 43,000 blacks in Kansas compared with 17,000 a decade earlier.[27] Most of those who arrived in Kansas were destitute and depended on the assistance of the relief associations which were hastily organized. Few blacks found opportunities there since they lacked the resources to purchase fertile land, and in the cities white animosity meant that few were employed. Most officials in Kansas did whatever they could to stem the in-migration, including the passage of ordinances which forbade paupers from entering a town or made it illegal for a black to be in a city after sundown. Eventually Singleton and other leaders recognized that blacks would not be welcome or economically successful in Kansas. The exodus stopped and there was no substantial movement of blacks away from the South until World War I.

Why did so few blacks move to the North? There was a tremendous need for labor as the United States became the world's leading industrial producer, and thousands of European immigrants were recruited to work in steel mills, on the railroads, or in the hundreds of manufacturing plants in the Northeast and Midwest. Employers often expressed their hostility toward foreign migrants, feeling that they were unproductive workers since they spoke no English and did not share the nation's culture.[28] Why did the entrepreneurs not turn to blacks who were, after all, native-born, English-speaking Protestants?

We also would expect northern migration because blacks had strong incentives to leave the South. During the first decade of Reconstruction, radical Republicans and the Union armies provided minimal protection for blacks and allowed them to exercise some civil rights. However, conservative southerners were determined to reestablish

[26]Painter (1976); Athearn (1978).
[27]U.S. Bureau of the Census, *Negro Population of the United States: 1790–1915* (1918), table 13.
[28]Zunz (1982), pp. 311–18.

white supremacy and were free to do so following the inauguration of President Hayes and his withdrawal of Union armies in 1877. Supreme Court decisions in the 1880s and 1890s defended the doctrine of states' rights, thereby nullifying the constitutional amendments and civil rights acts of the Civil War era. By 1900 Jim Crow became the rule of the South, and blacks lost both their civil rights and their economic apportunities.[29] What the white supremacist did not accomplish legally was achieved by violence; the number of blacks killed by lynching rose from 49 in 1882 to a peak of 161 a decade later.[30] Why did rural blacks not pack their belongings, take a train to New York, Detroit, or Chicago and get those jobs which were being filled by Italians, Poles, or Lithuanians?

Three factors account for the absence of this migration. First, blacks faced immense difficulty getting jobs in the North. Both white workers and their employers opposed the employment of blacks. Urban historians report that throughout the nineteenth century white immigrants in northern cities did whatever they could to keep blacks off the payroll, fearing they would lose their jobs if thousands of blacks came North.[31] Allan Spear observes that so long as there was an adequate supply of white labor, northern industrialists refused to hire blacks.[32] They assumed that white workers would object to working with blacks and believed that blacks were not qualified for urban employment. This was the age of Social Darwinism; popular scientific thought as well as the stereotypes of the day portrayed blacks as untalented, slow to learn, unwilling to work without close supervision, and totally unsuited for the modern industrial environment.

In a few cases employers made exceptions and imported trainloads of blacks to act as strikebreakers.[33] These blacks experienced the typical fate of strikebreakers—in the short run, they were subjected to physical attack and harassment by the white strikers. Indeed, the use of blacks to replace striking teamsters almost triggered a city-wide riot in Chicago in 1905.[34] Once the strikes were broken, most of the blacks lost their jobs as managements rehired the white workers they preferred.

The second reason for little northern migration by blacks is that

[29]DuBois (1935); Franklin (1961); Woodward (1957); Jaynes (1986).

[30]U.S. Bureau of the Census, *Historical Statistics of the United States: Colonial Times to 1970* (1975), series H-1170; Zangrando (1980), chap. 1; Williamson (1984), chap. 4.

[31]Curry (1981), chap. 2; Harris (1982), chap. 2; Zunz (1982), pp. 139 and 320–21; Pleck (1979), chap. 5; DuBois (1899), chap. 9; Lane (1986), pp. 36–42.

[32]Spear (1967), p. 34.

[33]Spero and Harris (1931), chap. 7; Bonachich (1976), p. 41; Wilson (1978), p. 64. For a discussion of the labor competition which helped lead to the 1917 riot in East St. Louis, see Rudwick (1966).

[34]Spear (1967), pp. 39–40.

southern leaders tried to retain their black populations, expecting that the cotton economy would once again boom and there would be a need for the cheap labor that blacks could provide. Many southern cities and states passed laws which made it difficult or impossible for labor agents to recruit local blacks.[35] Macon, Georgia, had among the most stringent regulations, charging a $25,000 license fee for recruiting and requiring forty-five character references.[36] Blacks were sometimes arrested for vagrancy as they waited to board northbound ships or trains.[37] Steinberg argues that federal policies throughout the post–Civil War period were also designed to keep blacks as a low-paid labor force in the rural South.[38] The Freedman's Bureau, for example, forced blacks into labor contracts with plantation owners and discouraged them from moving north. The federal government's efforts to settle the Great Plains and Far West were oriented to Europeans, not southern blacks, and few blacks obtained the land grants which the railroads and government were offering European migrants.

A third factor was the poverty and ignorance of blacks, along with the unfavorable experiences of some of the first blacks to leave. In the late nineteenth century the majority of blacks were illiterate and lived in rural areas, often remote places where there was little contact with the outside world.[39] Until the *Chicago Defender* was widely circulated after 1910, there was no voice in the North encouraging the migration of southern blacks. Indeed, many of those blacks who went to Kansas in the 1870s or who went north as strikebreakers in later decades returned to the South with reports of disappointment and failure.

Black Migration from World War I to 1970

Estimates of net interregional migration are shown in Table 5.1, along with an indication of how large the migration stream was in relationship to the size of the black population of the South. During the first decade of this century, for instance, the net number of blacks leaving the South was just under 200,000, or 2.4 percent of the southern black population at the middle of the 1900–10 period.

The World War I era was the first decade in which a substantial number of blacks moved away from the South. The volume and rate of black out-migration from the South increased in the 1920s, then declined during the Depression decade. During World War II and in the

[35]Harris (1982), p. 53; Johnson and Campbell (1981), p. 88.
[36]Scott (1920), p. 73; Johnson and Campbell (1981), p. 88.
[37]Scott (1920), p. 73.
[38]Steinberg (1981), chap. 7; Jaynes (1986), p. 313.
[39]U.S. Bureau of the Census, *Negro Population of the United States: 1790–1915* (1918), pp. 88 and 406; Kiser (1932), pp. 64–65.

TABLE 5.1

Black Out-Migration from the South, 1870–1970 (in thousands)

Decade	Black Population in South at Start of Decade	Estimated Net Out-Migration of Blacks	Out-Migrants as Percentage of Mid-Decade Population
1870–1880	4,421	71	− 1.4%
1880–1890	5,954	80	− 1.3
1890–1900	6,761	174	− 2.4
1900–1910	7,923	197	− 2.4
1910–1920	8,749	525	− 5.9
1920–1930	8,912	877	− 9.6
1930–1940	9,362	398	− 4.1
1940–1950	9,905	1,468	−14.6
1950–1960	10,225	1,473	−13.7
1960–1970	11,312	1,380	−11.9

SOURCES: U.S. Bureau of the Census, *Historical Statistics of the United States: Colonial Times to 1970 (1975)*, series A-176; *Current Population Reports*, series P-23, no. 80, table 8; Everett S. Lee et al., *Population Redistribution and Economic Growth: United States, 1870–1950*, vol. 1 (Philadelphia: American Philosophic Society), 1957.

two following decades, the exodus of blacks from the South continued, and in each period there was a net movement of almost 1.5 million blacks to the North or West. About one seventh of the South's total black popluation moved to other regions in each decade.

There were strong "push" factors encouraging blacks to leave the South. The cotton economy continued to prosper through the end of the nineteenth century, but thereafter it was troubled. Foreign competition led to tumbling prices. Furthermore, boll weevils gradually moved from Texas in the 1890s to the Carolinas in the 1920s. In many areas, production could be resumed after several years, but in the meantime the income of farmers, merchants, and those workers in the cotton economy declined abruptly.[40] The years before World War I were particularly bad since cotton production fell by more than 30 percent between 1914 and 1915.[41]

The need for farm labor was reduced in the 1930s by further agricultural changes. Prices for the South's three major crops—cotton, sugar, and tobacco—once again fell sharply between the late 1920s and the early 1930s. The New Deal's Agricultural Adjustment Act (AAA) sought to boost farm incomes by guaranteeing price supports through reduced acreage. When Franklin Roosevelt took office, there was a three-year inventory of cotton in storage so special efforts were made to

[40]Vance (1936); Daniel (1985), chap. 1; Wright (1986), chaps. 3 and 4.
[41]U.S. Bureau of the Census, *Historical Statistics of the United States: Colonial Times to 1970* (1975), series K- 553.

cut production. These programs benefited landowners but exacerbated the poverty of farm laborers and sharecroppers since they were laid off or forced out. In most of the South, especially in cotton areas, the owners were likely to be white while many of the unemployed laborers and evicted tenant farmers were black.[42] Weiss notes that the number of black tenant farmers fell by one third in the 1930s, while the number of sharecroppers fell by one quarter, changes she attributes to the AAA.[43]

Mechanization came slowly to southern farmers, primarily because they lacked the capital to buy the tractors and modern equipment used by the more prosperous farmers in the Midwest. In fact, Myrdal observed that the cultivation of cotton was just about as labor intensive in the 1930s as in the 1860s.[44] However, the New Deal's policies raised farmers' incomes, and in the subsequent decades expanding markets and generous price supports allowed southern farmers to amass capital. Tractors, mechanical harvesters, and similar equipment replaced mules and farmhands. Farm size increased, marginal farmers and agricultural laborers went to cities, and—in the course of three decades—blacks almost disappeared from American farming.[45] The number of black farmers—owners, tenants, or sharecroppers—peaked at about 900,000 in 1920 and declined very little before 1930, but by 1970 the number had fallen to 100,000.[46] In 1920, 49 percent of the nation's blacks lived on farms, in 1970, 4 percent, and in 1984, less than 1 percent.[47] Fligstein, who analyzed the out-migration of blacks from the South between 1900 to 1950, argues that the decreasing need for black labor in southern agriculture is the major reason underlying the redistribution of black population.[48] In his view, these factors far outweigh economic opportunities in the North or other factors such as the Klan violence directed against blacks.

Not only were there numerous push factors encouraging blacks to leave the South, but there were many pull factors attracting them to the North. During the period from 1914 to 1919, the gross national product increased by about one sixth, while manufacturing employment rose by almost one third.[49] Immigration to the United States averaged more than 1 million persons per year between 1910 and 1914, but this was

[42]Kirby (1980), p. 142; Mandle (1978), p. 79; Myrdal (1944), p. 258.

[43]Weiss (1983), p. 55; Wright (1986), p. 230.

[44]Myrdal (1944), p. 259.

[45]Beale (1966).

[46]U.S. Bureau of the Census, *Historical Statistics of the United States: Colonial Times to 1970* (1975), series K1-50.

[47]U.S. Bureau of the Census, *Current Population Reports*, series P-27, no. 58, (December 1985), table 1.

[48]Fligstein (1981), chap. 9.

[49]U.S. Bureau of the Census, *Historical Statistics of the United States: Colonial Times to 1970* (1975), series D-130 and F-3.

ended by German attacks on shipping in World War I and the restrictive immigration laws of the 1920s. Northern firms had to drastically increase their labor supply, but they were unable to tap a reserve of European or Canadian workers; instead, they hired those they previously avoided: blacks from the South. Simply stated, blacks who eked out a marginal living in the rural South could migrate to the North and find industrial jobs which paid much higher wages. Vickery, analyzing black migration between 1900 and 1960, contends that pull factors were primarily responsible for the population shifts, believing that blacks were responding to regional differences in wage levels.[50]

Few rural blacks had information about alternatives to their rural poverty, but after 1914 firms needing workers sent recruiters into the South. In one of the larger endeavors of this type, the Pennsylvania Railroad brought 12,000 blacks North to work on their tracks and equipment during World War I.[51]

Robert Abbott, editor of the *Chicago Defender*, used his paper to encourage blacks to come north. He called for a "Great Northern Drive" and established May 15, 1917, as the specific day of exodus. The newspaper stressed that high-paying industrial jobs were readily available in the North; that blacks in the North were not constantly at risk of being lynched, bulldozed, and terrorized by white racists; and that blacks had civil rights in such cities as Chicago. The public school system educated blacks, blacks were allowed to vote, and some even held influential positions in municipal government.[52]

When blacks began to move to cities in large numbers, many municipalities wished to confine them to a few neighborhoods. Ordinances were passed specifying where blacks and whites might live, but these infringed on property rights and often failed court tests. Eventually, the Supreme Court ratified the use of restrictive covenants to keep minorities out of a neighborhood.[53] Nevertheless, in many cities there were bloody conflicts over the issue of where blacks could live, until the post–World War II boom in suburban residential construction gave whites the option of moving away from racially mixed central cities. Struggles over residential areas were particularly important causes of two of the bloodiest urban riots—East St. Louis in 1917 and Chicago two years later.[54]

An equally contentious area was the workplace. During both World Wars I and II, white workers feared that prosperity would soon end and

[50]Vickery (1977).

[51]Franklin (1967), p. 472.

[52]Spear (1967), pp. 132–35; Chicago Commission on Race Relations (1922), pp. 87–92.

[53]Vose (1959).

[54]Chicago Commission on Race Relations (1922), pp. 73–74 and chap. 5; Rudwick (1966); Tuttle (1972), chap. 4.

that they would lose their jobs to blacks who would be willing to work for lower pay. In the 1920s most—but not all—unions strongly opposed black membership, and thereby made certain that few blacks would get skilled jobs. Indeed, some crafts unions continued their racial exclusion policies into the 1960s.[55] In many industries management and the workers tacitly agreed that certain low-skill and dirty jobs would be more or less reserved for blacks, while whites retained the skilled and supervisory positions.[56] As Myrdal observed, the majority of employed blacks worked in either domestic service or "Negro jobs"—that is, despised jobs at the bottom of the occupational ladder which carried a social stigma.[57]

A major change in the representation of blacks in industrial jobs occurred after President Roosevelt signed the National Labor Relations Act in 1935, strengthening the right of workers to organize. Some unions, primarily those affiliated with the Congress of Industrial Organizations (CIO), included blacks in their membership so as to preclude their use as strikebreakers. This development, along with the appointment of a Fair Employment Practices Committee (FEPC) in 1941 and the general labor shortage during World War II, undoubtedly opened new jobs to blacks. However, the basic pattern of occupational segregation was not changed.

The industrial history of Detroit has been studied more thoroughly than that of other cities. Throughout World War II there was a high level of racial conflict within the auto plants because white workers feared that blacks were being advanced into the jobs which should belong to them.[58] The 1943 riot which killed 34 Detroiters was preceded by a series of "hate strikes" in which whites refused to work if blacks retained the better jobs they had attained because of the labor shortage.[59] In many other cities blacks appealed to the FEPC for equitable treatment on the job. In some defense industries, blacks were advanced to better jobs, which often triggered walkouts by whites.[60]

Racial discrimination in employment was finally outlawed by Title VII of the Civil Rights Act of 1964, but by the time this law became effective most of the black migration to the North had occurred. Those millions of blacks who left the South may have escaped threats from a lynch mob and intimidation from the Klan, but they faced high levels of racial discrimination in housing and employment once they got to the North.

[55]Oates (1983), pp. 185–87.
[56]Zunz (1982), pp. 396–97; Foner (1981), chaps. 9–12.
[57]Myrdal (1944), chap. 13 and app. G.
[58]Meier and Rudwick (1979), chap. 3.
[59]Shogan and Craig (1964), chap. 1.
[60]Foner (1981).

The migration trends described in these pages have altered the racial composition of the nation's regions. Figure 5.2 shows that throughout the nation's history the South has had an unusually large representation of blacks. Until 1890 over one third of the southern population was black. The out-migration of blacks changed the racial make-up of the South, and the proportion black in the South fell during the twentieth century, reaching a record low of 19 percent in 1980. The Northeast and Midwest regions experienced increases in the representation of blacks because, once European migration was curtailed, black in-migration accelerated. Approximately 10 percent of the population in these regions is black. The West remains distinctive since blacks compose a small fraction of its population. Even with the in-migration to California which was encouraged by World War II, no more than 5 percent of the population in the West is black.

The Reversal of a Pattern: Migration Trends Since 1970

During the 1970s the traditional pattern of black out-migration from the South stopped, and that region gained in the interregional exchange of migrants. This can be seen in the Table 5.2, which shows the number of in-migrants and out-migrants for each of the four regions in the most recent period.

A region may experience a migration turnaround either because fewer people leave or because more people move into the region. With regard to blacks in the South, both changes have occurred. The number of out-migrants—that is, blacks living in the South at one date but enumerated in another region five years later—dropped sharply between the late 1960s and the late 1970s. A rapidly declining proportion of southern blacks are finding the move to Chicago, Detroit, or New York appealing. The South, however, is increasingly attracting blacks from the northeastern and midwestern states, and the number of blacks going from the North to the South more than doubled in the decade between the late 1960s and late 1970s.

The migration turnaround did not, at first, affect the black population of the western states, for they continued to attract blacks through the 1970s. However, the rate of net black immigration slowed down, and preliminary data from the 1980s suggest that the West is now also losing black population in the regional exchange. Between 1981 and 1983, for example, more blacks moved from the West to the South than in the opposite direction.[61]

[61]U.S. Bureau of the Census, *Current Population Reports*, series P-20, no. 384 (February 1984), table 42; no. 393 (October 1984), table 42.

TABLE 5.2

Black and White Interregional Migration, 1965–1980 (in thousands)

	Blacks			Whites		
	1965–70	1970–75	1975–80	1965–70	1970–75	1975–80
SOUTH						
Population at Start of Period	11,226	11,970	12,720	46,351	50,420	54,900
In-Migrants	+162	+302	+415	+2,954	+3,730	+3,787
Out-Migrants	−378	−288	−220	−2,084	−1,939	−2,176
Net Migrants	−216	+ 14	+195	+ 870	+1,791	+1,611
NORTHEAST						
Population at Start of Period	3,749	4,344	4,320	43,982	44,311	43,920
In-Migrants	+147	+118	+ 99	+1,115	+ 920	+ 996
Out-Migrants	−110	−182	−274	−1,893	−2,160	−2,312
Net Migrants	+ 37	− 64	−175	− 748	−1,240	−1,316
MIDWEST						
Population at Start of Period	4,321	4,572	4,800	49,734	51,641	51,240
In-Migrants	+204	+150	+170	+1,803	+1,569	+1,793
Out-Migrants	−111	−202	−221	−2,528	−2,714	−2,900
Net Migrants	+ 93	− 52	− 51	− 725	−1,145	−1,107
WEST						
Population at Start of Period	1,717	1,695	2,160	29,096	31,377	32,940
In-Migrants	+148	+153	+194	+2,126	+2,155	+2,558
Out-Migrants	− 62	− 51	−163	−1,523	−1,566	−1,740
Net Migrants	+ 86	+102	+ 31	+ 603	+ 594	+ 812

NOTE: "Migrants" are people who survived to the end of a five-year span, were age 5 or older and live in a different region at the end of the period than at the beginning. The population estimate for blacks in the West in 1965, developed from the *Current Population Survey*, slightly exceeded the count obtained in the 1970 census. International migration is excluded.

SOURCES: U.S. Bureau of the Census, *Census of Population: 1970*, PC(1)-B1, table 56; PC(2)-B2, table 42; *Census of Population: 1980*, PC80-1-B1, table 50; *Current Population Reports*, series P-20, no. 1! (September 27, 1966), table A; no. 285 (October 1975), table 30; no. 368 (December 1981), table 42; serie P-23, no. 80, table 5.

A racial lag in migration patterns is evident in Table 5.2. In the late 1960s, as well as during the 1970s, there was a substantial net flow of whites from the North to the South and West. It took about a decade for blacks to "catch up" with whites in this pattern of migration into the South.

Does the migration turnaround involve all age groups of blacks, or are young blacks still leaving the South while older blacks are moving into that region from the North? Figure 5.3 shows migration rates for specific age groups for the four regions during the 1960s and the 1970s. We began with the black and white populations by age as enumerated in 1960 or 1970 and used census survival rates to estimate what the population would have been a decade later in the absence of migration.[62] The estimated population was then compared with the census count in the later year and the difference was assumed to represent net migration. This figure shows the net number of migrants per 100 persons present at the start of a decade.

A historical change has occurred; for the first time since World War I, the cities of the North are no longer absorbing a stream of black migrants from the South. Among those blacks aged 10 to 19 in 1960, there was a net out-migration from the South equal to 28 percent of those present at the start of the period. But in the 1970s this was the only age group of southern blacks to experience out-migration, and the size of this stream was very small. Two percent of those southern blacks aged 10 to 19 in 1970 were gone by 1980. At all other ages there was a net in-migration of blacks to the South in the 1970s, whereas in the previous decade there was an out-migration at every age.

Migration rates for blacks in the Northeast and Midwest regions are the mirror images of those for the South. In the 1960s there was substantial in-migration as blacks left the South and sought jobs and opportunities in other regions. For example, for every 100 blacks aged 10 to 19 in 1960, there was a net in-migration to the Northeast of 58 additional blacks and to the Midwest of 30 additional blacks. In the 1970s, these numbers declined—for both regions—to a net in-migration of about 3 blacks per 100 present at the start of the decade.

The decennial change in migration patterns was much smaller among whites. In both decades the northern states lost white population at most ages while the South and West gained.

The Interregional Migrants: Who Are They?

What are the social and economic characteristics of people who migrate, and how do they compare with those of individuals who remain in their region of birth? Among the most consistent findings in demography are those concerning the selectivity of migration. Long-distance migrants are often higher in social and economic status than their peers who remained in their native area.[63]

To describe interregional migrants, we examined data from the census of 1980 for two areas—the South and all other states, a region we

---

[62]Because the census of 1980 classified as "other races" some 6.6 million persons who would have been classified as white had the 1970 coding rules been used, the only way to derive comparable estimates involved using data for whites and nonwhites for the 1960s and for blacks and nonblacks in the 1970s. For a description of the census survival estimation technique, see Shryock and Seigel (1971), chap. 21.

[63]Lee (1964).

FIGURE 5.3
Migration Rates for Blacks and Whites in the Four Regions by Age,
1960–1970 and 1970–1980

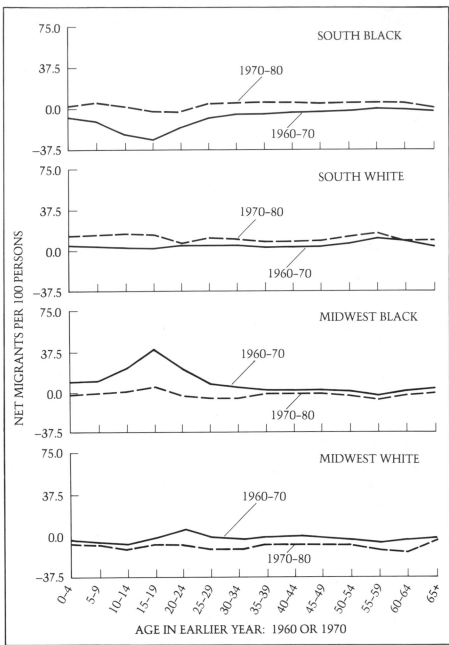

NOTE: Data for 1960 and 1970 pertain to whites and nonwhites; data for 1970 and 1980 pertain to blacks and nonblacks. This figure shows the estimated net number of migrants per 100 persons present in a region at the start of a decade.

FIGURE 5.3 (*continued*)

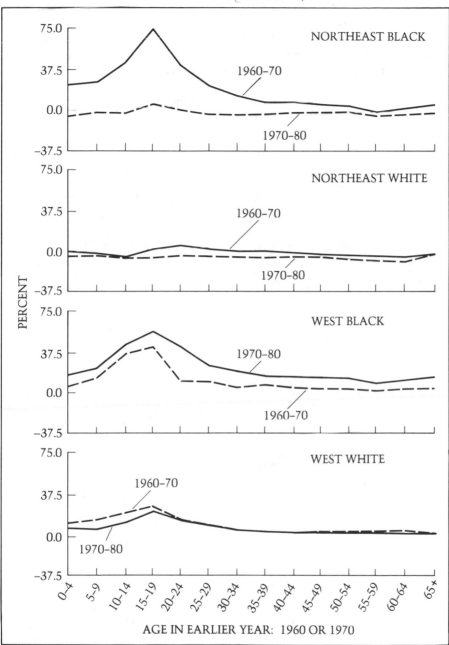

SOURCES: U.S. Bureau of the Census, *Census of Population: 1960*, PC(1)-1B, table 52; *Census of Population: 1970*, PC(1)-B1, table 57; *Census of Population: 1980*, PC80-1-B1, table 55.

will call the North. Individuals are classified as to whether they were living in their region of birth or another region in 1980. Lifetime residents are those who were living in the region where they were born. The 1980 census also posed a question about place of residence five years ago, and thus we can determine whether an in-migrant entered his present region before or after 1975. For our three-way classification we will use the terms "lifetime resident," "long-term migrant," and "recent migrant."[64]

Table 5.3 provides information about the lifetime residents of a region and the two streams of in-migrants. Since we are interested in social and economic status, we present findings for men aged 25 to 64 in 1980. For black men *living in the North,* for example, we find that 12 percent of the lifetime residents completed college compared with 8 percent of the long-term migrants from the South and 18 percent of the recent migrants.

Black men who moved away from the South before 1975 were less extensively educated than native-born black men in the North and held less prestigious jobs. For example, 43 percent of them compared with 34 percent of the native-born black men in the North worked as operatives, the job category which includes assembly line workers and laborers. However, a high proportion of these interregional migrants were employed. Their unemployment rate was lower than that of native-born men in the North, and they spent relatively many hours on the job. Their earnings were also considerably higher than those of black men who lived their entire lives in the North.

Previous investigators have found one other important way in which black migrants differ from nonmigrants: They are more likely to be married.[65] This is supported by evidence from the 1980 census; 79 percent of the men who moved north before 1975 were currently married compared with only 70 percent of those who were lifetime residents of the North.

Black men who moved north in the half decade before the 1980 census also worked a large number of hours; and, similar to earlier in-migrants, a high proportion of them were currently married. Although they were more extensively educated than previous migrants, their earnings were not unusually high. To the extent that wages are related to experience and seniority with an employer, these recent migrants are

[64]U.S. Bureau of the Census, *Census of Population and Housing: 1980,* Public Use Microdata Samples, Technical Documentation. We make use of the sample described in the previous chapter. Place of residence five years ago was coded for only one half of the individuals included in the public use sample. In this section, we are describing a 1 in 200 sample of blacks and a 1 in 2,000 sample of whites.
[65]Lieberson and Wilkinson (1976), table 6.

at a disadvantage since, at most, they spent five years on their present job.

During the 1950s and 1960s there was conjecture that the more liberal welfare regulations of northern states and their more generous benefits encouraged blacks to leave the South. Moynihan observed that in the mid-1960s families receiving Aid for Dependent Children were given $227 a month in New York, but only $62 in South Carolina. [66] He concluded that the migration of the rural poor to the urban centers was stimulated by this great regional difference in welfare payments. Empirical studies provided mixed support for this view. One analysis found a significant relationship between the level of welfare payments and black migration in the late 1960s. [67] But other investigations showed that southern-born blacks who moved to northern cities were less likely to obtain welfare benefits and less likely to be impoverished, but more likely to be employed than native-born northern blacks. [68]

If southern-born blacks moved north to obtain benefits, we would expect that an unusually high proportion of migrants would report transfer income. The census of 1980 asked whether a person received Social Security benefits from the federal government. If a man aged 25 to 64 received such benefits, they were probably from the Social Security's disability or health insurance program. It also asked about receipt of public assistance income such as the general assistance and disability programs administered by local agencies and subject to state control. [69]

The data reveal that black men who moved north before 1975 were no more likely to receive public assistance benefits than lifetime residents of the North: About 7 percent received such income. However, they were much more likely to obtain such benefits than blacks who remained in the South. Further investigations will be required to determine how much of this difference is attributable to migration and how much to a regional difference in the availability of welfare. In other words, we can neither confirm nor reject the hypothesis that migration from the South was stimulated by the availability of welfare. We note, however, that only 2 percent of recent black migrants to the North obtained these transfer payments.

Turning to information about blacks who were southern residents in 1980, we find that northern-born black men were more extensively educated, held more prestigious jobs, and reported higher earnings than life-

[66]Moynihan (1968), p. 28.
[67]Cebula (1974).
[68]Long (1974); Long and Heltman (1975).
[69]U.S. Bureau of the Census, *Census of Population and Housing: 1980,* Public Use Microdata Samples, Technical Documentation, p. K-23.

TABLE 5.3

Characteristics of Black and White Male Migrants and Nonmigrants Aged 25–64, 1980

	Blacks					
	North and West			South		
	Life-Time	In-Migrants		Life-Time	In-Migrants	
		Pre-1975	Post-1975		Pre-1975	Post-1975
EDUCATIONAL ATTAINMENT						
High School Graduates	69%	53%	77%	48%	75%	87%
College Graduates	12	8	18	7	15	23
Median Years	12.0	11.6	12.2	11.2	12.2	12.9
LABOR FORCE STATUS AND HOURS OF EMPLOYMENT						
Percent at Work	69%	71%	74%	75%	76%	78%
Percent of Labor Force Unemployed	13	10	11	7	7	9
Median Hours Worked, 1979	1,919	1,984	2,021	2,012	2,019	2,049
OCCUPATIONS OF EMPLOYED MEN						
Executive and Professional	16%	12%	16%	9%	23%	26%
Technical and Sales	7	5	6	4	10	13
Administrative Support	11	8	5	7	9	11
Service	17	16	10	13	10	9
Precision Production	15	16	17	17	18	19
Operatives	34	43	46	44	30	22
EARNINGS IN 1979						
Under $10,000	35%	27%	45%	48%	36%	48%
$20,000 or More	17	19	14	9	18	17
Median Annual Earnings	$10,079	11,511	9,528	8,052	10,000	8,712
Average Hourly Wage	$ 8.60	9.75	8.96	7.84	8.61	6.77
MARITAL STATUS						
Percent Currently Married	70%	79%	76%	62%	80%	81%
WELFARE RECIPIENCY IN 1979						
Social Security Benefits Only	4%	7%	1%	7%	5%	4%
Public Assistance Only	7	7	2	3	2	1
Both	1	1	<1	1	<1	<1

NOTE: The proportion of men who worked in agriculture is not shown but may be obtained by subtraction.

time residents of that region. The men who moved south were also distinctive in terms of a high proportion currently married and, compared with native-born southern blacks, few received transfer payments. To use the phrase of the economist, the "human capital" of blacks migrating to the South was greater than that of blacks migrating to the North.

Table 5.3 shows many substantial racial differences since whites complete more years of schooling, are less likely to be unemployed,

TABLE 5.3 *(continued)*

	Whites					
	North and West			South		
	Life-Time	In-Migrants		Life-Time	In-Migrants	
		Pre-1975	Post-1975		Pre-1975	Post-1975
EDUCATIONAL ATTAINMENT						
High School Graduates	80%	68%	82%	66%	86%	90%
College Graduates	25	20	37	18	35	35
Median Years	12.3	12.1	13.8	12.0	13.3	13.9
LABOR FORCE STATUS AND HOURS OF EMPLOYMENT						
Percent at Work	87%	83%	82%	85%	88%	84%
Percent of Labor Force Unemployed	5	5	6	4	2	4
Median Hours Worked, 1979	2,069	2,063	2,063	2,070	2,280	2,054
OCCUPATIONS OF EMPLOYED MEN						
Executive and Professional	29%	26%	32%	23%	40%	41%
Technical and Sales	12	10	11	13	18	19
Administrative Support	6	6	8	6	7	5
Service	6	7	13	6	6	7
Precision Production	21	22	22	27	19	15
Operatives	26	25	14	22	10	13
EARNINGS IN 1979						
Under $10,000	17%	18%	30%	25%	18%	25%
$20,000 or More	35	37	28	26	40	31
Median Annual Earnings	16,093	16,096	13,437	13,152	16,623	13,682
Average Hourly Wage	10.18	11.60	9.28	8.94	10.82	8.92
MARITAL STATUS						
Percent Currently Married	88%	86%	84%	88%	89%	87%
WELFARE RECIPIENCY IN 1979						
Social Security Benefits Only	5%	5%	2%	6%	5%	5%
Public Assistance Only	2	2	2	2	<1	1
Both	<1	1	1	<1	<1	<1

SOURCE: U.S. Bureau of the Census, *Census of Population and Housing: 1980*, Public Use Microdata Samples.

work at more prestigious jobs, earn more money, and are more likely to be married than comparable black men. Differences between migrants and nonmigrants among whites, however, are similar to those among blacks. Interregional white migrants tend to be more extensively educated and work at higher-status jobs than do those white men who remained in their regions of birth. White migrants who have been in their new region for more than five years earn more than the natives of that region and also earn more than their peers who remained in their region of birth.

A racial difference in the characteristics of migrants concerns marital status. Black men who moved away from their region of birth were more likely to be currently married than those who stayed, but this difference is not apparent among whites. Do stably married black men seek to maximize their income and thus migrate if higher-paying jobs are available in another region, or do personal values encourage both marriage and migration? If so, why are they apparent for blacks, but not for whites?

The Selectivity of Recent Interregional Migration

Between 1975 and 1980 about 2 percent of the blacks living in the South in 1975 moved to the North, while about 4 percent of those living in the North migrated to the South.[70] This migration was selective with regard to age, educational attainment, and region of birth. To determine more about who was moving from one region to another in this period we classified individuals by region of residence in 1975 and 1980. We then determined the proportion of people in a specific category in 1975 who were found in the *other region* five years later. The results, shown in Table 5.4, pertain to men aged 25 and over. We did not consider earnings, occupation, marital status, or welfare dependency, since those statuses may be confounded with moving; that is, they may change after arrival at the destination.

Of black men with an elementary education who lived in the South in 1975, only 0.5 percent moved to the North before 1980. Among college graduates, 6.4 percent left the South, indicating the educational selectivity of this migration stream. Black men aged 25 to 34 were much more likely to move away from the South than were older men. Region of birth also had an effect since the proportion going north between 1975 and 1980 was eight times as great for northern-born men living in the South in 1975 as it was for southern-born men. This is unambiguous evidence of return migration. The data reveal that education, age, and region of birth are related to migration in the same way for both whites and blacks.

Education and age are strongly related to each other since there has been a secular trend toward greater attainment. To better summarize the effects of these variables on migration, we fitted a model that had

---

[70]U.S. Bureau of the Census, *Current Population Reports*, series P-20, no. 368 (December 1981), table 42.

TABLE 5.4

*Proportion of Black and White Men in a Region in 1975
Who Lived in the Other Region in 1980,
Classified by Region of Birth, Educational Attainment, and Age*

	Proportion of Residents of South in 1975 Who Moved North or West Before 1980		Proportion of Residents of North or West in 1975 Who Moved South Before 1980	
	Black	White	Black	White
EDUCATIONAL ATTAINMENT				
Elementary	0.5%	1.5%	2.4%	2.2%
High School, 1–3 Years	1.8	2.0	1.7	2.5
High School, 4 Years	3.1	4.3	4.0	2.9
College, 1–3 Years	4.7	5.4	4.8	4.3
College, 4+ Years	6.4	8.0	6.0	5.0
AGE				
25–34 Years	3.2%	8.3%	6.0%	5.1%
35–44 Years	2.1	5.3	4.0	3.7
45–64 Years	0.8	1.1	1.8	2.2
65+ Years	0.6	1.4	2.1	3.0
REGION OF BIRTH				
South	1.8%	1.9%	6.0%	10.3%
North and West	14.3	13.4	2.3	2.9
Total	2.3%	4.3%	3.7%	3.5%

NOTE: Data exclude international migration and pertain to those men for whom a domestic state of residence in 1975 and state of birth were reported.

SOURCE: U.S. Bureau of the Census, *Census of Population and Housing: 1980*, Public Use Microdata Samples.

interregional migration between 1975 and 1980 as its dependent variable and region of birth, education, and age as independent variables. A model that called for additive effects of these variables—and no interactions—described the migration process for blacks. That is, the propensity to migrate varied by age, education, and region of birth, but an interaction effect of age and attainment was not needed to obtain a satisfactory fit.

Table 5.5 shows migration rates for different groups. For example, at the ages of peak migration—25 to 34—only 1.8 percent of the southern-born black men in the South with an elementary school education moved away compared with 10.7 percent of those with a college education. Among northern-born black men living in the South in 1975, the comparable proportions were 4.5 percent for those in the lowest attainment category and 23.5 percent in the highest. The effects of age on migration are seen very clearly when we control for attainment. Among black college graduates living in the South in 1975, 10.7 percent of the

TABLE 5.5

*Estimated Proportion of Black Men Who Lived in One Region in 1975
and Moved to the Other Region by 1980,
Classified by Education, Age, and Region of Birth*

	Proportion of Southern Residents Who Moved to North		Proportion of Northern Residents Who Moved to South	
	Born in South	Born Outside South	Born in South	Born Outside South
EDUCATION OF AGES 25–34				
Elementary	1.8%	4.5%	6.9%	4.0%
High School, 1–3 Years	4.4	10.6	5.7	3.3
High School, 4 Years	5.5	13.0	7.7	4.5
College, 1–3 Years	7.7	17.6	9.0	5.3
College, 4+ Years	10.7	23.5	12.3	7.4
1 TO 3 YEARS OF HIGH SCHOOL				
35–44 Years	2.5	6.2	3.1	1.8
45–64 Years	1.4	3.4	1.4	0.8
65+ Years	1.9	4.7	1.7	1.0
4 OR MORE YEARS OF COLLEGE				
35–44 Years	6.3	14.5	6.8	4.0
45–64 Years	3.6	8.4	3.3	1.9
65+ Years	4.8	11.3	3.9	2.3

NOTE: These estimates are derived from a logit model which treats the log of the odds of remaining in region as the dependent variable. It uses age, educational attainment, and region of birth as independe variables. Effects parameters are shown in Appendix Table 5.1. The two regions are South and No South.

SOURCE: U.S. Bureau of the Census, *Census of Population and Housing: 1980*, Public Use Microda Samples.

southern-born men aged 25 to 34 moved north compared with only 3.6 percent of those aged 45 to 64.

The factors influencing out-migration from the North are similar to those influencing migration from the South. Young college-educated men living in the North in 1975 were much more likely to move south than were older men who spent fewer years in school, and net of other factors region of birth had a strong effect on migration. Migration estimates for groups of black or white men with other characteristics may be calculated from the effects parameters shown in Appendix Table 5.1.

Interregional Migration: Its Consequences for Earnings

To examine economic aspects of migration, we considered the earnings of men in their prime working years—ages 25 to 64—and classified them by region of birth and region of residence. We fitted a model which specifies that a man's *earnings* during 1979—the year before the census—were determined by his age, his educational attainment, and how many hours he spent at work.[71] The means and coefficients for these earnings models are presented in Appendix Table 5.2.

Figure 5.4 presents information for men who were aged 35, college graduates, and full-time workers during 1979. Earnings for men with other characteristics may be estimated from the data shown in Appendix Table 5.2. College-educated black men who were born in the South and lived there in 1980 had average earnings of $12,411 (amounts in 1979 dollars). Southern-born black men with identical characteristics who moved north prior to 1975 earned an average of $13,951, revealing that migration was financially rewarding. Even those southern-born men who moved north after 1975 earned more than similar men who remained in the South. When black in-migrants to the North are compared with lifetime black residents of the North, we also find that they are successful economically once we control for education, age, and hours of employment. Southern-born, college-educated migrants with more than five years of residence in the North earned about $600 more than similar lifetime residents of the North.

Considering northern-born black men who moved into the South, we also find that migration was beneficial, despite the general pattern of lower wages in the South. College-educated men born in the North who moved to the South before 1975 had average earnings of $15,024, or about $1,700 more than similar men who stayed in the North. Even those northern-born men who moved to the South between 1975 and 1980 reported greater earnings than those who stayed in the North. Clearly, black men who move from one region to another—be it to the South or to the North—do well in economic terms when compared with "stayers."

Figure 5.4 also presents earnings information for men who graduated from high school and worked 35 hours per week for 50 weeks. Blacks who left their region of birth generally earned more than lifetime residents of either region. In particular, long-term migrants had much greater earnings than did those lifetime residents of either region who were similar in age, education, and hours of employment. For example, southern-born blacks who lived in the North for more than five years earned $10,296 compared with $9,048 for similar lifetime residents of the North and $8,376 for similar black men who remained in the South.

---

[71]The regression model takes the log of reported earnings in 1979 as its dependent variable. The independent variables are age, age-squared, years of schooling completed, and hours worked during 1979. Models were fitted separately for black and white men and for the six regions of birth by region of residence in 1980 groups. Men who reported earnings in 1979 of less than $1 were excluded as were men born outside the United States.

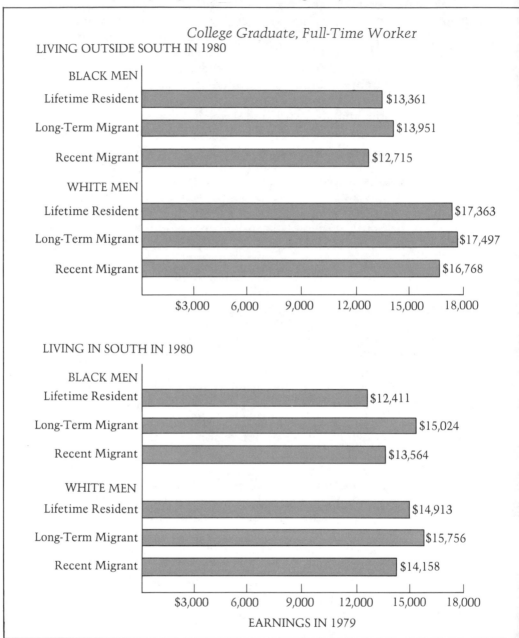

FIGURE 5.4
*Estimated Earnings in 1979 for Black and White Men Aged 35,
Classified by Region of Birth and Region of Residence in 1980*

College Graduate, Full-Time Worker

LIVING OUTSIDE SOUTH IN 1980

BLACK MEN
Lifetime Resident $13,361
Long-Term Migrant $13,951
Recent Migrant $12,715

WHITE MEN
Lifetime Resident $17,363
Long-Term Migrant $17,497
Recent Migrant $16,768

$3,000 6,000 9,000 12,000 15,000 18,000

LIVING IN SOUTH IN 1980

BLACK MEN
Lifetime Resident $12,411
Long-Term Migrant $15,024
Recent Migrant $13,564

WHITE MEN
Lifetime Resident $14,913
Long-Term Migrant $15,756
Recent Migrant $14,158

$3,000 6,000 9,000 12,000 15,000 18,000

EARNINGS IN 1979

NOTE: The coefficients used to estimate these earnings are shown in Appendix Table 5.2 along with th
means of the variables used in the 12 regression models. Earnings may be estimated for other groups
black or white men using the coefficients shown in Appendix Table 5.2. Amounts are shown in 19
dollars.

FIGURE 5.4 (*continued*)

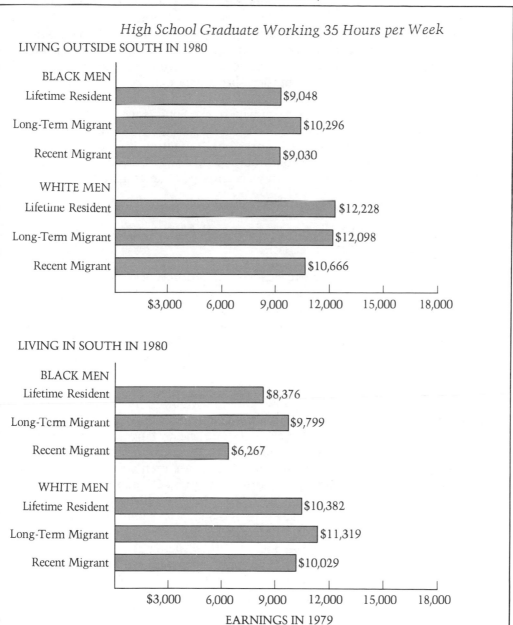

High School Graduate Working 35 Hours per Week

LIVING OUTSIDE SOUTH IN 1980

BLACK MEN
Lifetime Resident — $9,048
Long-Term Migrant — $10,296
Recent Migrant — $9,030

WHITE MEN
Lifetime Resident — $12,228
Long-Term Migrant — $12,098
Recent Migrant — $10,666

LIVING IN SOUTH IN 1980

BLACK MEN
Lifetime Resident — $8,376
Long-Term Migrant — $9,799
Recent Migrant — $6,267

WHITE MEN
Lifetime Resident — $10,382
Long-Term Migrant — $11,319
Recent Migrant — $10,029

EARNINGS IN 1979

SOURCE: U.S. Bureau of the Census, *Census of Population and Housing: 1980,* Public Use Microdata Samples.

131

Findings concerning the economic consequences of migration for whites are similar in some regards but very different in others. White men who were born in the South and moved north did well compared with similar white men who remained in the South. However, their earnings are just about equal to those of the white lifetime residents of the North. In other words, southern birth is not an asset with regard to the earnings of white men in the North, but it is an asset for black men. Northern-born black men who moved to the South reported larger earnings than those who remained in the North and earned much more than native-born southerners, but interregional migration did not have similar benefits for whites. Northern-born whites who moved south earned less than comparable white men who remain in the North.

The Causes and Consequences of Migration: A Search for Explanations

During the past two decades several conflicting perspectives about black migration have been popular. One view attributed problems of northern central cities to the influx of poorly educated southern blacks. High crime rates, the "explosion" of welfare rolls, the spread of urban poverty, and numerous other problems were thought to result from the fact that black in-migrants from the rural South did not have the abilities or skills to succeed in the urban North.[72] Demographic investigations disproved these speculations since they showed that southern-born blacks had relatively low unemployment rates compared with native blacks in the North, had more stable families, had lower poverty rates, and were hardly overrepresented on the welfare rolls. In particular, these studies consistently found that when education and occupation were taken into account, southern-born men who lived in the North earned more than native-born northern black men.[73]

National studies dealing with the total or white population consistently report that growing up in the South is a liability in terms of achievements as an adult.[74] After studying intergenerational social mobility, Blau and Duncan concluded that "being raised in the South equips a man poorly for occupational achievement."[75] Presumably, southern schools were less well-equipped, hired less competent teachers, and were in session fewer days per year than schools in the North. A high proportion of southerners—both black and white—

---

[72]Banfield (1968).
[73]Long and Heltman (1975); Hogan and Featherman (1977).
[74]Featherman and Hauser (1978), table 7.5.
[75]Blau and Duncan (1967), p. 214.

came from rural backgrounds, which encumbered those who entered a fast-paced competitive urban world. Furthermore, many southerners spoke with a drawl, which some northerners assumed connoted slow thought and ignorance. It is surprising that southern birth limits the achievements of whites outside the South but does not limit the achievement of blacks. Indeed, on several measures of economic achievement, southern birth appears to be an asset for those blacks who live in the North.

This led to the opposite speculation that, for blacks, growing up in a northern ghetto was much more of a liability than being raised in the rural South. Larry Long has argued that southern blacks have attitudes toward work and welfare that distinguish them from northern blacks. Supposedly, blacks raised in the rural South realized that their fathers and grandfathers performed manual labor from dawn to dusk and never had the option of depending on welfare. If they did not work at low-wage, low-skill jobs, they starved. Thus, southern-born blacks are presumed to be acclimated to back-breaking work, are willing to accept low-status jobs traditionally reserved for blacks, and have the southern view that adult men should support their own families and not depend on handouts from the government. Northern-born blacks, it is assumed, have greater educational attainments and much higher occupational aspirations. They also are more familiar with generous welfare systems. Thus, if they do not find rewarding jobs commensurate with their education and taste, they seek transfer payments which have traditionally been more available in the North.[76] Although this is an intriguing explanation, there have been few studies of differences in the attitudes of blacks by region of birth or the consequences these attitudes have for achievement.

It is also possible that the regional difference reported in this section came about because of the selectivity of migration. Analyses of earnings and occupational attainment take into account differences in years of education, length of employment, and even the characteristics of the family of origin.[77] However, persons who leave one region and go to another may be particularly ambitious or have other characteristics that enhance their chances for success but which are not measured in the census. In addition, the migrants who move to a new region and then remain there may be a selected subset of all who went there. It is highly probable that those men who are economically successful in their new location are more likely to remain than those who experience great difficulties.

[76]Long (1974), pp. 54–55; Long and Heltman (1975), pp. 1405–8.
[77]Hogan and Featherman (1977).

Urbanization and the Disappearance of the Black Belt

A description of minorities earlier this century would have been incomplete without an extensive discussion of the Black Belt. The Census Bureau's two statistical compendia about blacks devoted chapters to a description of population growth and change in this area.[78]

By the time of the Civil War, a group of counties extending from Chesapeake Bay in the Northeast to Texas in the Southwest had black majorities. In 1880 approximately one tenth of the nation's three thousand counties had larger black than white populations. A ratio of three or more blacks to each white resident was recorded in 55 counties in 1880, primarily in cotton-producing areas along the Mississippi River and in southern Alabama. During the period between the Civil War and the first decade of this century, most U.S. blacks lived in counties in which they were the dominant racial group. At this time, several states had majority black populations: Louisiana from 1810 through 1890; Mississippi from 1840 through 1930 and South Carolina for the century following 1820.[79]

An out-migration of blacks and an influx of whites altered this pattern so greatly that the term "Black Belt" has all but disappeared. By 1980 blacks outnumbered whites in fewer than 70 counties, almost all of them in the cotton areas of the Gulf Coast states. In recent decades, however, new counties have been added to the list: those counties or their equivalents which include such majority black cities as Atlanta, Baltimore, New Orleans, and St. Louis.

A fundamental demographic fact shaped race relations for many decades. Blacks were concentrated in isolated and impoverished areas of the rural South while many whites, including most of the immigrants from eastern and southern Europe, benefited from the expanding industrial economy of cities. Figure 5.5 describes the gradual incorporation of blacks into the modern economy, showing the proportion of total population living in urban places since 1870. The Census Bureau defines as urban all people who live in incorporated places of 2,500 or more, but caution is needed in interpreting change over time since this definition was altered in 1950 to include among the urban those people who lived in densely settled but unincorporated suburban locations.[80]

[78]U.S. Bureau of the Census, *Negro Population in the United States: 1790–1915* (1918); *Negroes in the United States: 1920–1932* (1935).
[79]U.S. Bureau of the Census, *Negroes in the United States: 1920–1932* (1935), table 12.
[80]U.S. Bureau of the Census, *Census of Population: 1980*, PC80-1-B1:A-2 and A-3. At present, the rural population should not be equated with the farm population since many rural people now work in cities. In 1980 only 6 percent of the rural black population lived on active farms. U.S. Bureau of the Census, *Census of Population: 1980*, PC80-1-C1, table 77; *Current Population Reports*, series P-27, no. 58 (December 1985), table 3.

FIGURE 5.5

Proportion of Black and White Population Living in Urban Places for the United States and its Regions, 1870–1980

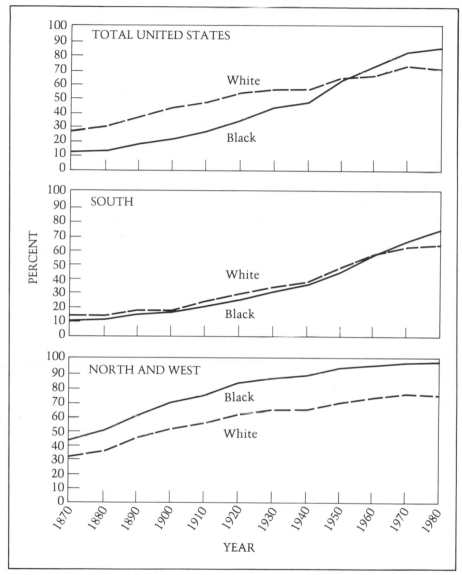

NOTE: Prior to 1950, the urban population included those who lived in incorporated places of 2,500 or more. Since 1950 the urban population has included those who live in densely settled but unincorporated suburban areas.

SOURCES: Daniel O. Price, *Changing Characteristics of the Negro Population*, (Washington, DC: U.S. Government Printing Office, 1969), table B-1; U.S. Bureau of the Census, *Census of Population: 1970*, PC(1)-B1, table 61; *Census of Population: 1980*, PC80-1-B1, table 50.

The urbanization of blacks lagged far behind that of whites at the turn of this century. For example, the proportion in cities was almost twice as great for whites (43 percent) as for blacks (23 percent). Whites became a predominantly urban population during World War I, while for blacks, this did not occur until World War II. Economic changes eventually eliminated this racial difference and since the mid 1950s the proportion urban has been higher for blacks than for whites. This is due in part to a new migration pattern since the recent movement of people away from large metropolises and into rural or remote areas has primarily involved whites.[81]

The lower panels of Figure 5.5 show that regional redistribution was a major factor in the urbanization of blacks. For over a century the majority of blacks in the North and West lived in cities, but in the South this has been the case for only a couple of decades. The migration of blacks away from the South was, in essence, a move from a region where cities were not so dominant into a highly urbanized region.

The shift of black population ended one type of geographic isolation and segregation but led to another. While the Black Belt pretty much disappeared, blacks continue to live in different areas than whites because of urban residential segregation.

Black-White Residential Segregation
The Emergence of Segregation

The patterns of black-white residential segregation which are found in metropolitan areas today date from the late nineteenth and early twentieth centuries. Urban historians who describe northern cities in the post–Civil War era note that blacks were one of many groups concentrated in low-income areas, but that blacks who wished to do so and were financially able could live throughout the city. Spear, for instance, asserted that a black ghetto did not exist in Chicago prior to the great migration of World War I.[82] Haynes, who thoroughly chronicled the growth of Detroit's black population, argued that no ghetto could be found in that city in 1908, and as late as 1915 blacks lived thoughout Detroit.[83] Green, in her history of black Washington, reported that for several decades after the Civil War blacks lived in all parts of the city, even in the northwestern quadrant.[84] Although many Philadelphia

---

[81]Long and DeAre (1981), p. 18. Lichter, Fuguitt, and Heaton (1985).
[82]Spear (1967), p. 7.
[83]Zunz (1982), pp. 374–375.
[84]Green (1967), p. 127.

blacks lived in the city's seventh ward, there was no ghetto at the turn of the century.[85]

The urban historians who describe southern cities distinguish an ante-bellum pattern from the residential segregation of the post-bellum era.[86] Blassingame, for example, argues that there was an absence of residential segregation in pre–Civil War New Orleans since blacks and whites lived side by side on the same streets or in the same rooming houses.[87] An analysis of census data for such southern cities as Augusta, Charleston, and Jacksonville in the decade after the Civil War reveals large racial differences in the type and quality of housing, but levels of segregation which are quite low compared with the present.[88] In 1890 blacks in Atlanta, Montgomery, Nashville, Raleigh, and Richmond were not concentrated into solidly black ghettoes. Rather, there were many pockets of blacks living in predominantly white neighborhoods.[89]

In the closing years of the nineteenth century, a Jim Crow system of segregation developed based on the premise that intimate social contact between the races was undesirable and would eventually weaken or destroy both races. Residential segregation was an important component of this systematic effort to isolate blacks from whites. It was accomplished in large part through a combination of real estate practices, intimidation, and legal regulations.

Real estate agents came to realize that their white clients did not want black neighbors so they turned black customers away. Green describes the situation in Washington.[90] Mary Terrell, a black woman who was a linguist, author, Oberlin graduate, member of the city's school board, and married to a *cum laude* graduate from Harvard, sought housing in the 1890s for her family in a white section of the city. No agent would sell or rent a house to them or to any other black unless the neighborhood already had a large black population. W. E. B. Dubois reports that by the 1890s Philadelphia real estate agents knew that blacks were confined to a segment of the housing market, so they charged them excessive rents.[91] Both Spear and Zunz describe the practices of real estate agents in Chicago and Detroit during World War I.[92] Realizing that a growing black population would be confined to limited areas, they took units occupied by whites and raised rents precipitously,

[85]Lane (1986), pp. 20–21.
[86]Rabinowitz (1978), chap. 5.
[87]Blassingame (1973), pp. 16 and 208.
[88]Taeuber and Taeuber (1965), pp. 45–53.
[89]Rabinowitz (1978), pp. 106–13.
[90]Green (1967), p. 127.
[91]DuBois (1899), pp. 348–49.
[92]Spear (1967), pp. 23–24; Zunz (1982), p. 375.

a technique which drove whites from these units, which were then subdivided into smaller units and rented to blacks. Similar strategies were used by several real estate firms during the first decades of this century to hasten the transition of Harlem from a Jewish to a black ghetto.[93]

Violence or intimidation was frequently directed toward those blacks bold enough to enter or remain in white areas. DuBois, for example, described the hostility directed toward a black former foreign service officer who moved into a white Philadelphia neighborhood and toward a bishop of the AME church who moved into a house owned by the Episcopalian diocese but located in a white section.[94] In Chicago quite a few blacks lived in the Hyde Park community on the south side at the turn of the century, but neighborhood organizations intimidated blacks and removed them from the area.[95] Similar strategies were used to confine blacks to the ghetto in Detroit during World War I.[96] In both Cleveland and Detroit some prosperous blacks tried to move into white communities during the 1920s, but they often met violence.[97]

The most publicized event of this type occurred in 1925. Dr. Ossian Sweet, a European-educated physician, moved his family into one of Detroit's white neighborhoods. After his house was surrounded by a hostile crowd, one of the occupants opened fire, killing one white. This led to charges of first-degree murder against the eleven blacks in the house. Clarence Darrow argued for the defendants and eventually won acquittal.[98]

Residential segregation was institutionalized by law, too. Jim Crow laws mandated segregation in most areas of public life, so it was only a small extension to legislate where people might live. In 1912 the Virginia legislature gave cities the right to designate neighborhoods as either black or white.[99] Several Virginia cities and Atlanta used such laws to isolate blacks and whites.[100] Baltimore and Greenville passed ordinances which designated individual blocks as available to only whites or blacks. Since many areas in southern cities were racially integrated, problems developed in drawing up these laws. Richmond and Winston-Salem, for example, effected *de jure* residential segregation when city councils defined the racial composition of a block on the basis of which race constituted the majority. The minority race did not

[93]Osofsky (1963), chaps. 7 and 8; Gurock (1979), pp. 145–50.
[94]DuBois (1899), p. 349.
[95]Spear (1967), pp. 21–23.
[96]Zunz (1982), p. 374.
[97]Kusmer (1976), p. 167.
[98]Conot (1974), pp. 300–3; Shogan and Craig (1964), pp. 20–21; Vose (1959), pp. 50–51.
[99]Johnson (1943), p. 173.
[100]Preston (1979), p. 96.

have to move, but no additional members of the minority group could move into the block.[101]

In 1917 the Supreme Court overturned laws of this type.[102] The litigation involved a Louisville statute which specified that only whites could live in a certain neighborhood. A white homeowner sold his property to a black, and the Court upheld the sale on the basis of property rights, not civil rights—that is, on the grounds that such ordinances denied owners the prerogative of disposing of their property as they wished.

Southern cities attempted to enact acceptable Jim Crow laws, but they generally failed to pass court tests. Eventually, the preferred method became the use of restrictive covenants. Around the turn of the century, as developers began to build several blocks or entire neighborhoods, they included clauses in deeds saying that the property could never be owned or occupied by blacks, Asians, Jews, or some other group deemed to be undesirable. The Supreme Court viewed these as private agreements between buyers and sellers and, in a 1926 ruling, claimed they involved no violation of civil rights.[103]

For several decades the NAACP and other civil rights organizations employed a litigation strategy to overturn residential segregation.[104] A major victory was achieved in 1948 when the Supreme Court ruled that neither federal nor state courts could enforce restrictive covenants, a precedent for other challenges to residential segregation.[105] In the 1960s the National Association of Real Estate Boards changed its code of ethics, which had previously supported residential segregation. During President Kennedy's Administration those regulations which called for residential segregation in federally funded housing were removed, and many municipalities adopted open housing laws.[106] The greatest legal change occurred in 1968 when Congress passed the Fair Housing Act, which bars racial discrimination on the part of any parties involved in the sale, rental, or financing of most housing units.[107]

Trends in Residential Segregation

The absence of racial data at the city block or census tract level makes it difficult to measure segregation trends precisely for many

---

[101]Johnson (1943), p. 174.
[102]*Buchanan* v. *Warley* 245 U.S. 60 (1917).
[103]*Corrigan* v. *Buckley* 271 U.S. 323 (1926).
[104]Vose (1959).
[105]*Shelley* v. *Kraemer* 334 U.S. 1 (1948).
[106]Helper (1969), chap. 3.
[107]Lamb (1984).

cities or metropolitan areas prior to 1940. Stanley Lieberson investigated trends in ten northern cities between 1910 and 1950, distinguishing the native-born white population, the foreign-born white population, and blacks.[108] In 1910 blacks were somewhat more segregated residentially from native-born whites than were foreign-born whites, but the difference was not great. Over time, the segregation of foreign-born whites from native-born whites decreased substantially, while blacks became much more segregated from both foreign- and native-born whites.[109] The Taeubers summarized trends for the first part of this century: "The most consistent findings in these historical investigations for various cities is a sharp increase in residential segregation between 1910 and 1930 in every city, both northern and southern, for which we have data."[110]

Since 1940 the decennial censuses have given us information about the racial composition of local areas, which permits a fine-grained analysis of trends in segregation. The findings reveal a high level of racial segregation with no more than modest changes in recent decades. Outside the South black-white residential segregation reached peak levels in 1950, and in the South in 1960. Since these dates, there have been small declines in segregation in most cities in all regions of the country.[111]

During the 1970s when the Fair Housing Act was effective for the entire decade, we might expect substantial reductions in residential segregation because of the improved economic status of some blacks,[112] more liberal racial attitudes on the part of whites,[113] and the continuing push by blacks for their civil rights.

Table 5.6 shows residential segregation scores for 1960, 1970, and 1980 for the 25 cities with the largest black populations in 1980. These cities included about three eighths of the nation's black population in 1980.[114] This measure of segregation—the index of dissimilarity—takes on its maximum value of 100 when all blacks and all whites live in racially homogeneous areas—an apartheid situation. Were individuals randomly assigned to their residence, this measure would approach its minimum value of zero. Its numerical value indicates the minimum proportion of either blacks or whites who would have to move from one area to another to eliminate residential segregation. The value of the index is unaffected by the relative size of the two racial groups.[115]

---

[108]Lieberson (1963).
[109]Lieberson (1963), table 38; Lieberson (1980), p. 291.
[110]Taeuber and Taeuber (1965), p. 54.
[111]Taeuber and Taeuber (1965), chap. 3; Sørensen, Taeuber, and Hollingsworth (1975), table 2; Van Valey, Roof, and Wilcox (1977), table 2; Taeuber (1983).
[112]Farley (1984).
[113]Taylor, Sheatsley, and Greeley (1978).
[114]U.S. Bureau of the Census, *Census of Population: 1980*, PC80-1-B1, table 69.
[115]Zoloth (1976).

TABLE 5.6

*Indexes of Racial Residential Segregation
for the 25 Central Cities with Largest Black Populations*

Central City	Black Population in 1980 (000)	Indexes of Black-White Residential Segregation			Change in Segregation Index	
		1960	1970	1980	1960 to 1970	1970 to 1980
ew York	1,784	80	77	75	−3	−2
hicago	1,197	89	93	92	+4	−1
etroit	759	85	82	73	−3	−9
iladelphia	639	87	84	88	−3	+4
s Angeles	505	82	90	81	+8	−9
ashington	448	80	79	79	−1	—
ouston	440	94	93	81	−1	−12
ltimore	431	90	89	86	−1	−3
ew Orleans	308	86	84	76	−2	−8
emphis	308	92	92	85	—	−7
tlanta	283	94	92	86	−2	−6
allas	266	95	96	83	+1	−13
leveland	251	91	90	91	−1	+1
. Louis	206	91	90	90	−1	—
ewark	192	72	76	76	+4	—
akland	159	73	70	59	−3	−11
rmingham	158	93	92	85	−1	−7
dianapolis	153	92	90	83	−2	−7
ilwaukee	147	88	88	80	—	−8
cksonville	137	97	94	82	−3	−12
ncinnati	130	89	84	79	−5	−5
ston	126	84	84	80	—	−4
olumbus	125	85	86	75	+1	−11
ansas City	123	91	90	86	−1	−4
chmond	112	95	91	79	−4	−12
verage for 25 Central Cities		88	87	81	−1	−6

NOTES: These measures are computed from census data for city blocks. The indexes for 1960 compare e distribution of the white and nonwhite populations; those for 1970 compare the distributions of the ite and black populations; those for 1980 compare the distribution of black and nonblack populations. djustments have been made for annexations or other changes in a central city's boundaries.

SOURCES: Annemette Sørensen, Karl E. Taeuber, and Leslie J. Hollingsworth, Jr., "Indexes of Racial sidential Segregation for 109 Cities in the United States, 1940 to 1970," *Sociological Focus* 8 (April 75): 125–43; Karl E. Taeuber, "Racial Residential Segregation, 28 Cities, 1970–1980," Working Paper . 83–12 (Madison: Center for Demography and Ecology, University of Wisconsin).

Between 1970 and 1980 black-white residential segregation decreased in 20 of the 25 cities. In Los Angeles, for example, the segregation score fell from 90 to 81, but in Washington there was no change. Overall, there was an average change of 6 points on this measure since the typical score fell from 87 to 81. In the 1960s segregation decreased in 17 of 25 cities, but the average decline was only 1 point. It is difficult to specify what constitutes a major decline, but between 1970 and 1980 drops of 10 points or more were recorded in six cities: Houston, Dallas, Oakland, Jacksonville, Columbus, and Richmond. On the other hand, blacks and whites became more residentially segregated in Philadelphia and Cleveland, while in three other cities there was no change in the level of segregation. There was no city showing a drop of 10 points in the 1960s.

Table 5.7 presents similar segregation scores for those 16 metropolitan areas which had black populations of 250,000 or more in 1980. A pattern of declining racial segregation is evident. In Chicago, for example, the metropolitan segregation score fell from 91 to 86; in Miami, from 86 to 77; and in St. Louis, from 87 to 82.

Only a few central cities or metropolitan areas experienced declines of 10 points or more in residential segregation. Nevertheless, the decreases were greater in the 1970s than in the 1960s. As Table 5.7 shows, segregation in metropolitan areas increased in the 1960s, but declined in the 1970s. Indeed, if segregation scores for central cities for the entire interval since 1940 are examined, the 1970s stand out as the decade in which black-white segregation declined the most. Despite these changes, the segregation scores for the largest cities and metropolises are around 80—a high level of segregation.

Between 1970 and 1980 the black population of central cities grew at a low rate, while in suburban rings the black population increased at a rate about three times that of the white population.[116] The proportion black in the suburban rings increased from 4.8 to 6.1 percent; in the cities, from 20.6 percent to 23.4 percent. Despite this change, there was still a high degree of racial isolation in 1980; 20 percent of the black population, compared with 42 percent of the nonblack, lived in the suburbs. Is this suburbanization of blacks leading to residential integration?

A definitive answer awaits more complete exploitation of the 1980 census data. There is considerable evidence showing that blacks who moved to the suburbs during this decade entered neighborhoods formerly occupied by whites.[117] However, an investigation of racial change in some 1,600 individual suburbs in 44 metropolitan areas found that

[116]Long and DeAre (1981), table 1.
[117]Spain and Long (1981).

TABLE 5.7

*Indexes of Racial Residential Segregation
for Metropolitan Areas with 250,000 or More Black Residents, 1980*

Metropolitan Area	Black Population in 1980 (000)	Indexes of Black-White Residential Segregation			Change in Segregation Index	
		1960	1970	1980	1960 to 1970	1970 to 1980
New York	1,941	74	74	73	—	−1
Chicago	1,427	91	91	86	—	−5
Los Angeles	944	89	89	76	—	−13
Detroit	891	87	89	87	+2	−2
Philadelphia	884	77	78	77	+1	−1
Washington	854	78	82	69	+5	−13
Baltimore	557	82	81	74	−1	−7
Houston	529	81	78	72	−3	−6
Atlanta	499	77	82	77	+5	−5
Dallas–Fort Worth	419	81	87	76	+6	−11
Newark	418	73	79	79	+6	—
St. Louis	408	86	87	82	+1	−5
San Francisco–Oakland	391	79	77	68	−2	−9
New Orleans	387	65	74	70	+9	−4
Cleveland	346	90	90	88	—	−2
Miami	280	90	86	77	−4	−9
Average for 16 Metropolitan Areas		81	83	77	+2	−6

NOTES: These indexes were calculated from data for census tracts. No adjustments have been made for changes in the definition of these metropolitan areas. These have been substantial in several metropolises including New York and Dallas–Fort Worth. The Memphis metropolitan area had a black population in excess of 250,000 in 1980, but segregation scores were not available. The indexes for 1960 compare the distribution of the white and nonwhite populations; those for 1970 compare the distribution of the white and black populations; those for 1980 compare the distribution of the black and nonblack populations.

SOURCES: Thomas L. Van Valey, Wade Clark Roof, and Jerome E. Wilcox, "Trends in Residential Segregation: 1960–1970," *American Journal of Sociology* 82 (January 1977): 826–44; Karl E. Taeuber, Arthur Sakamota, Jr., Franklin W. Monfort, and Perry A. Massey, "The Trends in Metropolitan Racial Residential Segregation," paper presented at the meeting of the Population Association of America, Minneapolis, May 5, 1984.

black-white residential segregation in suburban rings in 1980 was just about as great as it had been in 1970.[118]

These seemingly contradictory findings may be reconciled by considering the process of black suburbanization. A detailed analysis of changes in New Jersey in the 1970s found that many blacks were mov-

[118]Logan and Schneider (1984), table 1.

ing into the suburbs; however, they were generally entering neighborhoods which either already had black residents or were close to concentrations of black populations.[119] Thus, a process of racial transition was occurring in these suburbs as blacks—often of middle-class status—replaced whites, a process similar to that which occurred in many central cities after World War II.

The uniqueness of black-white residential segregation may be seen by analyzing segregation patterns for the two other racial or ethnic groups that have come to the United States recently: Hispanics and Asians. They differ from blacks in that their populations have grown more rapidly and they have entered most cities in large numbers since the 1960s. During the 1970s the black population grew 17 percent, while Hispanic population grew 61 percent and the Asian population 142 percent.[120] About one third of the Hispanic population was born outside the United States, while more than one half of the Asians were foreign-born; indeed, one quarter of the Asians counted in the census in 1980 had entered the country in the last five years. [121]

We might expect that many Hispanics and Asians would settle in immigrant enclaves and thereby be highly segregated from the non-Hispanic white population just as blacks are, but this is not the case. Levels of Asian-white and Hispanic-white segregation are quite low compared with that of blacks. This comparison is shown in Table 5.8. For this investigation, the non-Hispanic white population was defined as people who specified their race as white and then indicated they were not of Spanish origin. Hispanics are people who said they were white or "other" by race and that their origin was Mexican, Puerto Rican, Cuban, or other Spanish. Asians are people who selected an Asian or Pacific Islander response to the race question. These calculations were made with data for census tracts, which are urban areas containing about 5,000 people. Since they are larger than city blocks, they are more likely to include a heterogeneous population and, as a result, segregation indexes based on tract data are generally smaller in value than those calculated from data for city blocks.

In the nation's largest metropolis—New York—the residential segregation score comparing blacks and whites was 81, while that comparing Hispanics and whites was only 65 and that comparing Asians and whites was a much lower 49. In the Washington area the index of black-white segregation was 70, more than double the level of white-Hispanic or white-Asian segregation. In all 16 metropolises, blacks were

[119]Lake (1981).
[120]U.S. Bureau of the Census, *Census of Population: 1980*, PC80-1-B1, tables 38 and 41; *Census of Population: 1970*, PC(1)-D1, table 190.
[121]U.S. Bureau of the Census, *Census of Population: 1980*, PC80-1-D1-A, tables 253 and 254.

TABLE 5.8

Indexes of the Residential Segregation of Blacks, Hispanics, and Asians from Non-Hispanic Whites for Metropolitan Areas, 1980

	Blacks	Hispanics	Asians
Atlanta	77	31	39
Baltimore	74	38	44
Chicago	88	64	46
Cleveland	88	55	42
Dallas	79	49	43
Detroit	88	45	48
Houston	75	49	45
Los Angeles	81	57	47
Miami	78	53	34
New Orleans	71	25	54
New York	81	65	49
Newark	82	65	35
Philadelphia	79	63	47
St. Louis	82	32	44
San Francisco	74	41	47
Washington	70	32	31
Average	79	48	43

NOTE: These are indexes of dissimilarity which were calculated from census tract data. Data are shown for all metropolitan areas with 250,000 or more black residents in 1980, except Memphis.

SOURCE: U.S. Bureau of the Census, *Census of Population and Housing: 1980*, Public Use Samples, Summary Tape File 3.

much more residentially segregated from non-Hispanic whites in 1980 than were Asians or Hispanics.

These indexes also suggest that a continuation of the trends of the 1970s will leave blacks highly segregated in the foreseeable future. That is, if the average black-white segregation score declines by five points each decade, it will take about six decades for black-white residential segregation to fall to the current level of Asian-white or Hispanic-white segregation.

The Causes of Racial Residential Segregation

There are three popular explanations for the persistence of black-white segregation. One might be identified informally as the "birds of a feather" hypothesis. A second explanation focuses on economic differences between blacks and whites and contends that it is financially impossible for many blacks to share the same neighborhoods as whites. A third explanation argues that discrimination in the housing markets combines with racial differences in tastes to produce the continuing segregation of blacks and whites.

The Ethnic Homogeneity View

According to the homogeneity perspective, metropolitan communities are tesselations of ethnically identifiable subareas and the isolation of blacks from whites is typical, not unusual. Supposedly, ethnic groups prefer to live in homogeneous areas where they will find churches, social clubs, synagogues, bakeries, restaurants, and grocers serving their special needs.

In many cities we can identify areas in which an ethnic group once predominated or, in some cases, still predominates. The census of 1980 facilitates the analysis of ethnic patterns since it was the first to ask individuals about their ancestry. This was an open-ended question, allowing respondents to report any ethnic origin they wished, although they were encouraged not to give a religious response or answer that they were "American." This allows us to compare ethnic and racial residential segregation in 1980. Table 5.9 presents data for those metropolitan areas which had 250,000 or more black residents in 1980. The residential distribution of blacks and of the eleven largest ethnic groups are compared with that of people who said they were English; that is, the earliest of the European groups to arrive and the group which contributed most heavily to our political system and culture.

There were moderate levels of ethnic residential segregation in these metropolitan areas in 1980. Descendants of those groups coming to the United States prior to the Civil War were least segregated from the English, as illustrated by an average segregation score of 22 for Germans, 23 for Irish, 29 for French, and 30 for Scots. Descendants of groups who arrived later in the nineteenth century—Italians, Poles, and Hungarians—were more segregated from the English. In Los Angeles, for example, the segregation score comparing the Irish and the English was 17; that comparing the Poles and the English was 37. Apparently, the longer an ethnic group lives in the United States, the less its residential segregation from the English. The group most segregated from the English were the Russians, an ethnic group whose residential choices were once severely limited by restrictive covenants since many of them are Jewish.[122] Nevertheless, the Russians were much less segregated from the English than were blacks. In the San Franscisco area, for

---

[122]It is often assumed that a high proportion of people of Russian origin in the United States are Jewish. This is based, in part, on findings from the census of 1910, which found that 95 percent of the Americans born in Russia or with parents born in Russia had Yiddish or Hebrew as their mother tongue. Rosenthal (1975), p. 229.

TABLE 5.9

Indexes of the Residential Segregation of Blacks and Selected Ethnic Groups from the English Ethnic Group for Metropolitan Areas, 1980

	Blacks	Germans	Irish	French	Scots	Swedes	Dutch	Italians	Poles	Hungarians	Greeks	Russians
Atlanta	75	19	12	22	26	38	26	34	37	n.a.	n.a.	63
Baltimore	73	24	21	30	32	n.a.	37	34	45	48	56	73
Chicago	80	28	35	33	32	30	52	49	52	44	55	64
Cleveland	83	24	27	35	33	n.a.	n.a.	41	47	33	55	60
Dallas	77	16	14	23	27	33	28	33	37	n.a.	n.a.	n.a.
Detroit	85	21	20	27	28	36	37	45	42	44	52	66
Houston	73	17	17	22	31	35	33	29	35	n.a.	n.a.	n.a.
Los Angeles	78	14	17	24	28	25	34	25	37	41	46	55
Miami	71	18	17	27	29	32	37	29	50	48	44	61
New Orleans	63	27	23	31	n.a.	n.a.	n.a.	37	n.a.	n.a.	n.a.	n.a.
New York	67	39	43	40	n.a.	n.a.	n.a.	55	52	52	64	49
Newark	77	25	26	36	30	33	39	41	44	42	48	48
Philadelphia	77	27	32	35	32	n.a.	41	41	40	44	62	64
St. Louis	78	26	20	24	35	35	31	39	35	44	n.a.	75
San Francisco	71	15	21	26	25	25	32	30	28	41	45	43
Washington	68	15	17	25	27	36	33	25	29	41	46	51
Average	75	22	23	29	30	33	35	37	41	44	52	59

NOTES: These are indexes of dissimilarity calculated from census tract data. Data are shown for all metropolitan areas with 250,000 or more black residents in 1980 except Memphis. Each group is compared to the residential distribution of those who gave English as their only ancestry. Blacks are defined by the race question. Ethnic groups consist of individuals who reported one specific ancestry such as German or Irish.

    n.a. = Indexes not calculated if the group size was less than ten times the number of census tracts.

SOURCE: U.S. Bureau of the Census, *Census of Population and Housing: 1980*, Public Use Samples, Summary Tape File 3.

example, the segregation score comparing blacks and the English was 71; that comparing the Russians and the English was 43.

In many cities we now can locate neighborhoods in which the new immigrant groups—Hispanics and Asians—predominate. As we indicated in Table 5.8, these groups are residentially segregated from the non-Hispanic white population, but the degree of their segregation is approximately equal to the extent of English-Hungarian or English-Greek ethnic segregation.

These statistical measures suggest that residential segregation, to some extent, affects all racial and ethnic groups. There are, however, two distinctive aspects of black-white segregation. First, blacks are more isolated from whites than are the other major racial and ethnic minorities. Second, black-white segregation has persisted at high levels for decades, while the segregation of ethnic minorities from native whites has declined over time.[123] Even the newest minority groups to arrive in our cities in large numbers are less segregated from whites than are blacks.

The Economic Argument

It is often assumed that racial residential segregation is the result of the economic difference which distinguishes the races. Certainly, there are large differences in the financial status of blacks and whites. In 1985, 32 percent of the black population lived in households below the poverty line compared with 11 percent of the white households, and black families had median incomes only 58 percent of that of white families. Twenty percent of white families had incomes exceeding $50,000 compared with only 7 percent of black families.[124] The median net worth of white households in 1984—$39,000—was more than 11 times the median net worth of black households.[125] Whites typically have much greater economic resources than blacks, allowing them a wider array of housing opportunities.

If racial residential segregation were entirely dependent on a household's economic status, we would expect that poor blacks and poor whites would live together in some neighborhoods, middle-income blacks with middle-income whites in other neighborhoods, while rich blacks and whites would share the most exclusive and prestigious residential areas. Instead, we find that blacks of every economic level are highly segregated from whites of the same economic level.

[123]Lieberson (1963), p. 132; Lieberson (1980), chap. 9; Chaudacoff (1972), pp. 155–56.
[124]U.S. Bureau of the Census, *Current Population Reports*, series P-60, no. 154 (August 1986), table 5.
[125]U.S. Bureau of the Census, *Current Population Reports*, Series P-70, no. 7 (July 1986), table 6.

TABLE 5.10

*Indexes of Racial Residential Segregation, Controlling for
Income and Education, for Metropolitan Areas, 1980*

	Black-White Segregation in 16 Areas	Segregation in Three Metropolitan Areas	
		Black-White	Asian-White
FAMILY INCOME IN 1979			
Under $5,000	76	77	66
$5,000–$7,499	76	77	71
$7,500–$9,999	76	78	69
$10,000–$14,999	75	76	59
$15,000–$19,999	75	78	58
$20,000–$24,999	76	77	57
$25,000–$34,999	76	78	53
$35,000–$49,999	76	78	53
$50,000 or More	79	79	56
EDUCATIONAL ATTAINMENT OF PERSONS AGED 25 AND OVER			
Less than 9 Years	76	77	57
High School, 1–3 Years	77	79	56
High School, 4 Years	76	77	50
College, 1–3 Years	74	74	48
College, 4 Years or More	71	69	47

NOTES: The residential segregation scores are average values for the 16 metropolitan areas listed in the previous table and were computed from census tract data. For instance, the index shown for $20,000–$24,999—76—compared the residential distribution of black families in this income category with that of whites in the identical category.

The segregation scores in columns 2 and 3 are average values for the three metropolitan areas which contained both 250,000 blacks and 250,000 Asians: Los Angeles, New York, and San Francisco.

SOURCE: U.S. Bureau of the Census, *Census of Population and Housing: 1980,* Public Use Samples, Summary Tape File 3.

We computed measures of black-white residential segregation, controlling once for family income and then for educational attainment of people aged 25 and over. Average values for these segregation indexes for all 16 areas with black populations of 250,000 or more are shown in Table 5.10. For example, in Washington, the segregation score comparing black families with incomes of $10,000 to $14,999 with similar white families was 70; for families with incomes of $35,000 to $49,999 the segregation score was also 70.[126] For all 16 metropolitan areas, the average segregation score for families with incomes of $10,000 to $14,999 was 75; for families with incomes of $35,000 to $49,999 it was 76. This table presents similar residential segregation scores using educational attainment as the measure of economic status or social class.

[126]The census of 1980, conducted in April, asked about income received during 1979. Amounts are shown in 1979 dollars.

Blacks are thoroughly segregated from whites regardless of how much income they obtained or how many years they spent in school. The segregation score for families in the $50,000 and over range—79—is close to that for poverty-level families—76 for families with incomes under $5,000.

We might expect that the highly educated black elite would face few barriers in locating housing and would frequently live in the same neighborhoods as extensively educated whites, but the census of 1980 reports that they do not. The segregation scores comparing black and white college graduates were 80 in Detroit, 76 in Chicago, and 72 in New York. The corresponding residential segregation scores for blacks and whites who dropped out of high school were 77 in Detroit, 80 in Chicago, and 68 in New York.

The uniqueness of the black pattern is once again evident through an examination of Asian-white segregation. Three metropolitan areas— Los Angeles, New York, and San Francisco-Oakland—had both 250,000 black and 250,000 Asian residents in 1980. In these locations we compare the segregation of both blacks and Asians from whites, controlling for income and educational attainment. These segregation indexes are shown in Table 5.10.

At every income and educational level, black-white residential segregation was substantially greater than Asian-white segregation, even though many Asians arrived in the United States recently. For example, in Los Angeles, the score comparing the distributions of Asians and whites with more than $50,000 in family income was 58; for blacks and whites with similarly high incomes it was 83.

In contrast to the situation among blacks, as income or education increased, Asian-white residential segregation declined. This implies that social and economic factors account for some of the residential segregation of Asians since segregation levels varied by status. Asians with high incomes or extensive educations apparently could move into neighborhoods of similar whites much more easily than could blacks.

Racial Attitudes and Practices of Discrimination

A third explanation for persistent racial residential segregation focuses on the attitudes of whites and blacks and the discriminatory real estate practices that such attitudes may foster. Writing almost ninety years ago, W. E. B. Dubois asserted:

> The undeniable fact that most Philadelphia white people prefer not to live near Negroes limits the Negro very seriously in his choice of a

home and especially in the choice of a cheap home. Moreover, real estate agents knowing the limited supply, usually raise the rent a dollar or two for Negro tenants if they do not refuse them altogether. . . .[127]

Allan Spear's investigation of racial isolation in Chicago during the first decades of this century led him to conclude:

> The development of a physical ghetto in Chicago, then, was not the result chiefly of poverty; nor did Negroes cluster out of choice. The ghetto was primarily the product of white hostility. Attempts on the part of Negroes to seek housing in predominantly white sections of the city met with resistance from the residents and from real estate dealers. Some Negroes, in fact, who had formerly lived in white neighborhoods, were pushed back into black districts. As the Chicago Negro population grew, Negroes had no alternative but to settle in well-delineated Negro areas.[128]

Is it likely that racial animosity is responsible for the current high levels of residential segregation? On the one hand, we have convincing studies which demonstrate that almost all whites in all regions of the country endorse the idea that minorities should be able to live in whatever housing they can afford. In 1976, 88 percent of a national sample of whites said that whites did not have a right to keep blacks out of their neighborhoods. Furthermore, no more than a small fraction of whites claim that they would be disturbed if a black with an income and education similar to their own moved into their block.[129] These surveys show that white attitudes about racial mixing in neighborhoods are very different now from what they were forty or eighty years ago.

On the other hand, many whites apparently hold other perceptions which may have the consequence of encouraging segregation. We refer here to findings from an investigation of residential segregation in the Detroit area.[130] These results come from a location which is more polarized by race than many other areas (see segregation indexes in Tables 5.6 and 5.7). White residents, we found, generally held three beliefs about racial change in neighborhoods. First, they felt that stable interracial neighborhoods were rare. Once a few blacks entered an area, they thought that more would come and that, eventually, the neighborhood would become largely black. Second, many whites presumed that property values were lowered by the presence of black residents so it was

---

[127]DuBois (1899), p. 389.
[128]Spear (1967), p. 26.
[129]Schuman, Steeh, and Bobo (1985) table 3.3; Taylor, Sheatsley, and Greeley (1978); Pettigrew (1973), table 1.
[130]Farley et al. (1978); Farley, Bianchi, and Colasanto (1980).

seen as risky to hold property in an area undergoing racial change. Third, whites agreed that crime rates are usually much higher in black neighborhoods than in white ones. In particular, if whites are a minority in a black area, they may be exposing themselves to a high risk of victimization.

These attitudes have several consequences. Whites were extremely reluctant to purchase housing in neighborhoods that blacks were entering. In Detroit (and, we presume, in other metropolises) areas were clearly coded by color and individuals seeking or marketing housing knew with great certainty which neighborhoods were pretty much "open" to blacks and which were closed to them. [131]

A significant fraction of whites reported that they would be uncomfortable if blacks moved into their neighborhoods. If blacks made up as little as 7 percent of the population in an area, 25 percent of the whites said they would be uncomfortable and more than 25 percent said they would not enter such a neighborhood were they searching for housing. In a situation in which blacks composed 20 percent of the population, more than 40 percent of the whites would be uncomfortable and 25 percent would try to move away. In other words, the presence of even modest numbers of blacks in an area is very disturbing to a significant fraction of whites. [132]

This reluctance of whites to enter areas that are attracting blacks or even to remain in areas where blacks are represented provides motivation for real estate dealers to steer blacks and whites to distinct locations. Many of these violate the Fair Housing Law but, if commonly practiced, they help to account for the persistence of racial residential segregation.

Is racial steering still a common practice? The largest study of this question was conducted by the Department of Housing and Urban Development (HUD) in 1977. Prospective black customers were matched with prospective white customers, and they sought advertised housing in a sample of forty large metropolitan areas. The total number of tests for racial discrimination involved some 3,300 housing units. Despite presidential decrees, court rulings, local open housing ordinances, and federal laws, black customers were often treated differently from their white peers. Blacks who contacted four real estate agents had a 72 percent chance of experiencing discrimination if they wanted to rent and 48 percent if they wanted to buy. Another investigation of marketing practices in the 1970s made assumptions about the normal search process and concluded that 70 percent of the whites and blacks who wanted to rent were steered and among those who wanted to buy 90 percent were steered. [133]

[131]Molotch (1972); Rieder (1985), pp. 79–85; Taub, Taylor, and Dunham (1984).
[132]Farley et al. (1978), figure 7.
[133]Wienk et al. (1979), chaps. 2 and 3; Lamb (1984), p. 152.

Studies conducted in the 1980s in Boston and Denver show that there is a continuing high level of discrimination against blacks when they enter the housing market. The Boston investigation, for example, found that when matched whites and blacks sought housing, whites were given a larger array of houses from which to select and were invited to visit more locations. A black would have to visit nine real estate agents to get as many invitations as a white would receive in visiting five agents.[134]

These finding are consistent with earlier studies and experiences. In the late 1960s HUD supported a thorough investigation of the causes of racial residential segregation conducted by the National Academy of Sciences.[135] Their authoritative report contended that it was impossible to specify one cause for the persistent isolation of blacks from whites. Rather, there was a pervasive "web of discrimination" involving the actions and inactions of local governmental officials, federal agencies, financial institutions, and real estate marketing firms, which had the consequence of limiting housing opportunities for blacks and thereby creating the segregated patterns of metropolitan America. How that "web of discrimination" kept blacks and whites racially isolated in Chicago during the decades after World War II is thoroughly described in Hirsch's *Making the Second Ghetto*, a study which stresses the importance of both violence and governmental policies.[136]

Among the secretaries of HUD, George Romney was unusual in his attempts to seek equal opportunities for minorities in the housing market. He developed innovative programs which sought to open the suburbs to blacks through a carrot and stick approach that would reward communities that guaranteed equal opportunities and terminate funding in those that did not. Very strong opposition to such federal policies came from suburban communities, especially Warren, Michigan. This led President Nixon to withdraw his support for this program and it was subsequently terminated.[137]

The attitudes of blacks are also important in accounting for residential segregation. For more than two decades, national samples of blacks have been asked whether they prefer a racially mixed or largely black neighborhood.[138] Consistently, two thirds to three quarters of the black respondents have selected the integrated neighborhoods. In the Detroit study, we asked blacks if they would be comfortable in residential areas of different compositions and whether they would be willing to move into racially mixed areas. Most blacks would be comfortable in

---

[134]Yinger (1984), p. 14.
[135]Hawley and Rock (1973); National Academy of Sciences (1972).
[136]Hirsch (1983).
[137]Dimond (1985), pp. 183–84.
[138]Pettigrew (1973), table 5.

any neighborhood except those in which they were the only black resident. Almost all blacks were willing to be the third black to move into a formerly white area and many were willing to be the second black. However, blacks expressed great reluctance to be the first black on a white block. Some expressed a fear that crosses would be burned on their lawns or their windows would be stoned, but much more common was a feeling that their white neighbors would be unfriendly and critical of their behavior, and would make them feel unwelcome and out of place.[139]

These survey data from the Detroit area point out one of the serious problems which impedes residential integration. Whites strongly endorse the ideal of equal opportunities for blacks but would be uncomfortable if more than token numbers of blacks entered their neighborhoods. Blacks desire to live in mixed areas, but are reluctant to be the pioneers. It appears that whites are saying that integration is acceptable so long as black representation is minimal. Blacks, on the other hand, see integration as desirable but think the ideal neighborhood is one with a sizable black population—a number that will not only make whites uncomfortable, but will terminate white demand for housing in the neighborhood.

In the 1976 study of segregation in the Detroit metropolis, we found that blacks and whites were very knowledgeable about where they and the "other races" belonged. Real estate practices apparently reinforce and sharpen these judgments by steering blacks and whites to separate neighborhoods. Almost all Detroit area blacks knew about the strong antiblack reputation of one suburb, Dearborn, while whites defined the central city as appropriate for blacks and did not seek housing there even if it were available at attractive prices. We presented the following hypothetical situation to all respondents:

> I'd like you to imagine that you're going to move. You have a choice of buying two houses that are identical, except that one is located in northwest Detroit and the other is located in a desirable suburb. The house in the suburb costs $8,000 more than the house in Detroit. Which of the two houses would you choose to move into?

Given these alternatives, 90 percent of the whites selected the suburban house while 75 percent of the blacks selected the city house. Although endorsing the ideal of integration, many whites would actually spend large sums of money to avoid living in an area which they believe "belongs" to blacks. Blacks were knowledgeable about housing costs and the openness of various areas to them. They knew that whites in many suburbs would be somewhat upset if they moved there. For

[139]Farley et al. (1978), p. 332.

these reasons, many blacks who expressed a preference for integrated living may have avoided many suburbs and sought homes in those neighborhoods which were undergoing racial transition.[140]

The Consequences of Racial Residential Segregation

The consequences of residential segregation may be as difficult to assess as the causes. Certain aspects of the residential concentration of blacks may be beneficial. To the extent that voters prefer candidates of their own race and are hesitant to cast ballots for candidates of the other race, residential segregation has—at least in the short run—the effect of increasing black political representation. However, it may simultaneously permit white politicans to narrow their appeal if they concede the black vote to black candidates. Undoubtedly, the geographic concentration of blacks is beneficial to those businesses, social welfare organizations, and churches that focus on the specific needs of a black clientele.

Five adverse effects of racial residential segregation have been identified. First, if blacks are denied the opportunity to compete for a significant fraction of the housing market, they will be restricted to one portion of the market and may have to pay more than whites for equivalent housing. Courant argues that even if only a few whites discriminate against blacks, it will lead to a long-run stable equilibrium in which blacks pay more.[141] Investigations in the 1960s and early 1970s found that in many metropolitan areas blacks paid more than whites for comparable housing. Kain and Quigley, for example, showed that blacks in St. Louis spent more than whites to obtain housing similar in size and quantity.[142] King and Mieszkowski reported that New Haven blacks paid 6 to 13 percent more than whites for comparable housing.[143] There have been few recent investigations of these racial differences in housing costs, and the findings may differ because of the general decline in the demand for housing in many of the older metropolitan areas.

Second, blacks typically live in lower-quality housing than whites, occupy older housing, and are much less likely to be owners than whites. In 1980, for example, 68 percent of the white households owned their homes compared with 44 percent of the black households. Indeed, the proportion of black households who were owners in 1980 was lower than the proportion of white households who owned their homes in

[140]Farley et al. (1978), pp. 339–40.
[141]Courant (1978).
[142]Kain and Quigley (1975), table 7-1.
[143]King and Mieszkowski (1973).

1890—48 percent.[144] What remains open to dispute is whether blacks obtain lower-quality housing and have low rates of home ownership because they are poorer, on average, than whites or because they face discrimination in the housing market. Some analysts, such as Richard Muth, stress the racial difference in economic status and presume that this is the most important determinant of racial differences in housing quality and tenure.[145] Others demonstrate that if you take racial differences in income and demographic composition into account, you find a remaining net racial difference in quality or tenure.[146] For example, the 1977 Annual Housing Survey found that 52 percent of the black households and 68 percent of the white households lived in housing units built since World War II. Demographic and economic differences could account for 11 of the 16 point difference, suggesting that blacks were more likely than whites to live in old housing even after economic differences were taken into account. There was a 25 point racial difference in the proportion of householders who were homeowners. Economic and demographic factors accounted for 18 of those points, suggesting once again that there was a net racial difference in tenure, a difference which many believe reflects discrimination in the housing market.[147]

Third, several analysts have argued that racial residential segregation helps to explain the persistence of high unemployment rates among blacks. Supposedly, in the post–World War II period, manufacturers often shifted their production from older cramped plants in central cities to spacious new plants in the suburbs. Many shopping malls and service centers opened in the suburban ring, and the suburbs rapidly expanded their payrolls by hiring teachers, policemen, maintenance workers, and municipal officials. At this time, racial discrimination confined blacks to central city ghettoes, which put them at a great disadvantage with regard to jobs.[148] Empirical investigations have provided no more than mixed support for this hypothesis.[149] Racial residential segregation has measurable negative consequences for the occupational achievement and income of blacks,[150] but suburban blacks have unemployment rates almost as high as those of blacks who live in central cities.[151]

[144]U.S. Bureau of Census, *Census of Housing: 1960*, HC(1)-1, table H; *Census of Housing: 1980*,HC80-1-A1, table 7.
[145]Muth (1969, 1974).
[146]Jackman and Jackman (1980).
[147]Bianchi, Farley, and Spain (1982), table 2.
[148]Kain (1968).
[149]Mooney (1969); Masters (1975), chap. 5.
[150]Jibou and Marshall (1971).
[151]Westcott (1976), table 1.

Fourth, there can be no doubt that residential segregation is the major cause of racial segregation in public schools. If the constitutional mandate for integrated schools is to be fulfilled, it will be necessary either to integrate neighborhoods or to transfer a large proportion of metropolitan area students away from their neighborhood schools. Ann Schnare observed: "Residential segregation by race may then reduce the educational opportunities of blacks and perpetuate existing income and class differentials by depriving minority children of the chance to compete on an equal footing with white children."[152]

Finally, there are the psychological consequences of residential segregation. The isolation of blacks in American cities during the late nineteenth and early twentieth centuries was accomplished to fulfill the desire of whites that there be no social equality with blacks. What are the implications of this for blacks? W. E. B. DuBois argued that blacks in the United States were strangers in their own land. They were, he said, born with a veil because they had to perceive themselves and their world through the stereotypes imposed on them by whites, stereotypes that stressed their physical inferiority, their lack of morals, and the poverty of their culture.[153]

Segregated neighborhoods continue to limit the social contracts of whites and blacks, and high proportions of blacks and whites in metropolitan areas now reach adulthood without ever having a close friend of the other race. Many may complete their education without ever attending a school which enrolled students of the other race and without living in a neighborhood where the other race was well represented. This isolation may perpetuate stereotypes among both blacks and whites, stereotypes which reinforce the idea that one race is superior to the other.

[152]Schnare (1978), pp. 16–17.

[153]Toll (1979), p. 3.

APPENDIX TABLE 5.1

Effects Parameters from a Logit Model Which Treats the Log of the Odds of Interregional Migration as the Dependent Variable for Black and White Men Aged 25 and Over, Classified by Region of Residence in 1975

	Black		White	
	In South in 1975	Outside South in 1975	In South in 1975	Outside South in 1975
Total Sample Size	14,465	13,123	7,821	17,482
In South in 1980	14,137	482	7,485	607
In North/West in 1980	328	12,641	336	16,875
Percent Migrating	2.27%	3.67%	4.30%	3.59%
Ratio of Stayers to Movers	40.7::1	26.3::1	22.3::1	26.9::1
EFFECTS PARAMETERS				
Grand Mean	3.0992	3.5462	3.1308	2.8480
EDUCATION				
Elementary	+1.0916	+ .1768	+ .0862	+ .3740
High School, 1–3 Years	+ .1676	+ .3682	+ .2102	+ .1764
High School, 4 Years	− .0632	+ .0516	+ .0290	+ .0864
College, 1–3 Years	− .4154	− .1250	− .0526	− .2160
College, 4+ Years	− .7806	− .4716	− .2728	− .4208
AGE				
25–34 Years	− .6684	− .8268	− .9492	− .3706
35–44 Years	− .0804	− .1742	− .4670	− .0056
45–64 Years	+ .5380	+ .5958	+ .6844	+ .4362
65+ Years	+ .2108	+ .4052	+ .7318	− .0600
REGION OF BIRTH				
South	+ .4697	− .2835	+ .5100	− .3749
North or West	− .4697	+ .2835	− .5100	+ .3749
Likelihood Ratio χ^2	25.9	31.8	38.1	51.9
Probability of χ^2	>.50	.43	.18	.01
Degrees of Freedom	31	31	31	31

NOTE: This model treats the log of the odds of remaining in a region as its dependent variable. To estimate the proportion moving or staying for a specific group, it is necessary to sum the grand mean effect and add the effects for the appropriate categories of age, education, and region of birth. The sum estimates the log of the odds of staying in that region from 1975 to 1980. For college-educated black men aged 45–64 who were born in the South and lived there in 1975, the log of the odds of staying was 3.3263, implying an odds ratio of 27.84 stayers for each mover. In other words, 3.5 percent of the group moved away from the South between 1975 and 1980. For further details, see Davis (1978) or Swafford (1980).

SOURCE: U.S. Bureau of the Census, *Census of Population and Housing: 1980*, Public Use Microdata Samples.

Regression of Log of Earnings in 1979 on Age, Age-Squared, Educational Attainment, and Hours of Employment for Black and White Men Aged 25–64, Classified by Region of Birth and Region of Residence

	Black Men					
Birth →	N & W	South	South	South	N & W	N & W
1975 →	N & W	N & W	South	South	South	N & W
1980 →	N & W	N & W	N & W	South	South	South
ample Size	3,895	4,438	207	8,756	310	104
EGRESSION PARAMETERS						
Intercept	5.5154	6.3266	6.0557	5.9410	4.9683	3.8947
Age	.0773	.0639	.0611	.0585	.1152	.0600
Age Squared	−.0008	−.0006	−.0005	−.0006	−.0013	−.0003
Education	.0537	.0447	.0543	.0608	.0756	.1430
Hours of Work	.0007	.0005	.0005	.0006	.0005	.0008
R^2	.328	.191	.221	.242	.286	.274
IEANS						
Age	37.4	43.4	32.6	39.7	36.5	32.0
Age Squared	1,506	1,992	1,125	1,700	1,422	1,081
Education	12.4	11.3	12.9	10.7	12.6	13.8
Hours of Work	1.826	1,871	1,856	1,866	1,891	2,026
g of Annual Earnings	9.2316	9.3631	9.1665	9.0259	9.2292	8.9998
nnual Earnings	$13,681	15,008	12,647	11,276	13,771	13,573
g of Hourly Earnings	1.862	1.943	1.753	1.627	1.813	1.531
ourly Earnings	$ 8.60	9.75	8.96	7.83	8.61	6.77

	White Men					
Birth →	N & W	South	South	South	N & W	N & W
1975 →	N & W	N & W	South	South	South	N & W
1980 →	N & W	N & W	N & W	South	South	South
mple Size	10,672	829	96	3,958	942	324
EGRESSION PARAMETERS						
Intercept	5.9977	5.5643	5.2321	6.0967	5.2328	5.0523
Age	.0812	.0987	.1003	.0657	.1264	.1234
Age Squared	−.0008	−.0010	−.0010	−.0006	−.0014	−.0013
Education	.0564	.0610	.0881	.0593	.0577	.0612
Hours of Work	.0005	.0005	.0004	.0005	.0004	.0004
R^2	.230	.244	.356	.269	.219	.249
EANS						
Age	41.2	43.3	34.8	40.7	42.1	36.1
Age Squared	1,831	1,991	1,289	1,777	1,903	1,399
Education	13.3	12.6	14.1	12.2	14.0	14.5
Hours of Work	2,078	2,082	2,097	2,100	2,141	2,126
g of Annual Earnings	9.6406	9.6796	9.4938	9.4990	9.6960	9.5220
nual Earnings	$19,330	20,097	17,539	17,160	20,599	17,742
g of Hourly Earnings	2.079	2.117	1.946	1.932	2.100	1.944
ourly Earnings	10.18	11.60	9.28	8.94	10.82	8.92

URCE: U.S. Bureau of the Census, *Census of Population and Housing: 1980*, Public Use Microdata nples.

6

BLACK FAMILY, WHITE FAMILY:
A COMPARISON
OF FAMILY ORGANIZATION

I N THE span of 150 years, or five generations, black Americans were
emancipated, enfranchised, and empowered politically; engaged in a
massive migration from the rural South to the urban North;
realized dramatic economic gains; and moved from caste segregation to
social desegregation in the society. These massive changes in black
community and individual life, although ultimately to the better, had
disruptive consequences for many institutions in black society—black
families being only one of them.

The last 20 years have seen an explosion in the research literature
on black family life. During the 1970s over 50 books and 500 articles
were published. This production represented a five-fold increase over the
literature accumulated about black families in the half century between
W. E. B. DuBois's pioneering study of black family life and 1970,[1] the beginning of the most prolific period of research on black families.[2]
Despite this voluminous expansion in the literature, questions persist
about the true nature of black family life in the United States. Scholars
continue to debate whether black families are fundamentally comparable to or different from white families in the society. This chapter
evaluates empirical evidence of race differences in patterns and trends of

[1]DuBois (1970).
[2]Staples and Mirande (1980).

160

family life using data from the 1980 census. At the same time, we will be concerned with identifying shared characteristics in the organization and enactment of family life across race in the United States.

The Research Record:
Black and White Family Differences

In the past, many who have compared black and white families have advocated a pathology model. White family structure and processes were defined (either implicitly or explicitly) as normative. Thus, where black families departed from these patterns, they were defined as not normal—that is, pathological.[3] The origins of the pathology interpretation of black family life date back to E. Franklin Frazier's early studies.[4] However, recent impetus for these ideas was provided by Daniel Patrick Moynihan and Lee Rainwater. Merging key ideas from Gunnar Myrdal's concept of a "vicious cycle" and Frazier's notions of "family disorganization,"[5] Moynihan and Rainwater gave renewed currency to a view of black family life as fundamentally disorganized, unhealthy, and negative. To illustrate the point, Moynihan characterized black families as centerpieces in a community-wide "tangle of pathology,"[6] while Rainwater described black families as central actors in a process of "self-victimization" actively involving the black community as a whole.[7]

The thesis that black families are essentially pathological was soon vigorously challenged. Billingsley and Scanzoni criticized this model for its failure to recognize the diversity inherent in black family life.[8] They also criticized the model for failing to address the consequences that different contexts, values, and resources had for the structures and functions of black families. Hyman and Reed, Hill, and Heiss provided empirical documentation which refuted views of black family life as matriarchal, unstable, and welfare-reliant.[9]

Gutman demonstrated the relationship of changing historical circumstances to the evolution of black families,[10] while Stack carefully illustrated how different cultural values in black communities led to different conceptions of what constituted appropriate family forms.[11] By

[3]Staples (1971); Allen (1978); Engram (1982).
[4]Frazier (1932, 1939).
[5]Myrdal (1944); Frazier (1932, 1966)
[6]U.S. Department of Labor (1965), chap 4.
[7]Rainwater (1966), p. 175.
[8]Billingsley (1968); Scanzoni (1971), chap. 8.
[9]Hyman and Reed (1969); Hill (1972); Heiss (1975).
[10]Gutman (1976).
[11]Stack (1974).

1980 the pathological model had been resoundingly rejected. However, the task of identifying a model that sensitively and accurately portrayed important characteristics of black family life remained uncompleted.

The foundations of the "resilient-adaptive" model for interpretation of black families were laid during the 1970s by published challenges to the established pathology model. Simply stated, the "resilient-adaptive" model viewed black family organization as emerging from, and reflecting, differing historical circumstances, cultural values, and current socioeconomic settings. Thus, black and white family differences were taken as given, without the presumption of one family form as normative and the other as deviant. Rather, it became normative to expect the form and functioning of families to vary as their circumstances, resources, and cultural imperatives varied. [12] To some extent, the influence of this alternative model on interpretations of black-white family differences has been dampened by the relative scarcity of empirical research from this perspective. Such empirical research is especially sparse in regard to demographic studies, despite the historic importance of demographic statistics in the assessment of black family life in the United States. [13] This chapter uses 1980 census data to contribute to the literature on black and white family comparisons. Approaching observed family patterns from the "resilient-adaptive" or "culturally relative" perspective, we hope to enhance our understanding of the race factor in American family life.

Historical Perspectives On Black Family Life

Conventional wisdom presents a view of contemporary black family life as very much influenced by the period of slavery in this country. In particular, the higher incidence of marital disruption and female single-parent households among blacks (relative to whites) is seen as a product of this "slave heritage." For example, the historian Osofsky argued that *"slavery* initially destroyed the entire concept of family for American Negroes and the slave heritage, bulwarked by economic conditions, continued into the twentieth century to make family instability a common factor in Negro life."[14] In a similar vein, the sociologist Hauser wrote:

> Family disorganization and unstable family life among Negro Americans is a product of their history and caste status in the United States. During slavery and for at least the first half century after emancipation, the Negro never had the opportunity to acquire the patterns of sexual

[12]Elder (1985).
[13]DuBois (1909); Frazier (1966); Glick (1981).
[14]Osofsky (1963), p. 134.

behavior and family living which characterize middle-class white society.[15]

In large respect, these more contemporary interpretations of black families' historical roots in the United States draw from the prolific writings of Frazier about the black community and black families.

Frazier identified two traditions of black family life believed to emanate from the past.[16] The first was more widespread and seen as closely related to slavery and rural southern peasantry; its typical form was the "matriarchal," or female-headed, family. The second tradition of black family life identified was represented by the two-parent, male-headed household common to the minority of black Americans who were free, had independent artisan skills, and owned property. Frazier concluded that the family of the first and more common tradition "lacks continuity, and its roots do not go deeper than the contingencies of daily living."[17] This conclusion has been broadcast widely and contributes to the literature's negative interpretations of black family life.

Revisionist histories of black family life have taken issue with Frazier specifically and historical interpretations of black families generally. Gutman tells us that "much has been written about the history of the Afro-American family, but little, in fact, is really known about its composition and household at given historical moments, and even less is known about how and why it changed over time."[18] His findings from studies of black families in Buffalo, South Carolina, New York City, and Mobile disputed conventional viewpoints. Gutman's analyses of census data, courthouse records, and personal documents showed that most ante-bellum free Negroes *and* poor rural/urban freedmen and women lived in husband-wife headed nuclear families.[19] The fact that an average two thirds of black adults between 1850 and 1880 lived in double head and nuclear households raises serious questions about characteristics of slave families and the families of freed slaves as "a kind of cultural chaos."[20]

Numerous other studies also contest the conventional view of black families' historical development in this country. Furstenberg compared white and black family structure in Philadelphia for 1850 and 1880 using census data.[21] Seventy-three percent of white households and 75 percent of black households were nuclear. Among blacks, ex-slaves were *more* likely to reside in households headed by couples, lead-

[15]Hauser (1965), p. 854.
[16]Frazier (1932, 1966).
[17]Frazier (1949), p. 636.
[18]Gutman (1975), pp. 182–83.
[19]Gutman (1976).
[20]Stampp (1956), pp. 340–49.
[21]Furstenberg, Hershberg, and Modell (1975).

ing Furstenberg to conclude that urban and economic factors, more than cultural factors, accounted for observed increases in female-headed households. The conclusion that black family life during the late nineteenth century represented a stable form is supported by findings from Lammermeier,[22] who reported that over 60 percent of black families in the Ohio River Valley basin between 1850 and 1880 were headed by males. By the same token, Fogel and Engerman revealed two-parent households of long duration to be widespread among blacks across several slaveholding areas in the United States.[23] It would seem therefore, that the reputed link between the higher incidence of female-headed households and marital disruptions among contemporary black families and during the period of slavery is not supported by empirical facts.

Black and White Families in Historical Perspective

Since the beginning of the twentieth century, American family life has changed substantially. The evolution has been characterized by changes in the normative prescriptions governing family life, the roles assigned family members, the responsibilities of families to the larger society and their members, and the composition of modal family households, to name a few. Predictably, many of the changes observed in the general case have been duplicated by black families. In critical ways, however, changes in black families have resulted in either further differentiation from white families or in the creation of pronounced differences where none previously existed.

Between 1890 and 1984 households of both races increased in number; there was a 650 percent increase in black households (from 1.4 to 9.3 million) and a 660 percent increase in white households (from 11.3 to 74 million).[24] Until 1960 black households increased at a rate slower than or equal to the rate of white households. However, since 1960 black households have increased at almost twice the rate of white households (93 percent versus 55 percent). Higher growth rates in the black population among persons likely to form new households (for example, young adults) and higher marital disruption rates, which result in the formation of two separate households, largely explain race differences in the household growth rate. For both races, average household size declined from 1890 to 1984. The current average size of black

[22]Lammermeier (1973).
[23]Fogel and Engerman (1974a) pp. 84–85.
[24]U.S. Bureau of the Census, *Current Population Reports*, series P-23, no. 80 (1979), table 73; series P-20, no. 398 (April 1985), table B.

households (3.0) and white households (2.7) represents a decrease by two persons in each case.[25]

Fertility rates also reveal important historic differences between black and white women, discussed in detail in an earlier chapter. Although the general trend has been one of declining fertility for both races, significant differences persist in that black women's fertility continues to be higher by approximately 50 percent.[26] Black women are almost five times more likely to bear children out of wedlock (in 1984, 13 percent for whites versus 59 percent for blacks of all births);[27] and black teenage females bear children out of wedlock at four times the rate of whites (in 1984, 87 versus 19 per 1,000 unmarried women aged 15 to 19).[28]

The type and distribution of families by race reflect trends discussed above. By 1985 there were 6.8 million black families, or 250 percent more than in 1940; for the same period, white families grew by 90 percent, reaching 54.4 million. There has been a substantial increase in female-headed households among blacks. Between 1940 and 1985 the proportion of husband-wife black families declined steadily from 77 to 51 percent; white families remained essentially the same at 85 percent.[29] Since 1940 the proportion of black families headed by females has almost tripled, reaching 44 percent by 1985; again, white families were stable, hovering around 12 percent for the period.

Female heads of families differ drastically by race on several important dimensions. For instance, in 1984, 21 percent of the black women heads were separated as opposed to 13 percent of the white women. In the same year, 25 percent of black female family heads and 43 percent of whites were classified as divorced.[30] The latter pattern reflects a downturn since 1960 in the proportions of black women reported as married. In 1984, 28 percent of black women aged 15 to 44 lived with husbands, down from 52 percent in 1960. This compares with a drop of 14 percentage points among white women from 69 to 55 percent.[31]

---

[25]U.S. Bureau of the Census, *Current Population Reports*, series P-23, no. 80 (1979), table 73; series P-20, no. 398 (April 1985), table 21.

[26]U.S. Bureau of the Census, *Current Population Reports*, series P-23, no. 80 (1979) table 90.

[27]U.S. National Center for Health Statistics, *Monthly Vital Statistics Report*, vol. 34, no. 4, supplement (July 18, 1986), table 18.

[28]U.S. National Center for Health Statistics, *Monthly Vital Statistics Reports*, vol. 34, no. 4, supplement, (July 18, 1986), table 18.

[29]U.S. Bureau of the Census, *Current Population Reports*, series P-23, no. 80, (1979), table 74; series P-60, no. 149 (August 1985), table 1.

[30]U.S. Bureau of the Census, *Current Population Reports*, series P-23, no. 398 (1979), table 73; series P-20, no. 398 (April 1985), table 13.

[31]U.S. Bureau of the Census, *Census of Population: 1960*, PC(1)-1D, table 176; *Current Population Reports*, series P-20, no. 399 (July 1985), table 1.

The above patterns translate into a trend of decline in the numbers of children living with both parents. This trend was especially pronounced in black households where the proportion of children under age 18 living with both parents dropped from three quarters in 1960 to two fifths in 1984. For white households, the comparable figures were 93 and 81 percent.[32] Like many of the other observed patterns and trends, black children's living arrangements were strongly related to family economic status. The relationship was nearly linear in 1984; in families with annual incomes under $10,000, fewer than 15 percent of children lived with both parents; in families with incomes over $40,000, this proportion rose to 98 percent. A similar pattern held for whites: In families with incomes less than $10,000, 45 percent of children were in single-parent households. In white families with incomes over $40,000, 97 percent of the children lived with both parents.[33]

Family economic status has been shown to be a strong predictor of family structural and economic characteristics. At the same time, black Americans are known to be disproportionately represented among this country's poor. For this reason, it is important to examine historical trends in family income by race. Since we present a detailed discussion of black Americans' economic status elsewhere in this volume, our discussion here will be brief and focused on the family dimension. In 1935 incomes of two-parent black families in Atlanta, Georgia, were a third of their white counterparts' incomes. Similarly defined black families in New York City had incomes half those of white families.[34] Black families throughout the country had incomes about 60 percent those of white families. By 1985 these race and regional differences had moderated. Incomes of black families in the South were slightly over 50 percent of white incomes, while incomes of black families in the North were almost 60 percent of white incomes.[35]

From 1947 to 1985 the median income of black families, stated in constant dollars, more than doubled. Between 1964 and 1969 black family median income rose by over a third. Conversely, black family median income was stable during the recession years 1953–58 and actually *dropped* during the 1969–70, 1973–75, and 1979–81 recessions. By comparison, white families experienced mostly stable or steady growth

[32]U.S. Bureau of the Census, *Current Population Reports*, series P-23, no. 80 (1979), table 79; series P-20, no. 399 (July 1985), table 4.
[33]U.S. Bureau of the Census, *Current Population Reports*, series P-20, no. 399 (July 1985), table 9.
[34]U.S. Bureau of the Census, *Current Population Reports*, series P-23, no. 80 (1979), table 13.
[35]U.S. Bureau of the Census, *Current Population Reports*, series P-60, no. 164 (August 1986), table 1.

for the period from 1947 to 1985. By 1985 black family median income was 58 percent of white family income.[36]

The extent of the economic gulf between white and black families in this society is highlighted by two simple statistics from 1985. For this year, three times as many black families reported earnings of less than $7,500 (23 versus 7 percent); at the other extreme, nearly three times as many white families reported earnings in excess of $50,000 for the year (20 versus 7 percent). Concealed beneath the surface of these vast economic disparities, and most certainly related to them, are pronounced within-race differences in family economic well-being.

Female-headed families are at a profound disadvantage economically compared with families headed by couples—irrespective of race. In 1985 black female heads of families reported incomes 30 percent of what male heads of families with working wives reported (median income: $9,300 versus $30,500). For whites the figure was 40 percent (median income: $15,800 versus $37,000). So striking is the relationship between family structure and family income that some scholars have begun to speak of the "feminization" of poverty,[37] a reference to the overrepresentation of female-headed families among the nation's poor. To illustrate the point, 15 percent of black male-headed families fell below federally established poverty levels in 1985 compared with 53 percent of black female-headed families. For whites, the comparable figures were 7 and 30 percent.[38]

Race and Family Patterns in the 1980 Census

The 1980 census counted 69.0 million white, 8.2 million black, 4.0 million Hispanic, and 1.1 million Asian-Pacific Islander households. Of these households, 50.6 million white, 6.1 million black, 3.3 million Hispanic, and 0.8 million Asian-Pacific Islanders can be classified as family households.[39]

For purposes of the U.S. census, a household is defined as "the person or persons occupying a housing unit." A family is defined as "two or more persons, including the householder, who are related by birth, marriage, or adoption, and who live together as one household; all such persons are considered as members of one family."[40] In the discussion

[36]U.S. Bureau of the Census, *Current Population Reports*, series P-60, no. 154 (August 1986), table 11.
[37]Ross and Sawhill (1975).
[38]U.S. Bureau of the Census, *Current Population Reports*, series P-60, no. 154 (August 1986), table 16.
[39]U.S. Bureau of the Census, *Census of Population: 1980* (1983), PC80-1-C1, Tables 121 and 132.
[40]U.S. Bureau of the Census, *Current Population Reports*, series P-20, no. 398 (April 1985), pp. 227–29.

that follows, it is important to keep in mind the conceptual and analytical distinctions between family and household. For example, in using shared residence as the major criterion for defining a family, the census automatically omits extended families that transcend and link several different households, each containing a separate (or seemingly so) family.

Previous comparisons of family patterns by race using census data have looked at two major characteristics: household type and household composition. Household type refers to whether the household is headed by a married couple, a single male, or a single female; while household composition refers to whether the household consists only of parents and children or has additional members (the nuclear versus extended family distinction). For the most part, prior studies reveal pronounced differences in family patterns by race: Black families are less frequently headed by a married couple and more often extended.[41] Implicit in the paradigm which guides empirical study of race differences in family patterns is a normative statement. This society's norm of a monogamous, conjugal, nuclear family—that is, a family consisting of only the married parents and children in the household—influences how all family life is perceived and evaluated.

The 1980 census revealed striking differences by race in family patterns. Confirming findings from earlier studies, black families were considerably less likely than white families to be headed by married couples (42 versus 63 percent).[42] In fact, the proportion of married couple family households among blacks was lowest for all the various race/ethnic groups; 58 percent of Hispanic and 61 percent of Asian/Other family households were couple-headed. As Figure 6.1 illustrates, there were other important race differences in family patterns also. Compared with white households, black families were three times more likely to be headed by single females. Black families were also one third more likely than Hispanic families and twice as likely as Asian/Other families to be headed by women. Significant racial differences are also found in household composition. Thirty percent of black households in the 1980 census were extended compared with 18 percent of white households. The interesting fact is that Hispanic and Asian/Other households were closer to blacks than whites on this dimension; both race-ethnic groups had 28 percent of households with extended family membership. It seems, therefore, that nonwhite families are more likely than white families to have nonnuclear family members (that is, cousins and grandparents) living in the family residence.

Other general comparisons of black and white families using data

[41]Anderson and Allen (1984); Tienda and Angel (1982); Farley (1971).
[42]Allen (1979).

168

FIGURE 6.1
Types of Household by Race, 1980

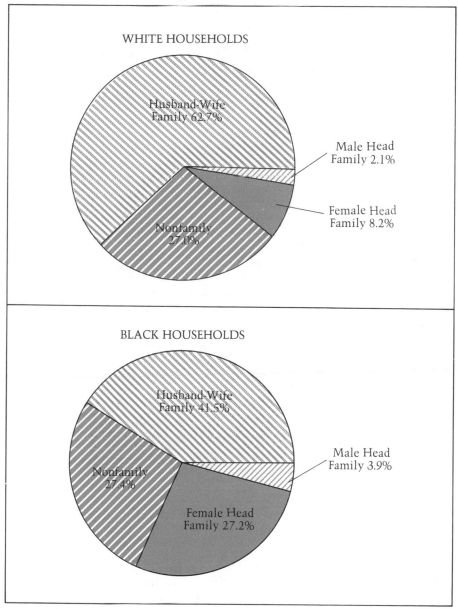

SOURCE: U.S. Bureau of the Census, *Census of Population and Housing: 1980,* Public Use Microdata Samples.

from the 1980 census are also quite revealing. Whites, for example, are one half times more likely to be married than blacks (59 versus 39 percent). Blacks, on the other hand, are one half times more likely to be single or never married (23 versus 15 percent). While the percentage widowed among whites is slightly below that for blacks (13.5 versus 14.6 percent), blacks are much less likely to be separated (2.3 versus 11.1 percent) or divorced (9.5 versus 12.4 percent).

The average age at first marriage for both blacks and whites reported in these data was about 23.5, but comparisons of sex ratios, the number of males to females in a given age group, suggest that marriage—or long-term monogamous relationships of any sort—pose more problems for blacks because black women outnumber black men in each age category during the critical dating, mating, and childbearing years of 20 to 49. At ages 20–24 there are 97 males for every 100 females; by age 30–34 the ratio drops to 96, and by ages 45–49 to 94. For whites, the ratios are much more favorable. At ages 20–24 there are 102 males for every 100 females; by ages 45–49 it is still a favorable 100.[43] A relative comparison of the four race-sex subgroups using data standardized for age reveals the proportion married with spouse present to be highest for white males (64 percent) and lowest for black females (37 percent). Fifty-seven percent of white females and 44 percent of black males were married with spouse present. The apparent conclusion is that women more than men, blacks more than whites, and black women most of all are disadvantaged in the "marriage market."

In male- or female-head, single-parent households, important differences are revealed by race in the age distribution of children. The observed patterns show black single-parent households to disproportionately contain younger children. Thus, of all the children residing in male-headed households, 25 percent of black children and 17 percent of white children were aged 3 or younger. Thirty-two percent of white children and 29 percent of black children in this category were aged 13 to 16. Likewise, of all the children living in female-headed, single-parent households, 18 percent of black children and 15 percent of white children were aged 3 or younger.

While the debate over whether black family life is pathological seems to have been resolved in favor of the "resilient-adaptive" position, we are still without clear understanding of *why* black and white families differ. Research findings have helped to correct the distorted impression that black-white family differences indicate black family deviance and pathology. However, the more vexing question of explaining race differences in family patterns persists.

[43]These sex ratios incorporate the estimates of net census undercount that appear in Passell and Robinson (1984), table 3.

Two possible explanations for black-white differences in family life are commonly advanced. One explanation attributes these systematic and persistent differences to cultural norms of black and white families. It is argued that black families are distinct from white families in patterns of organization and interaction because they operate on the basis of different cultural values, values in many respects derived from their African heritage of being a once enslaved, afterwards discriminated minority in American society.[44] The second explanation focuses on economic differences. Black families differ from white families, this position argues, because of their relative economic disadvantage, which is seen to have originated in slavery and to have been perpetuated by systematic economic discriminations and consequent economic deprivations since emancipation.[45]

Data from the 1980 census may prove useful in addressing questions of whether culture or economics better explains historically observed differences in black and white family organization and functioning. Using data from a national survey of over 150,000 households, Angel and Tienda found that "it is impossible to choose unequivocally between these two explanations."[46] Similarly, Allen concluded from analysis of 1970 census data that "questions surrounding the relative effects of class and culture on black family life in the urban United States remain largely unresolved."[47] It seems, therefore, that both explanations are valid to an extent; economic status and cultural values contribute independently (and one presumes jointly) to the explanation of black and white family differences.

In the analysis that follows, we shall explore family differences by race to establish the extent to which they are related to economic factors. Since specific information on cultural orientations is not available in census data, we will settle for a strategy that attributes race variations in family patterns unaccounted for by economic factors to other factors. We are inclined to believe that this "residual" category will likely include many cultural factors.

Economics and Race in the Explanation of Family Patterns

Not surprisingly, the economic disadvantage of blacks revealed by analyses elsewhere in this volume and in previous studies is reflected in comparisons of black and white households from the 1980 census. Black

[44]Sudarkasa (1981); Nobels (1979).
[45]Allen (1979); Angel and Tienda (1982).
[46]Angel and Tienda (1982), p. 526.
[47]Allen (1979), p. 310.

families are disproportionately poor; the proportion of black households with annual incomes below $5,000 is four times that of white households, while the proportion of black families with annual incomes above $25,000 is half that of whites. Bianchi and Farley argue that per capita family income (that is, income per member of the family) is a more sensitive and appropriate measure of family economic well-being.[48] When black and white families are compared on this measure, their stark differences in economic well-being persist; black families had per capita incomes only 58 percent of those of whites ($8,115 versus $4,731). Even more interesting is the fact that black family per capita income was lowest of the four race-ethnic groups. Asians/Others earned $6,772 per capita and Hispanics earned $4,868 per capita. (Amounts in this section are shown in 1979 dollars.)

The economic advantage of whites over blacks is also evident in comparisons between the two related aspects of socioeconomic status. White heads of household have completed on average over 1.5 years more of schooling than their black counterparts. Further, while 16 percent of white household heads have fewer than 9 years of schooling, 19 percent are college graduates. The comparable figures for blacks are 27 percent with fewer than 9 years and 8 percent college graduates. Blacks also do not fare as well occupationally. They are more often unemployed or outside the labor force. When employed, blacks hold jobs of lower prestige and power. Thus, 20 percent of white heads of households worked at jobs in the executive-managerial and professional specialty categories versus only 9 percent of black heads of household. At the other end of the occupational spectrum, 37 percent of blacks worked as operatives or service workers compared with 20 percent of whites.

We have seen that both household type and family income vary by race. The question arises as to whether a relationship exists between race, household type, and economic status. Is residence in a nontraditional household more likely for blacks because of their lower economic status or is this pattern the result of cultural preferences?

We see from Table 6.1 that traditional family structure, or headship of household by a married couple, is associated with income across race. Among whites, there is a linear increase in the percentage of households headed by couples as one moves from the low-income to high-income category. Forty percent of families in the $5,000–$9,999 a year category are headed by couples, 67 percent in the $15,000–$19,999 range, and 82 percent in the $25,000–$34,999 range. Just the opposite is true for female-headed families. Here, the proportion decreases with increases in economic level. Therefore, 13 percent of households in the

[48]Bianchi and Farley (1979).

$5,000–$9,999 a year category are headed by females, 9 percent in the $15,000–$19,999 range, and 5 percent in the $25,000–$34,999 range.

The pattern for black families is equally compelling; income is strongly associated with type of household. Among blacks, 73 percent of households with incomes in excess of $25,000 but less than $35,000 are headed by couples. This compares with 52 percent of households with incomes of $15,000–$19,999, and 29 percent with incomes of $5,000–$9,999. As with whites, the proportion of female-headed families among blacks runs in the reverse direction. These patterns confirm that household type varies by income for both races, households headed by couples are associated with higher income levels, and female-headed households are associated with lower income levels.

Race differences are apparent within income categories in the 1980 census data. A higher proportion of white households are headed by couples at each of the income levels in Table 6.1. Similarly, the proportion of female-headed households is higher for blacks at each income level. On average, the proportion of husband-wife households is 10 points higher for whites; the average difference between races on proportion of female-headed households is 14 points higher for blacks. It seems, therefore, that factors in addition to income may account for the observed race differences in household type. This residual may represent the so-called cultural effect, or that set of effects attributable to unspecified factors somehow correlated with race. Additional analyses convince us that these unspecified residual factors are not simply other aspects of socioeconomic status (for example, education, occupation) operating through race to indirectly influence household type. For instance, among whites with fewer than 12 years of education, the proportion of husband-wife households is 60 percent and the proportion of female-headed households is 9 percent. Among blacks with fewer than 12 years of schooling, 36 percent of households are headed by married couples and 27 percent are headed by single females. Among household heads with education beyond high school, 64 percent of white households and 44 percent of black households are headed by couples. Twenty-two percent of black families and 6 percent of white families have female heads.

Race differences in household headship persist across occupational categories and employment status. Thus, 70 percent of white household heads who were employed share family headship with a spouse; this compares with 50 percent of blacks (Table 6.2). Married-couple households are also more common among whites than blacks in the unemployed category (45 versus 27 percent). Consistent with this pattern, the proportion of households headed by females is higher in each employment status among blacks than among whites. Shifting attention to

TABLE 6.1

Household Type by Race and Household Income, 1980

Household Type	Under $4,999	$5,000–$9,999	$10,000–$14,999	$15,000 $19,999
WHITES				
Husband-Wife Heads, Family	18.6%	40.1%	54.5%	67.0%
Male Head, Family	1.2	1.8	2.1	2.3
Female Head, Family	12.4	12.8	11.7	8.7
Nonfamily Household	67.8	45.3	31.7	22.0
Total	100.0%	100.0%	100.0%	100.0%
BLACKS				
Husband-Wife Heads, Family	13.5%	29.0%	41.0%	51.5%
Male Head, Family	2.1	4.1	5.0	5.1
Female Head, Family	35.6	37.1	29.3	21.7
Nonfamily Household	48.8	29.8	24.7	21.7
Total	100.0%	100.0%	100.0%	100.0%

SOURCE: U.S. Bureau of the Census, *Census of Population and Housing: 1980*, Public Use Microdata Samples.

differences in household type by occupation of head, we see that 74 percent of white households in the executive-managerial category were husband-wife versus 52 percent of black households. At the lower end of the occupational hierarchy, 59 percent of black households headed by operatives were husband-wife versus 74 percent of white households. Taking income and other measures of socioeconomic status into account only slightly diminishes these powerful relationships between race and household type.[49]

Examination of the relationship between children's age and family income yields compelling results (Table 6.3). Life cycle and economic patterns predict that families with younger children will have lower annual incomes. This pattern is similar for both black and white families. There is a general shifting of age group proportions across income levels in Table 6.3 such that fewer families with incomes of more than $25,000 have children aged 3 or less (compared with those families who earn less than $15,000). Similarly, fewer families in the lower earnings categories have teenage children compared with those families in the higher income categories.

Tremendous racial differences exist, however, when one looks at

[49]These tabulations were made from the Public Use Microdata Sample Files of the Census of 1980 cited in Table 6.1.

TABLE 6.1 (*continued*)

	$20,000–$24,999	$25,000–$34,999	$35,000–$49,999	$50,000 and Over	Total
WHITES					
Husband-Wife Heads, Family	75.7%	82.4%	86.3%	86.1%	62.1%
Male Head, Family	2.8	2.3	2.5	2.5	2.3
Female Head, Family	5.8	4.7	3.5	2.4	8.3
Nonfamily Household	15.7	10.6	7.7	9.0	27.3
Total	100.0%	100.0%	100.0%	100.0%	100.0%
BLACKS					
Husband-Wife Heads, Family	62.0%	73.0%	80.2%	77.8%	40.5%
Male Head, Family	5.3	5.0	4.0	3.9	4.1
Female Head, Family	17.3	12.3	9.2	10.9	27.6
Nonfamily Household	15.4	9.7	6.6	7.4	27.8
Total	100.0%	100.0%	100.0%	100.0%	100.0%

the probability of living in a low-income household. Black children are significantly more likely to live in lower-income households; 59 percent of all black children in 1980 lived in the households with reported incomes of less than $15,000 compared with 28 percent of white children. In 1980, 42 percent of black children and 14 percent of white children lived with parents whose income did not exceed $10,000 for the year. At the other extreme, 17 percent of white children and 6 percent of

TABLE 6.2

Household Type by Race and Employment Status of Householder, 1980

	White			Black		
Household Type	Job, at Work	Job, Not at Work	Unemployed	Job, at Work	Job, Not at Work	Unemployed
Husband-Wife Heads, Family	70.0%	60.9%	44.8%	50.2%	36.3%	26.5%
Male Heads, Family	2.1	3.7	2.1	4.2	5.4	3.2
Female Heads, Family	6.9	11.1	11.4	22.2	34.8	35.0
Nonfamily Household	21.0	24.3	41.7	23.4	23.5	35.3
Total	100.0%	100.0%	100.0%	100.0%	100.0%	100.0%

SOURCE: U.S. Bureau of the Census, *Census of Population and Housing: 1980*, Public Use Microdata Samples.

TABLE 6.3
Child's Age by Race and Family Income, 1980

Age	Below $10,000		$10,000–$14,999		$15,000–19,999		$20,000–24,999		$25,000–49,999		$50,000 and over	
	White	Black	White	Black	White	Black	White	Black	White	Black	White	Black
3 Years or Less	27.3%	22.1%	26.8%	20.3%	25.8%	19.3%	19.6%	17.1%	12.2%	14.5%	12.0%	11.9%
4–6 Years	16.9	17.1	17.8	16.0	17.6	15.7	15.2	14.6	11.2	12.2	10.5	11.9
7–12 Years	34.0	35.8	35.1	36.5	35.2	37.5	38.2	38.9	37.5	37.2	37.8	35.5
13–16 Years	21.8	25.0	20.3	27.2	21.4	27.5	27.0	29.4	39.1	36.1	39.7	40.7
Total	100.0%	100.0%	100.0%	100.0%	100.0%	100.0%	100.0%	100.0%	100.0%	100.0%	100.0%	100.0%

SOURCE: U.S. Bureau of the Census, *Census of Population and Housing: 1980*, Public Use Microdata Samples.

black children lived in households where the annual income was above $35,000. One percent of black children compared with 6 percent of white children lived in the most privileged of households: those where annual income exceeded $50,000 during 1980.

The conclusions are clear and undeniable; a majority of black children find themselves in low-income households. Moreover, significant proportions of black children, at rates three times those for whites, live out their lives in households characterized by poverty and deprivation. The implications of these statistics for the quality of life experienced and the future prospects of black children are apparent.

We saw above that the proportion of extended households, like income and household type, varied systematically with race in the 1980 census. Blacks were more likely than whites to live in households that included members other than the nuclear group of parents and children. In assessing this pattern, it is important to determine whether race–cultural preferences or income–economic necessity provides the better explanation for these observed differences. Stack has documented the economic incentives that encourage low-income groups to form extended families.[50] Such arrangements are, on balance, beneficial in that they allow for the pooling of economic resources and ensure greater continuity in the availability of these resources over time. One wonders, therefore, if the observed race differences in proportion of extended households would disappear if the income differences of blacks and whites are taken into account.

The statistics in Table 6.4 suggest that race is related to household extendedness independent of household income. At each level of income, a greater proportion of white households includes only parents and children. Extended households are, on average, twice as common among black households, the one exception being for families in the no-income category, where race differences in household extendedness are not significant. Contrary to our expectations and findings from the literature, the proportion of extended households did not decrease as family income increased. Indeed, whites in the categories below $10,000 and above $50,000 were equally likely to live in extended households; the same was true for blacks in the two extreme income categories.

There are relationships between household extendedness and other measures of socioeconomic status. Among household heads with less than a high school education, 33 percent of blacks and 19 percent of whites were in extended households; for high school graduates, the comparable figures were 30 and 17 percent; and for college graduates, 19

[50]Stack (1974).

TABLE 6.4
Household Extendedness by Race and Family Income, 1980

Household Extendedness	Below $10,000	$10,000– $14,999	$15,000– $19,999	$20,000– $34,999	$35,000– $49,000	$50,000 and over
WHITE,						
Nuclear	87.3%	88.3%	89.8%	90.4%	89.5%	87.7%
Extended	12.7	11.7	10.2	9.6	10.5	12.3
Total	100.0%	100.0%	100.0%	100.0%	100.0%	100.0%
BLACK						
Nuclear	71.2%	69.9%	72.0%	74.1%	74.3%	69.6%
Extended	28.8	30.1	28.0	25.9	25.7	30.4
Total	100.0%	100.0%	100.0%	100.0%	100.0%	100.0%

SOURCE: U.S. Bureau of the Census, *Census of Population and Housing: 1980*, Public Use Microdata Samples.

and 13 percent. Among whites, the difference between the lowest and highest educational category was minimal (5.5 percentage points); among blacks, however, this difference was much larger (13.2 percentage points).[51] When we look at occupation of household head, the following patterns are evident. Greater proportions of black than white households are extended across occupational categories, and the proportion of extended households drops for both races with increased status or prestige of occupation. Thus, 12, 16, and 30 percent, respectively, of white households whose heads work as executives-managers, operatives, and service workers resided in extended households. For blacks, 22 percent of executives-managers, 27 percent of operatives, and 33 percent of service workers headed extended households. The difference between employed and unemployed household heads in the proportion of extended households was negligible for both races.

Race and Household Extendedness: A Cultural Link?

The link between race and household extendedness in these data is clear and decisive. Across various characteristics, blacks are significantly more likely than whites to live in households that include members other than the immediate, nuclear family. The association of race with household extendedness persists even after the acknowledged economic differences between blacks and whites are taken into account. This finding suggests that factors other than economic necessity may explain the greater tendency of blacks to opt for this particular nontraditional sort of family arrangement. Blacks may indeed more often organize themselves into extended family households for reasons of cultural preference. The tendency of Asians and Hispanics, two other non–Anglo-American groups, to live in extended over nuclear family arrangements relative to whites adds further weight to this cultural hypothesis.

Economic factors were seen as clearly predictive of household types, but less so of household extendedness. We are led to speculate whether other, noneconomic factors might better explain the greater proportion of extended households among blacks. Studies by McAdoo, Shimkin et al., and Martin and Martin provide a clue on this point.[52] This research indicates that the benefits to family members from residence in an extended household often go far beyond the economic. Thus, the major lure of such arrangements may have more to do with

[51]These tabulations were made from the Public Use Microdata Sample Files of the Census of 1980 cited in Table 6.1.

[52]McAdoo (1978); Shimkin, Shimkin, and Fratc (1978); Martin and Martin (1978).

the needs of family members for companionship, assistance with child care, help around the house, and so forth. Census data do not lend themselves to the investigation of these kinds of motives for location in extended households. However, we do know that certain family statuses and life stages are associated with demands likely to exceed an individual's available resources. Therefore, one wonders whether the elderly more than the young, those divorced or separated more than those who are married, and parents with infants or young children more than those with older or no children would be more likely to join extended households as a way to increase the pool of available social and emotional resources on which to draw in efforts to meet life's demands.

Examining the hypothesis that needs for additional emotional or social resources encourage the formation of extended households (as much as if not more than needs for additional economic resources), we first look at differences in household extendedness by marital status of household head. The observed patterns lend some credence to our hypothesis that people who are more self-sufficient economically, emotionally, and socially will be less likely to live in extended households. In both races the proportion of adults living in extended families is lowest for married people and highest for those who are single. Likewise, people who are widowed, divorced, or separated are significantly more likely than those who are married to be members of extended households. Social, cultural, *and* economic considerations apparently create a situation such that individuals in nontraditional marital statuses (that is, single, widowed, divorced, or separated) are also more greatly represented among those in nontraditional housing arrangements.

Race, Household Composition, and the Economic Well-Being of Families

To this point our analyses have produced a variety of findings—some contradictory, others revealing, and many confusing. Multivariate analyses are helpful in such situations for their ability to clarify the complex relationships between multiple variables. In essence, these procedures allow researchers to examine the relationship between two variables while controlling for how each of these variables may be influenced by others. We focus on economic well-being as the outcome variable because it holds profound implications for the quality of life experienced by blacks and whites in American society.

Our multivariate analysis relates per capita income to seven predictor variables: household type, household extendedness, region, area type, age, and education. The analysis attempts to determine the relative importance of each of these factors in the determination of the household's economic status. We use Multiple Classification Analysis (MCA) as the multivariate statistical procedure because of its ability to handle predictor variables that have no better than nominal or categorical measurement. The dependent variable, however, is measured on a numerical scale. The dependent variable equals the per capita income of households classified by characteristics of the household head. This is not equal to the per capita income of persons since household size is not taken into account in our calculation. Because households with many children have lower per capita incomes, the per capita incomes for households shown in Table 6.5 are larger than other indexes of per capita income.

Results from our MCA analysis of income are presented in Table 6.5. The ETA squared values provide an estimate of the explanatory power of a set of variables. In this respect, the statistic is analogous to multiple R-squared in regression analysis and may be interpreted as the percentage of variance in the outcome variable accounted for by the predictor variables. Together, the seven predictor variables in this analysis account for a sizable 33 percent of the total variance in per capita household income among blacks and 32 percent among whites.

In relative terms, household type is the predictor variable most strongly correlated with household income. Among blacks 18 percent of the variance in this income measure is attributable to the type of household. Among whites, it is 15 percent. Households headed by married couples have significantly higher per capita incomes; for example, households headed by a married black couple were an average of $6,025 above the mean for all black households. With other contributing factors controlled, the net deviation from the mean for married couple households was still $6,122 above the mean. By contrast, incomes in female-headed and nonfamily households were far *below* the mean.

Educational attainment of the household head provides the second most powerful explanation of household income in this analysis. Education accounts for 9 percent of the variance in levels of household per capita income among blacks and 12 percent among whites. Black households whose heads graduated from college averaged about $11,000 more in income than those headed by persons who failed to complete high school.

The other predictor variables—area type, household extendedness, region, and age—manage to account individually for less than 5 percent of the total variance observed in the per capita income of households. Despite this, several revealing findings are apparent. For instance, the economic advantage of extended households is revealed. Compared by

TABLE 6.5

Multiple Classification Analysis of Per Capita Income of Households in 1980, by Race (in 1979 dollars)

	White Households				Black Households			
	Mean Income	Gross Deviation	Net Deviation	ETA Squared	Mean Income	Gross Deviation	Net Deviation	ETA Squared
HOUSEHOLD TYPE								
Married Couple	$24,709	+4,313	+4,407	.145	$19,807	+6,025	+6,122	.182
Male Head	22,027	+1,632	−1,413		15,573	+1,792	−115	
Female Head	14,368	−6,028	−6,629		10,038	−3,744	−4,198	
Nonfamily	12,267	−8,128	−7,911		8,657	−5,124	−4,538	
HOUSEHOLD EXTENDEDNESS								
Nuclear	19,979	−416	−931	.005	13,006	−775	−1,186	.013
Extended	23,297	+2,902	+6,482		16,256	+2,475	+3,787	
REGION								
Northeast	21,095	+699	+633	.004	14,881	+1,100	+1,033	.011
Midwest	19,572	−823	−663		13,444	−337	−222	
South	19,203	−1,192	−763		11,996	−1,785	−1,463	
West	21,854	+1,458	+704		15,605	+1,824	+739	

AREA TYPE							
Central City	19,591	−805	+140	.018	13,885	+104	+135	.012
Ring	21,855	+1,459	+795		14,771	+990	+493	
Outside SMSA	16,823	−3,573	−2,462		10,712	−3,069	−1,815	
AGE OF HOUSEHOLD HEAD							
Under 20 Years	8,489	−11,906	−8,298	.040	6,266	−7,515	−5,375	.033
20–24 Years	13,766	−6,629	−6,221		9,267	−4,513	−4,504	
25–29 Years	18,376	−2,020	−4,062		12,857	−924	−2,351	
30–34 Years	21,826	+1,430	−1,785		15,288	+1,506	−331	
35–44 Years	25,180	+4,784	+2,059		17,112	+3,330	+1,840	
45 Years and Over	20,087	−308	+1,601		13,168	−613	+827	
EDUCATIONAL ATTAINMENT OF HEAD							
0–8 Years	12,758	−7,638	−7,868	.120	10,086	−3,695	−4,193	.0850
9–11 Years	16,556	−3,839	−4,230		12,064	−1,718	−1,534	
12 Years	20,042	−353	−315		14,672	+891	+1,149	
13–15 Years	21,790	+1,394	+2,203		16,853	+3,072	+3,376	
16+ Years	29,231	+8,836	+8,546		22,526	+8,746	+8,487	
Total Adjusted R^2	$20,396			.318	$13,781			.325

NOTE: The dependent variable refers to per capita income and equals household income divided by household size. These data are *not* weighted by household size since each household is counted once in these calculations. Weighting by household size typically reduces per capita income since households with many children tend to have low per capita incomes.

SOURCE: U.S. Bureau of the Census, *Census of Population and Housing: 1980*, Public Use Microdata Samples.

region, per capita household incomes were highest in the West and lowest in the South. Predictably, mean household income was lower in the central city than in the suburbs, but households located outside metropolitan areas had lower incomes than those in central cities. The per capita income of households rose steadily with head's age up to age 45, when it then drops off.

In summary, our multivariate analysis reveals an interesting pattern of effects. Apart from household type and education of household head, no single variable exerts substantial influence on our measure of household income. However, there is clear evidence that the cumulative effects of membership in certain categories exert profound influence over household income levels. Thus, young, black, single female heads of households, without high school diplomas, who live in southern central cities were the poorest of the poor. The combined cumulative negative effects of their membership in each of these categories offsets and overwhelms any positive effects that might accrue from their greater likelihood to be members of extended households.

Summary, Discussion, and Conclusions

In this chapter we examined the differences between black and white families as revealed in the 1980 census. Family organization is a critical feature of modern society, and families are expected to play a central role in smoothing the adjustment of individuals to life's stresses and strains. Many would argue for a view of families as the central determinants not only of individuals' emotional well-being, but of their social and economic well-being as well. It is in this latter connection that black families are often discussed. The published literature is replete with characterizations of their supposed pathology and their negative influences on the development of black communities and individuals. Although revisionists prefer to view black families as culturally different, rather than pathologically deviant, negative views of black family life still predominate in the literature.

There is little of a direct, definitive nature that our analyses can contribute to this debate over black family wellness or illness. Demographic data, by their very nature, simply do not allow researchers to concisely ascertain motives or to systematically reveal individual differences within categories. Thus, we cannot say with absolute certainty where cultural imperatives begin and economic necessity ends in the explanation of race differences in the proportions of families headed by women. Nor, for that matter, can we fully elucidate the experiences of atypical types (for example, female-headed households in the top 5 percent income bracket).

These data do allow us, however, to effectively compare patterns in the organization and conduct of family life across race. Moreover, we are able to make these comparisons using large-scale, nationally representative data. We found the results from our comparisons of black and white families to be revealing in several important respects.

Generally speaking, white families were more often "conventional" in their organization, as defined by the societal norm of conjugal, monogamous, nuclear households, than are black families. This is not to say that no black families conformed to this norm; a sizable proportion did. Rather, this model of family life was less characteristic of black families than of white families.

Several consequences or patterns were associated with observed differences by race in family organization. For instance, blacks were more likely to be separated and divorced. Similarly, a significantly higher proportion of white youngsters were residing with both parents. Most noteworthy, however, was the dramatic correspondence between family organization and family economic status. The proportion of households headed by women, the proportion of children living with one parent, and the incidence of separation and divorce were all higher for lower-income individuals. Since Black Americans were clearly disadvantaged relative to whites on various measures of economic well-being (for example, occupational attainment and income), it is not surprising that black families differ from white families.

In attempting to account for observed differences in black and white family organizational patterns, the nearly overwhelming temptation is to accept the vast economic differences between blacks and whites as the prime determinants. Certainly, there is much evidence to support this view, the strongest correlation (controlling for the influence of other variables) in our analysis was between household income and household headship. By the same token, there were near linear decreases in the proportions of households headed by women, households where children reside with a single parent, and extended households with increases in economic status.

One expects race differences in family organization to diminish as economic differences by race are lessened, which is indeed the case. As one moves up the economic ladder, black-white differences in family structure are reduced. Upper middle-class black and white families are more alike than different on these factors when compared with lower-income families. It is worth noting the plethora of historical and structural factors which contribute to observed race differences in family organization as we struggle to understand the origins of these differences. Economic explanations are not wholly satisfying; much of the variation by race in family organization seems to be attributable to other factors.

Cultural factors—that is, family preferences, notions of the appropriate and established habits—also help explain race differences in family organization. It is necessary to lend some credence to the cultural explanation because of the compelling link between race and family organization, a link equal in force to that between family organization and economic status. For each indicator of family organization, there persists a race difference that is sustained even after extensive statistical controls for race differences in economic status. Thus, where blacks and whites of equal education, income, and occupational status are compared, blacks consistently have higher rates of female-headed households, extended households, and children living with one parent. The question to be posed, therefore, asks whether it is indeed reasonable to speculate that there are norms in the black community which encourage marital breakup and family disruption.

Since the late nineteenth and early twentieth centuries, there has been a more or less steady shifting away from the "conventional" family model among black families. This shift accelerated after 1960. So pronounced was the pattern, and so consistent across the years, that both Frazier and Moynihan were prompted to comment on what they interpreted as the "disorganization" of black family life.[53] What, in fact, was happening was a redefinition and reorganization of black family life. Frazier (but not Moynihan) saw this as the logical next stage in the evolution of black American family life. This reorganization was necessitated by the rapid social, economic, political, and cultural changes that black people underwent in this society.

These large-scale historical, social, and economic factors would seem to provide more reasoned explanation of the evolution that has brought black family organizational patterns to their current state. We say this not to dismiss cultural factors totally, but rather to place them in their proper perspective. Questions of cultural preference in the determination of black family patterns are best considered after these other societal-level factors have been taken into account. In a succinct statement of the position pointing to structural factors as the best explanations for black family patterns, Frazier concludes:

> The character of the Negro family during the various stages of its development has been affected by the social isolation of Negroes in American society. The lack of opportunity for the Negro male to parti-

---

[53]Frazier (1949, 1966); U.S. Department of Labor (1965).

cipate freely in the economic organization and his subordination to whites as well as the general exclusion of Negroes from political activities have all affected the organization and the function of the Negro family.[54]

Thus, the best single category of explanations for race differences in the organization of family life in contemporary U.S. society would seem to be those that attribute importance to race differences in functional relationships with the larger society and its institutions. Black-white differences in economic well-being, political power, and social standing—more so than differences in values and predispositions—explain the observed race differences in family organization and process.

[54]Frazier (1949), p. 84.

7

THE SCHOOLING OF AMERICA: BLACK-WHITE DIFFERENCES IN EDUCATION

E DUCATION is accorded a special status in this country. Decisions of where to buy a home, which politician to vote for, and how income is allocated are routinely based on educational considerations. Americans view education as the route to upward mobility. As a society, we firmly believe that any citizen, no matter how humble his or her beginnings, can with sufficient industry and intelligence climb the educational ladder to a better life.

It is against this backdrop of beliefs about education that we compare the relative educational statuses of black and white Americans. Historically, blacks have lagged behind their white fellow citizens on the major indicators of educational status. Many of the reasons for this discrepancy are apparent; other reasons are more obscure. Many have speculated about the causes of these persistent educational differences between blacks and whites. Working from the assumption that education represents a setting for truly egalitarian competition, they wonder why black Americans have failed to take full advantage of the opportunities provided. It is all the more perplexing when they note the widely trumpeted educational successes of Vietnamese, Asian Indians, and other new immigrant groups to the United States.

Race and Education in the United States: A Brief History

Black-white differences in education are rooted in the historical realities that shaped the relations between the races in the United States. Therefore, to compare the educational statuses of black and white Americans apart from their histories is to risk serious misinterpretation. During the period of slavery the overwhelming majority of blacks were denied access to formal schooling. Except for the infrequent pockets of opportunity represented by cities, the North, Freedman status, and sympathetic whites, blacks were not allowed to gain even the rudiments of an education. For example, laws were passed that made it illegal to teach slaves to read and write.[1]

Few scholars are unaware of the severe constraints placed on black education during slavery; however, many incorrectly see those constraints as having ended with Emancipation. In fact, the impediments to black education persisted under the feudal agrarian system of "sharecropping" that replaced slavery in the South. Education for children was secondary to the contingencies of raising cotton, the chief agricultural crop of the South.

> When time comes to break the sod, the sod must be broken; when the time comes to plant the seeds, the seeds must be planted, and when the time comes to loosen the red clay from about the bright green stalks of the cotton plants, that, too, must be done even if it is September and school is open. Hunger is the punishment if we violate the laws of Queen Cotton.[2]

Beyond the rural South were other barriers to black achievement in education. Black Americans in cities and in the North confronted systematic race discrimination which denied them educational progress. State laws and local ordinances, supported by the Supreme Court ruling in *Plessy* v. *Ferguson*, consigned blacks to inferior schooling under an institutionalized system of "separate and unequal." Further, black Americans were denied access to employment opportunities usually available to similarly educated whites. "An inferior education, for instance, typified Irish peasants and southern Italians as much as blacks; still, a European immigrant who had just arrived in Boston had a far better chance of securing a well-paying job than a black laborer whose ancestors had been in Boston for generations.[3]

[1]Anderson (1984).
[2]Wright (1941), p. 64.
[3]Anderson (1984), p. 107.

Despite the historic barriers, blacks have continued to cling to their belief in the American ethic of education for self-advancement. They have approached education as a transformational power capable of changing in fundamental ways the conditions of their lives and the lives of their children. Thus, an examination of the historical record reveals steady, though at times slow, gains in the educational statuses of black Americans. In 1880, 15 years after Emancipation, only 20 percent of blacks were literate (that is, could write). By 1930, 84 percent of blacks were literate.[4] In 1860, 2 percent of school-aged black children attended school; by 1880, 33 percent were in school; by 1930, 60 percent of black children attended school.[5] The proportions of blacks completing high school and college also increased sizably. In 1940 the median number of years of schooling completed for blacks was 5.7; by 1980 it was 12. The proportion of high school graduates among blacks grew from 8 percent in 1940 to 51 percent in 1980; the proportion of college graduates increased from 1.3 to 8 percent for these same years.[6] A revolution in learning had been wrought: in 1980 blacks were barely distinguishable from whites in literacy and school enrollment rates.

At the same time that black Americans were advancing educationally, education was a growth industry for the United States as a whole.[7] School enrollment rates rose and illiteracy rates dropped for whites as well. In 1890, 60 percent of white children were enrolled in school; by 1940 that figure was 72 percent. The white illiteracy rate dropped from 8 percent to 1 percent between 1890 and 1969. Between 1940 and 1980 the median number of years of schooling for whites aged 25 and over increased from 8.8 to 12.5 years. In the same period the proportion of whites who completed high school grew from 26 percent to 69 percent; the proportion of whites who had completed college increased from 5 to 17 percent.[8]

From the late nineteenth century to the present, several major sociohistorical events combined to influence the educational statuses of both black and white Americans. With the age of technology came increased need for and compensation of educated Americans; at the same time, educational programs multiplied as a society hungry for educated people sought to expand the pool of talent.[9]

[4]U.S. Bureau of the Census, *Negroes in the United States 1920–1932* (1934), p. 231.
[5]Anderson (1984); Reid (1982); U.S. Bureau of the Census, *Negroes in the United States 1920–1932* (1934), chap. 11.
[6]Reid (1982); Current Population Reports (1979): series P-23, no. 80, table 70; series P-20, no. 390 (August 1984), table 1.
[7]Hurn (1985).
[8]U.S. Bureau of the Census, *Current Population Reports*, series P-23, no. 80 (1980), tables 69 and 70; *Census of Population: 1980*, PC80-1-C1, table 123.
[9]Hurn (1985); Brookover (1979).

While the educational status of blacks has come to more closely approach that of whites with time, sizable differences persist. The dream of equality of education between the races, fed by the Supreme Court ruling of 1954 striking down racially segregated public schools, has yet to be achieved. On various indicators of educational achievement, black Americans lag well behind their better-educated white peers.

Race and Education in 1980

This country's race relations have been likened to a caste system by several observers.[10] In such systems the relations between a disenfranchised minority group (in power terms) and a dominant majority group are structured hierarchically. The society's power relationships, patterns of interaction, and belief systems coalesce to institutionalize the inferior position of the minority group.

Educational outcomes are greatly influenced by the constraints of caste membership. In a systematic cross-cultural study of how caste influences education, Ogbu observed several consistent features for Japan, Israel, Britain, New Zealand, and the United States.[11] Each society shared the belief that education determines a person's position in adult life. Historically, caste minorities in the societies studied were either first denied formal education and later given inferior education or were given inferior education from the beginning. As a result, a wide gap in educational attainment between the minority group and the dominant group was characteristic for each society.

Ogbu identified four typical explanations for the lower school performance of minority group members in the societies he examined.[12] Minority group members and members of the majority group offered competing explanations for the observed gap in education. Majority group theories usually dominate the literature on group differences in educational status. Where available, minority group literature adopts a diametrically opposed position, and theories of minority group members attribute the sources of their educational problems to inequities in the prevailing caste stratification system and in the legal-extralegal discriminatory policies and practices of the dominant group. By contrast, dominant group explanations of the educational problems of the minority locate the sources in the social, cultural, familial, or biological inadequacies of the minority group.

In the 1980 census blacks and whites were found to differ by educational status in many important ways. Although black Americans had

[10]Cox (1945).
[11]Ogbu (1978).
[12]Ogbu (1978), pp. 343–49.

made phenomenal educational progress since 1900, they continued to be disadvantaged in relative terms. As Table 7.1 shows, blacks aged 16 and over have lower educational attainments than whites. For example, 40.5 percent of blacks and 28.1 percent of whites aged 16 or over had completed 8 years or less of formal schooling. White high school graduates outnumbered blacks by 29.6 to 22.3 percent. Whites were more than twice as likely to have graduated from college or to have completed advanced college degrees.

The educational profile for Hispanic Americans was very similar to that of blacks in 1980. Hispanics were more likely than blacks to have completed fewer than 8 years of formal schooling (49.9 versus 40.5 percent); nevertheless, the two groups had similar proportions of college graduates and advanced degree recipients. A slightly higher proportion of blacks than Hispanics were high school graduates.

Asian Americans provided an interesting contrast to black Americans in their patterns of educational attainment. (The "Other" category in Table 7.1 includes Native Americans, but is predominantly composed of Asian Americans). Unusually higher proportions of Asians graduated from college or held postgraduate degrees. The unusual shape of the Asian population's educational distribution results, in large part, from the more recent immigration of this population to the United

TABLE 7.1

Educational Attainment by Race
for the Population Aged 16 and Over, 1980

Educational Attainment	White, Non-Hispanic	White, Hispanic	Black	Asian/Other
Primary School (0–8 Years)	28.1%	49.9%	40.5%	33.9%
Some High School (9–11 Years)	15.9	17.8	21.6	13.9
High School Graduate (12 Years)	29.6	18.6	22.3	21.8
Some College (13–15 Years)	14.2	9.2	10.6	15.1
College Graduate (16 Years)	6.7	2.2	2.8	7.2
Post B.A. (17+ Years)	5.5	2.3	2.2	8.1
Total	100.0%	100.0%	100.0%	100.0%

SOURCE: U.S. Bureau of the Census, *Census of Population and Housing: 1980,* Public Use Microdata Samples.

TABLE 7.2

Educational Attainment by Race and Sex
for the Population Aged 16 and Over, 1980

Educational Attainment	Black		White	
	Female	Male	Female	Male
Primary School (0–8 Years)	38.3%	43.2%	26.9%	29.4%
Some High School (9–11 Years)	21.9	21.2	16.3	15.5
High School Graduate (12 Years)	23.6	20.8	32.5	26.5
Some College (13–15 Years)	11.0	10.1	14.4	13.9
College Graduate (16 Years)	3.0	2.5	6.1	7.4
Post B.A. (17+ Years)	2.2	2.2	3.8	7.3
Total	100.0%	100.0%	100.0%	100.0%

SOURCE: U.S. Bureau of the Census, *Census of Population and Housing: 1980,* Public Use Microdata Samples.

States. The Asian population is in the unique position of having large numbers in both the *lowest* and the *highest* educational categories.

Table 7.2 reveals sexual differences in years of attainment. For both blacks and whites, women were less likely to have only primary school educations and were more likely to have graduated from high school. Among whites, a sizable male advantage was clear for more advanced years of schooling—white men were more likely to have completed college (7.4 versus 6.1 percent) and to have continued college beyond the bachelor's level (7.3 versus 3.8 percent) than white women. Among blacks, similar proportions of women and men completed college or went on to advanced degree programs.

Comparison of black and white educational attainment by age conveyed empirically some of the historical effects on race and education discussed above (see Table 7.3). While blacks had lower educational attainment in all categories, these differences were considerably more extreme among older cohorts. The proportion of young blacks who had graduated from high school was comparable to that for whites aged 26 to 35. Whites in the oldest cohort (56 years or older) were more than twice as likely to have graduated from college (5.4 versus 2.0 percent). Whites generally were two to three times as likely to have graduated

TABLE 7.3
Educational Attainment by Race and Age for the Population Aged 16 and Over, 1980

Educational Attainment	Black					White				
	16–25 Years	26–35 Years	36–45 Years	46–55 Years	56+ Years	16–25 Years	26–35 Years	36–45 Years	46–55 Years	56+ Years
Primary School (0–8 Years)	6.6%	6.9%	14.4%	30.0%	58.4%	3.4%	3.6%	6.9%	12.9%	31.8%
Some High School (9–11 Years)	40.0	19.2	26.1	27.7	19.0	30.8	8.5	12.6	17.1	18.6
High School Graduate (12 Years)	33.4	40.4	35.3	25.1	13.5	37.2	38.8	41.7	39.8	28.9
Some College (13–15 Years)	16.9	21.8	14.8	9.7	4.9	21.4	23.2	17.9	14.1	10.8
College Graudate (16 Years)	2.5	6.8	4.4	3.5	2.0	5.6	14.1	9.7	8.1	5.4
Post B.A. (17+ Years)	.6	4.9	5.0	4.0	2.2	1.6	11.8	11.2	8.0	4.5
Total	100.0%	100.0%	100.0%	100.0%	100.0%	100.0%	100.0%	100.0%	100.0%	100.0%

SOURCE: U.S. Bureau of the Census, Census of Population and Housing: 1980, Public Use Microdata Samples.

from college or pursued advanced degrees; this held true no matter the age groups. Together, these patterns illustrate the effects of historical race discrimination. As noted above, such discrimination diminished for younger cohorts of blacks.

Our review of historical factors in the education of black and white Americans revealed regional effects of considerable import. The system of legal bondage and the remnants of agrarian feudalism combined to disadvantage blacks raised in the South. It should also be noted that, on average, lesser development of the South as a region resulted in a disadvantage to all residents in terms of educational attainment (Table 7.4). Thus, we see that the proportion of people aged 16 and over who have completed only primary school is highest in the southern region, and this holds true for both blacks and whites. Forty-five percent of blacks and 31 percent of whites living in the South completed no more than 8 years of formal schooling. This pattern is repeated when we look at the proportions of residents in the South with high school diplomas.

Educational attainment is greatest in the West, where 7.4 percent of whites and 3.5 percent of blacks have completed college (Table 7.4). By the same token westerners had the lowest proportion of residents with no more than a primary school education. Predictably, whites outdistance blacks in educational attainment in all regions. Comparison of black-white differences within regions shows these differences to be extreme only in the case of completion of primary school. In the South 44.6 percent of blacks compared with 30.7 percent of whites have only primary school educations. Given the number of black colleges concentrated in the South and the disproportionate share of black college graduates these schools account for, we expected to find a regional black advantage in college attainment. This was not the case. Blacks in the South are no closer to whites in the proportions completing college or advanced degrees than in any other region of the country.

Household type was also associated with the educational attainment of the head of household (Table 7.5). Prior research, like the findings reported in chapter 6 of this book, reveals female-headed households to be more common among blacks than among whites. Heads of husband-wife households are half again as likely as female heads to have graduated from college and twice as likely to hold advanced degrees. However, heads of nonfamily households were by far the most extensively educated household type among blacks. Over 5 percent of this group held college degrees, while an additional 24.2 percent were high school graduates.

The educational differences by household type observed among whites were more in line with the expected (Table 7.5). These differences, unlike those for blacks, effectively presage the known economic

TABLE 7.4
Educational Attainment by Race and Region for the Population Aged 16 and Over, 1980

Educational Attainment	Northeast		Midwest		South		West	
	White	Black	White	Black	White	Black	White	Black
Primary School (0–8 Years)	27.5%	35.9%	29.2%	38.3%	30.7%	44.6%	22.8%	31.0%
Some High School (9–11 Years)	15.4	22.9	15.9	22.6	17.4	21.2	13.9	18.7
High School Graduate (12 Years)	30.9	25.0	31.3	23.0	27.0	20.6	29.6	25.2
Some College (13–15 Years)	12.9	11.0	12.9	11.5	13.4	8.9	19.1	18.4
College Graduate (16 Years)	7.3	2.8	6.0	2.5	6.5	2.8	7.4	3.5
Post B.A. (17+ Years)	6.0	2.4	4.7	2.1	5.0	1.9	7.2	3.2
Total	100.0%	100.0%	100.0%	100.0%	100.0%	100.0%	100.0%	100.0%

SOURCE: U.S. Bureau of the Census, Census of Population and Housing: 1980, Public Use Microdata Samples.

TABLE 7.5

Educational Attainment for Household Heads Aged 16 and Over, 1980

Educational Attainment	Black				White			
	Husband-Wife Household	Female-Headed Household	Male-Headed Household	Non-family Household	Husband-Wife Household	Female-Headed Household	Male-Headed Household	Non-family Household
Primary School (0–8 Years)	39.8%	39.7%	47.1%	32.4%	29.0%	31.1%	37.1%	17.6%
Some High School (9–11 Years)	20.7	23.2	23.8	18.7	16.2	19.8	19.0	12.4
High School Graduate (12 Years)	23.0	23.3	19.0	24.2	30.1	27.4	26.3	28.9
Some College (13–15 Years)	10.7	10.5	7.8	15.0	12.9	12.0	11.1	20.9
College Graduate (16 Years)	3.2	2.1	1.4	5.4	6.5	5.1	3.7	10.8
Post B.A. (17+ Years)	2.6	1.2	.9	4.5	5.3	4.8	2.8	9.4
Total	100.0%	100.0%	100.0%	100.0%	100.0%	100.0%	100.0%	100.0%

SOURCE: U.S. Bureau of the Census, *Census of Population and Housing: 1980*, Public Use Microdata Samples.

disadvantages of single female heads of household compared with males who head husband-wife households. Among whites, female heads of household were disadvantaged relative to heads of husband-wife households at each level of educational attainment above high school, but in no case were these differences more than a few percentage points.

There is a strong relationship between educational attainment and employment status. Americans expect those who are better educated to have better jobs and to be above the ravages of unemployment. To be well educated, we assume, is to be "recession proof." Our examination of employment status and occupation by educational attainment is intended to test this premise (Tables 7.6 and 7.7). Table 7.6 shows that unemployment was less common among high educational achievers in both races. Labor participation rates were also higher for persons with more education. The proportion of blacks not in the labor force was higher for high school graduates than for college graduates. White high school graduates were also more likely than college graduates or advanced degree holders to be nonparticipants in the labor force.

Educational attainment seems to be highly correlated with occupation in both races, although slightly less so among black Americans (Table 7.7). Generally speaking, educational attainment is higher for the more powerful, more prestigious, better-paying occupations (for example, executive-administrative, professional specialty). In terms of modal characteristics, the white labor force may be occupationally classified as follows: workers in the executive-administrative and professional specialty categories are characterized by college education. Workers in the administrative support, services, precision production, and operative categories are most frequently high school graduates, some having spent additional time in college without earning the degree. Workers in the private household and farming-fishing categories for the most part have primary or secondary school educations.

The educational profiles of blacks by occupation are very similar to those for whites. Black Americans with college degrees tend to concentrate in the executive-administrative and professional specialty categories. Characteristically blacks in the occupational categories of technical, sales, administrative support, services, and manufacturing (that is, precision production and operative) had either graduated from high school or attended college without obtaining a degree. As was true for whites, blacks in the private household and farming-fishing categories were substantially less educated than their peers.

Income is another area where we expect to see the returns on higher education manifested (see Table 7.8). It is an article of faith that better-educated individuals in this society earn more money than those who are less educated. A clear pattern emerges for both blacks and

TABLE 7.6
Educational Attainment by Race and Employment Status, 1980

Educational Attainment	Black			White		
	Employed	Unemployed	Not in Labor Force	Employed	Unemployed	Not in Labor Force
Primary School (0–8 Years)	32.8%	4.1%	63.1%	29.2%	2.9%	67.9%
Some High School (9–11 Years)	41.3	8.3	50.4	46.0	5.0	49.0
High School Graduate (12 Years)	63.4	8.1	28.5	64.0	4.0	32.0
Some College (13–15 Years)	67.9	6.6	25.5	66.8	2.9	30.3
College Graduate (16 Years)	80.1	3.6	16.3	75.1	2.0	22.9
Post B.A. (17+ Years)	82.4	2.3	15.2	82.4	1.4	16.1
Total	52.6	6.8	40.6 100%	58.9	3.5	37.6 100%

SOURCE: U.S. Bureau of the Census, *Census of Population and Housing: 1980*, Public Use Microdata Samples.

TABLE 7.7

Educational Attainment by Race and Occupation for Persons Aged 16 and Over 1

Educational Attainment	Executive, Administrative	Professional Specialty	Technical	Sales	Administrati Support
BLACKS					
Primary School (0–8 Years)	0.7%	0.7%	0.3%	0.9%	1.8%
Some High School (9–11 Years)	1.2	1.4	0.6	3.4	5.7
High School Graduate (12 Years)	2.8	2.7	1.9	5.0	17.0
Some College (13–15 Years)	6.4	8.7	4.3	6.9	27.5
College Graduate (16 Years)	13.3	36.0	3.5	6.1	15.9
Post B.A. (17+ Years)	15.3	54.6	2.4	2.7	7.2
Total	3.2	5.6	1.7	4.0	11.8
WHITES					
Primary School; (0–8 Years)	1.7	0.7	0.4	3.0	2.2
Some High School (9–11 Years)	2.5	1.1	0.5	8.2	6.6
High School Graduate (12 Years)	6.0	2.7	1.8	9.2	19.3
Some College (13–15 Years)	10.4	9.8	4.7	11.0	19.5
College Graduate (16 Years)	19.0	29.3	3.6	11.5	11.0
Post B.A. (17+ Years)	16.6	55.2	3.2	5.1	4.8
Total	7.4	9.2	2.2	8.5	13.2

SOURCE: U.S. Bureau of the Census, *Census of Population and Housing: 1980*, Public Use Microd Samples.

whites: as the income level increases, the proportion of people at successively higher levels of income with primary school or secondary school education declines. Interestingly, this linear relationship between educational attainment and household income is much stronger for blacks than for whites.

Education in the United States: Separate and Unequal?

Educational attainment was shown to vary on many important dimensions in the United States. We identified clear regional, income, oc-

TABLE 7.7 *(continued)*

Private household	Protection Services	Other Services	Farming, Fishing	Precision Production	Operative	No Work	Total
4.6%	0.5%	13.1%	3.7%	5.1%	17.8%	50.8%	100.0%
2.3	0.7	18.6	1.9	5.6	21.3	37.6	100.0%
1.3	1.6	16.2	1.1	7.4	24.8	18.2	100.0%
0.5	2.6	11.8	0.6	6.9	15.2	8.6	100.0%
0.3	1.8	4.0	0.4	3.6	5.7	9.4	100.0%
0.4	0.9	2.9	0.3	1.8	2.8	8.7	100.0%
2.2	1.3	14.8	1.8	6.1	19.7	27.8	100.0%
0.5	0.6	7.2	4.2	8.5	16.2	54.8	100.0%
0.5	0.8	14.0	3.2	10.0	18.5	34.1	100.0%
0.3	1.2	8.8	2.4	11.9	15.7	20.7	100.0%
0.1	1.7	8.0	1.8	8.7	8.5	15.8	100.0%
0.1	1.0	2.7	1.2	3.9	2.5	14.2	100.0%
0.0	0.5	1.4	0.6	2.0	1.3	9.3	100.0%
0.3	1.1	8.5	2.5	9.3	13.0	24.8	100.0%

cupation, and sex differences, to name a few, and the conclusion is inescapable: There exists an independent race effect which ultimately disadvantages black Americans in terms of educational attainment. To be sure, much of this effect is historical and cumulative, and the continued educational disadvantage thus can be partly explained as little more than the legacy from generations of educational deprivation. For as Anderson reminds us, "Any student of black achievement behavior should recognize at the outset that much discordance has reigned historically between blacks' motivation to achieve and their opportunities to achieve."[13]

A paradox is posed, however, by the educational progress of black

[13]Anderson (1984), p. 104.

TABLE 7.8

Educational Attainment of Household Head by Race and Annual Household Income, 1980

Educational Attainment	Under $5,000	$5,000-$9,999	$10,000-$14,999	$15,000-$19,999	$20,000-$24,999	$25,000-$34,999	$35,000-$49,999	$50,000 and Above	
BLACK									
Primary School (0–8 Years)	25.2%	23.5%	16.6%	12.0%	9.2%	9.8%	4.3%	1.4%	100%
Some High School (9–11 Years)	18.1	21.6	17.4	13.5	10.4	11.5	5.4	2.1	100%
High School Graduate (12 Years)	11.6	17.2	18.4	15.2	12.8	15.2	7.2	2.4	100%
Some College (13–15 Years)	8.9	14.5	16.4	15.1	13.5	18.4	9.8	3.4	100%
College Graduate (16 Years)	6.1	9.1	14.1	15.1	13.9	22.2	13.8	5.7	100%
Post B.A. (17+ Years)	4.8	7.0	10.5	12.8	13.5	22.6	19.0	9.5	100%
Total	17.0	20.0	16.9	13.5	11.0	12.9	6.4	2.3	
WHITE									
Primary School (0–8 Years)	9.8%	14.8%	14.4%	14.8%	14.4%	18.2%	9.0%	4.5%	100%
Some High School (9–11 Years)	7.8	13.9	14.6	14.5	13.8	18.8	11.4	5.2O	100%
High School Graduate (12 Years)	4.5	10.5	14.0	16.0	15.6	21.9	12.3	5.3	100%
Some College (13–15 Years)	4.5	9.0	12.8	14.2	15.2	21.3	14.4	8.6	100%
College Graduate (16 Years)	3.1	5.7	10.2	12.1	14.0	23.3	18.4	13.2	100%
Post B.A. (17+ Years)	2.7	4.8	7.5	10.2	13.3	23.4	20.9	17.2	100%
Total	6.4	11.5	13.5	14.6	14.7	20.0	12.5	6.8	

SOURCE: U.S. Bureau of the Census, *Census of Population and Housing: 1980*, Public Use Microdata Samples.

Americans since 1900. Over this period, the increase in enrollment rates was greater for blacks than whites. By 1980 blacks at most ages were just as likely to be enrolled as whites, and the racial gap in attainment among those in their 20s was very small. Many living in the United States during the Reconstruction never imagined that this might occur.

School enrollment rates are an important indicator of educational status for several reasons. On the one hand, these figures gauge the accessibility of formal schooling to various groups in a society. At the same time enrollment figures presage the future, indicating the proportion of different subgroups enrolled at different levels of the schooling sequence.

Black Americans have made progress in closing the enrollment gap with whites over the last century, and by 1980 black school enrollment rates were close to those for whites (Table 7.9). Black and white school attendance rates for children aged 4 to 15 were essentially the same: 93 percent. Ninety-two percent of "other" and 91 percent of Hispanic children aged 4 to 15 attended school.

Careful scrutiny of these comparable patterns reveals important underlying differences, however. For instance, black children were more likely than their white peers to attend public schools; in fact, blacks were more likely than any of the various racial-ethnic groups to enroll their children in public schools. More whites aged 4–15 than other groups attended church-affiliated private schools. Blacks were least likely to attend such schools.

The implications of these simple differences in school enrollment patterns are manifold. They reveal that critical differences are camouflaged beneath apparent similarities; thus, while it may appear that

TABLE 7.9

*School Attendance by Race and Type of School
for Persons Aged 4–15, 1980*

	Currently Attending School	Type of School		
		Public	Church-Affiliated	Other Private
Blacks	92.4%	85.7%	4.9%	1.8%
Whites	92.9	79.1	10.3	3.5
Hispanics	90.8	80.2	8.7	1.9
Others	91.7	79.8	7.1	4.8

SOURCE: U.S. Bureau of the Census, *Census of Population and Housing: 1980*, Public Use Microdata Samples.

blacks and whites are receiving comparable educations, the actual facts may be the opposite. These enrollment patterns also hold implications for questions about the relative quality of education received by blacks and whites. Since private schools often operate on a smaller scale and are generally more responsive to student-parent desires, they can be expected to provide superior educational experiences. Research using more sensitive data than census statistics has demonstrated the validity of the expectation that blacks and whites generally receive different educational experiences which then reverberate by manifesting different educational outcomes.[14]

Despite the educational progress of black Americans to date, many find reasons for concern. There is reason to believe that it will be several generations before blacks achieve complete parity with whites educationally.[15] Moreover, serious questions continue to be raised about the relative quality of education received by the two races. It has been suggested that although blacks are approaching whites in the quantity of education (that is, years of schooling), they in fact continue to be far behind in the quality of education (that is, academic content of schooling).

Debates about the quality of education and race are long-standing. More often than not such debates take as their point of reference the stubborn differences between blacks and whites on nationally administered, standardized tests. Tests such as the California Achievement Test (given during the primary school years in many areas); the Scholastic Aptitude Test (required of students seeking admission to college), and the Law School Aptitude Test (used to select students for admission to law schools) consistently reveal a white advantage. At times, blacks score an average 1 to 2 standard deviations below the mean. Similar patterns are observed on various other standardized tests used in this country to evaluate applicants for entry-level positions in the Armed Services, building trades, service industries, and professions. In each of these instances as well, the average scores of blacks are significantly lower than those of whites.

While there is much discussion over the causes and implications of black-white differences in performance on standardized tests, the fact of these substantial differences remains uncontested.[16] Table 7.10 provides details on the Scholastic Aptitude Test scores for 1983 college-bound seniors to illustrate the point. Whites scored highest as a group on the three tests, while blacks scored lowest. Performance differences on standardized tests have fueled discussions of differential quality of

[14]Hare and Levine (1984).
[15]Hill (1971).
[16]Taylor (1980).

TABLE 7.10

Scholastic Aptitude Test Scores by Race, 1983

	Verbal	Math	Writing
Black	339	369	34.1%
White	443	484	44.2%
Mexican-American	375	417	38.2%
Asian/Pacific Islander	395	514	38.6%

SOURCE: L. Ramist and S. Arbeiter, *Profiles, College-Bound Seniors, 1983* (New York: College Entrance Examination Board), 1984.

schooling by race. It is important to note the strong effects of socioeconomic background on test scores. For instance, the median scores for all students on the verbal and mathematical sections of the SAT test vary substantially by family socioeconomic status.[17] The median scores increase proportionately with parental education and family income; the better-educated and more affluent a student's parents, the higher that student scores on the examination (Table 7.11). The effects of socioeconomic status, however, should not be overstated. Black students whose fathers graduated from college scored lower, on average, than white students whose fathers had only an elementary school education.

The racial segregation of public education is another factor often cited in discussions of race and the quality of schooling. Demographic change in central cities has implications for the school districts that serve these areas. For one thing, whites as a relatively more affluent segment of the population drain financial support when they move out of the district in sizable numbers; the few whites and sizable black and minority populations that remain are generally less able to fund the public school system through property taxes and special tax levies. With the erosion of the school district's financial base comes a diminution in the educational services it is able to provide. Thus, student-teacher ratios may rise, it becomes increasingly difficult to attract more qualified teachers, the physical plant deteriorates, and the problems associated with poorly financed urban school districts proliferate.

White representation in the urban school districts of this country has been declining since the mid-1960s. The percentage white enrollment declined substantially between 1967 and 1978 in each of the nation's twenty largest urban school districts.[18] Table 7.12 shows the reality and impact of racial change in the nation's urban school dis-

[17]Ramist and Arbeiter (1984).
[18]Farley (1984), pp. 22–33; Orfield (1983).

TABLE 7.11

Scholastic Aptitude Test Scores by Race and Family Socioeconomic Status, 1983

	Father's Education			Annual Parental Income			
	Primary School	High School Graduate	College Graduate	Below $6,000	$18,000–23,999	$30,000–39,999	Over $50,000
Black	300V 330M	313V 338M	366V 384M	287V 322M	340V 358M	362V 383M	406V 423M
White	395V 429M	413V 451M	453V 500M	411V 438M	432V 476M	439V 484M	463V 513M
Mexican-American	338V 381M	362V 396M	408V 443M	320V 360M	370V 407M	393V 431M	422V 452M
Asian/Pacific Island	314V 487M	350V 478M	394V 529M	261V 454M	378V 507M	420V 526M	464V 569M

NOTE: Median scores for verbal and mathematics subtests.

SOURCE: L. Ramist and S. Arbeiter, *Profiles, College-Bound Seniors, 1983* (New York: College Entrance Examination Board), 1984.

TABLE 7.12

Changes in Enrollment and Racial Segregation in School Districts of Largest U.S. Cities, by Region, 1967–1978

	Change in Enrollment 1967–78		Index of Dissimilarity			Percentage White in School of Typical Black			Percentage Black in School of Typical White		
	White (1)	Black (2)	1967 (3)	1978 (4)	Change (5)	1967 (6)	1978 (7)	Change (8)	1967 (9)	1978 (10)	Change (11)
NORTH AND WEST											
New York	−45%	+16%	62	70	+ 8	26%	14%	−12%	16%	19%	+ 3%
Chicago	−56	− 1	90	90	—	6	4	− 2	7	11	+ 4
Los Angeles	−54	− 4	91	71	−20	7	15	+ 8	3	13	+10
Philadelphia	−32	− 7	75	79	+ 4	14	10	− 4	21	20	− 1
Detroit	−74	+ 8	75	60	−15	15	11	− 4	21	62	+41
San Diego	−21	+27	80	51	−29	24	39	+15	4	10	+ 6
Indianapolis	−48	− 1	78	46	−32	22	36	+14	10	32	+22
San Francisco	−67	−30	54	35	−19	27	18	− 9	15	25	+10
Milwaukee	−48	+31	86	39	−47	16	38	+22	6	32	+26
Cleveland	−48	−23	89	89	—	7	5	− 2	9	11	+ 2
SOUTH											
Houston	−57%	+ 7%	93	75	−18	5%	12%	+ 7%	3%	18%	+15%
Dallas	−55	+39	93	64	−29	7	17	+10	3	25	+22
Baltimore	−52	− 7	82	67	−15	10	12	+ 2	18	40	+22
San Antonio	−63	−15	86	67	−19	8	11	+ 3	4	14	+10
Memphis	−50	+29	95	60	−35	4	16	+12	4	45	+41
Washington	−62	−25	75	85	+10	4	2	− 2	44	40	− 4
New Orleans	−66	+ 2	85	74	−11	9	8	− 1	17	45	+28
Jacksonville	−23	+ 6	92	40	−52	7	49	+42	3	26	+23
Nashville	−29	+ 6	83	40	−43	19	53	+34	6	25	+19
Atlanta	−84	+ 2	94	75	−19	5	6	+ 1	7	55	+48

NOTE: The census of 1980 showed that Phoenix, San Jose, and El Paso were among the twenty largest cities. Phoenix is not listed here because its central city contains many school districts; San Jose and El Paso, because less than 3 percent or their enrollment was black in 1967. The school district for Indianapolis does not include the entire central city. Data for Houston and Dallas refer to 1968 and 1978. The school district for Jacksonville includes all of Duval County, that for Nashville, all of Davidson County.

SOURCE: Reynolds Farley, *Blacks and Whites: Narrowing the Gap?* (Cambridge, MA: Harvard University Press), 1984, table 2.1.

tricts. School districts in Detroit, San Francisco, San Antonio, Washington, New Orleans, and Atlanta experienced a 60 percent or more decline in the white student population between 1967 and 1978. It is reasonable to assume that these districts suffered some negative effects in terms of funding patterns, strength of students' academic backgrounds, and so on, as a result of this massive racial redistribution.

In our search for answers to the paradox of black progress and retardation in education in the United States, we return to Ogbu's cross-national study of minority education. Clearly there are elements of truth in both the minority- and majority-group explanations of the current educational statuses of blacks. American society has historically discriminated against blacks, and blacks have over time come to expect such discrimination as normative. The result, Ogbu suggests, is the creation of a mutually reinforcing system that undercuts black educational progress. Structural barriers set up in the society and psychological barriers within the individual interact to restrict the educational development of blacks. This pattern is repeated internationally, across societies.

> None of the societies studied has a history of rewarding minority-group members equally for equivalent training and ability. In almost every case, the minorities have become disillusioned with the prevailing belief that the way to get ahead is through hard work and success in school. This disillusionment contributes significantly to their academic retardation and lower educational attainment.[19]

Anderson noted in the specific case of black Americans that

> those who deplore low academic achievement among black children and attribute this situation to deficiencies in the will to learn fail to understand the quality of oppression and motivation in black life and culture and how the former has eaten away at the latter for 363 years.[20]

[19]Ogbu (1984), p. 346.
[20]Anderson (1984), p. 120.

8

EMPLOYMENT

T
HE FIRST census to gather data about work activity was that of 1850, when census marshals recorded the occupation of all free males aged 16 and over.[1] Leonard Curry, a historian who analyzed these data concluded, "The most obvious point demonstrated by the census figures is the extent to which legal and societal restrictions and prohibitions were effective in limiting the employment opportunities of urban free blacks."[2]

After acquainting himself with the conditions of blacks living in Philadelphia in the 1890s, W. E. B. DuBois observed:

> For a group of freedmen the question of economic survival is the most pressing of all questions. . . . But when the question is complicated by the fact that the group has a low degree of efficiency on account of previous training; is in competition with well-trained, eager and often ruthless competitors; is more or less handicapped by a somewhat indefinite but existent and wide ranging discrimination; and, finally, is seeking not merely to maintain a standard of living but steadily to raise it to a higher plane—such a situation presents baffling problems to the sociologist and philanthropist.[3]

---

[1]U.S. Bureau of the Census, "Population and Housing Inquiries in U.S. Decennial Censuses, 1790–1970," Working Paper no. 39 (1973).

[2]Curry (1981), pp. 21 and 33.

[3]W. E. B. DuBois (1899), p. 47. For a new analysis of employment opportunities for blacks in Philadelphia in the late nineteenth century, see Lane (1986), pp. 36-42.

After the lapse of another four decades, Gunnar Myrdal devoted five chapters of *An American Dilemma* to the economic problems of blacks. Blacks were poor, he contended, because they were unable to get high-paying jobs. Most were forced to work as farm hands, sharecroppers, unskilled laborers, or domestic servants. Blacks had little incentive to get extensive educations, he thought, since employment opportunities for the highly trained blacks were restricted to the black community. Myrdal blamed the shortage of opportunities directly on racial discrimination. White employers were unwilling to hire blacks for anything but low-paying manual jobs. White workers feared blacks as competitors, would not accept them as colleagues, and certainly would not have them as supervisors. In Myrdal's view, whites were more willing to provide welfare benefits for blacks than to remove the job restrictions which kept blacks impoverished and underemployed.[4]

Although many racial changes have occurred in the last century, we are reminded on the first Friday of every month that the employment problems described by Curry, DuBois, and Myrdal, still affect the black community. On that day, the Bureau of Labor Statistics announces the unemployment rate for the previous month. Since the mid-1950s, the proportion of blacks out of work has been twice that of whites in both prosperous and lean times.[5] Among black men under age 25, the statistics indicate that a remarkably high proportion of those who seek work cannot find it—30 percent in the early 1980s.[6]

To describe trends in labor force participation and employment, we begin with a review of the concepts which are currently in use.[7] The *labor force* consists of persons who are either employed or unemployed. All other persons are not participants in the labor force; that is, they are neither working nor seeking a job. This includes many students, homemakers, retirees, institutionalized persons, and those who are unable to work.

Persons are classified as employed if they worked at any *paid* job during the survey week or if they spent 15 or more hours as an unpaid worker in a family business or on a family farm. People who had jobs but were temporarily absent because of a strike, sickness, or vacation are also considered employed. People who only do unpaid housework or perform volunteer work are not classified as employed no matter how difficult their tasks or how many hours they spent in this activity. Men

[4]Myrdal (1944), p. 301.
[5]Killingsworth (1968), p. 2.
[6]U.S. Bureau of Labor Statistics, *Employment and Earnings*, vol. 32, no. 1 (January 1985), table 3.
[7]U.S. Bureau of the Census, *Census of Population and Housing: 1980*, Public Use Microdata Samples, Technical Documentation K-25 and K-26.

and women on active service in the Armed Forces are considered employed regardless of their duties.

Individuals are counted as *unemployed* if they satisfy three criteria: (1) they must not be working at a job; (2) they must have made efforts to get a job within the previous month; and (3) they must be able to do some type of work. Almost any efforts to find employment—such as reading help-wanted ads—are considered as job-seeking activities. The unemployed also includes people on temporary layoff who are awaiting recall by their employer.

These definitions do not depend on how a person spends most of his or her time; rather, they depend on paid employment and the search for work. A farrier who would like a full-time job at a racetrack but can only locate a job parking cars for a few hours one evening a week is classified as employed. A medical student who spends 60 hours a week studying but who is actively looking for a one-night-per-week job as a go-go dancer is considered unemployed.

There has been much debate about the appropriateness of these concepts. Should the definition of unemployment be tightened to exclude persons who are not seriously looking for employment? An individual who is willing to work only in a specific job with specific hours— perhaps a part-time job—is considered unemployed. Furthermore, almost any activity—even asking friends about openings—counts as a search. Undoubtedly, the unemployment rate would *decrease* if more stringent definitions were used.[8]

There is also the issue of underemployment. A person who wants to work full time but can locate only a part-time job is considered employed even if the job is far below his or her skill level or desired hours of work. The term "underemployment" refers to these situations. The unemployment rate would *increase* were it to include a measure of underemployment.[9]

Discouraged workers are also not included among the unemployed. Suppose that a woman wants to work but believes that no jobs are available. If there is an interval of four or more weeks in which she makes no effort to find work, she will be classified as out of the labor force and will not contribute to the unemployment rate. Because many feel that discouraged workers should be included among the unemployed, the Census Bureau attempts to gather data on nonparticipants in the labor force. They are asked what they are doing and whether they want a job. In 1985 about 80 percent of the black nonparticipants, male

[8]U.S. National Commission on Employment and Unemployment Statistics (1979), p. 52.
[9]U.S. National Commission on Employment and Unemployment Statistics (1979), p. 51.

and female, said they did not want jobs. The remaining 20 percent of those out of the labor force who wanted jobs were asked why they had not searched for one in the last four weeks. About two thirds of them gave reasons such as school attendance, home responsibilities, or health problems. The other one third said they were not looking because they thought no jobs were available.[10] These people are currently defined as discouraged workers by the federal statistical system. Including these persons raised the unemployment rate for black men in 1985 from 15.3 to 17.3 percent. For black women, the increase was from 14.9 to 17.5 percent.[11]

These are conservative estimates of discouraged workers. In recent years, the National Urban League has conducted a Black Pulse Survey, a major aim of which has been to estimate the number of blacks who gave up the search for employment and thereby became nonparticipants in the labor force. Using a more liberal definition of discouraged workers, Robert Hill contends that if they were completely counted, the true unemployment rate for blacks in 1980 would have been almost double the 13 percent rate shown by the Bureau of Labor Statistics.[12]

Problems in defining unemployment and discouraged workers have led analysts to another measure of labor force activity: the *employment-population ratio*. Because this equals the proportion of a group who held jobs at a specific point in time, it is useful for assessing the extent of employment.

The data and trends described in this chapter were gathered by the Census Bureau's Current Population Survey and released by the Bureau of Labor Statistics. These trend data refer to the noninstitutionalized population, and thus exclude prisoners or those living on military posts. Since the mid-1940s a monthly survey has asked a sample of the population about their employment status. Those holding jobs are asked about their occupations, their employer's product, and the number of hours they worked in the previous week. Those unemployed were asked when they last worked, whether they were able to work, and whether they were searching for jobs. All adults in the sample were asked questions about the number of weeks they worked in the previous year, their weeks of unemployment in that year, and their usual hours of employment while working. The most widely cited unemployment statistics are based on these figures. They are known as household data since

[10]U.S. Bureau of Labor Statistics, *Employment and Earnings*, vol. 33, no. 1 (January 1986), tables 3 and 36.

[11]Among whites, the corresponding percentages unemployed were 6.1 for men and 6.4 for women before adjustments for discouraged workers and 6.6 percent for men and 7.4 percent for women after adjustments; U.S. Bureau of Labor Statistics, *Employment and Earnings*, vol. 33, no. 1 (January 1986), tables 3 and 36.

[12]Hill (1981), p. 17.

they are obtained from those people who live in a sample of the nation's residences. Data from this source may differ from those obtained from the state employment agencies or from employer records. Until the mid-1970s the Bureau of Labor Statistics generally released data for the nonwhite and white population.[13] As a result, we frequently use the term "nonwhite" in this chapter and the next. We begin with an analysis of unemployment trends, move on to labor force participation, and then describe the employment-population ratio.

Unemployment Trends

This analysis of the labor force divides the population into three age groups: persons aged 16 to 24, many of whom are enrolled in school or just starting their careers; persons aged 25 to 54, of whom the majority of men and many women are fully employed, often because they are supporting families; persons aged 55 and over. We will be looking at trends among those who are just beginning to work, those at the ages of peak employment, and those who are nearing retirement or have retired.

Figure 8.1 shows changes from 1950 to 1985 in the proportion of labor force unemployed. Looking first at trends among young nonwhite men, we observe that unemployment rates were at a low of about 6 percent in the early 1950s, rose in the recessionary period just before 1960, sank to low levels during the "guns and butter" era of the late 1960s, and then increased to high levels during the 1973–75 recession and again in the early 1980s. Peaks and troughs in the unemployment rate for young nonwhites occur at the same times as those for whites.[14] Trend lines for other age groups of men and for most groups of women moved in a parallel fashion; that is, unemployment rates fell during periods of economic expansion and rose during recessions.

An easy way to summarize the relationship of economic growth to

[13]The nonwhite population, as defined for Bureau of Labor Statistics tabulations, is more nearly equivalent to the black population than is the nonwhite population as defined for the census of 1980. In brief, many individuals who reported "Hispanic" as their race in the 1980 census enumeration were classified as "other races." In the Bureau of Labor Statistics tabulations, many of them would have been classified as "white." In the census of 1980, 69 percent of the nonwhite population 16 and over was black. In the 1980 Bureau of Labor Statistics tabulations for persons aged 16 and over, 85 percent of the nonwhite population was black. U.S. Bureau of the Census, *Census of Population: 1980*, PC80-1-B1, table 38; U.S. Bureau of Labor Statistics, *Employment and Earnings*, vol. 28, no. 1 (January 1981), table 44.

[14]Since unemployment rates vary by age, Figure 8.1 shows age-standardized rates developed from information about the component age groups within the three broad age categories. Age standardization removes the confounding effects of changes in age composition.

FIGURE 8.1

Proportion of Labor Force Unemployed by Race, Sex, and Age, 1950–1985

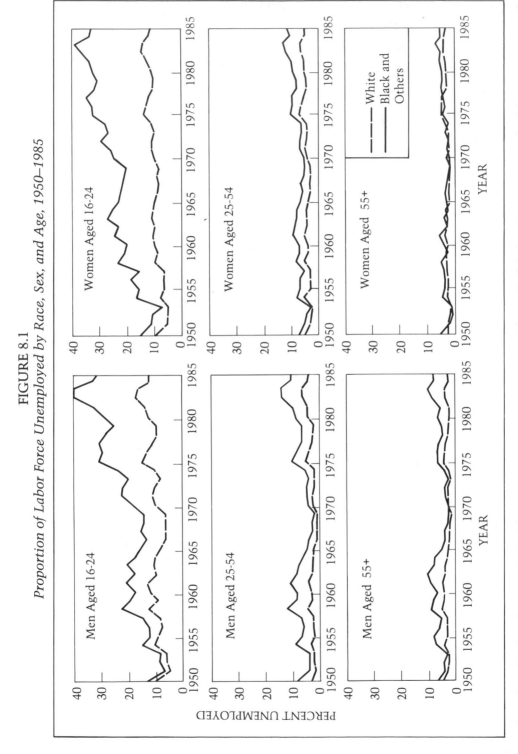

NOTE: These data have been standardized for age using component age groups within each of the broad age categories. The data are average annual rates for the entire year and refer to the civilian population. Data for 1950 through 1971 refer to whites and nonwhites. Data for later years refer to whites and blacks.

SOURCES: U.S. Bureau of Labor Statistics. *Handbook of Labor Statistics 1975. Reference Edition. Bulletin 1865 (1975), tables 4 and 63. Handbook*

unemployment is to ask what is the change in the unemployment rate associated with a 1 percent real change in the Gross National Product (GNP).[15] The peak growth rate for the GNP was about 8 percent between 1950 and 1951 while the biggest decline was a drop of 2 percent between 1981 and 1982. On average, the GNP grew 3.4 percent each year.

Table 8.1 presents the change in the unemployment rate associated with a 1 percent change in the GNP and a measure of the strength of this relationship. If unemployment rates and economic growth were so closely linked that there was a one-to-one relationship between the two, the correlation coefficient would take on its maximum value of 1.00. If changes in unemployment were unrelated to changes in GNP, the correlation coefficient would have a value of 0.00.

A 1 percent increase in real GNP typically reduced the unemployment rate of nonwhite men aged 25 to 54 by .63 points; that of similarly aged white men, only .34 point. Stated differently, if the GNP grew by 5 percent from one year to the next, the unemployment rate for nonwhite men of prime working ages would drop 3.2 percent; while among white men, it would drop 1.7 percent. For other age groups of men and for women of all ages, nonwhites also suffer more than whites when the economy declines, but also benefit more when the economy expands.

Basing their conclusions on observations such as this, some contend that rapid economic growth is the ideal way to improve the status of blacks, perhaps a more effective and quicker solution than implementing affirmative action or upgrading central city schools. Presumably, during rapid economic growth, less-skilled workers will be hired and employers will not have the luxury of discriminating against blacks. Although rapid economic growth is recommended to solve problems of poverty, racial differences in unemployment are persistent and will likely remain large. For example, in 1985 the unemployment rate among black men aged 25 to 54 was 10 percent, while among similar white men it was only 5 percent. To eliminate this racial difference within a year would require a growth rate of the GNP of 17 percent or five times the average rate of growth for the post–World War II era.

Table 8.1 shows that the impact of economic changes is greatest among young workers and least among older people. Indeed, since World War II unemployement rates at ages 55 and over have not been closely tied to changes in the economy. This age group is not adversely affected by an economic turndown—at least in terms of unemploy-

[15]The term "real change in the Gross National Product" is used to indicate that we analyzed the change after adjustments were made for inflation. Growth rates which include inflation would be much higher than those used in this investigation. For a discussion of previous findings, see Tobin (1965), p. 881.

TABLE 8.1

Change in Unemployment Rate and Change in Proportion Out of the Labor Force Associated with a 1 Percent Change in the Gross National Product, 1950–1951 to 1984–1985

Age	Nonwhites		Whites	
	Change in Rate or Proportion	Correlation Coefficient Squared	Change in Rate or Proportion	Correlation Coefficient Squared
CHANGE IN UNEMPLOYMENT RATE				
MEN				
16–24 Years	−1.01	.71	−.77	.85
25–54 Years	−.63	.77	−.34	.82
55+ Years	−.39	.34	−.20	.50
WOMEN				
16–24 Years	−.44	.72	−.67	.35
25–54 Years	−.30	.76	−.39	.57
55+ Years	−.15	.40	−.24	.31
CHANGE IN PROPORTION OUT OF THE LABOR FORCE				
MEN				
16–24 Years	−.16	.15	−.15	.18
25–54 Years	−.05	.06	−.01	.05
55+ Years	−.07	.02	.08	.12
WOMEN				
16–24 Years	−.09	.04	−.12	.12
25–54 Years	.07	.05	.05	.03
55+ Years	.00	.00	−.02	.02

NOTE: Coefficients in the upper panels are taken from models which regress year-to-year percentage change in the age-standardized unemployment rates year-to-year changes in the constant dollar gap. Coefficients in the bottom panel relate annual changes in the age-standardized proportions out of the labor force on year to year changes in the GNP.

SOURCES: U.S. Bureau of Labor Statistics, *Handbook of Labor Statistics* (December 1983), bulletin 2175, tables 4 and 26, *Employment and Earnings*, vol. 32, no. 1, table 3; vol. 33, no. 1 (January 1986), table 3; U.S. Bureau of the Census, *Historical Statistics of the United States: Colonial Times to 1970*, (1975), part 1, series F-3; *Statistical Abstract of the United States: 1984*, table 735.

216

ment—nor does it benefit much from an expansion. Older employees have the seniority to avoid a layoff during a recession, but an economic boom, on the other hand, does little to cut their unemployment rates.

Is the racial gap in unemployment rates expanding or is it contracting? To investigate racial disparities in unemployment, we assumed that the unemployment rate of white men aged 25 to 54 was a standard by which to judge the rates of other men and women. Presumably, adult white men do not suffer from the lack of experience or prejudice which may diminish employment opportunities for minorities, women, and young workers. The model we used related a group's unemployment rate to that of white men aged 25 to 54 and also included a time trend. If the difference between a group's unemployment rate and that of white men aged 25 to 54 grew larger, the coefficient associated with the time trend will be positive. Indeed, its value will tell us the average annual increase in the gap separating a group's unemployment rate from that of white men aged 25 to 54. If the gap grew smaller, the coefficient associated with the time trend will be negative. A coefficient for the time trend of zero would suggest that the unemployment rate of a group moves in tandem with that of adult white men.

For nonwhite men aged 16 to 24, the ratio of unemployment rates is 3.6 (see Appendix Table 8.1); that is, once the time trend is eliminated, the proportion unable to find jobs among nonwhite men aged 16 to 24 was, on average, 3.6 times that of white men aged 25 to 54. Stated differently, young nonwhite men typically had an unemployment rate almost four times that of adult white men. The coefficient associated with the time trend was +.49, showing that, on average, the disparity in unemployment rates between nonwhite men aged 16 to 24 and white men aged 25 to 54 increased by almost 0.5 percent annually between 1950 and 1985. This is a substantial trend and clearly indicates that employment opportunities declined for young blacks relative to those of adult white men.

The unemployment rates of young black women and young whites of both sexes also rose compared with those of the reference group: white men aged 25 to 54. The changes, however, were very much greater for young nonwhites than whites. This examination of thirty-five year trends suggests that the deterioration in the employment situation for the young is not just a consequence of one recession in the mid-1970s and another at the start of the 1980s.[16] Rather, there was a secular trend toward higher unemployment rates for young workers of both races and both sexes.

[16]For more detailed analyses, see Mare and Winship (1979, 1984); Freeman and Wise (1982); Cogan (1982).

The unemployment rate for nonwhite men aged 25 to 54 was much higher than that of similar white men: Its average level was 2.5 times that of whites. Unlike the situation for the young, the secular trend for this age group implies a slight decline in racial difference, since the racial gap in unemployment at these ages declined slightly. Compared with adult white men, the unemployment rate is not getting higher among adult nonwhite men.

This analysis of unemployment trends leads to three conclusions. First, throughout the post–World War II era, the unemployment rates of nonwhites have been much higher than those of whites, frequently on the order of two to one. Second, year-to-year fluctuations in unemployment rates are greater among nonwhites than among whites. Blacks benefit more than whites from economic expansions but suffer more during recessions. Third, if racial progress is indicated by the elimination of the racial gap in unemployment, there is mixed evidence of improvement. The employment situation unambiguously worsened for young nonwhites in the labor force, but among men aged 25 to 54 there are signs of a possible racial convergence in unemployment rates.

Labor Force Participation

There is consensus that unemployment is a problem since a person who cannot find a job may be unable to support himself or his family, and his self-esteem may be diminished. There is less agreement about nonparticipation in the labor force. If an adult spends many years outside the labor force, quite likely his or her income and savings will be reduced. However, it is necessary to consider the reason for nonparticipation. If a woman is not working because she devotes all her efforts to raising children or inherits so much money that she can retire at age 43, nonparticipation may not be a problem. If she is not participating because she is discouraged about employment prospects, then nonparticipation may be a social and economic problem.

Recent trends in participation reveal an increasing racial disparity among men, which helps to explain why racial gaps in income remain large. Figure 8.2 shows the proportions of total population who are out of the labor force for the six age-race groups. Trends in this figure contrast sharply with those shown in Figure 8.1. Unemployment rates fluctuated from one year to the next in response to economic conditions, but there are clear trends for the nonparticipation rates. A steadily increasing proportion of men aged 25 and over of both races are nonparticipants, and among nonwhites there has also been a rise in the proportion of young men who neither work nor seek work. Since women in-

FIGURE 8.2
Trends in Proportion of Population Out of the Labor Force by Race, Sex, and Age, 1950–1985

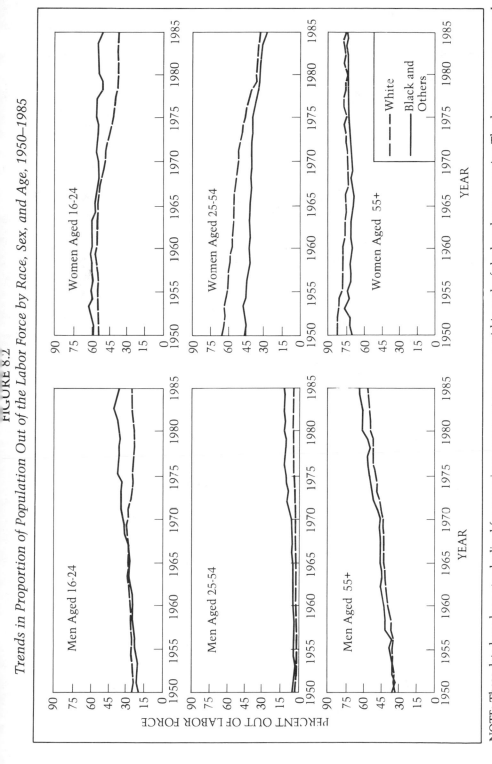

NOTE: These data have been standardized for age using component age groups within each of the broad age categories. The data are average annual rates for the entire year and refer to the civilian population. Data for 1950 through 1971 refer to whites and nonwhites. Data for later years refer to whites and blacks.

SOURCES: U.S. Bureau of Labor Statistics, *Handbook of Labor Statistics 1975—References Edition*, Bulletin 1865 (1975), tables 4 and 63; *Handbook of Labor Statistics*, Bulletin 2217 (June 1985), tables 5 and 27; *Employment and Earnings*, vol. 32, no. 1 (January 1985), table 3; vol. 33, no. 1 (January 1986), table 3.

creasingly spend their adult lives at work, a declining share of those aged 25 to 54 are nonparticipants, a trend that is also evident among young white women.

The difference between unemployment trends and labor force participation trends is also illustrated in Table 8.1, which relates annual changes in unemployment rates and the proportion out of the labor force to changes in the GNP. If there are many discouraged workers who withdraw from the labor force because they think work is not available, they should be pulled into the labor force by rapid economic growth; thus we would expect a strong inverse relationship between changes in nonparticipation and changes in the GNP. Among men—and among young women—there is some evidence that this happens, but the relationship of economic growth to labor force participation is very weak. For example, a 1 percent rise in the GNP lowered the unemployment rate of nonwhite men aged 25 to 54 by .63 percent while it reduced their nonparticipation rate by only .05 percent. Economic upturns and recessions apparently played a very small role in altering labor force participation rates in the 1950s, 1960s, and 1970s. Presumably, even rapid economic growth in the near future will not lead to drastic rises in labor force participation.

Changes in nonparticipation may be readily described for most groups by fitting a trend line to the information shown in Figure 8.2. Appendix Table 8.2 shows the average annual change in the proportion of the population out of the labor force for 1950–85. There are substantial racial and sexual differences in these trends. Among adult men aged 25 to 54, the proportion who are out of the labor force has been going up, but the annual increase has been about three times as fast among nonwhites as among whites. At the younger ages, the proportion out of the labor force has been going up rapidly among nonwhite men, but there has been an actual increase in labor force participation for young white men and for women of both races. Improvements in pensions, the indexing of Social Security payments, and reductions in retirement age help to account for the growing proportion of men aged 55 and over who are not members of the labor force.

What do people do if they are not participating in the labor force? Since 1967 the Census Bureau has asked nonparticipants about their *major* activity. Their answers were grouped into four categories: attending school, keeping house, unable to work, and other reasons. This last category is a broad one that includes retirees, discouraged workers, and people who are nonparticipants for many other reasons. Table 8.2 presents age-standardized data for blacks and whites in two age groups. Data for nonparticipants aged 55 and over, mostly retirees, are not shown.

TABLE 8.2

Reported Activities of Persons Aged 16–24 and 25–54 Who Were Outside the Labor Force, by Sex and Race, 1967 and 1985

Reasons for Nonparticipation in Labor Force	Blacks				Whites			
	16–24 Years		25–54 Years		16–24 Years		25–54 Years	
	1967	1985	1967	1985	1967	1985	1967	1985
MEN								
In School	79%	69%	9%	8%	83%	77%	15%	11%
Keeping House	<1	2	1	8	<1	<1	2	5
Unable to Work	2	1	39	25	2	1	36	24
Other Reasons	19	28	51	59	15	22	47	60
Total	100%	100%	100%	100%	100%	100%	100%	100%
WOMEN								
In School	42%	52%	1%	4%	44%	54%	<1%	3%
Keeping House	48	31	94	78	48	31	92	87
Unable to Work	1	<1	2	4	<1	<1	<1	2
Other Reasons	9	17	3	13	8	15	6	8
Total	100%	100%	100%	100%	100%	100%	100%	100%

NOTES: These data have been standardized for age to remove the confounding effects of changes in the age compositions. Data in both years refer to the noninstitutionalized population. Data for 1967 refer to nonwhites; those for 1985 refer to blacks.

SOURCES: U.S. Bureau of Labor Statistics, *Employment and Earnings*, vol. 32, no. 1 (January 1986), table 3; *Employment and Earnings*, vol. 14, no. 7 (January 1968), table A-1.

A comparison of reasons given by blacks and whites shows that the racial differences are small. About three quarters of the young men who are nonparticipants attend school, while among young women both attending school and keeping house were frequently reported as reasons for being out of the labor force. Few adult men attend school or keep house, but about one quarter of those out of the labor force report that they are unable to work. Keeping house is the major activity of women out of the labor force.

Is there any evidence that the proportion of discouraged workers is increasing? Unfortunately, we lack the data needed to answer the question unambiguously. For every group of both races there was a rise in the proportion who reported that they were out of the labor force for "other reasons." Much of this change, especially at the younger ages, may be attributable to more discouraged workers.

Two conclusions come from this analysis of labor force participation. First, participation was not strongly related to economic changes; rather, there are long-run trends that are little affected by booms or recessions. Second, racial differences in labor force participation are increasing among men under age 55 and among young women. Among adult women, however, the racial gap vanished. Formerly, nonwhite women participated to a much greater extent than whites, but this is no longer the case.

The Employment-Population Ratios

Do unemployment rates best describe trends, or should we analyze labor force participation rates? These two measures gauge different aspects of the employment situation. Many analysts believe that a third measure—the employment-population ratio—may provide more useful information, that is, the proportion of population actually holding jobs. Figure 8.3 presents age-standardized information for the six age-race groups.

Among the young in 1950, about 70 percent of the men of both races held jobs. The trend toward greater school attendance delayed entry into the labor force, and by 1985 a much lower proportion of young men were at work. The drop-off in employment was greater among nonwhite men, partly because white men are now more likely than black men to combine work and school. Just after the Korean War, the racial difference in the proportion of 16-to-24-year-olds employed was only 2 percentage points. This increased to 14 points in 1973, jumped to 20 points in the 1973–75 recession, and then remained at that level.

Although racial differences in employment are much smaller

FIGURE 6.3

Proportion of Total Population at Work by Race, Sex, and Age (employment-population ratios), 1950–1985

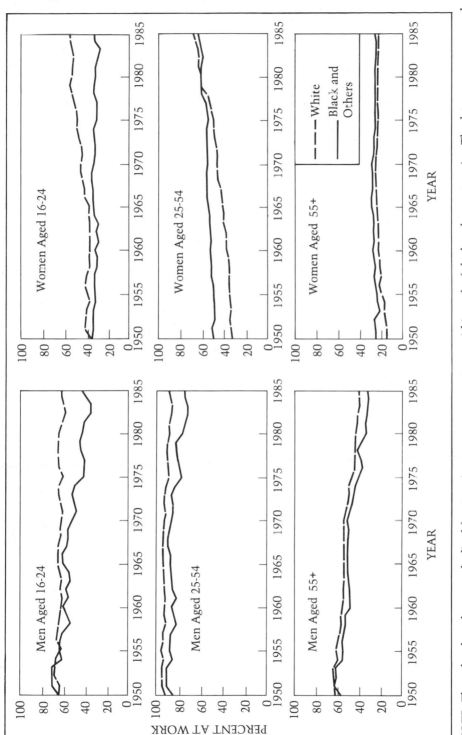

NOTE: These data have been standardized for age using component age groups within each of the broad age categories. The data are average annual rates for the entire year and refer to the civilian population. Data for 1950 through 1971 refer to whites and nonwhites. Data for later years refer to whites and blacks.

SOURCES: U.S. Bureau of Labor Statistics, *Handbook of Labor Statistics 1975—Reference Edition*, Bulletin 1865 (1975), tables 4 and 63; *Handbook of Labor Statistics*, Bulletin 2217 (June 1985), tables 5 and 27; *Employment and Earnings*, vol. 32, no. 1 (January 1985), table 3; vol. 33, no. 1 (January 1986), table 3.

among men aged 25 to 54, trends are similar: a gradual increase in the racial disparity until 1973–75, a sharper increase in the racial disparity during that recession, and no contraction thereafter. By 1985 only 78 percent of the black men in this group held jobs compared with 90 percent of the white men. Stated differently, despite the economic recovery, one adult black man in five is either unemployed or not in the labor force. Among men aged 55 and over, there has been a decline in employment, but the racial gap has not grown larger since older men of both races are retiring rather than working.

Throughout the post–World War II era, there has been a persistent increase in the employment of young white women aged 16 to 24 but very little change among nonwhite women, and thus the racial difference has grown much larger. In 1950, 40 percent of the white women and 35 percent of the nonwhite women aged 16 to 24 were at work; in 1985 the racial disparity was 22 points since 57 percent of the white women and only 35 percent of the black women had jobs.

The proportion of adult women at work was once higher for blacks than for whites. Largely because of the increasing employment of white women, this has changed and in the early 1980s, for the first time, the proportion of women aged 25 to 54 at work was greater for whites than blacks.

Racial Differences in Employment: Are They Limited to Certain Groups?

The trend lines shown in this chapter suggest that the racial gap in employment has grown larger since World War II. Bowen and Finegan report that racial differences in male labor force participation were increasing even during the 1940s.[17] If whites are increasingly likely to hold jobs while blacks are increasingly likely to be out of the labor force, racial differences in income and wealth will undoubtedly increase.

Perhaps the employment problems of blacks are concentrated among individuals with certain demographic characteristics, such as those with limited educations. If we control for the social and demographic differences that distinguish the races, we might expect to find that racial differences in employment and labor participation disappear.

We start by considering the population aged 25 to 54 and use five indicators of employment:

[17]Bowen and Finegan (1969), table 3–4.

1. the proportion of the 25 to 54 age group holding a job when the census was conducted in April 1980 (the employment-population ratio);
2. the proportion of the *labor force* unemployed in April 1980 (the conventional unemployment rate);
3. the proportion of the *total population* aged 25 to 54 unemployed at some time in 1979 (the prevalence of unemployment);
4. the average number of weeks of unemployment in 1979 for those who were out of work at some point during the year (the duration of unemployment);
5. the average hours of employment in 1979 for people who worked during that year.

April 1980 is an appropriate date for gauging employment differences, for this was a time of neither economic boom nor bust. For example, during the economic expansion of the late 1960s, the nation's overall unemployment rate fell to a low of 3.5 percent in 1969. In the recession of 1982 it climbed to 9.7 percent, while in April 1980 it was 7.0 percent.[18]

Adult Men

Analysts of labor force activity and unemployment often observe that differences are associated with age, educational attainment, marital status, and health status.[19] In Table 8.3, we show indicators of employment for blacks and whites classified by these variables. Data for men reveal that racial differences in employment are not restricted to any particular demographic group and that no matter how we classify men blacks were less likely to be at work, more likely to be unemployed, and worked fewer hours than did comparable whites. It is erroneous to assume that the employment problems of black men are restricted to a poorly educated underclass.

For example, among men aged 25 to 29 and those aged 50 to 54, the proportion holding a job in 1980 was 87 percent for whites but only 73 percent for blacks. Unemployment varied by age, of course, but for all age groups the proportion out of work was about twice as great for blacks as for whites. In addition, those blacks who were out of work in 1979 spent an average of four more weeks unemployed than did those white men who were out of work, regardless of their characteristics.

For both races, the proportion employed increased with educational

[18]U.S. Bureau of Labor Statistics, *Employment and Earnings*, vol. 27, no. 5 (May 1980), table A-1; vol. 32, no. 1 (January 1985), table 1.
[19]Bowen and Finegan (1969), chap. 3; Bancroft (1958).

TABLE 8.3

Employment Information for the Black and White Population Aged 25–54, 1980

	Percent of Total Population at Work in 1980			Percent of Labor Force Unemployed in 1980			Percent Unemployed at Some Time in 1979		Average Weeks Out of Work for Those Unemployed in 1979		Hours of Employment for Those Working in 1979	
	Black	White	Racial Difference	Black	White	Racial Difference	Black	White	Black	White	Black	White
MEN												
Age												
25–29 Years	73%	87%	−14%	12%	6%	+6%	24%	20%	18	13	1,761	2,003
30–34 Years	76	91	−15	10	4	+6	20	14	18	14	1,838	2,140
35–39 Years	79	92	−13	9	4	+5	18	12	18	15	1,909	2,163
40–44 Years	80	91	−11	8	3	+5	15	10	17	13	1,901	2,189
45–49 Years	78	89	−11	7	4	+3	14	9	18	16	1,887	2,178
50–54 Years	73	87	−14	7	4	+3	13	9	19	16	1,883	2,136
Educational Attainment												
Elementary	66	76	−10	11	8	+3	17	18	17	16	1,769	1,970
High School, 1–3 Years	69	84	−15	13	7	+6	21	19	19	16	1,770	2,035
High School, 4 Years	79	90	−11	9	5	+4	19	15	18	13	1,880	2,145
College, 1–3 Years	82	91	−9	8	4	+4	20	13	18	14	1,874	2,132
College, 4 Years	89	95	−6	4	2	+2	14	9	17	12	1,959	2,177
College, 5+ Years	89	94	−5	3	2	+1	12	7	17	13	1,957	2,164
Current Marital Status												
Single	60	77	−17	16	8	+8	23	19	20	16	1,655	1,864
Married-Spouse-Present	87	93	−6	6	4	+2	16	11	16	13	1,937	2,187
Widowed, Divorced, Spouse Absent	66	81	−15	13	8	+5	21	19	20	16	1,767	1,981
Place of Residence												
Metropolitan	77	90	−13	9	4	+5	19	13	18	14	1,856	2,113
Non-Metropolitan	74	87	−13	9	5	+4	17	14	16	15	1,808	2,172
Region of Residence												
Northeast	74	90	−16	10	5	+5	19	13	21	16	1,830	2,091
Midwest	71	90	−19	15	5	+10	22	13	19	14	1,833	2,178
South	79	90	−11	7	3	+4	17	12	16	13	1,862	2,139
West	76	89	−13	8	5	+3	20	15	18	14	1,852	2,064
Migration Stream												
Born North, Lives North	70	90	−20	13	5	+8	22	13	20	14	1,814	2,119
Born North, Lives South	80	92	−12	7	3	+4	18	12	18	11	1,909	2,169
Born South, Lives South	79	89	−10	8	4	+4	17	12	16	14	1,866	2,126
Born South, Lives North	75	88	−13	11	6	+5	19	15	19	15	1,875	2,074
Presence of Work-Limiting Disability												
Yes	33	53	−20	18	9	+9	18	16	24	19	1,467	1,775
No	82	92	−10	9	4	+5	19	13	17	14	1,875	2,143

WOMEN

Age												
25–29 Years	62%	63%	– 1%	12%	6%	+6%	22%	15%	18	12	1,543	1,533
30–34 Years	66	59	+ 7	9	5	+4	18	12	18	13	1,611	1,484
35–39 Years	66	61	+ 5	8	5	+3	16	10	17	13	1,621	1,515
40–44 Years	65	62	+ 3	6	5	+1	13	10	16	14	1,655	1,545
45–49 Years	62	59	+ 3	6	4	+2	11	8	18	16	1,646	1,581
50–54 Years	55	55	—	6	4	+2	10	8	18	15	1,566	1,615
Educational Attainment												
Elementary	43	40	+ 3	11	10	+1	13	11	19	16	1,467	1,517
High School, 1–3 Years	51	48	– 3	13	8	+5	17	12	19	16	1,502	1,512
High School, 4 Years	66	59	+ 7	9	5	+4	17	11	17	13	1,632	1,544
College, 1–3 Years	74	64	+10	7	4	+3	17	11	17	12	1,669	1,540
College, 4 Years	85	69	+16	3	3	—	16	10	14	11	1,552	1,496
College, 5+ Years	87	81	+ 6	3	2	+1	13	11	13	13	1,529	1,597
Current Marital Status												
Single	61	80	–19	11	4	+7	20	14	18	16	1,594	1,811
Married-Spouse-Present	66	55	+11	7	5	+2	14	9	17	13	1,596	1,449
Widowed, Divorced, Spouse Absent	60	73	–13	10	6	+4	17	17	18	16	1,609	1,694
Place of Residence												
Metropolitan	63	61	+ 2	9	5	+4	16	11	18	13	1,610	1,546
Non-Metropolitan	60	56	+ 4	9	6	+3	17	10	18	14	1,531	1,500
Region of Residence												
Northeast	62	60	+ 2	9	5	+4	15	11	18	15	1,627	1,494
Midwest	58	59	– 1	12	5	+7	19	10	19	14	1,590	1,503
South	65	60	+ 5	8	4	+4	16	10	17	12	1,591	1,606
West	65	61	+ 4	8	5	+3	16	13	16	13	1,624	1,528
Migration Stream												
Born North, Lives North	61	61	—	10	5	+5	18	11	19	14	1,587	1,496
Born North, Lives South	71	61	+10	7	3	+4	20	11	17	10	1,593	1,571
Born South, Lives South	65	60	– 5	8	4	+4	16	9	17	12	1,539	1,616
Born South, Lives North	60	55	– 5	9	7	+2	16	13	19	15	1,615	1,564
Presence of Work-Limiting Disability												
Yes	22	29	– 7	18	13	+5	13	11	23	17	1,162	1,233
No	68	62	+ 6	8	5	+3	17	11	17	13	1,624	1,550
Overall Mean	63	60	+ 3	9	5	+4	16	11	18	13	1,601	1,538

SOURCE: U.S. Bureau of the Census, *Census of Population and Housing: 1980*, Public Use Microdata Samples.

attainment while the proportion unemployed was inversely related to schooling. Nevertheless, at all attainment levels, black men fell behind white men on these indicators of employment. Even among those who completed five or more years of college, black men were at a disadvantage.[20] They were more likely to be unemployed and spent more time out of work in 1979, and those who held jobs worked about 10 percent fewer hours than comparable white men.

There has been much discussion in recent years about the relationship of marital status to the employment and occupational achievement of black men. In 1965 Daniel Patrick Moynihan described a deterioration of black families in urban ghettos involving illegitimacy, divorce, desertion, and female-headed households.[21] Young men raised in such families, Moynihan asserted, had difficulties in school, were prone to delinquency, and were often rejected for military service because their test scores were so low. He stressed that the lack of employment opportunities for black men was a major cause of the deterioration of black families. Because of discrimination, limited education, and slow economic growth, black men could not often find good jobs and so they were unable to support their wives or raise their children.

There has been much controversy about the direction of causality. Some accuse Moynihan of implying that blacks were responsible for their own travails.[22] That is, blacks allowed their families to deteriorate, which doomed the life chances of young blacks. Others read Moynihan's report differently and believed that he was arguing that employment problems were a fundamental cause of the deterioration of black families.

This debate has not been resolved in the twenty years since Moynihan offered his views, but there have been detailed analyses of the history of black families,[23] as well as empirical studies which show that black men who are raised by both parents spend more time in school and have greater occupational achievement as adults than do black men raised by their mothers only.[24] In addition, there have been several excellent accounts of how unemployment affects black men and women which suggest that the cause and effect relationship between black family structure and black employment is complex.[25]

Although we know little about the exact relationship of employ-

[20]For an analysis of the effects of attainment on labor force participation by race at earlier dates, see Bancroft (1958), table 41; Bowen and Finegan (1969), table 3-5.
[21]U.S. Department of Labor (1965).
[22]Rainwater and Yancey (1967); Billingsley (1968), pp. 198–201.
[23]Gutman (1976); Litwak (1979).
[24]Duncan and Duncan (1969); Duncan, Featherman, and Duncan (1972), p. 65; Featherman and Hauser (1978), tables 6.7 and 6.10.
[25]Liebow (1967); Stack (1975); Willie (1976).

ment to family stability, marital status is strongly linked to labor force activities. In 1980, 87 percent of the black men aged 25 to 54 living with wives held jobs and only 6 percent of them were unemployed. For never-married black men, just 60 percent were at work, and their unemployment rate was 16 percent. In addition, black and white men who lived with wives were much more comparable in their labor force activity than were men in other marital statuses. Indeed, this is one of the few groups in which the unemployment rate for blacks was not double that of comparable whites. Black men with wives reported short durations of unemployment during 1979 and many hours of employment compared to black men who were never married or were divorced.[26]

Perhaps racial differences in health and disability help account for black-white differences in employment. The census of 1980 was the first since 1910 to inquire about health; individuals were asked if they had physical, mental, or other health conditions which lasted at least six months and limited the amount or kind of work they could do. There was a substantial racial difference: 12 percent of the black men aged 25 to 54 and 8 percent of the white men reported such limitations. As expected, the proportion at work and the number of hours on the job were relatively low for men whose health was impaired.

To further explore racial differences in employment, we fitted models using information about labor force participants in 1980. The independent variables were those that were strongly related to unemployment: age, educational attainment, marital status, region, and presence of a work-limiting disability. The log of the odds of unemployment was the dependent variable, which was related to the independent variables separately for blacks and whites. Parameters from these models are shown in Appendix Table 8.3 and may be used to estimate unemployment rates for any groups of black or white men. Selected findings are shown in Figure 8.4.

Figure 8.4 demonstrates the strong relationship of marital status, education, and region to unemployment. It also indicates that racial differences do not disappear after black and white men are equated with regard to these variables. For example, in the North, black college graduates aged 25 to 34 who were married had a low unemployment rate of 3.5 percent, but it was still twice that of comparable white men—1.7 percent. For both races, greater educational attainment meant less unemployment. However, the educational disparity in unemployment is clearly indicated in that black men who completed one to three years of college typically had unemployment rates close to those of white men who failed to finish high school. In both the North and the South,

---

[26]For an analysis of differences in 1960, see Bowen and Finegan (1969), chap. 3.

FIGURE 8.4

Estimated Unemployment Rates in April 1980 for Blacks and Whites
Aged 25–34, by Education and Marital Status

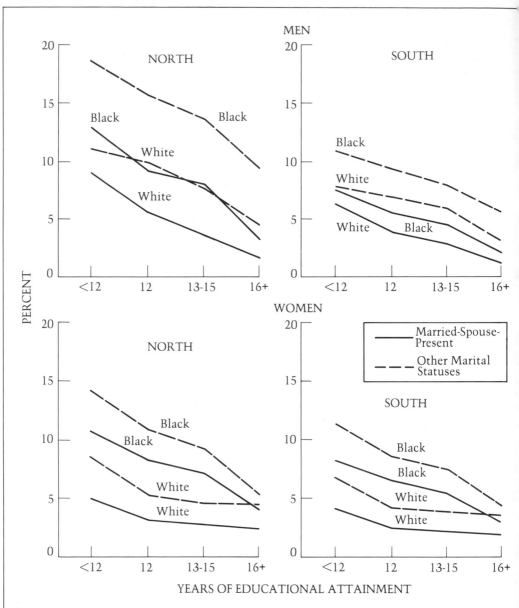

NOTE: Estimates assume no work limitation for men and no child under age 6 for women. North refers the entire country except the South.

SOURCE: U.S. Bureau of the Census, *Census of Population and Housing: 1980*, Public Use Microda Samples.

married black men living with spouses had lower unemployment rates than black men in other marital statuses, but they were higher than those of comparable whites.

Studies of racial attitudes have traditionally reported that on racial matters southern whites were more conservative—or prejudicial—than northern whites.[27] If unemployment results from discriminatory practices such that blacks are the last hired and first fired, we might expect higher unemployment rates for blacks in the South than for those in the North. This is not the case. Figure 8.4 reports that unemployment rates for blacks in the South are consistently 2 to 5 percentage points below those of similar blacks in the North. Perhaps by 1980 attitudes had changed such that southern white employers were even more willing to hire blacks than were northern employers. More likely, the regional difference in unemployment reflects differences in economic growth and industrial structure, since white unemployment rates are also lower in the South than in the North. Regional differences in the benefit levels of state-run transfer programs may also help to account for these regional differences in unemployment.

One of the most consistent racial differences concerns hours of employment. Black men aged 25 to 54 averaged 275 fewer hours of work in the year than whites, or an average of about six fewer hours on the job every week. This racial difference in hours of work is a consistent one. Among men aged 25 to 29, the racial difference was about the same as it was among men in their 50s. Black men with elementary school educations worked 201 fewer hours than did white men, while among men with five or more years of college the difference was 207 hours. As we will see in Chapter 11, this is an important reason for the lower earnings of blacks.

Additional information about the net racial difference in hours of employment is shown in Figure 8.5. We considered individuals who were employed at some point during 1979 and then fitted a model which assumed that the number of hours they worked was influenced by their educational attainment, age, region of residence, marital status, and the presence or absence of a work-limiting disability. The parameters of these regression models are listed in Appendix Table 8.4 and may be used to estimate hours of work for persons with any set of characteristics.

Black men aged 30 in 1980 who had completed only junior high school and were living with wives spent an average of 1,880 hours on the job. Similar black men who were single, separated, or divorced worked an average of 1,700 hours, revealing once again the strong rela-

[27]Schuman, Steeh, and Bobo (1985), chap. 3.

FIGURE 8.5
*Estimated Hours of Employment in 1979 for Black and White Men
Living in the South by Educational Attainment, Age, and Marital Status*

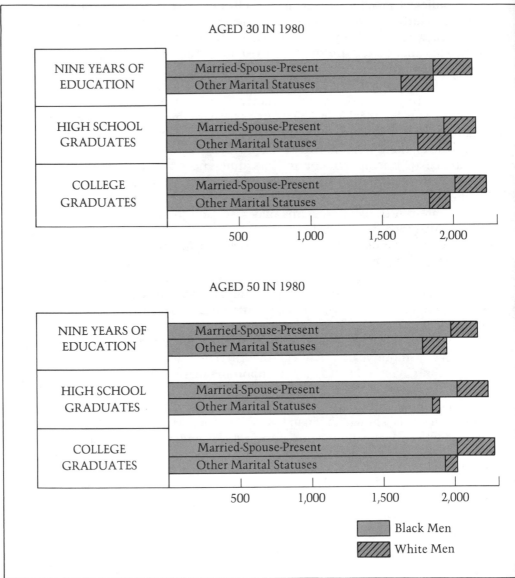

SOURCE: U.S. Bureau of the Census, *Census of Population and Housing: 1980,* Public Use Micr
Samples.

tionship of marital status to employment.[28] The hours of employment for similar white men were 2,115 for men living with wives and 1,875 for those who were single, separated, or divorced.

Educational attainment was closely linked to hours of employment since those who spent many years in school worked more hours. College-educated black men aged 30 who lived with wives worked in excess of 2,000 hours in 1979, which was still more than 200 hours below the employment figure for comparable white men. At age 50 the similar racial difference was 150 hours. As this figure demonstrates, racial differences in hours of employment cannot be attributed to differences in age, marital status, or education since white men typically worked about 200 hours, or 4 hours per week, more than comparable black men. There appears to be a net racial difference both in unemployment itself and in hours of employment.

Adult Women

On all indicators of employment, black men fall behind similar white men, but the situation is more complicated among women. Unemployment rates are higher for black women than for white women, but the proportion of women aged 25 to 54 holding jobs hardly differs by race. This seeming paradox occurred because labor force participation rates were also higher for black women.

Table 8.3 shows five indicators of employment for black and white women. Demographic characteristics were related to employment in much the same way among women as among men. Educational attainment, for example, was directly related to employment for both black and white women and inversely related to the unemployment rate or length of unemployment. Among women—just as among men—those in the South had higher rates of employment and lower rates of unemployment than those who lived in the Northeast or Midwest.

There are both sexual and racial differences in the way marital status relates to employment. Among men of both races, those who were married and lived with wives were more likely to be at work than were single, divorced, or separated men, but the reverse relationship was found among white women. White women who lived with husbands had low rates of employment, perhaps because of their household responsibilities and husband's income. For black women, the proportion *at work* was highest for those who lived with husbands.[29]

---

[28]For similar findings for earlier years, see Bianchi (1981), table 5.4; Bowen and Finegan (1969), chaps. 5–7.

[29]For possible explanations, see Cain (1966), p. 80; Bancroft (1958), pp. 52–57.

Black women in 1980 not only had higher unemployment rates than white women but a much greater proportion were unemployed at some point during 1979. Furthermore, black women out of work in that year averaged about five weeks more unemployment than white women similar to the situation among black men.

Finally, there was a racial difference in hours of employment. Recall that this analysis of hours is restricted to those who worked at some point during 1979. Given the greater prevalence and longer duration of unemployment for blacks, we might expect that black women would spend fewer hours on the job than white women, but this is not the case. Black women who held a job in 1979 averaged about one more hour of work per week than white women—31 hours per week for blacks compared with 30 for whites.

Are the determinants of unemployment pretty much the same among adult women as among men? Our investigation assumed that whether a woman was unemployed depended on her educational attainment, age, region of residence, and marital status. Work-limiting health conditions were reported less frequently by women than by men so this variable was supplanted with one indicating the presence in the household of a child under age 6.[30] This analysis is restricted to those black and white women who were labor force participants in 1980. Undoubtedly, there was a great deal of selectivity involved since women who were obligated to care for children, who had no marketable skills, or who suffered from serious health problems remained out of the labor force.

Figure 8.4 summarizes findings while the coefficients from the models are shown in Appendix Table 8.3. Educational attainment was strongly linked to unemployment among women; 11 percent of the married black women in the North with husbands present who had less than 12 years of education were unemployed compared with only 4 percent of those who completed college. The net effect of living with a spouse was a reduction in unemployment. There may, of course, be additional selectivity here, since married women who cannot locate jobs may drop out of the labor force more rapidly than single or divorced women. The net effects of region were the same for both races: Unemployment rates of women in the North generally exceeded those of the South. Having a child under age 6 in the household had no more than a small net effect on unemployment rates among women of either race.

The most striking findings in Figure 8.4 are the persistent and substantial racial differences. Black women in every comparison had higher

[30]Bowen and Finegan (1969), pp. 108–27; Bancroft (1958), pp. 63–64; Sweet (1973), chap. 4.

unemployment rates than comparable white women. Indeed, the ratio of black to white unemployment rates for most groups of women exceeded 2 to 1. Black women who had college degrees typically had unemployment rates about the same as those of white women with just a high school diploma, leading us to conclude that unemployment was not just a problem for those black women with limited educations.

Despite their much higher unemployment rates, employed black women put in more hours in a year than white women. Is this true for most groups of women or is it true of only some groups? To answer this question, we used a model similar to that for men. Hours of employment in 1979 were presumed to be influenced by education, age, region, marital status, work limitation, and the presence of a preschool child in the household. The statistical models are described in Appendix Table 8.4.

There is only a modest relationship of educational attainment to hours of employment among both black and white women. The major racial difference concerned the effects of marital status and children under age 6. Among white women, the net effect of living with a husband was a reduction in annual employment of about 250 hours while among black women it was an insignificant reduction of only 14 hours. Having a young child in the home reduced the hours of employment by about 6 hours per week for white women but less than 3 hours for black women. Married black women and those who had young children worked more hours than comparable white women. Among those who were not currently married and those without young children, white women worked the greater number of hours.

Why should having a young child or a husband reduce the work efforts of white women much more than those of black women? There is no simple answer. One might speculate that since black families are more often extended than white families, there is frequently a grandmother or other relative who can care for the child. Ethnographic accounts of low-income black neighborhoods suggest that older black women are quite willing to care for or raise young children;[31] and an investigation based on Current Population Survey data found that in about 5 percent of the black families and 3 percent of the white families, there was a nonworking adult relative who could provide child care. The presence of such potential babysitters significantly increased the hours of employment of both black and white women.[32]

Another reason for this difference may be the limited earnings capacity of black men, especially husbands. A black family that wishes to attain economic security will have to rely on the earnings of both

[31]Stack (1975); Willie (1976).
[32]Bianchi (1981), tables 5.3 and 5.4.

spouses more often than comparable white families because the earnings of black men continue to lag behind those of white men.[33] Many black women apparently do not have the option of reducing their employment when they become wives and mothers.

Employment and Joblessness Among the Young

Employment opportunities for young people have attracted the nation's attention as civil rights organizations stressed the exceptionally high unemployment rates of young black men.[34] If many youths are unemployed or have to accept low-wage jobs which offer no opportunities, they may turn to delinquency or illegal activities. Their chances for successful careers will be diminished and they may be unable to fulfill their responsibilities as husbands and fathers.

This section focuses on people aged 16 to 24 in 1980. Table 8.4 shows four indicators of employment for young men and women of both races. These are then related to age, educational attainment, enrollment, marital status, and place of residence. We turn first to the employment differences by age and educational attainment.

Labor force participation and employment, of course, increase with age. Only 17 percent of the blacks aged 16 and 17 in 1980 held jobs, compared with 64 percent of those aged 22 to 24. It is important to note, however, that the racial disparity in employment did not diminish with age. At all of these youthful ages, and among both men and women, a much higher proportion of whites than blacks held jobs. Unemployment rates (shown in the middle columns of Table 8.4) decline with age, but at all ages the unemployment rates of black youth were double or triple those of white youth.

In this analysis of youth employment, respondents were classified simultaneously by educational attainment and enrollment, revealing that black and white students differ in the way they mix education and work. A consistently smaller proportion of blacks held jobs while attending school or college; for example, among men enrolled in the first year of college, 34 percent of the blacks compared with 52 percent of the whites were working. The one exception to this pattern involves male graduate students since, in this group, blacks had the higher employment rate.

Among those youth who were not going to school, we find a familiar pattern: the proportion with a job in April 1980 was much higher for whites than for blacks, while the proportion unemployed was about

[33]Farley (1984), chap. 5.
[34]Hill (1981), p. 7.

twice as great among blacks. These census data also reveal that if racial differences in attainment were eliminated, there would still be large racial differences in youth employment and unemployment.

The racial differences by marital status were much the same for young persons as for older people. Young men of both races who lived with wives had high rates of employment and low rates of joblessness. Among black women, an unusually large proportion of those with husbands held jobs, but being married had no such effect for young white women.

There is now a large racial difference in the way young women begin their job careers and families. Since 1960 the fertility rates of white women aged 16 to 24 (see Chapter 4) have fallen while their enrollment rates and rates of labor force participation increased, suggesting that many of them are spending these years as students and employees rather than as wives and mothers. Among black women, the fall in fertility has been more modest and there has been little, if any, increase in employment.

Table 8.4 shows that among young men, there were substantial regional differences; the proportion of black youth employed was greater in the South and West than in the Northeast or Midwest. Since the employment rates of whites do not vary by region, racial differences in employment were also greatest in the Northeast and Midwest.

If we consider several of these variables related to employment simultaneously, will we discover that racial differences disappear? What factors actually determine the unemployment rates of young people? To answer such questions, we fitted models which specified that the unemployment rate of young people—that is, those out of work as a percentage of the total labor force—was influenced by educational attainment, current enrollment, region of residence, marital status, and age. Figure 8.6 presents a summary of major findings while the statistical details are shown in Appendix Table 8.5. The trend lines in this figure refer to people aged 20 and 21 living outside the South.

Education and enrollment had substantial consequences for unemployment rates. As this figure illustrates, in every comparison those attending school had much lower rates of unemployment than similarly educated persons who were not enrolled. Furthermore, unemployment was inversely related to attainment, except among blacks enrolled in high school. Racial differences are evident since in every comparison unemployment rates were higher among blacks than among comparable whites. Out-of-school blacks who completed some college training, for example, had an unemployment rate similar to that of out-of-school whites who did not finish high school.

This analysis of youth employment leads to two conclusions. First,

TABLE 8.4
Employment Information for the Black and White Population Aged 16–24, 1980

	Percent of Total Population at Work in 1980			Percent of Labor Force Unemployed in 1980			Percent Unemployed at Some Time in 1979		Average Weeks Out of Work for Those Unemployed in 1979	
	Black	White	Racial Difference	Black	White	Racial Difference	Black	White	Black	White
MEN										
Age										
16–17 Years	17%	36%	−19%	28%	14%	+14%	15%	15%	17	13
18–19 Years	39	60	−21	23	13	+10	30	29	17	13
20–21 Years	55	71	−16	20	10	+10	33	29	18	12
22–24 Years	64	80	−16	17	9	+ 8	30	27	17	12
Educational Attainment and Current Enrollment										
Currently Enrolled										
Completed Less Than 12 Years	19%	38%	−19	25%	12%	+13	15	15	18	13
Completed 12 Years	34	52	−18	21	7	+14	31	25	14	9
Completed 1–3 Years College	42	50	− 8	15	6	+ 9	30	22	13	7
College Graduate	58	51	+ 7	6	6	—	28	24	8	7
Not Currently Enrolled										
Completed Less Than 12 Years	44%	67%	−23	28%	18%	+10	34%	37%	20	16
Completed 12 Years	68	84	−16	17	10	+ 7	31	30	18	13
Completed 1–3 Years College	74	89	−15	15	7	+ 8	34	28	16	11
College Graduate	84	90	− 6	7	5	+ 2	27	24	12	10
Current Marital Status										
Single	40%	57%	−17	22%	12%	+10	27%	25%	18	13
Married-Spouse-Present	84	91	− 7	10	6	+ 4	26	26	13	10
Widowed, Divorced, Spouse-Absent	61	75	−14	17	13	+ 4	31	36	18	14
Place of Residence										
Metropolitan	45%	64%	−19	21%	10%	+11	28%	26%	18	12
Non-Metropolitan	45	62	−17	17	11	+ 6	24	24	16	13
Region of Current Residence										
Northeast	35%	59%	−24	26%	13%	+13	26%	25%	19	14
Midwest	39	64	−25	30	12	+18	32	27	20	12
South	48	66	−18	16	7	+ 9	25	23	16	11
West	52	65	−13	15	11	+ 4	27	29	17	12
Overall Mean	45%	64%	−19	20%	11%	+ 9	27%	25%	17	12

WOMEN

Age										
16–17 Years	14%	33%	−19%	29%	15%	+14%	12%	12%	16	11
18–19 Years	30	54	−24	27	10	+17	27	26	17	11
20–21 Years	43	62	−19	21	8	+13	30	25	18	10
22–24 Years	54	67	−13	17	6	+11	29	21	17	10
Educational Attainment and Current Enrollment										
Currently Enrolled										
Completed Less Than 12 Years	16%	36%	−20	25%	12%	+13	14%	12%	17	11
Completed 12 Years	34	49	−15	20	8	+12	29	22	13	8
Completed 1–3 Years College	41	55	−14	14	4	+10	27	21	13	7
College Graduate	56	63	−7	5	4	+1	24	17	11	7
Not Currently Enrolled										
Completed Less Than 12 Years	26%	36%	−10	36%	21%	−15	27%	25%	20	15
Completed 12 Years	51	69	−18	20	9	−11	30	25	18	11
Completed 1–3 Years College	66	80	−14	13	4	−9	32	22	15	9
College Graduate	81	90	−9	8	3	−5	33	28	13	8
Current Marital Status										
Single	34%	56%	−22	22%	8%	+14	24%	20%	17	10
Married-Spouse-Present	54	55	−1	15	9	+6	29	21	17	10
Widowed, Divorced, Spouse-Absent	43	55	−12	23	15	+8	30	27	16	13
Place of Residence										
Metropolitan	38%	58%	−20	21%	8%	+13	25%	21%	17	10
Non-Metropolitan	33	48	−15	24	11	+13	24	20	17	12
Region of Current Residence										
Northeast	32%	55%	−23	23%	9%	+14	23%	20%	18	12
Midwest	35	59	−24	26	9	+15	27	21	18	10
South	39	53	−14	20	9	+11	25	20	16	10
West	42	57	−15	17	9	+8	25	25	16	10
Overall Mean	37%	56%	−19	21%	9%	+12	25%	21%	17	11

SOURCE: U.S. Bureau of the Census, *Census of Population and Housing: 1980, Public Use Microdata Samples.*

FIGURE 8.6
Estimated Unemployment Rates for Persons Aged 20 and 21 by Race, Sex,
Educational Attainment, and Enrollment, 1980

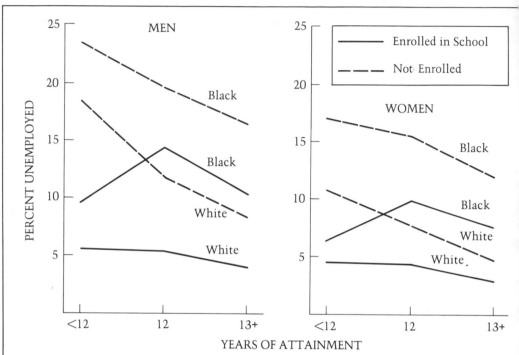

NOTE: Estimates are derived from the logit model which is summarized in Appendix Table 8.5. Estimat
are for persons 20–21 years old, living in North and *not* married-spouse-present.

SOURCE: U.S. Bureau of the Census, *Census of Population and Housing: 1980,* Public Use Microda
Samples.

racial differences in unemployment among the young are not due to racial differences in demographic characteristics. They do not come about
just because black youths drop out of school at early ages or are concentrated in regions of high unemployment. Joblessness is consistently
higher among blacks than among comparable whites, and young blacks
not only experience more unemployment than whites, but they are
unemployed for longer periods. Second, educational variables—
attainment and enrollment—have the strongest independent effects on
the unemployment rates of young people. But at all levels of attainment, blacks differ from whites.

Racial Differences in Employment: Why Do They Remain So Large?

Whether we consider labor force participation, unemployment, or the proportion at work, we find that racial differences are large and that many of them have increased since the civil rights revolution of the 1960s. Among the young, the employment picture has grown steadily worse for blacks.[35] In the period since World War II we might have expected a convergence of employment characteristics for four reasons: (1) racial differences in years of schooling were drastically reduced;[36] (2) white attitudes about the employment of blacks have changed;[37] (3) there have been important changes in federal laws, especially the adoption of Title VII of the Civil Rights Act of 1964; what were common hiring or promotion practices in the 1940s or 1950s are now criminal offenses;[38] (4) blacks have continued their struggle for equal opportunities.

The persistence of racial differences in employment is an intriguing question which has great consequences for race relations. Commentators and social scientists have never been reluctant to offer their views, but often they deal with only one facet of the issue or present speculations which are not bolstered by evidence.

Welfare as an Alternative to Work

The most popular and influential of the current views contends that the nation has created so many benefit and welfare programs that there is now an attractive alternative to work. As Martin Anderson observes: "Why should someone work forty hours a week, fifty weeks a year for, say, $6,000? People on welfare may be poor but they are not fools. Any rational calculation of the net returns from working by someone on welfare would discourage any but the most doggedly determined."[39]

George Gilder, whose ideas provided a rationale for the reductions in social spending advocated by the Reagan Administration, argues that welfare programs, although created with good intentions, have destroyed the work ethic: "Welfare now erodes work and family and thus

---

[35]Mare and Winship (1984); Cogan (1982).

[36]U.S. Bureau of the Census, *Sixteenth Census of the United States: 1940, Population*, vol. 4, pt. 1; table 18: Characteristics by Age; *Current Population Reports*, series P-20, no. 390 (August 1984), table 1.

[37]Schuman, Steeh, and Bobo (1985), table 3.1.

[38]Rodgers (1984).

[39]Anderson (1978), p. 47.

keeps poor people poor. Accompanying welfare is an ideology—sustaining a whole system of federal and state bureaucracies—that also operates to destroy their faith."[40]

Governmental benefits, it is assumed, reduce the incentive to work among both races, but their effects are proportionally greater among blacks because blacks typically have a lower earnings potential than whites. Gilder also contends that welfare agencies discriminate in favor of blacks because, to increase their own budgets and bureaucracies, they make special efforts to look for potential beneficiaries where they are most numerous—in minority communities.[41]

Charles Murray speculates that were the government to cut back on welfare benefits and eliminate other transfer programs, there would be an increase in employment because men and women would accept the low-paying jobs they now spurn.[42] Walter Williams believes that this would be particularly beneficial for black men since, even if they had to start at the bottom, they would develop the work habits they now lack.[43]

These authors also stress that the expansion of the welfare system in recent decades is a leading cause of the changing structure of black families. As Gilder states: "Most men make the sacrifices necessary to reach the higher reaches of the American economy chiefly to support their wives and families."[44] Supposedly, the welfare system permits men to abandon their wives or children, knowing that they will not starve. It also permits wives to terminate marriages hastily; indeed, in some circumstances, the income from transfer payments may exceed the financial contributions of the former husband. These views are often presented without corroborating evidence but, as this chapter indicates, there is a strong link between marital and employment status in the black community.

Moving beyond these speculations, Census Bureau surveys report that after income and type of family are taken into account, blacks are more likely than whites to obtain transfer benefits. Among households with incomes of under $7,500 in 1984, 26 percent of the white and 57 percent of the black received food stamps. In 1983, 47 percent of the black female-headed households received food stamps compared with 21 percent of the white female-headed households.[45] If Anderson, Gilder, and Murray are correct, the expansion of welfare since the mid-1960s may have reduced employment more among blacks than whites.

[40]Gilder (1981), p. 127.
[41]Gilder (1981), p. 133.
[42]Murray (1984), pp. 227–35.
[43]Williams (1982), p. 49.
[44]Gilder (1981), p. 151.
[45]U.S. Bureau of the Census, Current Population Reports, series P-60, no. 150 (November 1985), tables 1 and 5.

Some aspects of the effects of governmental transfer programs on labor supply have been studied in detail. For example, in the 1960s and 1970s impaired workers were increasingly able to obtain long-term disability benefits from the Social Security system.[46] Apparently, this is responsible for some of the decreasing employment of men over age 50. The declines in employment have been larger among older blacks than among whites, seemingly because disability benefits were an attractive alternative to earnings for people who would have earned little had they stayed on the job.

At the end of the 1960s, President Nixon proposed replacing welfare with a guaranteed annual income, often called the negative income tax.[47] Congress did not adopt this plan but authorized large-scale income maintenance experiments which were carried out in New Jersey, Denver, Seattle, Gary, and other locations. They were designed to assess the possible effects of a federal program which would guarantee all households or adults a modest income. One of the most important questions concerned the consequences of such a policy on hours of employment. If one guarantees individuals a modest income, will they reduce their time on the job and, if so, by how much? Samples of low-income families were selected for this test. Some adults were guaranteed modest incomes for a specified period of time regardless of their work effort or marital status, while others served as a control group and got no such guarantee, receiving only a small payment for participating in the study.

There has been much discussion of the way these experiments were conducted, the accuracy of the investigations, and the generalizability of findings.[48] The New Jersey study found that guaranteed incomes did little to reduce the work effort of husbands, but that they had an effect on the hours of employment of wives. In particular, the hours of employment for black husbands with a guaranteed income did not significantly differ from those of husbands in the control group, but black wives with guaranteed incomes reduced their work effort.[49]

The Seattle and Denver experiments found larger effects on labor supply. Both husbands and wives participating in those programs reduced their employment: about two hours per week for husbands and about five hours for wives.[50] Racial differences in the effects of a guaranteed income were not large for wives, but black husbands reduced their time at work by about twice as much as white husbands.[51]

These experiments provide limited support for the view that

[46]Parsons (1980).
[47]Marmor (1971); Moynihan (1973).
[48]Anderson (1978), chap. 5.
[49]Rees (1977), p. 23.
[50]Robins (1980), p. 63.
[51]Robins and West (1980), p. 98.

transfer benefits reduce labor supply. The Denver and Seattle experiments found that guaranteed incomes reduced work effort, but the New Jersey results were ambiguous. As a result, we still lack evidence which would permit us to conclude that the expansion of welfare programs in the last three decades is the primary cause of the decreasing labor force participation of black men.

The Impact of New Governmental Regulations

A second popular view contends that blacks have made little progress in employment because of new governmental laws and regulations. Minimum wage laws are frequently cited as governmental interference in the labor market which reduces employment opportunities for young workers or those who lack advanced training. Throughout the post–World War II era, the federal minimum has been raised—although often at a slower rate than inflation—and coverage has been extended to more industries. Many believe that individuals with few skills would now be at work if employers could hire them at a wage appropriate for their productivity.[52] Banfield argued that during the 1950s and 1960s increases in the minimum wage did not reduce the employment of skilled workers, but that rises in unemployment among the unskilled—such as black youth—matched increases in minimum wages.[53]

The quantitative investigations of economists provide only a little support for the view that minimum wages keep unemployment rates high among blacks.[54] To be sure, minimum wage laws depressed the employment of young blacks, but there is a secular trend toward lower proportions of black men at work which cannot be accounted for by these laws, and there is no reason to expect that the elimination of minimum wage laws would lead to a racial convergence in employment.

Credentialism, certification, and licensing are cited by several investigators as new developments, often supported by governmental bureaucracies, which reduce the employment of minorities.[55] Supposedly, many employers and state agencies demand that individuals have special and unrealistic credentials, such as two years of college to be a policeman, completion of a formal course in cosmetology to trim fingernails, an expensive state-issued medallion to drive a hack, or a federal license to take a delivery truck across a state line. While these

[52]Williams (1982), chap. 3.
[53]Banfield (1968), p. 197.
[54]Mincer (1976); Ragan (1977); Freeman (1979), p. 23.
[55]Williams (1982); Sowell (1981b), pp. 40 and 123; Gilder (1981), pp. 145–52.

regulations may seem racially neutral, they effectively discriminate against those who are unlikely to have the specialized training, the finances, and the political contacts needed to obtain a license or special certificate.

Finally, several scholars believe that the Civil Rights Act of 1964 had the latent consequence of making it more difficult for blacks to get on payrolls.[56] Presumably, employers in the past were free to hire blacks without regard to their credentials and then promote or fire them depending on their performance. Now black workers are a special protected class with prerogatives ordinarily denied their white colleagues. If they are quickly fired or moved up the ladder too slowly, blacks can appeal to the EEOC or even go into federal court. This may lead employers to carefully scrutinize the applications of blacks and hire only those who are unusually qualified. This would improve the career chances of blacks who are hired but make it difficult for other blacks to get their first good job. Thomas Sowell argues:

> As the government makes it more dangerous to fire, demote, or even fail to promote, members of minority groups, this tends to increase the demand for the more demonstrably able among them and reduce the demand for the average or below average, or those with too little experience to provide a reassuring track record.[57]

Changes in Economic and Demographic Structure

A third explanation for the persistence of racial differences in employment focuses on changes in the nation's industrial structure and its demographic characteristics. Prior to mechanization, southern farms were able to employ large numbers of blacks if the economy boomed, and even illiterate men could find jobs whenever there was a shortage of labor.[58] When southern agriculture became capital-intensive, employment opportunities were eliminated for a large segment of the black population. Thus, industrial change is seen as the cause of the trends we described. This argument is based, in essence, on the assumption that although the skill level of blacks has obviously improved because of greater educational attainment, employers' expectations have risen even more rapidly, and thus many blacks face a difficult time getting jobs.

The effects of this change have been analyzed most thoroughly for young workers. Cogan points out that between 1950 and 1970 the pro-

[56]Sowell (1983), p. 134; Williams (1982), chap. 2.
[57]Sowell (1983), p. 134.
[58]Fligstein (1981), chap. 7; Mandle (1978), Piven and Cloward (1971), chap. 7.

portion of white teenage males at work remained about 40 percent, but the proportion of blacks employed fell from 47 to 27 percent.[59] He shows that this racial differential was entirely attributable to a declining need for low-skill workers in southern agriculture.

A different structural explanation was offered by Kain.[60] He observed that after blacks migrated to the North, they were concentrated in central-city ghettos while jobs moved to suburban shopping centers and industrial parks. He predicted that employment opportunities would get worse for blacks as the chocolate city—vanilla suburb pattern of residential segregation persisted.

This explanation is highly plausible, but it assumes that jobs are deconcentrating more rapidly than the black population and that suburban workplaces are inaccessible to central-city blacks. Several investigations have challenged this explanation for the decline in the proportion of blacks at work and have discovered, at most, a weak relationship between the residential segregation of blacks and their economic status.[61] For young blacks, unemployment rates in the suburbs are just about as high as those in central cities, suggesting that a movement to the suburbs will not eliminate the racial difference in employment.[62] The black population in the suburbs grew rapidly throughout the 1970s, but the proportion of black men at work slowly declined, casting further doubt on this hypothesis.[63]

Several other demographic changes are more important in accounting for the increasing racial difference in the employment of young men. Mare and Winship demonstrate that between the 1960s and the 1980s, an increasing share of young black men attended school, while the fraction in the Armed Forces also rose relative to that of white men.[64] Those going to school or in military service were, apparently, more talented and employable than other black men. Thus, enrollment and enlistment culled the more able blacks from the labor force, leaving those with less skill and thereby swelling the unemployment rate of blacks.

The Increased Competition Hypothesis

James Tobin, a Nobel Prize economist, argues that blacks make economic gains only when there is a shortage of labor.[65] Basing these

[59]Cogan (1982), table 1.
[60]Kain (1968).
[61]Masters (1975), chap. 4; Jiobu and Marshall (1971).
[62]Westcott (1976), table 1.
[63]Long and DeAre (1981); Logan and Schneider (1984).
[64]Mare and Winship (1984).
[65]Tobin (1965), p. 879–87.

246

ideas on a model which sees the labor market as a queue, it is presumed that employers select workers of one racial or ethnic group rather than others because they feel some workers are more productive or, perhaps, because of their own prejudices.[66] Hiestand, for example, argues that the only way blacks made employment gains in this century was by filling vacancies created by a shortage of whites.[67] When white men shifted from blue collar jobs to offices, black men could move from farm labor to the manual jobs vacated by whites. Rees and Schultz observed, in their thorough study of the Chicago labor market in the 1960s, that employers selectively sought new workers with an awareness of their race or ethnicity, a factor thought to be related to productivity.[68] Steel mills, for example, advertised for labor in foreign-language newspapers and in those circulating in selected white neighborhoods, but not in the black-oriented *Chicago Defender.*

The table below shows average annual growth rates of the adult population, the civilian labor force, and civilian employment for five intervals between 1940 and 1984.[69]

	1940–50	1950–60	1960–70	1970–80	1980–84
Population 16 to 64 Years	+0.4%	+0.3%	+1.5%	+1.9%	+1.1%
Civilian Labor Force	+1.1	+1.1	+1.7	+2.6	+1.5
Civilian Employment	+1.6	+1.6	+1.8	+2.3	+1.4

Between 1940 and 1960 the adult population grew slowly compared with both the labor force and civilian employment. Indeed, employment increased twice as rapidly as population, implying a general shortage of labor. Since 1970 the situation has been quite different; the labor force has grown more rapidly than the adult population but it has also increased more rapidly than civilian employment. As a result, there was a labor surplus and unemployment rates have risen.

The rapid growth of the labor force in recent decades is attributable to greater labor force participation by white women and white teenagers, and to a substantial influx of Asians and Latin Americans, some of them immigrating specifically to find jobs in the United States. Do

[66]Lieberson (1980), pp. 296–304.
[67]Hiestand (1964), chap. 9.
[68]Rees and Schultz (1970), p. 206.
[69]U.S. Bureau of the Census, *Census of Population: 1980*, PC80-1-B1, table 45; U.S. Bureau of Labor Statistics, *Labor Force Statistics Derived from the Current Population Survey: A Databook*, vol. 1 (Bulletin 2096), tables A-3, A-4 and A-10; *Employment and Earnings*, vol. 32, no. 1 (January 1985), table 1.

these demographic changes help to explain the persistently high unemployment rates of blacks? If blacks make progress only when labor is in short supply, and employers place black men at the bottom of a queue, the answer appears to be yes. However, similar to other single-factor explanations, this one has deficiencies when examined more thoroughly. There is a high degree of job segregation by sex (see Chapter 9). If men and women either select or are slotted into different occupations, the increase in the female labor force will not restrict job opportunities for men unless employers shift occupations from male to female.

Migration into the United States has increased since the Cellar-Hart Act became effective in 1968, and legal migration now accounts for one quarter of total population growth.[70] In addition, there is a substantial flow of undocumented migrants from Mexico who work on a short-term basis.[71] Since the nineteenth century, there has been strident opposition to immigration on the basis of fears that newcomers will take jobs which rightfully belong to natives.

It is difficult to know how current immigration affects the labor force participation and unemployment rates of native-born blacks and whites. Because of differences in geographic locations and skills, blacks and the rapidly growing Asian and Hispanic population may compete in separate labor markets. The census of 1980 and recent Current Population Surveys report that Asians, on average, are more extensively educated than blacks or whites and work at more prestigious jobs because they are overrepresented in the professional specialty and technical occupations. Hispanics, on the other hand, report less educational attainment than blacks and work at less prestigious jobs since they are overrepresented among the ranks of construction laborers, machine operators, and other manual workers.[72] A recent investigation found no significant relationship between the 1970-80 growth rate of the Hispanic population in a labor market and its unemployment rate for blacks in 1980.[73] The authors of another new study of this topic summarize: "The fear that immigration brings 'cheap labor' to disemploy native-born workers is without foundation as a general proposition . . . immigration can lead to increased output, increased employment and an improved economic climate for the native population as a whole."[74] In brief, it is difficult to assert with any confidence that the black unemployment rates remain high either because of immigration or be-

[70]U.S. Bureau of the Census, *Current Population Reports*, series P-25, no. 990 (July 1986), table 1.
[71]Passel (1985).
[72]U.S. Bureau of the Census, *Census of Population: 1980*, PC80-1-C1, tables 123, 125, 133, and 135.
[73]Espenshade (1985).
[74]Cafferty et al. (1983), p. 18.

cause of the rise in labor force participation among white women and white youth.

Social-Psychological Explanations

A different explanation for the low employment rates of black men is based on the findings of urban ethnographers and directs attention to the attitudes of these workers and the expectations of potential employers.[75] Black youths may often aspire to careers for which they are not qualified, at least in the view of those who might hire them. The youngsters may feel that a high school diploma or some training at a community college allows them to escape manual labor and makes them good candidates for supervisory jobs. Employers, on the other hand, disparage the abilities and training of young blacks. They distrust their educational credentials and believe that many black high school graduates are unable to write coherent paragraphs or solve simple problems in algebra. They expect that they will work irregularly, often be absent on Mondays or Fridays, come to work with "ghetto blasters" or "dew-rags," and use drugs or liquor while on the job. In addition, they are seen as crime-prone. Both Anderson and Liebow, who carried out ethnographic studies in Philadelphia and Washington, claim that employers adjust the wage rates of black men downward to compensate for the thievery which, they believe, inevitably occurs when they are on the payroll.[76] Employers certainly do not see those workers as candidates for advancement within the firms. If the urban ethnographers are correct, the elimination of the minimum wage will do little to increase the employment of young blacks since, in the eyes of management, few of them have the skills or industriousness needed to be productive. Indeed, they are employees of last resort.

Although they desire good jobs, these young black men may anticipate no upward mobility during their working lives. They observe their fathers or relatives who are in their 40s and 50s and are now filling the same mundane jobs available to teenagers. They look for attractive alternatives. They realize that a few young blacks become extraordinarily rich participating in professional entertainment or sports or by marketing women and drugs, and some enter these lines of work, but the success rate is not high. Others realize that they can survive by sporadically working and then making use of unemployment compensation, welfare benefits, and food stamps. As an outcome, the proportion of black men at work remains low while jobless rates stay high.

[75]Anderson (1978); Hannerz (1969); Liebow (1967).
[76]Anderson (1978), p. 77; Liebow (1967), p. 39.

Long argues that these attitudes emerged primarily among black men who were recently raised in northern cities.[77] Supposedly, four or five decades ago black men who grew up in the rural South had limited aspirations and saw no alternative to backbreaking jobs in the cotton, rice, or tobacco fields. Growing up in the urban North with constant exposure to both a thriving underground economy and extensive welfare payments is seen as much more of a liability than being raised in the rural South.

Summary and Conclusions

When we consider trends in the major indicators of the economic status of blacks, we find that employment trends differ greatly from those in educational attainment, occupational achievement, and earnings. There is no evidence of any racial convergence on most indicators of employment during the post–World War II era. In some important manner, the increasing educational attainment of blacks, the more liberal racial attitudes of whites, and civil rights legislation have failed to narrow the significant gaps in employment and unemployment. Indeed, the evidence in this chapter convincingly demonstrates that racial disparities in employment are growing larger among young women and among men at all ages except the oldest. If present trends continue, blacks will fall further behind whites in terms of employment.

Since the actual employment trends differ so drastically from the firmly held expectations of scholars and hopes of civil rights advocates, many explanations have been offered focusing on such things as the availability of federal transfer payments, basic macro-economic trends, or the emergence of new attitudes toward work on the part of black men. We reviewed and, to a limited degree, evaluated these conflicting explanations for the persistent employment problems of blacks. Each may contribute to understanding, but none provides a succinct or convincing explanation for the puzzling trends described in this chapter.

[77]Long (1974), p. 55.

APPENDIX TABLE 8.1

Ratio of a Group's Unemployment Rate
to That of White Men Aged 25–54 and Time Trend, 1950–1985

Age	Ratio of Unemployment Rate to That of White Men 25–54 Years (1)	Time Trend in Relative Unemployment Rates (2)	Correlation Coefficient Squared (3)
NONWHITE MEN			
16–24 Years	3.59	+.49	.96
25–54 Years	2.48	−.06	.93
55+ Years	2.10	−.05	.65
NONWHITE WOMEN			
16–24 Years	3.56	+.58	.79
25–54 Years	1.86	+.06	.86
55+ Years	1.11	.00*	.60
WHITE MEN			
16–25 Years	2.62	+.10	.93
55+ Years	1.08	−.03*	.57
WHITE WOMEN			
16–24 Years	2.00	+.17	.80
25–54 Years	1.12	+.03*	.83
55+ Years	.83	+.02*	.47

Coefficient not statistically significant at the .05 level.

NOTES: Coefficients in this table result from the regression of a group's unemployment rate in a given year on the unemployment rate of white men aged 25 to 54 in that year and the number of years elapsed since 1950. The regression coefficient associated with the unemployment rate of white men aged 25 to 54 is identified as the ratio of unemployment rates while the regression coefficient associated with the number of years since 1950 is identified as the time trend. Age-standardized data were used.

SOURCES: U.S. Bureau of Labor Statistics, *Handbook of Labor Statistics* (December 1983), bulletin 2175, tables 4 and 26; *Employment and Earnings*, vol. 32, no. 1 (January 1985), table 3; vol. 33, no. 1 (January 1986), table 3; U.S. Bureau of the Census, *Historical Statistics of the United States: Colonial Times to 1970*, series F-3; *Statistical Abstract of the United States: 1984*, table 735.

APPENDIX TABLE 8.2

Average Annual Change in the Proportion of the Population Out of the Labor Force, 1950–1985

Age	Nonwhite		White	
	Average Annual Change	Correlation Coefficient Squared	Average Annual Change	Correlation Coefficient Squared
MEN				
16–24 Years	+.64	.96	−.01	.00
25–54 Years	+.29	.93	+.09	.79
55+ Years	+.83	.98	+.74	.99
WOMEN				
16–24 Years	−.32	.83	−.78	.87
25–54 Years	−.46	.92	−.98	.95
55+ Years	+0.4	.07	−.17	.50

NOTE: These coefficients result from the regression of the average annual proportion of a group out of the labor force on the number of years since 1950. Age-standardized data were used.

SOURCES: U.S. Bureau of Labor Statistics, *Handbook of Labor Statistics* (December 1985) bulletin 2175, table 4; *Handbook of Labor Statistics* (December 1980), table 4; *Employment and Earnings*, vol. 31, no. 1 (January 1984), table 3; vol. 32, no. 1 (January 1985), table 3; vol. 32, no. 1 (January 1985), table 3; vol. 33, no. 1 (January 1986), table 3.

APPENDIX TABLE 8.3

Analysis of Determinants of Unemployment in April 1980 for Black and White Men and Women Aged 25–54

	Men		Women	
	White	Black	White	Black
Intercept	−3.0656	−2.7846	−3.3744	−3.0658
EDUCATIONAL ATTAINMENT				
Less Than 12 Years	+.6026	+.4628	+.4482	+.4430
12 Years	+.2816	+.1772*	−.0484*	+.1596*
13 to 15 Years	−.0854*	−.0030*	−.1260*	−.0174*
16 Years or More	−.7988	−.6370	−.3706*	−.5852*
AGE				
25–34	+.2702	+.2462	+.1542*	+.4572
35–44	−.0901*	+.0120*	+.0558*	−.0362*
45–54	−.1801	−.2582	−.2100*	−.4121

APPENDIX TABLE 8.3 *(continued)*

	Men		Women	
	White	Black	White	Black
REGION OF RESIDENCE				
North	+.1252*	+.3032	+.0738*	+.1356*
South	−.2860	−.2980	−.1560*	−.1352*
West	+.1608*	−.0052*	+.0822*	−.0004*
MARITAL STATUS				
Married-Spouse-Present	−.3464	−.3468	−.2766	−.1726*
Other Marital Statuses	+.3464	+.3468	+.2766	+.1726*
WORK LIMITATION OR PRESENCE OF CHILD UNDER AGE 6				
Yes	+.1150*	−.0728*	−.0408*	−.0688*
No	−.1150*	+.0728*	+.0408*	+.0688*
EDUCATION BY MARITAL STATUS INTERACTION				
<12, Married-Spouse-Present	+.2200	+.1372		
Other Marital Statuses	−.2200	−.1372		
12, Married-Spouse-Present	−.0338	+.0392		
Other Marital Statuses	+.0338	−.0392		
13–15, Married-Spouse-Present	−.0092	+.0418		
Other Marital Statuses	+.0092	−.0418		
16+, Married-Spouse-Present	−.1772	−.2182		
Other Marital Statuses	+.1772	+.2182		
L.R. χ^2	148.89	185.37	143.01	140.49
Degrees of Freedom	131	131	134	134
Probability	.14	<.01	.28	.33

*Coefficient is not significantly different from zero at .05 level.

NOTES: These are logit models which use the log of the odds of unemployment as their dependent variable. The proportion unemployed in any group may be estimated by summing the intercept and the effects of parameters associated with that group. For example, to estimate unemployment among black women aged 35 to 44, who lived in the West, who finished 12 years of schooling, and who had neither a husband nor a child under age 6, one would sum: −3.0658, −.0362, −.0004, +.1596, +.1726 and +.0688. The sum, −2.7014, is the log of the odds of unemployment for these women. Converting to a proportion leads to an estimate of 6.29 percent as the unemployment rate for this group.

"North" includes Midwest and North East regions.

The "work limitation" variable was used for men; the "presence of child under age 6" variable for women.

The use of "education by marital status interaction" term significantly improved the fit of equations for men but not for women.

SOURCE: U.S. Bureau of the Census, *Census of Population and Housing: 1980*, Public Use Microdata Samples.

APPENDIX TABLE 8.4

Regression of Hours of Employment in 1979 on Years of Education, Age, Age-Squared, Region, Marital Status, Work Limitation, and Presence of Child Under Age 6 for Blacks and Whites Who Were Employed in 1979

	Whites		Blacks	
	Means of Variables	Coefficients From Regression Equation	Means of Variables	Coefficients From Regression Equation
MEN				
Hours Worked in 1979	2,125		1,854	
Years of Education	13.1	+ 17.0	11.7	+ 19.1
Age	37.6	+ 48.8	36.9	+ 32.7
Age Squared	1,488	− 0.6	1,434	− 0.3
Region	.32	+ 22.3	.51	+ 41.0
Marital Status	.77	+228.8	.61	+182.0
Work Limitation	.05	− 326.0	.06	− 374.2
Intercept		749		810
R²		.05		.05
Sample Size		33,963		34,553
WOMEN				
Hours Worked in 1979	1,542		1,612	
Years of Education	12.9	+ 3.9	12.1	+ 17.3
Age	37.2	− 27.9	36.7	+ 35.8
Age Squared	1,461	+ 0.4	1,417	− 0.4
Region	.31	+116.5	.52	− 2.3*
Marital Status	.69	− 246.8	.48	− 14.4*
Work Limitation	.04	− 355.3	.05	− 432.7
Presence of Child<6	.18	− 325.4	.22	− 138.8
Intercept		2,208		760
R²		.07		.05
Sample Size		25,437		34,669

*Regression coefficient *is not* significantly different from zero at the .05 level.

NOTES: Region: Residence in the South is coded as one.
  Martial Status: Individuals who are married-spouse-present are coded as one.
  Work Limitation: Individuals reporting a physical, mental or other health condition which lasted for six months or more and which limited the kind or amount of work they could do are coded a one.

SOURCE: U.S. Bureau of the Census, *Census of Population and Housing: 1980*, Public Use Microdat Samples.

APPENDIX TABLE 8.5

Analysis of Determinants of Unemployment Among
Men and Women Aged 16–24, 1980

	Men		Women	
	White	Black	White	Black
Intercept	−2.8298	−2.2720	−2.9934	−2.3510
AGE				
16–19	+.0788*	−.1086*	+.2208*	+.0066*
20–21	−.0324*	+.0650*	+.0406*	+.0638*
22–24	−.0462*	+.0436*	−.2614*	−.0704*
REGION				
North	+.1816*	+.3230*	+.0076*	+.0850*
South	−.3688*	−.1990*	−.0756*	−.0254*
West	+.1872*	−.1240*	+.0680*	−.0596*
MARITAL STATUS				
Married-Spouse-Present	−.2004*	−.1444*	−.0708*	−.0806*
Other Marital Statuses	+.2004*	+.1444*	+.0708*	+.0806*
EDUCATIONAL ATTAINMENT AND ENROLLMENT				
<12 Years, Not Enrolled	+.9862	+.5580	+.7636	+.5346
<12 Years, Enrolled	−.3940*	−.4942*	−.1816*	−.5826*
12 Years, Not Enrolled	+.4646	+.3252*	+.3932	+.4068
12 Years, Enrolled	−.4254	−.0612*	−.2222*	−.0886*
13+ Years, Not Enrolled	+.0588	+.1044*	−.1216*	+.1100*
13+ Years, Enrolled	−.6902	−.4322*	−.6314*	−.3802*
L.R. χ^2	106.87	147.57	113.95	129.76
Degrees of Freedom	97	97	97	97
Probability	.23	<.01	.12	.01

*Regression coefficient is not significiantly different from zero at the .05 level.

NOTE: This table presents the parameters from logit models which take the log of the odds of unemployment as their dependent variable. The unemployment rate for any group of persons may be estimated by summing the intercept for that group and the appropriate effects parameters.

SOURCE: U.S. Bureau of the Census, *Census of Population and Housing: 1980*, Public Use Microdata Samples.

9

RACIAL DIFFERENCES
IN OCCUPATIONAL ACHIEVEMENT

G IVEN that the unemployment situation for blacks was as bleak in the 1980s as it had been in the 1950s, we might expect that blacks made no progress in moving into those prestigious jobs traditionally reserved for whites. This is not the case. During the last several decades barriers have been removed and racial differences in the occupations of employed men have substantially decreased, and among women, a continuation of recent trends implies an elimination of racial differences in occupations.

The census of 1940 was the first to include a question about what a person did while he or she was at work. Thirty-six percent of the employed white men and only 6 percent of the employed black men held white collar jobs; 37 percent of the black men and 9 percent of the white men worked as domestic servants or laborers.[1] Quite clearly, blacks were concentrated at the bottom of the occupational ladder.

When we examine the findings of historians who describe the labor movement and those who analyze the urbanization of blacks, we learn that racial differences in occupation did not come about solely because blacks had limited educations or because they lived in the rural South. Rather, there were widely held social values which led to the color coding of jobs.

[1]U.S. Bureau of the Census, *Sixteenth Census of the United States: 1940, Population*, vol. 3, pt. 1, table 62.

Long before Emancipation, white laborers in both the South and the North recognized that their wages and standards of living would be lowered if blacks left the rural South in great numbers and competed for their jobs.[2] There was no large-scale migration of blacks to northern cities after the Civil War, but the investigations of historians reveal that blacks were driven from some of the crafts and trade niches they had once occupied in these cities. Pleck reports that racial prejudice forced blacks out of laboring, carpentry, and wainwrighting in Boston, while in Philadelphia blacks were driven from the ranks of artisans.[3] In Cleveland blacks just about disappeared from the skilled trades between 1870 and 1915.[4] In 1860 blacks were well represented among skilled tradesmen in New Orleans, but by the turn of the century their numbers were greatly reduced.[5] Although there are some notable exceptions, union leaders and members often viewed blacks as threatening as management, so they excluded blacks from their ranks. Some unions, particularly the rail operating brotherhoods, were especially effective in eliminating black employees, and thereby reserved for whites the occupational slots which were formerly filled by blacks.[6] Although employers occasionally used blacks as strikebreakers to thwart the power of unions, it seldom led to substantial occupational gains for blacks.[7]

The "scientific" evidence of the Social Darwinists stressed the innate inferiority of blacks, and so there emerged an image of a Sambo-type black man who was ill-prepared for the rigors of serious work in the modern shop or office, a stereotype which justified reserving the better jobs for white men. Lieberson reports that northern employers often felt that blacks worked too slowly to maintain production and thus European immigrants were almost universally preferred.[8]

DuBois thoroughly analyzed the employment opportunities available to blacks in Philadelphia at the turn of the century.

No matter how well-trained a Negro may be, or how fitted for work of any kind, he cannot in the ordinary course of competition hope to be much more than a menial servant.
He cannot get clerical or supervisory work to do save in exceptional cases.
He cannot teach save in a few of the remaining Negro schools.

[2]Curry (1981), chap. 6.
[3]Pleck (1979), pp. 144–47.
[4]Kusmer (1976), p. 75.
[5]Woodward (1957), p. 361; Myrdal (1944), pp. 280–93; Blassingame (1973), pp. 59-66.
[6]Harris (1981), chap. 2; Foner (1981).
[7]Wilson (1978), p. 72; Spear (1967); Spero and Harris (1968), chap. 7; Kusmer (1976), p. 69; Dickerson (1986), pp. 8–10.
[8]Lieberson (1980), p. 348; see also Chicago Commission on Race Relations (1922), chap. 8.

He cannot become a mechanic except for small transient jobs, and cannot join a trades union.
A Negro woman has but three careers open to her in this city: domestic service, sewing or married life.[9]

DuBois gives dozens of specific examples of highly trained blacks who could find only menial work, such as a man who graduated from the University of Pennsylvania with a degree in mechanical engineering and was employed as a waiter. In his view, occupational segregation came about solely because of color prejudice. Whites, even those without hostile feelings for blacks, accepted the view that intimate social contact on the job with a lower race was undesirable and would ultimately weaken both races.

During World Wars I and II the severe shortage of labor forced employers to hire more blacks and some blacks were advanced into better jobs, but these were periods of urban racial violence, and those who describe it claim that it was fostered by white beliefs that blacks were taking jobs and moving into neighborhoods which rightfully belonged to whites.[10] As Bonacich observed, a split labor market developed in the United States between the Civil War and World War I. Both management and workers agreed that blacks should be excluded from the prestigious and high-paying jobs. A variety of customs and institutional practices—often written into law or labor contracts—ensured that white men would be atop the occupational distribution.[11]

The occupational segregation that emerged was not easily changed, and the exceptionally talented or highly trained black could not hope to compete with whites on the basis of merit. Blacks who obtained a professional education at Howard, Maherry, or Atlanta University could only serve a black clientele. A small number of blacks secured federal civil service appointments, but southern congressmen challenged them, and, especially during the Wilson Administration, efforts were made to limit the number of blacks and ensure that they never supervised whites.[12] Civil rights organizations and liberal whites could not alter this pervasive system of occupational stratification.

[9]DuBois (1899), p. 323. For additional information about the denial of employment opportunities to blacks in Philadelphia in the late nineteenth century, see Lane (1986).
[10]Chicago Commission on Race Relations (1922); Shogan and Craig (1964); Rudwick (1964); Trotter (1984).
[11]Bonacich (1972, 1976).
[12]Williamson (1984), chap. 12; Green (1967), pp. 170–74. Jim Crow policies in federal employment affected the Census Bureau. In 1918 it published *Negro Population: 1790–1915*. The introductory page reports that the tabulations were made by Negro clerks who were supervised by Negroes (p. 14). The 1935 volume, *Negroes in the United States: 1920–32*, indicates that the tabulations were made by a corps of Negro clerks. U.S. Bureau of the Census (1935), p. iii.

One of the most clearly documented exclusions of black employees was that in professional baseball.[13] In the late nineteenth century a few blacks played for major league teams, but white players and team owners objected to their participation, and by 1900 blacks were eliminated from professional baseball. A few dark-hued Caribbeans were accepted, but American blacks were banned. Civil rights organizations occasionally questioned why blacks were excluded from the national game. Some team owners contended that blacks lacked the stamina, the ability, and the dedication to play baseball with whites, but this was difficult to maintain when squads of all-star black players regularly defeated similar teams of whites in exhibition games. Others claimed that their white players—especially those from the South—would quit if forced to play alongside blacks. Some team owners argued—contrary to factual evidence—that their clientele would not pay to see blacks compete with whites or that fights on the field between black and white players would trigger racial riots.

After World War II civil rights organizations put increased pressure on the sport's leadership, and in 1946, for the first time in fifty years, a black was permitted to earn his living by playing baseball with whites. The racial integration of the major leagues, which occurred the next year, has been heralded as one of the important symbolic victories for blacks. Nevertheless, the first black players were harassed in numerous ways, and many owners strongly resisted hiring blacks. As late as 1960 some major league teams had never employed black players.

The Beginnings of Occupational Change

The occupational distribution of blacks has been upgraded more than that of whites in recent decades. Four factors contribute to this change. First, passage of the Wagner Act in 1935 greatly strengthened the position of unions in their negotiations with managements, and industrial unions became effective bargaining agents for workers in steel, rubber, and automobile plants. These unions foresaw the possible use of blacks as strikebreakers and, much more than the traditional crafts unions, accepted black members. To be sure, accounts of labor history report that black workers were often treated shabbily by their white colleagues, and wildcat strikes were frequently called by whites to protest the hiring or promotion of blacks, but the emergence of industrial unions improved occupational opportunities for blacks.[14]

[13]Tygiel (1983); Petersen (1984).
[14]Meier and Rudwick (1979).

The expansion of employment during World War II and the continuation of high rates of economic growth that followed meant that there was a higher demand for labor, and blacks who were displaced by the mechanization of southern agriculture could often move into higher-paying industrial or service jobs. The migration of blacks to northern cities and their rising incomes also created opportunities for more black entrepreneurs and professionals.

Another factor in the improved occupational distribution of blacks is that a higher proportion were qualified for better jobs. Throughout the post–World War II period, there was a tremendous expansion of educational facilities. A secondary school education became typical, and many young people had the opportunity to enroll for even more education. The 1954 *Brown* decision benefited blacks particularly. Southern states were certainly not willing to racially integrate their schools in the 1950s or early 1960s, but they recognized numerous violations of the separate-but-equal principle and sought to rectify this by spending more on black schools. By the early 1980s, blacks aged 25 to 29 averaged 12.6 years of schooling, just three tenths of a year less than similar whites, implying that the racial difference in years spent in school had just about disappeared.[15]

There have also been changes in the attitudes of whites about the employment of blacks. Four decades ago, the majority of whites felt that it was acceptable to restrict the employment of blacks. For example, a national sample of whites in 1944 was asked whether Negroes should have as good a chance as whites to get any kind of job. Only 45 percent said that blacks should have an equal chance. When the question was last asked in 1972, 97 percent of the whites endorsed the ideal of equal employment opportunities for blacks.[16]

Finally, there have been changes in civil rights laws, undoubtedly reflecting the growing consensus that racial discrimination is un-American. In 1946 Congress was unwilling to extend the life of a toothless Fair Employment Practices Committee. Eighteen years later they enacted an encompassing law which simultaneously banned all racial discrimination in the labor market and established powerful enforcement mechanisms. The Equal Employment Opportunity Commission (EEOC) was created to ensure nondiscrimination by private employers who have 25 or more workers. This agency hears complaints from workers, requires employers to report the number of workers in each occupational category classified by sex and race, and, after the 1972 re-

---

[15]U.S. Bureau of the Census, *Current Population Reports*, series P-20, no. 390 (August 1984), table 1.

[16]Schuman, Steeh, and Bobo (1985), table 3.1.

visions to the law, obtained the power to initiate suits. Litigation involving equal employment opportunities is now well known, especially that concerning American Telephone and Telegraph, which eventually led to the award of $80 million in back pay and wage adjustments.[17] Suits accusing fire and police departments of discrimination in 35 major cities resulted in specific hiring quotas for women and minorities.[18]

Federal courts upheld the powers of EEOC and issued numerous rulings which should lead to greater occupational achievement by blacks, including the *Griggs* decision, which eliminated unfair employment tests;[19] The *Weber* decision, which upheld voluntary affirmative action programs;[20] and the *United Paperworkers* decision, which challenged seniority systems if they had a discriminatory effect.[21] The courts also strengthened the negotiating powers of EEOC by accepting—at least in some circumstances—statistical disparities in employment as evidence of discrimination.[22]

Approximately one third of the nation's employees work for governmental agencies or for firms with governmental contracts, and in 1965 the Office of Federal Contract Compliance Programs was established to guarantee equal employment opportunities in this section of the economy. Indeed, they have established employment guidelines for contractors and may terminate federal funding if the underrepresentation of minorities and women on an employer's payroll reveals discrimination.[23]

The situation for black workers is very different now from what it was decades ago. If they believe there is racial discrimination in hiring, in promotions, or in setting wage rates, there are a variety of governmental agencies to contact and an array of supportive federal court decisions.

The Challenges of Assessing Occupational Change

Blacks have made occupational progress, but specifying exactly how much is a challenge. The analysis should be made with a long time series showing the number of black and white men and women working

[17]Wallace (1976).

[18]Rodgers (1984); Burstein (1979).

[19]*Griggs* v. *Duke Power Company*, 401 U.S. 424 (1971).

[20]*United Steel Workers* v. *Weber*, 443 U.S. 193 (1979).

[21]*United States* v. *Local 189, United Paperworkers*, 397 U.S. 919 (1970). The retroactive creation of seniority for minority workers who experience discrimination has been questioned by a ruling involving the Memphis fire department. *Firefighters Local Union No. 1784* v. *Stotts*, 467 U.S. 561 (1984); *Wygant* v. *Jackson Board of Education*, 476 U.S., 90 L.Ed.2d 260 (1986); *Local 93, International Association of Firefighters, AFL-CIO C.L.C.V. City of Cleveland*, 478 U.S., 92 L.Ed.2d 405 (1986); *Local 28 of Sheet Metal Workers' International Association* v. *E.E.O.C.*, 478 U.S., 92 L.Ed.2d 344 (1986).

[22]Rodgers (1984), p. 97.

[23]Rodgers (1984), pp. 94–95.

at specific jobs, but it is difficult to do this for four reasons. First, there are literally thousands of different jobs, and it is impossible to present occupational trends in such elaborate detail. It is necessary to group jobs into categories or use average prestige scores, even though this leads to a loss of information.

Second, there is no agreement about which jobs are the better ones. Most would agree that corporate executives have more prestigious jobs than numbers runners, but does the manager of a bank have a better or worse job than a junior high school principal? Is it more prestigious to be a carpenter or an electrician? These questions led many investigators to develop socioeconomic index scores for occupations. Otis Dudley Duncan was among the first to do this,[24] using data from a survey in the 1940s which asked a national sample of the population to evaluate the status of 45 specific jobs. He then related this evaluation to the earnings and educational attainment of men working at those occupations in 1950. Using the relationship among these variables, he then estimated a socioeconomic score for every occupation. In subsequent decades, many other investigators have tried to adapt and improve this system for developing occupational status scores.[25] Although there is a continuing debate about whether it is legitimate to estimate occupational status in this way, the use of such scores allows us to readily summarize occupational status.[26]

A third difficulty results from changes over time in the way occupations have been classified. Although questions about work activity have appeared on the census since the nineteenth century,[27] the Census Bureau did not adopt a modern classification scheme until 1940. Between 1940 and 1970 the job of each worker was classified into one of approximately 475 detailed occupational codes. These detailed occupations were then grouped into approximately 160 intermediate level categories and then into 13 broad categories.[28] The occupational codes were changed moderately in this interval, but it is possible to assemble roughly comparable data for 11 broad occupational groups from the censuses of 1940–70. Similar occupational data are available by race from the Current Population Survey for 1958–82.

A major change in occupational classification occurred in 1980.[29]

[24]Duncan (1961).
[25]Stevens and Featherman (1981); Nam and Powers (1968); Featherman and Stevens (1982); Stevens and Cho (1985); Hauser and Featherman (1977), app. B.
[26]Featherman and Hauser (1976).
[27]Conk (1978).
[28]U.S. Bureau of the Census, *Census of Population: 1950*, P-C1, pp. xix-xxiv; *Census of Population: 1960*, PC(1)-1D, pp. xxvii-xxxiv; *Census of Population: 1970*, PC(1)-D1, app. 17.
[29]Stevens and Cho (1985); Rytina and Bianchi (1984); Bianchi and Rytina (1984).

More than 500 detailed occupational titles were grouped into 13 broad categories, but because there are many new and basically different occupational titles, the classification system is not comparable to that used previously.[30] In addition, old titles cannot easily be matched to new ones. For example, a woman who was classified as a nurse in 1970 might be any one of six occupational categories in 1980.[31] If we wish to study the changing occupational status of blacks, we need to examine one time series based on the previous occupational classification system and another based on the new categories.

Fourth, occupational tabulations from the Current Population Survey generally refer to whites and nonwhites rather than blacks and whites.[32] In this chapter the term, "nonwhite" refers to all people who are not classified as white, while "black" specifically refers to people identified as black or Negro.

Occupational Change: 1940 to 1980

Changes in the occupations of employed nonwhites may be analyzed using data for 11 broad occupational categories shown in Table 9.1. Information for 1940 and 1950 came from decennial censuses and refer to blacks and whites, while that for 1960 to 1980 are from the Current Population Survey and refer to nonwhites.

Looking at employment data for black men in 1940, we find that about two fifths of those who held jobs worked on farms, while another one third were machine operators or laborers in factories. The biggest change in the following decades was the exodus from agriculture. During World War II, black men moved into the blue collar occupations, becoming laborers, operatives, or crafts workers, while in the 1960s and 1970s there was an increasing representation of blacks in the higher occupational categories: professional, managerial, and clerical positions. Of course, there was an upgrading of the jobs of whites, especially before 1960, as they shifted from the factories to offices, but occupational change among whites has been much less than among blacks, partly because whites were never so highly concentrated in farming and laboring jobs.

Employment changes among black women can best be understood by examining the decline in domestic service. On the eve of World War II,

[30]U.S. Bureau of the Census, *Census of Population: 1980*, PC80-1-D1-A app. B.
[31]Bianchi and Rytina (1984), p. 10.
[32]In the decennial census, persons who placed themselves in the "other races" category and subsequently indicated they were Hispanic were classified as nonwhite by race. In the Current Population Survey, most such individuals were classified as white by race. See chap. 8, footnote 13.

TABLE 9.1

Occupational Distributions by Race and Sex, 1940–1980

	Black or Nonwhite					White				
	1940	1950	1960	1970	1980	1940	1950	1960	1970	1980
MEN										
Professional	1.8%	2.2%	3.8%	7.8%	10.7%	5.9%	7.9%	11.4%	14.6%	16.1%
Proprietors, Managers, Officials	1.3	2.0	3.0	4.7	6.7	10.7	11.7	14.5	15.3	15.3
Clerical	1.1	3.1	5.8	7.4	8.4	7.1	6.8	7.2	7.1	6.2
Sales	1.0	1.1	1.2	1.8	2.7	6.8	7.0	6.5	6.1	6.4
Craftsmen	4.4	7.8	9.5	13.8	17.1	15.7	20.0	19.8	20.7	21.4
Operatives	12.6	21.4	24.3	28.3	23.4	19.0	20.3	19.0	18.6	16.1
Domestic Service	2.9	1.0	0.4	0.3	0.2	0.2	0.1	0.1	0.1	<0.1
Other Service	12.4	13.5	14.9	12.8	15.8	5.9	5.2	5.6	6.0	7.9
Farmers, Farm Managers	21.2	13.5	4.8	1.7	0.6	14.1	10.1	6.1	3.6	2.6
Farm Laborers	19.9	10.4	9.5	3.9	2.4	7.0	4.2	3.3	1.7	1.5
Nonfarm Laborers	21.4	24.0	22.8	17.5	12.0	7.6	6.7	6.5	6.2	6.5
Total	100.0%	100.0%	100.0%	100.0%	100.0%	100.0%	100.0%	100.0%	100.0%	100.0%
Mean Socioeconomic Index	16	18	21	27	31	30	33	36	39	40

WOMEN

Professional	4.3%	5.7%	6.0%	10.8%	14.8%	14.9%	13.5%	13.1%	15.0%	17.0%
Proprietors, Managers, Officials	0.7	1.4	1.8	1.9	3.7	4.4	4.8	5.4	4.8	7.4
Clerical	0.9	4.0	9.3	20.9	29.3	25.0	31.1	32.9	36.3	36.0
Sales	0.5	1.4	1.5	2.5	3.1	8.2	9.6	8.5	7.7	7.3
Craftsmen	0.1	0.7	0.5	0.8	1.4	1.1	1.6	1.1	1.2	1.9
Operatives	6.2	14.9	14.1	17.6	14.9	20.5	20.3	15.1	14.1	10.1
Domestic Service	60.0	42.0	35.2	17.5	6.5	11.1	4.1	6.1	3.4	1.9
Other Service	10.5	19.1	21.4	25.7	24.3	11.6	11.5	13.7	15.3	16.0
Farmers, Farm Managers	3.0	1.7	0.6	0.1	0.1	1.1	0.6	0.5	0.3	0.4
Farm Laborers	13.0	7.6	9.0	1.5	0.5	1.2	2.2	3.3	1.5	0.8
Nonfarm Laborers	0.8	1.5	0.6	0.7	1.4	0.9	0.7	0.3	0.4	1.2
Total	100.0%	100.0%	100.0%	100.0%	100.0%	100.0%	100.0%	100.0%	100.0%	100.0%
Mean Socioeconomic Index	13	18	21	29	36	36	39	39	40	43

INDEXES OF OCCUPATIONAL DISSIMILARITY BY RACE

	Men	Women
1940	43	63
1950	37	53
1960	37	43
1970	31	28
1980	24	18

NOTE: Data for 1940 and 1950 are from decennial censuses. Data for other years are from the Current Population Survey. They are estimates of annual averages rather than specific months. Data for 1940 and 1950 refer to blacks; for the other years, to nonwhites. Data for 1940 and 1950 refer to the employed population aged 14 and over; for other years, employed population aged 16 and over.

SOURCES: U.S. Bureau of the Census, *Sixteenth Census of the United States: 1940, Population*, vol. 3, pt. 1, table 63; *Census of Population: 1950*, P-C1, table 126; P-E, no. 3B, table 9; U.S. Bureau of Labor Statistics, *Handbook of Labor Statistics 1975-Reference Edition*. Bulletin 1865 (1975), table 19; *Employment and Earnings*, vol. 28, no. 1 (January 1981), table 22.

6 out of 10 employed black women were cleaning floors, cooking meals, or washing windows in other people's homes. During the war years openings developed for black women to take other service jobs or to run machines in factories. During the 1950s they moved into clerical positions and in the 1960s into the professional ranks. Once again, we observe that occupational change in the last four decades was much greater among nonwhites than whites, largely because the proportion in domestic service was never great for white women.

To summarize the occupational status of a group, socioeconomic index scores developed by Duncan from 1950 census data are presented in Table 9.1. They range from a high of 74 points for professionals to a low of 7 points for domestic servants and nonfarm laborers.[33] That is, if a group of men all held professional jobs, their average socioeconomic score would be 74 points; if they were all day laborers, 7 points. These scores summarize the occupational status of a group in much the same way average income summarizes an income distribution.[34]

A look at these scores reveals both the upgrading of the occupations of nonwhites and the large remaining differences. For minority men, the average socioeconomic score rose from 16 to 31 points or, roughly, the change from the status of a machine operator to that of a craftsman. Nevertheless, in 1980 the average status of a nonwhite man's job was equivalent to that of the typical white man four decades earlier. As this table indicates, the proportion of men working as laborers or machine operators was much greater for nonwhites in 1980 than for whites in 1940.

The shift away from domestic service led to a rapid increase in the socioeconomic status of black women, but we once again observe a substantial racial lag. Because many nonwhite women are service workers, including domestic service, their average socioeconomic score in 1980 equaled that of white women forty years earlier.

Are blacks and whites becoming more alike in their occupations? A good way to answer this question is to compare the job distributions of blacks and whites using indexes of occupational dissimilarity. If both races were identically distributed across job categories, this index would take on its minimum value of zero. If all job categories were either exclusively black or exclusively white, the index would equal 100, its maximum value.

---

[33]Duncan (1961), p. 155.

[34]These socioeconomic indexes are based on 1950 census data and are not adjusted for any changes in occupational coding. The indexes developed for employed males are applied to females even though there are questions about the appropriateness of this procedure. The average socioeconomic scores are no more than a crude index of a group's status.

Table 9.1 shows indexes of occupational dissimilarity for men and women. They were calculated from data for 11 broad occupational categories, so they do not give a fine-grained picture of racial differences, and indexes computed from a greater number of occupational categories would have larger values than those shown here.

These measures report a sharp decline in occupational segregation by race. At the end of the Depression decade, their values were 43 for men and 63 for women; by 1980 these values had fallen to 24 for men and 8 for women. The decades of greatest change were the 1960s and 1970s when nonwhites moved into the higher-ranking occupations. These measures clearly suggest that employed nonwhites are gradually coming to resemble whites in terms of the jobs they hold.

Occupational changes since 1958 may be monitored more thoroughly because data are available from the Current Population Survey for an intermediate-level system of 28 job categories such as salaried manager, carpenter, motor vehicle driver, and construction laborer. Figure 9.1 shows the proportion of employed workers in the two most prestigious broad categories: professional or technical workers and managers or administrators. The percentage of white men holding such jobs changed only a little in the last 25 years, but the proportion of nonwhite men in such occupations more than tripled from 6 percent to 21 percent. The shift of nonwhite women away from domestic service and farm labor was matched by a rise in the proportion in professional and managerial jobs. As a result, nonwhite women are gradually catching up with white women in terms of occupational prestige.

The figure also indicates a large racial difference in the allocation of jobs by sex. Among whites, the proportion of employed workers in professional and managerial jobs has always been much greater for men than for women. Among nonwhites, there is no such sexual difference, and employed nonwhite women are as likely as nonwhite men to be working at top jobs. There is no satisfactory explanation for this finding. One possibility is that the occupational achievements of white women are "depressed" because many of them are wives who place family responsibilities ahead of their own career achievements. As a result, they may be qualified for jobs that are more prestigious than those they actually hold. Another possibility is that the occupational achievements of black men are "depressed," perhaps because employers are reluctant to promote them to higher-ranking positions, especially if it involves the supervision of whites, particularly white women.

The occupational distributions of whites and nonwhites are becoming more alike, suggesting that some of those barriers which once confined blacks to the least desirable jobs have been removed. Decreas-

FIGURE 9.1

Proportion of Employed Workers Holding Professional, Technical, Managerial, and Administrative Jobs, 1958–1982

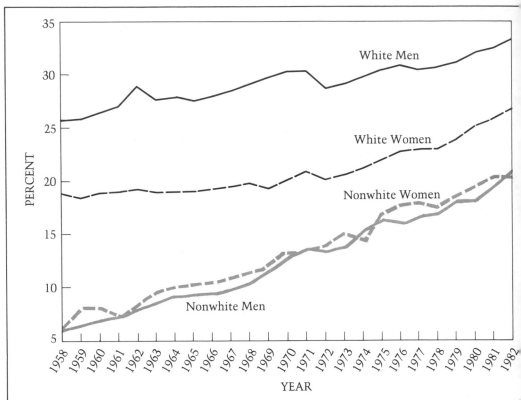

SOURCES: U.S. Bureau of Labor Statistics, *Handbook of Labor Statistics*, bulletin 2000, table 18; bulletin 2070, table 20; *Employment and Earnings*, vol. 28, no. 1, table 22; vol. 30, no. 1, table 22; vol. 29, no. table 22.

ing racial differences are reported in Figure 9.2, which shows indexes of occupational dissimilarity based on 28 job categories. The upper panel compares white and nonwhite workers, controlling for sex. In 1958 the index for men was 40, implying that 40 percent of the men in either race would have to shift from one job classification to another to eliminate racial differences. Through both prosperous and lean times, this index steadily declined, falling to 25 in 1979, the last year for which this time series is available. There has been an even more rapid racial convergence in the occupations of women since this index of racial disparity declined from 47 in 1958 to 18 in 1979.

If we look at specific occupations, we find substantial gains in the

FIGURE 9.2

Measures of the Similarity of the Occupations of Whites and Nonwhites and of Men and Women, 1958–1979

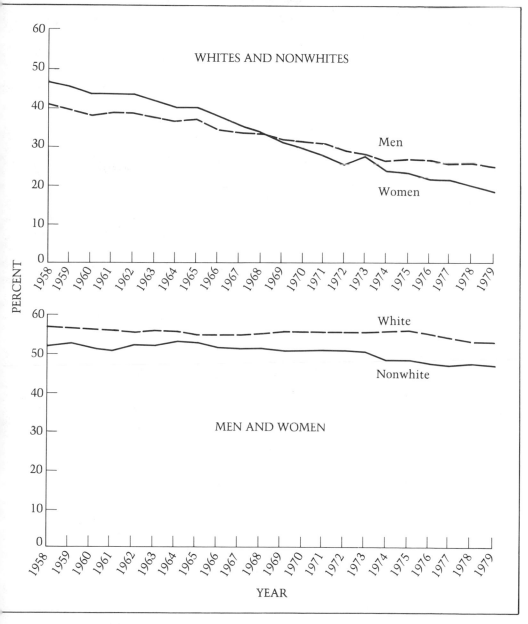

URCES: U.S. Bureau of Labor Statistics, *Handbook of Labor Statistics*, bulletin 2000, table 18; bulletin 70, table 20.

representation of minorities in better jobs between 1958 and 1979. Among men working as professionals in medicine and health, the proportion nonwhite doubled from 4 to 8 percent; for salaried managers, from 1 to 5 percent; and for construction craftsmen, from 6 to 10 percent. Among women, there has been a similar upgrading. The proportion of nonwhite retail sales clerks tripled from 2 to 6 percent; for stenographers and typists, from 3 to 9 percent; and for women who worked as salaried managers, from 2 to 6 percent.

The magnitude of this change may be seen by comparing trends in racial differences to trends in sexual differences. Whites and nonwhites are gradually becoming more alike in their occupations, but the sexual segregation of jobs is disappearing much more slowly, if at all. This is shown in Figure 9.2. Using the same 28 occupational categories, we calculated measures of the similarity of the occupational distributions of men and women. Despite economic growth and civil rights legislation which prohibits employment discrimination on the basis of sex, men and women were almost as differentiated with regard to occupations in 1979 as they had been in 1958. Occupational differences between blacks and whites are declining, but those employment practices and social pressures which assign men and women to different jobs have hardly changed.[35]

An Analysis of Change
Using the New Occupational Categories

To analyze occupational change in the most recent period, it is necessary to use the new occupational categories developed for the census of 1980. The Census Bureau recoded a sample from the census of 1970 into the new codes and, since 1983, the new occupational categories have been used by the Current Population Survey. Table 9.2 presents information about the occupations of black and white workers between 1970 and 1986.

To provide some information about the relative standing of the new occupational categories, it shows the median earnings for full-time, year-round workers by race and sex. In terms of the earnings and educational levels of their incumbents, the broad occupational categories may be sorted into three groups. People who hold executive, administrative, or managerial jobs or who are professional specialty workers have the greatest earnings and educational attainment.[36] Average earnings and

---

[35]Beller (1984); Blau and Hendricks (1979); Bianchi and Rytina (1984).

[36]U.S. Bureau of the Census, *Census of Population: 1980*, PC80-1-D1-A, tables 281 and 282.

educational attainment are roughly similar for an array of white and blue collar categories, including technicians, sales, administrative support, precision production and crafts, machine operators, and transportation occupations. Finally, there are four broad occupational categories distinguished by the low earnings and limited educations of their occupants: handlers, helpers, and laborers; farming, forestry, and fishing; other service; and private household. The rankings of the broad occupational groups by average earnings is much the same for blacks and whites, but a sexual difference stands out. Among men, the earnings of blacks fall behind those of whites in every category, but in 7 of the 13 occupational categories, black women earned more in 1979, on average, than white women.

Table 9.2 provides information about the jobs of employed blacks and whites. Racial differences are striking. In 1980 about 12 percent of the black men compared with 25 percent of the white men had jobs in the top two occupational categories. Half of the black men compared with one quarter of the white men worked at the less skilled blue collar jobs: other service; machine operators; transportation; and handlers, helpers, and laborers. In 1980 more black women were employed as household servants than as professionals, but among whites the number of professionals was ten times the number of maids, cooks, and servants.

Despite these large racial differences, there has been an upgrading of the occupations of blacks relative to whites, as indicated in Table 9.2, which shows average annual changes in the numbers employed in the occupational categories between 1970 and 1980 and 1980 and 1986. If we look at the most prestigious occupations, we find that black employment increased much more rapidly than white, especially during the 1970s. The number of black men holding executive, administrative, or managerial jobs rose 8 percent each year; the number of white men, only 3.5 percent. Among women, the corresponding growth rates were 13.6 percent for blacks and 10 percent for whites. In other occupational categories with above-average earnings, we also find that black employment grew more rapidly than white between 1970 and 1980.

The data for 1986 were collected by the Census Bureau and are basically comparable to those of 1980, although the sample size is much smaller.[37] The interval since 1980 differs from the previous decade in that rates of employment change in many occupations were about the same for both races. Blacks apparently had less of an advantage in mov-

---

[37]In the census of 1980, occupational data were gathered from a 19.4 percent sample of the total population. The March 1985 Current Population Survey (CPS) gathered occupational information from persons in approximately 60,000 households. The CPS excludes the institutional population, including members of the military living on base.

TABLE 9.2

Earnings by Occupation in 1979, Occupational Distributions, and Employment by Occupation by Race and Sex, 1970–1986 (employed persons aged 16 and over)

| | Median Annual Earnings in 1979 of Full-Time Workers (in thousands) | | Distribution by Employment (April Data) | | | | | | Average Annual Change in Number Employed | | | |
| | | | Blacks | | | Whites | | | 1970–1980 | | 1980–1986 | |
	Black	White	1970	1980	1986	1970	1980	1986	Black	White	Black	White
MEN												
Executive, Administrative, Managerial	$17.1	$23.0	2.9%	5.7%	6.7%	10.8%	13.5%	14.1%	+8.3%	+3.5%	+4.9%	+2.2%
Professional Specialty	16.9	22.5	4.3	5.9	6.3	10.6	11.5	12.2	+4.6	+2.0	+3.3	+2.5
Technicians and Related Supports	15.4	18.1	1.1	2.0	1.7	2.6	3.1	2.9	+7.3	+3.1	-0.4	+0.4
Sales	13.2	18.1	2.9	3.9	5.2	10.4	9.8	11.9	+4.4	+0.7	+6.9	+4.7
Administrative Support	14.1	16.7	7.8	9.3	9.1	7.2	6.6	5.4	+3.2	+0.5	+1.9	-2.0
Private Household	6.4	8.0	0.5	0.2	0.1	<0.1	<0.1	<0.1	-7.5	-3.4	-9.8	—
Protective Service	14.0	16.7	1.8	3.1	4.2	2.1	2.3	2.4	+7.1	+2.3	+7.3	+2.1
Other Service	9.5	11.3	14.0	13.6	13.0	5.3	5.9	6.1	+1.2	+2.4	+1.5	+1.9
Farming, Forestry, and Fishing	7.1	10.5	6.6	3.4	3.6	5.4	4.3	5.0	-5.3	-1.1	+3.4	+4.0
Precision Production and Crafts	14.2	17.4	15.5	15.6	15.9	21.6	21.3	20.4	+1.4	+1.2	+2.7	+0.6
Machine Operators and Assemblers	12.8	15.5	15.6	14.7	11.0	10.4	9.1	7.4	+0.8	-0.1	-2.5	-2.0
Transportation	12.9	16.5	11.6	11.1	9.7	7.2	6.9	6.5	+1.0	+0.9	0.0	+0.4
Handlers, Helpers, and Laborers	10.6	13.1	15.4	11.5	13.5	6.4	5.7	5.7	-1.6	-0.0	+5.0	+1.6
Total	$12.8	$17.6	100.0%	100.0%	100.0%	100.0%	100.0%	100.0%	+1.4%	+1.4%	+2.3%	+1.4%
Average Socioeconomic Score			26	29	29	35	36	36				

WOMEN

Executive, Administrative, Managerial	$13.1	$12.8	1.7%	4.7%	5.9%	4.0%	7.8%	9.8%	+13.6%	+10.0%	+6.2%	+6.5%
Professional Specialty	13.1	13.8	9.8	11.8	10.8	13.7	14.6	14.9	+5.3	+4.0	+0.8	+3.0
Technicians and Related Supports	11.4	11.6	2.5	3.3	3.7	2.1	3.1	3.2	+6.1	+2.1	+4.2	+3.2
Sales	8.1	9.0	4.4	6.1	8.7	11.8	12.0	13.6	+6.8	+3.5	+8.3	+4.7
Administrative Support	10.3	10.0	19.2	25.9	26.1	34.0	32.1	30.1	+6.4	+2.7	+2.6	+1.6
Private Household	4.8	4.4	17.8	5.0	4.0	2.0	0.8	1.7	-9.3	-5.3	-1.3	+14.4
Protective Service	11.4	11.2	0.3	0.7	0.7	0.2	0.4	0.4	+12.4	+8.5	+3.4	+3.6
Other Service	7.5	7.4	23.8	23.6	23.8	14.9	15.1	15.1	+3.4	+3.5	+2.5	+2.7
Farming, Forestry, and Fishing	6.7	6.1	1.3	0.5	0.4	0.8	1.0	1.1	-5.7	+5.3	-2.9	+4.5
Precision Production and Crafts	9.7	10.6	2.4	2.3	2.6	2.7	2.3	2.2	+3.0	+3.4	+4.2	+2.1
Machine Operators and Assemblers	8.4	9.0	13.1	12.3	10.1	10.9	8.0	5.7	+2.8	+0.2	-0.9	-2.9
Transportation	9.8	9.7	0.5	0.9	1.2	0.5	0.8	0.9	+9.5	+7.5	+6.2	+4.0
Handlers, Helpers, and Laborers	8.5	9.1	3.2	2.9	2.0	2.4	2.0	1.3	+2.5	+1.4	-3.9	-4.1
Total	$9.5	$10.2	100.0%	100.0%	100.0%	100.0%	100.0%	100.0%	+2.0%	+5.3%	+2.4%	+2.7%
Average Socioeconomic Score			29	34	33	37	38	37				

SOURCES: U.S. Bureau of the Census, *Census of Population: 1980*, PC80-1-C1-A, table 89; PC80-1-D1-A, table 281; U.S. Bureau of Labor Statistics, *Employment and Earnings*, vol. 33, no. 5 (May 1986), table A-23. Amounts shown in 1979 dollars.

ing into the executive and professional specialty occupations in the 1980s than in the 1970s. Socioeconomic status indexes are also shown in Table 9.2 using a new set of scores for the 1980 occupational categories.[38] They report a modest upgrading of the jobs of blacks in the 1970s but very little change in the 1980s.

Once again, we use the index of occupational dissimilarity (shown below) to determine whether blacks are becoming more like whites in terms of their jobs. These were calculated from data for the 13 broad occupational categories, so they are smaller in value than would be indexes computed for the more than 500 detailed occupational categories.

	Men	Women
1970	30	29
1980	27	19
1986	27	18

These indexes demonstrate that between 1970 and 1980 occupational differentiation by race declined, but this trend may not be continuing into the 1980s. At this time, we cannot determine whether this is a temporary pattern associated with the economy of the 1980s or whether the long-run trend toward the racial convergence of occupations has ended.[39]

[38]Stevens and Cho (1985).

[39]Data from surveys conducted in the 1980s are consistent in reporting that racial differences in occupational distributions declined little, if at all, after 1980. Listed below are indexes of occupational dissimilarity calculated from the new 13 broad occupational categories.

	Men	Women
Census: April 1980	26.9	18.5
Current Population Survey		
Annual Average for 1983	26.2	18.5
Annual Average for 1984	26.7	18.7
Annual Average for 1985	26.7	17.9
April 1983	27.2	19.1
April 1984	26.8	19.2
April 1985	28.7	17.3
April 1986	27.0	17.6

SOURCES: U.S. Bureau of Labor Statistics, *Employment and Earnings*, vol. 31, no. 5 (May 1984), table A-23; vol. 32, no. 1 (January 1985), table 21; vol. 33, no. 1 (January 1986), table 21; vol. 33, no. 5 (May 1986), table A-23; U.S. Bureau of the Census, *Census of Population: 1980*, PC80-1-C1, table 89.

274

Occupational Differences in 1980

We might expect that by 1980 racial differences would be small among the highly educated and that blacks and whites who entered the job market after the Civil Rights Act banned discrimination would be much more alike in their occupational distributions than people who entered the labor market decades earlier. To examine the magnitude of racial differences, data from the census of 1980 were examined. As an indicator of status, we used the proportion who held jobs in the top two categories: executive, administrative, or managerial and professional specialty occupations. Within categories of the independent variable, blacks and whites were also compared using the index of occupational dissimilarity. The analysis was restricted to persons of prime working ages: 25 to 64.

Turning first to information about workers classified by age, we find that blacks who entered the labor force recently were somewhat better represented in the top job categories than were older blacks. Although these are cross-sectional data, they suggest a diminution of discrimination and an improvement in employment opportunities for blacks. Nevertheless, the percentage of young black men in the top-ranked occupations—18 percent—is far below the percentage of white men of any age.

The likelihood of a person getting an administrative or professional job was strongly related to his schooling, and after controlling for education, we found small racial differences. As Table 9.3 shows, those people who did not attend college seldom got the top-ranked jobs, but the majority of both black and white college graduates were administrators or professionals. Black women who obtained college degrees were slightly more likely than similar white women to be in the top occupational categories, but among men blacks were at a slight disadvantage in 1980. A look at the entire occupational distribution shows that white men who held college degrees were more likely than black men to work as executives or as sales officials, while black college graduates were more likely than whites to fill administrative support positions or work as craftsmen.

Regional differences in the employment of blacks and whites are apparent. Blacks lagged furthest behind whites in the South, whereas in the West the job distributions of blacks and whites were most alike and the proportion of blacks with high-ranking jobs was greatest.

Region of birth was related to occupational prestige since northern-born blacks were considerably more likely to hold administrative or

TABLE 9.3

Percentage of Employed Workers in Executive, Administrative, Managerial, and Professional Specialty Occupations and Index of Occupational Dissimilarity for Black and White Men and Women, Aged 25–64, 1980

	Men			Women		
	Percent Executive, Administrative, Managerial, and Professional Specialty		Index of Occupational Dissimilarity	Percent Executive, Administrative, Managerial, and Professional Specialty		Index of Occupation Dissimilar
	Black	White		Black	White	
Total	15%	30%	28	19%	26%	24
AGE						
25–34 Years	18	29	25	19	27	24
35–44 Years	17	33	28	19	27	24
45–54 Years	13	30	31	18	24	33
55–64 Years	10	28	35	14	20	38
EDUCATIONAL ATTAINMENT						
Elementary	3	6	19	4	5	32
High School, 1–3 Years	5	9	23	6	7	23
High School, 4 Years	10	15	26	9	13	24
College, 1–3 Years	21	29	24	23	29	12
College, 4 Years	50	57	20	67	62	7
College, 5+ Years	75	80	12	86	83	4
REGION OF RESIDENCE IN 1980						
Northeast	17	31	27	19	28	18
Midwest	14	28	25	20	24	15
South	14	29	32	18	26	29
West	25	35	23	21	28	13
REGION OF BIRTH BY REGION OF RESIDENCE						
Born North, Lives North	19	31	28	22	27	11
Born North, Lives South	32	42	22	31	34	16
Born South, Lives South	13	25	30	18	24	27
Born South, Lives North	12	32	30	18	25	18
MARITAL STATUS						
Currently Married	16	30	27	20	27	23
Widowed, Separated or Divorced	14	29	28	18	24	22
Never Married	12	31	29	14	26	25

NOTE: "North" refers to all areas outside the South.

SOURCE: U.S. Bureau of the Census, *Census of Population and Housing: 1980*, Public Use Microd Samples.

professional jobs than were southern-born blacks. Northern-born blacks as well as northern-born whites who moved into the South were most likely to be holding the high-paying jobs.

Occupational differences by marital status are small for both races.

However, those who were married and lived with a spouse in 1980 were somewhat more likely to hold administrative or professional jobs.

New Opportunities for Black Employees: Have Racial Differences Disappeared?

Richard Freeman analyzed long-run trends in the occupations of black workers and found that a great change occurred in the 1960s. He concluded that for the first time blacks were competing successfully for good jobs.

> Black Americans have historically been concentrated in low-level occupations as a result of little and poor-quality education, unfavorable family backgrounds, and job market discrimination. The highly-educated were found in teaching or segregated professional services, but rarely in managerial or professional jobs in major corporations. From the period of slavery until the 1950s or 1960s, the occupational standing of blacks relative to that of whites improved little, if at all. . . . In the new market of the 1960s, the historic pattern of little or no occupational progress was broken. Blacks moved up the occupational hierarchy rapidly, with the highly-educated breaking into previously "closed" managerial and professional occupations.[40]

If Freeman is correct about occupational change, and if those factors which once kept blacks at the bottom of the ladder are no longer effective, we should expect that in 1980 blacks and whites who had similar qualifications worked at similar jobs.

To investigate this, we considered men and women who were participants in the labor force in 1980. They were divided into two groups: those in executive, administrative, managerial, and professional specialty occupations—the top categories in the new occupational system—and all others. This latter group includes all lower-ranking occupations as well as unemployed persons, since we are interested in assessing the equality of occupational achievement. The proportions with prestigious occupations were 13 and 28 percent for adult black and white men, respectively; 18 and 26 percent for black and white women.

At first glance, it appears that a substantial racial difference remains, but blacks and whites differ on many characteristics that influence their occupational achievement. To take such differences into account, we estimated models which said that the probability that an individual held a top-ranking job was influenced by his or her age, educational attainment, region of residence, and marital status. Coef-

[40]Freeman (1976) p.2.

ficients for these models are shown in Appendix Table 9.1. Once we equate blacks and whites with regard to these factors, we may find that equivalent proportions of blacks and whites hold good jobs, which would suggest an absence of discrimination.

In Figure 9.3 persons in the labor force are classified by age and sex, and the proportion of blacks and whites holding administrative and professional jobs is indicated for each of six educational attainment levels. We expect the smallest racial differences among those young people who benefited from improvements in the educational system after World War II and who entered the job market after the Civil Rights Act banned discrimination.

Looking at the upper half of Figure 9.3, we find that at every attainment level, the proportion of men aged 25 to 34 with prestigious jobs was greater among whites. The differences, however, were quite small. Among those young men who completed five or more years of college, the model estimates that about 75 percent of the whites and 70 percent of the blacks held administrative or professional jobs. Among young women, the proportion with prestigious occupations was also greater for whites than for blacks at all educational levels except among college graduates.

It is desirable to determine if these patterns have changed and if the importance of race as a factor influencing occupation has declined. Ideally, we would go back to the 1960 or 1970 census and fit similar models to see if racial differences among young people in 1980 were smaller than they had been at previous dates, but this cannot be done since the occupational categories are not comparable. As an alternative, we looked at individuals aged 45 to 54 in 1980 to find out if racial differences are larger for them than for those aged 25 to 34 in 1980. The racial pattern for these older workers is quite similar to that for younger ones. At every attainment level, black men lagged behind white men with regard to employment in the top jobs. The actual racial differences were only slightly smaller than those among younger men, suggesting, at best, a modest improvement for blacks. For women, the pattern also did not vary by age. Older black women who graduated from college were more likely to hold managerial or professional jobs than their peers.

Although this analysis describes only some aspects of occupational achievement, it suggests that one should be cautious in drawing conclusions about the elimination of racial discrimination in the job market, even if the focus is on college-educated men. Undoubtedly, many of the traditional barriers have been removed, and a black man who obtains a college diploma is no longer restricted to the public school system, the civil service, or the pulpit of a black church. However, after taking relevant factors into account, the models show that a black man is less

FIGURE 9.3
Proportion of Labor Force Participants
Holding Executive, Managerial, or Professional Specialty Occupations
by Race, Sex, Age, and Educational Attainment, 1980

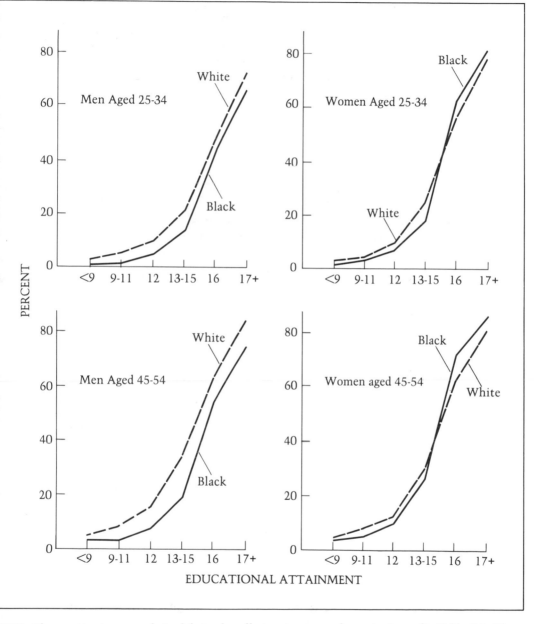

NOTE: These estimates were derived from the effects parameters shown in Appendix Table 9.1. They pertain to married-spouse-present people living in the South in 1980.

SOURCE: U.S. Bureau of the Census, *Census of Population and Housing: 1980*, Public Use Microdata Samples.

likely than a white man to get an executive or professional position. The same pattern of racial difference is observed among those who entered the labor force before and after passage of civil rights legislation.

Freeman did not describe discrimination trends in the employment of highly educated women. Here it is much easier to draw conclusions about the limited impact of racial discrimination, since these models report that black women of any age who complete college are more likely than comparable white women to work as administrators, executives, managers, or professional specialists.

Conclusion

There have been improvements since 1940 in the occupational status of blacks, attributable in large part to the exodus of black men from farm labor and black women from domestic service. Racial differences in occupational distributions are declining and differences are much smaller within specific educational groups than in the aggregate. Nevertheless, blacks are more concentrated at the bottom of the occupational hierarchy than whites, and the only blacks who have attained occupational parity with similar whites are college-educated women.

Will racial differences in occupational status continue to decrease? Will there be a time when blacks, on average, work at jobs that are just as prestigious and financially rewarding as those of whites? Answers to these questions may depend on three trends. First, whites have shifted away from the racial views which were common at the turn of the century—namely, that blacks were unqualified for most jobs, that close social contact with black workers was inappropriate, and that whites had a right to an elevated position in the occupational structure. Nevertheless, the numerous Title VII lawsuits which are settled in favor of plaintiffs prove that some firms and governmental agencies continued to assign blacks and whites, men and women, to different jobs many years after passage of the Civil Rights Act of 1964.[41] Opposition to affirmative action in employment may be based on the belief that individual competition is the best way to fill all openings, but it may also be related to the economic interests of white men who have benefited from discrimination in the past. Quite likely, there is a legacy of the traditional racial views and some employers may still be reluctant to hire blacks or promote them.

Second, there are questions about the supply of qualified black workers. In terms of formal schooling, blacks and whites are approach-

[41]Burstein (1979); Rodgers (1984); Smith, Craver, and Clark (1982).

ing parity, but there are persistent racial differences in achievement test scores,[42] and black college students enroll in different programs than white students.[43] In addition, there may be persisting or even increasing racial differences in the attitudes and social skills employers consider relevant when making hiring or promotion decisions. Several writers suggest that black youth, especially males from the ghetto, lack the work habits, speech patterns, and dress standards which employers find appealing.[44]

Third, a minority group may have difficulty moving up the occupational ladder when jobs are in short supply. Between 1974 and 1985 the adult labor force expanded by 26 percent, but the number of employed workers grew only 23 percent.[45] Several recent studies provide empirical evidence of racial and ethnic queues in the American labor market.[46] As discussed in Chapter 8 when explaining the persistently high unemployment of blacks, employers may feel that some workers are more productive and qualified for top-ranking jobs than others. Traditionally, they have assigned white men to the most prestigious jobs, and they need not change this practice now that labor is in great supply.

Finally, there has been a fundamental change in federal policy with regard to affirmative action. By the end of the 1970s judicial decisions and pressures from governmental agencies, especially from the Equal Employment Opportunity Commission, had led employers to adopt hiring practices which guaranteed that some minorities and women would be on the payroll. There is lively debate, of course, about the fairness of such policies, but there is evidence that the occupational distribution of blacks was upgraded more than that of whites during the 1960s and 1970s when the policies received federal backing.[47] Since 1982 the Department of Justice has challenged the constitutionality of these agreements and rulings, contending that they are a type of reverse discrimination. At present, there are no rigorous investigations of the consequences of the Justice Department's new approach to affirmative action, but this change may strongly influence future trends in the occupational status of blacks.

[42]U.S. Bureau of the Census, *Social Indicators III*, (1980), table 6/17; Burton and Jones (1982); Thomas (1986).

[43]U.S. Bureau of the Census, *Current Population Reports*, series P-20, no. 351 (May 1980), table 1.

[44]Murray (1984), p. 188; Williams (1982), chap. 3.

[45]U.S. Bureau of Labor Statistics, *Employment and Earnings*, vol. 33, no. 1 (January 1986), table 1.

[46]Hout (1986); Lieberson (1980) chaps. 10 and 11.

[47]Burstein (1985); Smith and Welch (1986), pp. 85–94.

APPENDIX TABLE 9.1

Effects Parameters for Logit Model Which Treats the Odds of Employment in Executive, Administrative, Managerial, and Professional Specialty Occupations for Labor Force Participants, by Race and Sex, 1980

	Men		Women	
	Black	White	Black	White
Sample Size	33,586	33,249	33,992	23,787
Percent with Selected Jobs	12.5%	28.1%	17.7%	25.9%
Odds of Holding such a Job	.14:1	.39:1	.22:1	.35:1
	EFFECTS PARAMETERS			
Intercept	− 1.7810	− 1.2344	− 1.2974	− 1.1184
EDUCATIONAL ATTAINMENT				
Less Than 9 Years	− 1.8659	− 1.9414	− 2.1612	− 2.0445
9–11 Years	− 1.6725	− 1.4494	− 1.7144	− 1.4973
12 Years	− .9752	− .7972	− 1.0792	− .8420
13–15 Years	+ .1362	+ .2018	+ .0402*	+ .2122
16 Years	+ 1.7016	+ 1.4342	+ 1.9782	+ 1.5958
17+ Years	+ 2.6758	+ 2.5520	+ 2.9366	+ 2.5758
MARITAL STATUS				
Married-Spouse-Present	+ .0612	+ .1046	+ .0696	− .0056*
Other Marital Statuses	− .0612	− .1046	− .0696	+ .0056*
AGE				
25–34 Years	− .2458	− .3476	− .2506	− .1208
35–44 Years	+ .0992	+ .0832	+ .0880	+ .0300*
45–54 Years	+ .1466	+ .2644	+ .1626	+ .0908*
REGION				
South	− .0388*	− .0066*	+ .0026*	+ .0534*
North and West	+ .0388*	− .0066*	− .0026*	− .0534*
Likelihood Ratio, χ^2	126.6	115.2	121.7	74.0
Probability of χ^2	<.01	<0.1	<.01	.14
Degrees of Freedom	62	62	62	62

*Coefficient *not* significantly different from zero at .05 level.

NOTE: This model treats the log of the odds of being employed as an executive, administrative, managerial, or professional worker as its dependent variable. To estimate the proportion of a given group holding such jobs, it is necessary to add the intercept to the appropriate effects parameters for educational attainment, marital status, age, and region of residence. The sum is the log of the odds of being employed in one of these prestigious occupations. For black men who completed 16 years of education, who lived with a wife, were between ages 25 and 34, and resided in the South, the sum of the effects parameters is − .3028, implying an odds ratio of .73 to 1 which implies that 42.5 percent of the men in this group held such jobs.

SOURCE: U.S. Bureau of the Census, *Census of Population and Housing: 1980*, Public Use Microdata Samples.

10

PERSONAL INCOME

B LACK men who received income in 1985 reported an average of $10,400 while white men reported $16,500, which means blacks had incomes only 63 percent of those of whites.[1] Black women, on the other hand, had incomes 85 percent of those of white women: $6,100 versus $7,100.[2] This chapter describes personal income and seeks to determine if racial differences have declined over time. There has been a trend toward a racial convergence in educational attainment, the attitudes of whites regarding the employment of blacks have changed, and federal laws now prohibit discrimination in pay rates and promotions. In addition, black men have moved out of low-paying jobs in southern agriculture, and black women are no longer concentrated in domestic service. These changes should reduce racial differences in income.

There are numerous challenges involved in the analysis of income trends. First, the tradition of asking individuals about how much money they receive is not a long one. Because of the economic crisis of the 1930s, a question about income was proposed for the 1940 census.

[1]Amounts shown in this chapter are in dollars that have the purchasing power of 1984 dollars. The Consumer Price Index was used to adjust for inflation. These income data refer to persons aged 15 and over.
[2]U.S. Bureau of the Census, *Current Population Reports*, series P-60, no. 154 (August 1986), table 8.

Many people were strongly opposed, feeling that their constitutional right to privacy would be compromised by an intrusive federal agency.[3] As a result, the Census Bureau limited its inquiry to one question about the amount of wage and salary earnings in 1939 for persons earning less than $5,000. Individuals who were reluctant to tell an enumerator about their earnings could fill out a form *in camera* and send it to Washington.

By 1950 attitudes had changed, and the census asked about wage, salary, and self-employment earnings as well as cash received from interest, dividends, pensions, and rent in the previous year.[4] Over the years additional questions have been added. At present the major source of income data is the Census Bureau's Current Population Survey, when in March of each year the residents of about 60,000 households are asked an array of income questions. Changes in the sources of income as well as the evolution of income questions mean that it is difficult to assemble a long time series that accurately reports racial differences in income.

What should be included in a definition of personal income? We can identify several components:

1. the wage, salary, and self-employment earnings that people receive for doing their jobs;
2. property income such as interests on deposits, income from the rental of a house, dividends, and capital gains;
3. cash transfer payments obtained from the government including Social Security benefits, unemployment compensation, and veteran's benefits;
4. fringe benefits associated with employment such as health and life insurance, contributions to retirement programs, subsidized cafeterias, free parking, and the use of company telephones or cars;
5. noncash transfer benefits such as food stamps, school lunches, public housing, Medicare, Medicaid, and government-backed loans for education.

The magnitude of racial differences, of course, depends on the definition of income. In recent years one of the most repeated arguments has been that the poor are really not so impoverished because most tabulations of income omit food stamps, school lunches, and Medicaid programs.[5] Estimates of income which include noncash transfer payments lead to

---

[3]Eckler (1972), pp. 192–95; Halacy (1980), p. 143; Scott (1968), pp. 45–46.
[4]Miller (1955), p. 129.
[5]Murray (1984), p. 63.

lower rates of poverty than those based on cash income alone.[6] However, at this time there is little consensus about which noncash benefits should be included as income or how they should be evaluated.

Another challenge is that respondents are often not precise in reporting their incomes because it is difficult to remember some sources and exact amounts. Independent estimates may be developed for some types of income. For example, the Internal Revenue Service scrutinizes tax forms to determine wage and salary earnings, and their figures can be compared with the reports individuals make on their census schedules or in the March Current Population Survey. Governmental agencies know how much money is dispersed in Social Security checks or as unemployment compensation, and these amounts can be compared with what people report. Since Census Bureau data are confidential, findings of this type are published only for broad statistical categories.

The decennial census and the Current Population Survey do a good job of gathering data about wage and salary earnings, Social Security payments, and federal retirement pensions. In the aggregate, the amount reported to the Census Bureau from these sources is just about equal to the independent estimates.[7] But welfare payments and unemployment compensation are less completely reported: Recipients report about 75 percent of what the agencies disperse. Interest, dividends, and worker's compensation are still less accurately enumerated since the independent estimates suggest that only 45 percent of the income from these sources is reported.

The Census Bureau estimates that the recent March surveys underestimate aggregate income by about 10 percent. Studies from the 1960 census suggest an undercount of about 13 percent of income and from the 1950 census of 16 percent.[8] This implies that the income figures discussed in this chapter substantially underestimate individuals' actual money income, but there may be a trend toward improved reporting of income. Unfortunately, we have no information about racial

[6]Smeeding (1982), chap. 6; Anderson (1978), pp. 19–25.
[7]U.S. Bureau of the Census, *Current Population Reports*, series P-60, no. 151 (April 1986), table A-2. Over time the Census Bureau has increased the number of questions that respondents are asked about earnings or income. This has been matched by an increase in nonresponse. In the census of 1940, only 2.5 percent of the sample failed to answer the one question about wage and salary earnings; in 1982, 26.6 percent of the sample respondents failed to answer one or more of the earnings and income questions. At present, the Census Bureau imputes an amount to individuals who fail to answer these questions, but there is evidence that the current imputation procedure leads to underestimates of income since high income persons are apparently more likely to be nonrespondents. For further information see Lillard, Smith and Welch (1986).
[8]U.S. Bureau of the Census, *Current Population Reports*, series P-60, no. 142 (April 1985), table A-2; Miller (1955), app. B; Miller (1966), app. A; National Bureau of Economic Research (1958).

differences in reporting. The income figures described in this chapter are pretax amounts.[9]

Although the census and Current Population Survey guarantee confidentiality and do not distinguish between legal and illegal income, we know little about the completeness of the reporting of illegally obtained income. Nor do we know about the frequency with which income is concealed from both the Internal Revenue Service and the Census Bureau.

If we are interested in the economic status of a group, however, there is good reason to examine personal income. This chapter describes racial differences in this important measure. Since much of the civil rights struggle sought to eliminate discrimination in pay rates, there is also good reason to assess racial differences in earnings as well. Do similarly qualified blacks and whites earn the same amounts or is there a penalty associated with having a black skin? The next chapter describes the earnings of employed workers.

Racial Differences in Sources of Income

Do whites have greater incomes than blacks just because they earn more or do whites get more income from other sources? Suppose that racial differences in employment and earnings were eliminated. Would blacks receive as much income as whites? Quite likely they would not, because blacks and whites differ greatly in their sources of income. In brief, whites are more likely than blacks to report substantial amounts of property income. Blacks, on the other hand, get a larger share of their total income from governmental transfer payments.

Table 10.1 indicates the proportion of black and white men and women aged 15 and over who obtained cash income from each of eight

[9]In recent years, the Census Bureau has used a simulation model to estimate the distribution of posttax income. The taxes included are federal income taxes, state income taxes, and Social Security. Sales taxes, city income taxes, and license or user fees are excluded.

In 1984 whites living in households had a per capita income of $10,826 pretax and $8,469 posttax, implying that they paid 22 percent of their income in taxes on average. For blacks, the pretax per capita income was $6,071 and posttax was $5,001, implying a tax rate of 18 percent. The pretax racial difference in per capita income was $4,800; the posttax difference was $3,500. In other words, more than one quarter of the white advantage in per capita income was taken away by the higher taxes whites typically pay. Before taxes, blacks in households had per capita incomes 56 percent of those of whites; after taxes, 59 percent. This implies that studies which adjust for taxes will report smaller racial gaps than those which do not. U.S. Bureau of the Census, *Current Population Reports*, series P-23, no. 147 (July 1986), table 1.

TABLE 10.1

Sources and Amounts of Personal Income for Blacks and Whites Aged 15 and Over by Sex, 1984 (in 1984 dollars)

	Blacks				Whites			
		Amount				Amount		
	Percent Receiving	Per Capita	Per Recipient	Distribution Of Income by Source	Percent Receiving	Per Capita	Per Recipient	Distribution of Income by Source
MEN								
Earnings								
Wage and Salary	64%	$ 8,465	$13,226	81%	72%	$14,669	$20,450	76%
Nonfarm Self-Employment	3	291	9,922	3	9	1,353	15,349	7
Farm Self-Employment	<1	2	$ 500	<1	2	105	4,921	<1
Property Income	24%	$ 179	$ 740	2%	60%	$ 1,149	$ 1,900	6
Transfer Payments—Total	32%	$ 1,463	$ 4,516	14%	32%	$ 2,090	$ 6,475	11%
Retirement and Annuities	6	362	6,311	3	10	787	7,919	4
Social Security	14	602	4,288	6	16	897	5,573	5
Public Assistance and Welfare	3	59	2,017	<1	<1	20	2,197	<1
Supplemental Income	4	97	2,400	<1	1	28	2,621	<1
Total	86%	$10,400	$12,119	100%	96%	$19,366	$20,259	100%
WOMEN								
Earnings								
Wage and Salary	55%	$5,680	$10,284	77%	55%	$5,812	$10,489	66%
Nonfarm Self-Employment	1	47	4,478	<1	4	209	5,425	2
Farm Self-Employment	<1	<1	<1	<1	<1	8	3,041	<1
Property Income	23%	$ 154	$ 663	2%	60%	$1,241	$ 2,057	14%
Transfer Payments—Total	42%	$1,472	$ 3,474	20%	36%	$1,507	$ 4,221	17%
Retirement and Annuities	3	136	4,628	2	6	245	4,353	3
Social Security	16	553	3,516	8	21	859	4,134	10
Public Assistance and Welfare	14	396	2,888	5	3	77	2,930	<1
Supplemental Income	7	156	2,334	2	2	42	2,217	<1
Total	85%	$7,353	$ 8,622	100%	91%	$8,777	$ 9,682	100%

NOTE: "Transfer Payments" include veteran's benefits, unemployment compensation, and worker's compensation, which are not shown separately. "Social Security" includes Railroad Retirement benefits.

SOURCE: U.S. Bureau of the Census. *Current Population Reports*, series P-60, no. 151 (April 1986), tables 32 and 35.

sources and shows the dollar amounts they received on both a per recipient and a per capita basis.[10]

A higher proportion of whites than blacks had cash income. That is, 96 percent of the white men compared with 86 percent of the black men reported receiving money. Among women, the corresponding proportions were 91 percent for whites and 85 percent for blacks. The group without income includes some students, adult dependents, and people who are cared for by friends or relatives.

There are large racial differences in the sources of income. The proportion of adults with earnings is greater among whites than blacks, especially for men, which is attributable to the higher unemployment rates of black men and their much lower rates of labor force participation. The proportion with self-employment income among whites is three times that among blacks, reflecting the fact that white men are more likely to be in business for themselves. The largest racial difference in recipiency, however, pertains to property income: monies received as interest on savings accounts, from dividends, or from rental properties.

Because of the tradition of low incomes in the black community, there is now a great racial difference in wealth holdings. In 1983 the Census Bureau initiated the new Survey of Income and Program Participation (SIPP), designed to provide more detailed information about income sources, recipiency of transfer payments, and poverty. In 1984 the survey began to collect information about asset holdings. A comparison of households reveals, first, that those headed by blacks were much less likely to have valuable assets than those headed by whites.

Table 10.2 shows the percentage of black and white households that own each of ten types of assets. Three quarters of the white house-

[10]Decennial enumerations and the Current Population Survey (CPS) both gather data about income during the year before the census. Although the CPS is a large survey, its income data differ from those of the census for two reasons. First, it is primarily an income and employment survey conducted by trained interviewers. Since it includes many more income questions than the census, it should obtain a more complete coverage of income. Second, the CPS excludes some groups, such as residents of institutions and barracks, which are included in decennial enumerations. Since this is only a small fraction of the population, it should have little influence on the income estimates.

Appendix Table 10.1 shows summary income statistics for 1979 obtained from the March 1980 CPS and the April 1980 census. As expected, the proportion of adults reporting any cash income was higher in the CPS than in the census, but the amounts of income are greater in the census. Presumably, many people with only a few dollars of income are apt to overlook them when filling out their census forms, but are prodded to remember such income by CPS interviewers. The censuses of 1950 and 1960 similarly estimated larger median incomes per recipient than did the comparable CPS. Miller (1966), pp. 197–205. Per capita incomes, however, are quite similar in the two sources.

Whenever possible, data in a specific figure or table in this chapter come from either decennial censuses *or* Current Population Surveys.

TABLE 10.2

Households Owning Assets and Mean Value of Holdings by Race, 1984

Type of Asset	Percentage of Households Owning Assets		Mean Value of Holdings for Asset Owners	
	Black	White	Black	White
Interest Earning Assets at Financial Institutions	44%	75%	$ 3,135	$16,865
Checking Accounts	32	57	599	947
Stocks and Mutual Funds	5	22	2,813	27,694
Own Home*	44	67	29,914	51,939
Rental Property*	7	10	38,142	73,831
Other Real Estate*	3	11	14,423	35,102
Own Business or Profession*	4	14	33,997	64,495
Motor Vehicles*	65	89	3,446	5,707
U.S. Savings Bonds	7	16	550	2,624
IRA/Keogh Accounts	5	21	3,441	9,047
Mean Value for Asset Holders			20,241	86,332

*Shows amount of net equity.

SOURCE: U.S. Bureau of the Census, *Household Economic Studies*, P-70, no. 7 (July 1986), tables 1 and 3.

holds but less than one half of the black households had interest-bearing deposits at financial institutions. One fifth of the white households but only one twentieth of the black households owned shares of stock or mutual funds. The table also reveals very large racial differences in home ownership, checking accounts, and IRA/Keogh plans.

Table 10.2 also reports the mean value or equity of assets for the households holding them. The cash deposits of whites exceed those of blacks by a factor of five; for stock holdings, the ratio is ten to one.

The SIPP survey reports that at all income levels and for all types of households the assets of blacks are a small fraction of those of similar whites. It is not just that female-headed black households or low-income black households have many fewer assets than whites. This is shown in Table 10.3, which presents the median value of assets for households classified by their monthly income in the fourth quarter of 1984 and by their type. Assets, of course, increased with household income, but even in the highest income category—$4,000 or more in monthly income—blacks holdings were small compared with those of whites.

Finally, racial differences in wealth are very much greater than racial differences in current income. As indicated in Table 10.3, black

TABLE 10.3

Median Net Worth of Households by Race, Monthly Income,
and Type, 1984

	Black Households	White Households	Black as a Percentage of White
Monthly Income in Fourth Quarter, 1984			
<$900	$ 88	$ 8,443	1%
$900–1,999	4,218	30,714	14
$2,000–3,999	15,977	50,529	32
$4,000 or More	58,758	128,237	46
Type of Household			
Married Couple	13,061	54,184	24
Female-Headed	671	22,500	3
Other Male Household	3,022	11,826	26
For Total Households			
Median Net Assets	3,397	39,135	9
Median Monthly Income	1,088	1,760	62

SOURCE: U.S. Bureau of the Census, *Household Economic Studies,* P-70, no. 7 (July 1986), table G.

households in late 1984 had monthly incomes which were 62 percent of those of white households, but their median assets were only 9 percent of those of whites. Given these large discrepancies in assets, it is not surprising to find that three fifths of adult whites in 1984 received property income compared with only one quarter of adult blacks.

Table 10.1 also shows that some types of income were obtained more frequently by blacks. These include public assistance (that is, state payments to the indigent) and supplemental income (that is, benefits from that component of the Social Security system which supports the blind, the disabled, and the impoverished elderly).[11] One black woman in 7 reported public assistance or welfare income compared with one white woman in 35, a consequence of the frequency with which black women maintained their own families.

The races differ substantially in their total income. On a per capita basis, the earnings of white men are more than 70 percent greater than those of black men, while the earnings of white women exceed those of black women by just 2 percent. Very large differences are found in property income since, on a per capita basis, white men receive six times as much as black men and white women eight times as much as black women. Those incomes that are tied to earnings—namely, Social Secu-

[11]Levitan (1980), pp. 35–38; Patterson (1986), pp. 197–98.

rity payments and retirement or annuity payments—are also much greater for whites.

In 1984 white men had a total per capita (mean) income of $19,400, and black men had $10,400. If there had been no racial difference in earnings, black men would still have fallen about $2,500 behind white men in average earnings because of their lower property incomes and their smaller Social Security payments. Equating black and white women with regard to earnings would not eradicate racial differences because the elevated economic status of white women results, in large measure, from their high property income, their annuities, and their larger Social Security checks.

Income statistics—such as those in Table 10.1—omit several kinds of benefits. Does this result in an overestimate or an underestimate of racial differences? It is necessary to consider the types of income which are excluded. First, Census Bureau estimates of income omit capital gains or losses. For many families, a major source of capital gains—if they have any—is appreciation of the value of their home, and the majority of whites own their homes while the majority of blacks rent. This tenure advantage is found for whites at all levels of current income. [12] Thus, blacks benefit less from appreciation in the value of their property. The Census Bureau studied changes in the value of housing by combining findings from the 1973 Annual Housing Survey and the 1980 Census of Housing, estimating market values for both years for a large sample of homes. The median worth of homes owned and occupied by whites in both years rose by 8 percent (in constant dollars) in this seven-year span. For those owned and occupied by blacks, there was an actual decrease of 3 percent in their market value. [13] This may not be surprising since the majority of black owners live in central cities where the demand for housing is often weak. Presumably, if the federal statistical system counted capital gains and inheritances, racial differences in income would be larger than those shown in Table 10.1.

Second, income statistics exclude employer-paid benefits be they the use of a car, day care, life insurance, or a company cafeteria. Since 1979 there has been enumeration of employer-paid health insurance and pension plans. At first glance, it appears that there is only a small racial

---

[12]U.S. Bureau of the Census, *Census of Housing: 1980*, HC80-1-B1, pt. 1, tables 88 and 89; *Annual Housing Survey, 1981*, series H-150-81, pt. C, tables A-1 and A-4.

[13]The Components of Inventory Change Study was based on data for about 49,000 housing units included in the 1973 Annual Housing Survey and 64,000 units from the 1980 Census of Housing. Shown on bottom of page 292 are data pertaining to the estimated market value of houses owned by whites and blacks in both 1973 and 1980. These data exclude houses in which there was a change in either tenure or the race of the owner, as well as houses built since 1973. The data pertain to single-family dwelling units and exclude farm houses. Amounts are in 1984 dollars.

difference in some of the most important benefits. In 1984, 57 percent of the black wage and salary workers had access to an employer- or union-provided group health plan. Forty-one percent of the black workers and 43 percent of the white workers had access to an employer- or union-provided pension plan. However, the benefits whites receive are apparently more generous. For example, 43 percent of the whites with a health care plan have all costs born by their employer or union compared with just 33 percent of the black workers who have health coverage.[14] If employer benefits were included, racial differences would also be larger than they appear on the basis of cash income only.

Third, cash welfare benefits such as Aid to Families with Dependent Children and Public Assistance are included in income tabulations, but noncash benefits are excluded, and these are much more frequently obtained by blacks than by whites. In recent years the Census Bureau has gathered data about the receipt of means-tested welfare benefits—that is, benefits which may be obtained only if income and wealth fall below certain cut-off points. The most common of these are food stamps, public housing or subsidized rent, Medicaid, and reduced-price or free school lunches. Since these benefits are often provided on a household rather than a personal basis, it is difficult to link them to an individual's income. Indeed, the data are obtained for households rather than for persons. Shown on page 293 are the proportions of black and

[13](continued)

	Owned by Blacks in Both Years			Owned by Whites in Both Years			Distribution of 1980 Owners by Place of Residence	
	1973 Value	1980 Value	% Change	1973 Value	1980 Value	% Change	Black	White
Total United States	$38.6	$37.5	−3%	$59.4	$64.0	+ 8%	100%	100%
Within Metropolitan Areas in 1980								
Central Cities	$39.5	$38.2	−4	$55.7	$61.9	+11	53	23
Suburban Rings	$49.8	$47.9	−4	$69.4	$72.6	+ 9	23	45
Outside Metropolitan Areas	$27.1	$29.5	+9	$45.6	50.9	+12	24	32

NOTE: Amounts shown in thousands.

SOURCE: U.S. Bureau of the Census, *Census of Housing: 1980*, HC80-1, table SA-3B.

[14]U.S. Bureau of Labor Statistics, *Employment and Earnings*, vol. 30, no. 1 (January 1983), table 46; U.S. Bureau of the Census, *Current Population Reports*, series P-60, no. 150 (November 1985), tables 4, 14, and 17.

white households who had one or more members benefiting from four of the largest noncash programs in 1984.[15]

There are large racial differences in who benefits from these programs. More than one quarter of the black households obtain food stamps, and half of those with children make use of subsidized school lunches. For each program, the proportion of blacks getting noncash transfers is three to four times that of whites. Were it possible to estimate the cash value of these benefits, racial differences in income might be reduced.[16]

Programs	Black	White
Total Households		
Food Stamps	25%	6%
Publically Subsidized Housing (Renters only)	25	9
Medicaid	27	7
Households with One or More Members Aged 5 to 18		
Free or Reduced-Price School Lunch	51	16

The absence of data about capital gains, the limited information about employer-paid benefits, and difficulties in estimating the value of noncash transfers prevent us from concluding whether the income statistics used in this chapter basically overstate or underestimate racial differences. Undoubtedly, income from capital gains and inheritance raise levels more among whites than blacks, but food stamps, Medicaid, and free school lunches benefit proportionately more blacks than whites.

Trends in the Incomes of Individuals

Have blacks been catching up with whites in terms of personal income, or are the racial differences just as large in the 1980s as they were at the end of World War II?

[15]U.S. Bureau of the Census, *Current Population Reports,* series P-60, no. 150 (November 1985), table B.

[16]The Census Bureau has not published estimates of personal income adjusted for the value of these noncash transfer payments. However, the Census Bureau has estimated poverty rates adjusted for such benefits. In 1979, using the money-only income concept, 29.9 percent of the black population and 9.9 percent of the white population were below the poverty line. Attributing a market value to food stamps, subsidized school lunches, and housing leads to revised poverty rates of 24.3 for blacks and 8.7 percent for whites. Smeeding (1982), table F-1. (This adjustment excludes medical or insurance benefits from Medicare or Medicaid.)

Before drawing conclusions about racial differences, we note several aspects of the income data used for our analysis. First, no one index perfectly summarizes something as complex as the income distribution, but the choice of a measure has important substantive implications. The two indexes most frequently used are the *median* and the *mean*. The median, which separates the upper half of the income distribution from the lower half, is often preferred to the mean because it does not give undue weight to those with very high incomes, and thus gives a better picture of the typical income. But, because the highest income categories in the United States include many white men but few women or blacks, racial and sexual differences in income appear much smaller when the median is used. In 1985, for example, median income figures for black and white men were $10,400 and $16,500, respectively, a difference of $6,100. The mean income of black men was $12,900 and of white men was $20,800, a much larger difference of $7,900.[17]

The amounts shown in this chapter are in constant dollars which have the purchasing power of 1984 dollars, adjusting for inflation and allowing us to conclude that increases over time in income represent real gains in purchasing power. These adjustments were made using the Consumer Price Index (CPI), calculated by the Bureau of Labor Statistics. Estimated on a monthly and annual basis, it represents the retail price of a package of goods and services for an unchanging array of consumer needs, such as food, shelter, clothing, and transportation. If the same items which cost $100 in one year are priced at $200 five years later, inflation is said to have halved the value of a dollar.[18]

Caution is needed when interpreting income figures using the CPI. For one thing, people change their tastes over time and in 1984 spent their income for a different mix of food, shelter, medical items, transportation, and clothing than they had three decades earlier.[19] It is also likely the economic progress during the 1970s is seriously understated by the use of the CPI, since it gives great weight to the cost of purchasing a home. As interest rates rose rapidly in the 1970s, the CPI also escalated, and while the cost of living certainly increased for that small fraction who became homeowners for the first time, most consumers were less affected by the rising interest rates.[20] Nevertheless, while some distortion is introduced by any index used to control for inflation, these distortions do not greatly affect the relative income position of blacks and whites.

[17]U.S. Bureau of the Census, *Current Population Reports*, series P-60, no. 154 (August 1986), table 8.

[18]Triplett (1981).

[19]Callahan (1981); Cagan and Moore (1981).

[20]Jencks (1986).

If blacks had been incorporated into the economic mainstream and if racial discrimination declined, we would expect that their incomes would approach those of whites. In Figure 10.1 we see a remarkable parallel in the income trend lines for black and white men. Increases in income for black men were more than matched by increases for white men; the racial difference in income grew wider through the mid-1960s and has contracted only a little since then. In 1948 the racial difference in median income for men was $5,000. This rose to $8,000 in 1965 and then fell to $6,100 in 1985. Prosperous and lean times affected the races similarly, but the racial gap in personal income hardly narrowed in the post–World War II era.

Changes for women are not at all similar to those for men. The median income of black women rose much more rapidly than that of white women, leading to smaller racial differences. In 1948 black women were $2,800 behind white women in median income, but in the late 1970s this difference declined to about $500. For most of the post–World War II era, black women were catching up with white women in terms of personal income.

Two decades ago Herman Miller described income trends as revealed by the census of 1960.[21] He stated his expectation that he would find declining racial differences in both the 1940s and 1950s, primarily because blacks moved into industrial jobs and because of their increased educational attainment. He limited his analysis to men but expressed surprise at his findings: Blacks made progress during the 1940s but the racial gap in income actually widened in the 1950s. Writing twenty years after Miller, we can be no more optimistic. Figure 10.1 suggests that, in terms of income, black men have gained very little on whites since the end of World War II, and the racial difference in purchasing power was actually greater in 1985 than it had been at the end of World War II. Should we conclude that the 1940s were the only years in which progress was made in reducing racial differences in the income of men? Have the geographic changes, the improvements in education, and the civil rights laws of the post–World War II era failed to reduce the advantage white men enjoy in income?

Table 10.4 provides answers by showing median income figures for 1939–85. The amounts for 1939 came from the census of 1940 and refer to wage and salary earnings only, while data for more recent years were gathered in the Census Bureau's March surveys.

Before answering questions about the income status of blacks, we need to define progress. Should we look at differences in amounts of in-

[21]Miller (1966), pp. 40–43. For another analysis of income trends for black men in the 1950s, see Batchelder (1964).

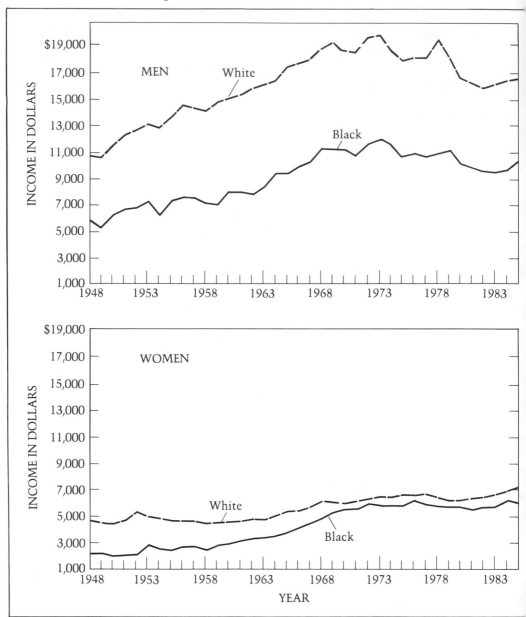

FIGURE 10.1

*Median Income for Black and White Men and Women Aged 14 and Over
Who Reported Income, 1948–1985 (in 1984 dollars)*

NOTE: Data for years since 1979 refer to people aged 15 and over; data for years through 1966 pertain
whites and nonwhites.

SOURCE: U.S. Bureau of the Census, *Current Population Reports,* series P-60, no. 151 (April 1986), ta
29; no. 154 (August 1986): table 8.

come, or should we focus on the relative status of blacks? Suppose that an employer has a talented sales manager who earns $25,000 per year and an efficient secretary who earns $10,000. At the end of the year, he decides that both of them deserve generous raises since they are key employees. He gives the sales manager a 20 percent increase, raising his salary by $5,000 to $30,000. He is even more impressed by the productivity of his secretary and gives her a 25 percent raise, increasing her salary by $2,500 to $12,500.

If we look at absolute differences in income, we conclude that this was a bad year for the secretary since her purchasing power fell further behind that of the sales manager. In the earlier year he earned $15,000 more than she did; in the next year, $17,500 more. The gap in income widened and the secretary made no progress in catching up with the highly paid sales manager.

On the other hand, the percentage increase in pay was greater for the secretary than for the sales manager: 25 percent versus 20 percent. In terms of her salary during the earlier year, she did better than the sales manager and if she continues to get 25 percent raises while he gets only 20 percent, her salary will equal his in eleven years. In relative terms, she also did well. At the beginning of the year her salary was 40 percent as large as his; at the end of the year, 42 percent.

If the secretary focuses on absolute differences in income, she will be disappointed with her pay increase and may think her boss does not appreciate her work. If she focuses on her income relative to that of the sales manager or her rate of increase compared with his, she may be quite happy. There is no consensus about whether the emphasis in income comparisons should be on relative differences or absolute differences. If a low-income group receives greater percentage increases than a high-income group, their incomes in the long run will equal or exceed those of the high-income group. In the short run they will often fall further behind in amounts of income since a small percentage increase for a high-income group leads to a bigger absolute change than does a large percentage increase for a low-income group.

If we look at income changes among blacks and whites, we often find that analyzing absolute differences and analyzing relative differences lead to opposite conclusions. Returning to Table 10.4, we find that black men made income progress relative to white men in three decades: the 1940s, the 1960s, and the 1970s. During the 1940s the black median as a percentage of the white rose from 41 to 48 percent, in the 1960s from 47 to 58 percent, and in the 1970s another 5 points. Despite these relative gains for black men, the racial difference in absolute amount—that is, the racial difference in purchasing power—

TABLE 10.4
Median Income for Black and White Adults Who Received Income, 1939–1985 (in 1984 dollars)

	1939	1949	1959	1969	1979	1985
MEN						
Median Income						
Black	$ 3,400	$ 5,200	$ 7,100	$11,200	$11,100	$ 10,400
White	8,300	10,800	15,000	19,200	17,600	16,500
Racial Gap	−4,900	−5,600	−7,900	−8,000	−6,500	−6,100
Black as Percentage of White	41%	48%	47%	58%	63%	63%
Percentage with Income						
Black	n.a.	85%	88%	86%	87%	87%
White	n.a.	89%	92%	93%	96%	96%
WOMEN						
Median Income						
Black	$ 1,800	$ 2,200	$ 3,000	$ 5,200	$ 5,800	$ 6,100
White	5,100	4,700	4,700	6,200	6,300	7,100
Racial Gap	−3,300	−2,500	−1,700	−1,000	−500	−1,000
Black as Percentage of White	36%	46%	62%	84%	92%	85%
Percentage with Income						
Black	n.a.	54% *	63%	73%	83%	85%
White	n.a.	46%	52%	65%	90%	91%

NOTES: Data for 1939 refer to wage and salary earnings only. Data from 1939 and 1949 were obtained from the decennial censuses. Data for subsequent years refer to all monetary income and were obtained from the Current Population Survey. Figures for 1939, 1949, and 1959 refer to whites and nonwhites; for later years, to whites and blacks. Through 1969, income questions were asked of persons aged 14 and over; since then, of persons aged 15 and over.

SOURCES: U.S. Bureau of the Census, *Census of Population: 1950*, P-C1, table 137; *Current Population Reports*, series P-60, nos. 80, 90, 127, 132, and 154.

increased throughout the 1940–70 period. In relative terms, black men moved closer to white men in income, but in absolute terms their purchasing power fell further behind.

Trends among women are more readily summarized. The incomes of black women increased more rapidly than those of white women in both absolute and relative terms throughout the 1940 to 1980 interval. Black income, as a percentage of white, rose from 36 percent to 92 percent.

It is important to observe that sexual differences in income increased during this span: Men of both races widened the income gap which separated them from women of their race. In 1939 black women had a median income $1,600 less than that of black men; in 1985, $4,300. Among whites, the sexual gap in median income tripled, from $3,200 to $9,400. When interpreting the meaning of these sexual disparities, we note that an increasing proportion of women report income and that the majority of men and women pool their incomes by sharing a household.

We have used the median income of people who obtained income as a measure of well-being. This is a changing pool of individuals. On the one hand, an increasing proportion of young whites and of women of both races are at work, many of them holding part-time jobs. Also, more elderly people now receive cash benefits from the governmental transfer programs. These changes tend to minimize increases over time in median income because the pool of recipients now includes many people who get small amounts of income. On the other hand, discussions of the black underclass suggest that an increasing percentage of black men—especially those in central city ghettoes—have no attachment to the labor force. If this is occurring, a declining percentage of black men might receive income, and an analysis that is limited to the median income of recipients will present an overly optimistic picture of the well-being of blacks.

Table 10.4 also shows the proportion of men and women who received income in each year and reveals that changes are quite modest. A better way to investigate economic well-being might be to use a per capita measure. Table 10.5 presents data about the per capita income of adults from 1949 through 1985. The income received by all persons in a group, such as all black men, was divided by the size of the group.

Trends in per capita income are similar to those in the income of recipients. The incomes of black men rose more rapidly than those of white men for much of the post–World War II period. In terms of the actual difference in per capita income, black men made no progress in reducing the large gap which separated them from white men.

THE COLOR LINE AND THE QUALITY OF LIFE IN AMERICA

TABLE 10.5

*Per Capita Income for the Adult Population by Race and Sex,
1949–1985 (in 1984 dollars)*

	1949	1959	1969	1979	1985
MEN					
Black	$ 5,100	$ 7,600	$10,900	$11,800	$11,300
White	11,000	15,3000	19,800	20,600	19,900
Racial Gap	− 5,900	− 7,700	− 8,900	− 8,800	− 8,600
Black as Percentage					
of White	46%	50%	55%	57%	57%
WOMEN					
Black	$ 1,700	$ 2,800	$5,100	$6,900	$7,400
White	2,600	3,500	5,500	7,800	8,400
Racial Gap	− 900	− 700	− 400	− 900	− 1,000
Black as Percentage					
of White	66%	79%	91%	88%	88%

NOTES: Data for 1949 were obtained from the 1950 census; for subsequent years, from the Current Population Survey. Data for 1959 refer to whites and nonwhites; for other years, to blacks and whites. Data through 1969 refer to the population aged 14 and over; for later years, aged 15 and over.

SOURCES: U.S. Bureau of the Census, *Census of Population: 1950*, P-C1, table 137; *Current Population Reports*, series P-60, no. 75 (December 14, 1970), table 45; no. 129 (November 1981), table 51; no. 154 (August 1986), table 8.

Once again, we find that the picture of change is more optimistic when the income trends of women are considered. During the 1950s and 1960s, the incomes of black women rose more rapidly than those of white women, but since 1970 the relative income position of black women has not improved. Even though black women in the mid-1980s had incomes that were 88 percent of those of white women, the racial gap in median income more than doubled between 1969 and 1985.

Income Trends by Region

Blacks in the North and West traditionally had larger incomes than those in the South for several reasons. Few northern or western blacks worked at the low-paying jobs in agriculture, which employed the majority of southern blacks. Racial attitudes regarding equal treatment for blacks have been more conservative in the South than elsewhere, presumably leading to more discrimination in the job market.[22] In addition, the state-funded components of transfer programs such as Aid to

[22]Schuman, Steeh, and Bobo (1985), chap. 3.

300

Families with Dependent Children were slower to develop in the South and continue to provide less support than those in the North or West.[23] There is a long history in the South of both keeping blacks off state-funded support programs and making their benefits smaller than those of whites.[24]

As we have seen, black men have made little progress in narrowing the income gap that separates them from white men. Does this apply to all regions or has there been rapid convergence in some places and no change in others? Information about regional trends is shown in Table 10.6. For 1939 we are once again restricted to information about wage and salary income, while for the other years income from all sources is reported.

Blacks in the South have always had smaller incomes than those in other regions, but the regional disparity has decreased. In 1939 black men in the South had a median income less than half of that of black men in the Northeast, the region where they were most prosperous. By 1984 the median income for black men in the South was about 80 percent of that of black men in the West, the most prosperous region. Among women, there was an even more rapid regional convergence. In terms of income blacks in the South are gradually catching up with blacks in the rest of the nation.

An examination of trends over time within each of the regions reveals differences in the rate of black progress. Black men made relative gains in income during the 1940s in all regions. During the 1950s there were small gains for black men in the Northeast and West; in the 1960s there were gains in all regions. The South was distinguished from other regions in the 1970s because black men made relative gains in that region but in no other. The years following 1979 were not prosperous for black men in any region since their incomes fell more precipitously than those of whites in all regions but the West.

The Midwest is unique in its pattern of change. During the 1940s the incomes of black men rose more rapidly than those of white men, and by 1950 racial differences in income were relatively small. This is attributable to industrialization during and after World War II, which attracted many blacks to the region's steel mills, automobile factories, and rubber plants. Employers offered good wages, and the presence of industrial unions may have curtailed racial discrimination. However, the following three decades were not prosperous for midwestern blacks since their incomes rose at a slower rate than those of whites. Indeed, the median income of black men in this region was greater, in constant

[23]Smeeding (1984), table 11.
[24]Myrdal (1944), pp. 359–60; Piven and Cloward (1971), pp. 115 and 133; Sitkoff (1978); Weiss (1983), pp. 58–59.

TABLE 10.6

Median Income by Region for Black and White Men and Women Who Received Income, 1939–1984 (in 1984 dollars)

	Men						Women					
	1939	1949	1959	1969	1979	1984	1939	1949	1959	1969	1979	1984
NORTHEAST												
Median Income												
Black Median	$5,900	$9,000	$11,800	$15,500	$12,800	$10,500	$3,500	$5,100	$6,100	$8,600	$7,700	$6,500
White Median	8,800	13,000	16,600	20,600	18,600	16,900	5,300	6,100	6,200	7,300	7,700	7,000
Racial Gap	−2,900	−4,000	−4,800	−5,100	−5,800	−6,400	−1,800	−1,000	−100	+1,300	<100	−500
Black as Percentage of White	67%	69%	72%	75%	69%	62%	67%	83%	99%	118%	103%	94%
MIDWEST												
Median Income												
Black Median	$5,700	$9,500	$12,400	$16,200	$13,500	$8,600	$2,900	$4,100	$4,800	$7,300	$7,300	$5,900
White Median	8,500	11,900	16,100	20,200	19,400	16,300	4,900	4,700	4,900	6,200	7,400	6,600
Racial Gap	−2,800	−2,400	−3,700	−4,000	−5,900	−7,700	−2,000	−600	−100	+1,100	−100	−700
Black as Percentage of White	67%	80%	77%	80%	70%	52%	60%	86%	97%	117%	98%	90%
SOUTH												
Median Income												
Black Median	$2,800	4,500	$5,900	$9,200	$10,000	$8,800	$1,400	$1,900	$2,600	$4,400	$5,700	$5,800
White Median	6,500	9,000	12,600	16,500	17,000	15,600	4,700	4,100	4,700	6,500	7,600	6,900
Racial Gap	−3,700	−4,500	−6,700	−7,300	−7,000	−6,800	−3,300	−2,200	−2,100	−2,100	−1,900	−1,100
Black as Percentage of White	43%	50%	47%	56%	59%	56%	31%	47%	56%	68%	75%	84%
WEST												
Median Income												
Black Median	$5,200	$8,100	$12,100	$14,400	$12,800	$12,000	$3,400	$3,900	$5,500	$7,600	$8,300	$7,600
White Median	8,900	11,900	17,000	20,000	19,500	17,500	5,700	4,400	5,700	7,000	8,200	7,600
Racial Gap	−3,700	−3,800	−4,900	−5,600	−6,600	−5,500	−2,300	−500	−200	+600	+100	<100
Black as Percentage of White	56%	69%	71%	72%	65%	69%	60%	88%	97%	109%	101%	101%

NOTES: Data for 1939 refer to the wage and salary earnings of people who were in the labor force in April 1940; for subsequent years, to the total income of all persons reporting income. Figures for 1939 and 1949 are for whites and nonwhites; for subsequent years, for whites and blacks. Data for 1984 were obtained from the Current Population Survey; for other years, from the decennial census. Figures through 1969 refer to the population aged 14 and over; for later years, aged 15 and over.

SOURCES: U.S. Bureau of the Census, *Sixteenth Census of the United States: 1940, Population, The Labor Force, Wage and Salary Income in 1939; Census of Population: 1950, P-C1, table 162; Census of Population: 1960, PC(1)-1D, table 262; Census of Population: 1980, PC80-1-C1, table 95; Current Population Reports*, series P-60, no. 151 (April 1986), table 30.

dollars, in 1949 than in 1984. Miller pointed out this trend two decades ago when he observed that even during the booming post-World War II era, the incomes of blacks in the Midwest failed to move closer to those of whites.[25] By 1984 black men in the South had larger incomes than black men in the Midwest.

Looking at the racial gap in purchasing power among men—as measured by the difference in median income—suggests even less progress. In every region the census of 1980 found a larger racial difference separating white men from black than did the censuses of 1940 and 1950.

Once again, trends among women differ from those among men. Incomes of black women rose more rapidly than those of white women in all regions between 1940 and 1970. Not only was there relative progress but actual racial differences in median incomes declined, and by 1970 black women outside the South reported larger incomes than white women. In the Northeast and Midwest, for example, black women averaged more than $1,000 per year in income than white women. It is rare to find that blacks are ahead of whites on any important indicator of economic status. The advantage of black women in 1969 is explained, in part, by their greater work effort. In the Northeast, for example, 36 percent of the black women who had income worked full time for the entire year compared with 31 percent of the white women.[26] During the 1970s the employment patterns of white women came to resemble those of blacks and the monetary advantage of blacks declined or disappeared in all regions since the median incomes of white women rose much faster than those of black women.

Income Trends by Educational Attainment

If discriminatory practices are outlawed and if the attitudes of whites change, we might anticipate that the incomes of highly trained blacks would be the first to converge with those of whites. An employer might be apprehensive about hiring a black who had dropped out of high school, but if a black applicant had an actuarial degree or an MBA, the risks might seem small. The census does not measure talent, ask about specialized training, or determine who holds an occupational certificate or license. However, it obtains information about educational attainment. If there is a racial convergence in economic

[25]Miller (1966), pp. 85–88. For a further analysis of regional trends in relative income, see Reich (1981), pp. 46–49.

[26]U.S. Bureau of the Census, *Current Population Reports,* series P-60, no. 75 (December 14, 1970), table 59.

status, many would expect it to happen first among the college educated.[27]

Table 10.7 shows income trends for educational groups. Since many people do not complete their schooling until they are in their 20s, this information refers to persons aged 25 and over. There has, of course, been a great change over time in the selectivity of these educational groups. In the past, college graduates were a very small fraction of the black population, but this is no longer the case. In 1950 college graduates made up just 2 percent of the adult black population but in 1984, 11 percent. While the college graduate category has become less selective, the elementary education category represents a rapidly declining fraction since the proportion of adult blacks with less than 8 years of education fell from 61 percent in 1950 to 15 percent in 1984.[28]

The income figures reveal, on the one hand, some evidence of racial progress and, on the other, very persistent and large race differences. In 1949 black men who graduated from college had incomes about equal to those of white men who dropped out after 8 years of elementary school.[29] Since the late 1970s black men with a college degree have had incomes as large as those of white men who attended but did not graduate from college.

On measures of relative income, black men at each educational level gained on whites, particularly during the 1960s. This tendency toward a racial convergence apparently ceased in the first half of the 1980s, and, at most levels, the incomes of black men fell more rapidly than those of white men.

Racial differences in relative incomes hardly vary by how long men attend school. In 1949 black men at all educational levels reported incomes about 64 percent of those of comparable white men; in 1984, about 71 percent. Differences in absolute amounts of income, of course, are directly related to educational attainment. In 1984 black men with four or more years of college training received $8,900 less than comparably educated white men, while for those with only one to three years of secondary school, the racial difference in median income was only $3,800. The more years a black man spends in school, the further his purchasing power falls behind that of white peers.

Racial progress is considerably less when the racial gap in income is taken as the measure of economic change. Black male college graduates had incomes $8,100 less than those of white college graduates in 1949, a

[27]Freeman (1976), chap. 1; Wilson (1978), pp. 99–104.
[28]U.S. Bureau of the Census, *Census of Population: 1950*, P-C1, table 115; *Current Population Reports*, series P-60, no. 151 (April 1986), table 33.
[29]Miller argues that the situation in 1950 was an improvement over 1940 since in the earlier year black men with college degrees earned less, on average, than white men who did not complete elementary school. Miller (1955), p. 46.

Median Income by Educational Attainment for Black and White Men and Women Who Received Income, 1949–1984 (in 1984 dollars)

	Men					Women				
	1949	1959	1969	1979	1984	1949	1959	1969	1979	1984
COLLEGE GRADUATES										
Black Median	$11,400	$17,300	$25,000	$24,300	$21,900	$ 9,000	$13,300	$18,100	$18,100	$18,400
White Median	19,500	27,700	36,300	32,200	30,800	10,200	13,400	16,500	15,600	15,700
Racial Gap	−8,100	−10,400	−11,300	−7,900	−8,900	−1,200	−100	+1,600	+2,500	+2,700
Black as Percentage of White	59%	62%	69%	76%	71%	88%	99%	109%	116%	118%
1 to 3 YEARS OF COLLEGE										
Black Median	$ 9,400	$14,400	$20,200	$18,300	$16,300	$ 5,500	$ 7,700	$12,100	$12,400	$11,600
White Median	15,600	21,800	27,200	25,100	22,100	7,400	8,600	10,300	11,000	10,300
Racial Gap	−5,700	−7,400	−7,000	−6,800	−5,700	−1,900	−900	+1,700	+1,400	+1,300
Black as Percentage of White	63%	66%	74%	73%	74%	74%	89%	118%	114%	113%
HIGH SCHOOL GRADUATES										
Black Median	$ 9,800	$13,300	$18,000	$16,200	$12,900	$ 4,800	$ 6,100	$ 9,600	$ 9,800	$ 8,400
White Median	14,500	19,700	24,700	23,400	19,600	7,000	8,000	9,500	9,300	7,800
Racial Gap	−4,700	−6,400	−6,700	−7,100	−6,700	−2,200	−1,900	+100	+500	+600
Black as Percentage of White	68%	68%	73%	69%	66%	67%	76%	101%	104%	108%
1 TO 3 YEARS OF HIGH SCHOOL										
Black Median	$ 8,700	$11,600	$15,200	$12,600	$ 9,600	$ 3,500	$ 4,300	$ 6,800	$ 6,500	$ 5,000
White Median	13,100	17,900	21,400	17,400	13,400	5,000	6,000	7,200	6,700	5,700
Racial Gap	−4,400	−6,300	−6,200	−4,800	−3,800	−1,500	−1,700	−400	−200	−700
Black as Percentage of White	66%	65%	71%	72%	71%	70%	71%	94%	96%	89%
8 YEARS OF ELEMENTARY SCHOOL										
Black Median	$ 8,100	$10,400	$12,700	$10,200	$ 7,300	$ 3,200	$ 3,500	$ 4,900	$ 5,100	$ 5,000
White Median	11,300	14,200	15,900	13,000	10,600	4,100	4,000	4,800	5,400	5,200
Racial Gap	−3,200	−3,800	−3,200	−2,800	−3,300	−900	−500	+100	−300	−200
Black as Percentage of White	72%	73%	80%	79%	69%	79%	86%	101%	95%	95%
LESS THAN 8 YEARS OF ELEMENTARY SCHOOL										
Black Median	$ 5,200	$ 6,500	$ 8,400	$ 7,600	$ 6,100	$ 2,100	$ 2,600	$ 3,600	$ 4,500	$ 4,000
White Median	9,100	9,700	10,900	9,900	8,000	3,200	3,100	4,100	5,000	4,600
Racial Gap	−3,900	−3,200	−2,500	−2,300	−1,900	−1,100	−500	−500	−500	−600
Black as Percentage of White	57%	67%	78%	76%	76%	63%	84%	88%	90%	88%

NOTE: Data for 1949 and 1959 refer to whites and nonwhites. Data for 1984 were obtained from the Current Population Survey; for earlier years, from the decennial census. Income figures refer to persons aged 25 and over.

SOURCES: U.S. Bureau of the Census, *Census of Population: 1950,* P-E, no. 5B, table 12; *Census of Population: 1960,* PC[1]-1D, table 223; *Census of Population: 1970,* PC[2]-5B, tables 7 and 8; *Census of Population and Housing: 1980,* Public Use Microdata Samples: *Current Population Reports,* series P-60, no. 151 (April 1986), table 33.

gap that grew to \$11,300 in 1969 and then declined to \$8,900 in 1984. At most educational levels the racial gap in median income of men was just about as great in 1984 as in 1950.

Among women, we find the expected pattern of racial change since extensively educated black women caught up with white women at an early date. By 1960—half a decade before Title VII banned discrimination in employment—black women with college degrees had average incomes similar to those of college-educated white women. A look at the full array of information for women shows that racial differences declined at all attainment levels throughout the post–World War II era in both absolute and relative terms. Since the late 1960s black women with high school diplomas have received more income than comparable white women, attributable in part to their greater hours of employment.

Income Trends by Age Groups

There has been a slight improvement in the relative income of black men and a large decline in the racial gap in income of women. Do these changes occur on a cohort basis or a period basis? One might assume that young blacks and whites who enter the labor force are more alike in both their educational attainment and their aspirations than were blacks and whites who began their careers many decades ago. If so, we would anticipate that racial differences in income would get smaller with each new entering cohort. A process such as this would gradually lead to an elimination of racial differences as younger cohorts replace older ones.

It is also plausible that racial change occurs in a given *period* rather than on a cohort basis. The 1960s witnessed many civil rights activities and the enactment of more laws than did the previous or the following decades. Perhaps during this decade employers terminated those practices that either kept blacks off the payroll or unfairly assigned them to low-wage jobs. If this decade—or any other decade—was one of momentous racial change, we might expect that income differences contracted for all age groups simultaneously, that is, a *period* pattern of change.

Table 10.8 shows the median income of black and white men classified by age and allows us to examine changes on a period basis for specific age groups. Recall that we have been looking at both the relative income of blacks and the absolute racial difference in income.

Among men, there is clear evidence of a *period* pattern of change. Black incomes—as a percentage of whites—rose more during the 1960s than in the previous decade across the range of ages. During the 1970s the incomes of black and white men in most age groups rose at about

TABLE 10.8

Median Income by Age for Black and White Men and Women Who Received Income, 1949–1984 (in 1984 dollars)

	Men					Women				
	1949	1959	1969	1979	1984	1949	1959	1969	1979	1984
AGES 15–24										
Black Median	$3,400	$3,800	$5,100	$5,200	$3,300	$1,900	$2,700	$4,200	$4,200	$3,100
White Median	5,200	5,600	5,600	7,000	5,000	4,100	3,500	4,400	4,900	3,700
Racial Gap	−1,800	−1,600	−500	−1,800	−1,700	−2,200	−800	−200	−700	−600
Black as Percentage of White	66%	71%	91%	75%	67%	45%	79%	95%	85%	83%
AGES 25–34										
Black Median	$7,100	$10,400	$15,900	$14,500	$11,500	$3,300	$4,600	$8,900	$10,100	$8,700
White Median	12,600	18,000	23,200	21,500	19,100	6,300	7,100	9,500	10,900	9,500
Racial Gap	−5,500	−7,600	−7,300	−7,000	−6,600	−3,000	−2,500	−600	−800	−800
Black as Percentage of White	57%	58%	68%	69%	60%	52%	65%	94%	93%	91%
AGES 35–44										
Black Median	$7,400	$11,500	$16,600	$18,100	$15,900	$3,200	$4,700	$8,600	$10,700	$10,100
White Median	14,000	20,200	27,000	28,000	25,500	6,600	7,800	9,800	10,400	9,400
Racial Gap	−6,600	−8,700	−10,400	−9,900	−9,600	−3,400	−3,100	−1,200	+300	+700
Black as Percentage of White	53%	57%	62%	65%	63%	49%	60%	89%	103%	107%
AGES 45–54										
Black Median	$7,000	$10,000	$15,200	$16,800	$15,300	$2,800	$3,600	$7,100	$8,800	$9,000
White Median	13,600	19,000	26,000	28,600	25,700	6,400	8,400	10,600	10,800	8,900
Racial Gap	−6,600	−9,000	−10,800	−11,700	−10,400	−3,600	−4,800	−3,500	−2,200	+100
Black as Percentage of White	51%	53%	59%	59%	60%	44%	43%	67%	80%	101%
AGES 55–64										
Black Median	$5,500	$8,000	$11,500	$12,400	$9,800	$2,200	$2,900	$4,700	$5,700	$5,800
White Median	11,600	16,400	21,600	23,000	20,700	4,800	6,200	8,800	8,800	7,000
Racial Gap	−6,100	−8,400	−10,100	−10,400	−10,900	−2,600	−3,300	−4,100	−3,100	−1,200
Black as Percentage of White	47%	49%	53%	54%	47%	46%	47%	54%	65%	83%
AGE 65 AND OVER										
Black Median	$2,600	$3,400	$4,900	$5,900	$6,200	$1,700	$2,200	$3,100	$4,000	$4,300
White Median	5,300	6,500	8,500	10,600	10,900	2,700	2,900	4,200	5,600	6,300
Racial Gap	−2,700	−3,100	−3,600	−4,700	−4,700	−1,000	−700	−1,100	−1,600	−2,000
Black as Percentage of White	49%	53%	57%	56%	57%	62%	78%	73%	73%	69%

NOTE: Data for 1949 and 1959 refer to whites and nonwhites. Data for 1949 to 1969 refer to persons aged 14 to 24 who reported income; for later years, aged 15 and over. Data for 1984 were obtained from the Current Population Survey; for earlier years, from the decennial census.

SOURCES: U.S. Bureau of the Census, *Census of Population: 1950*, P-C1, table 139; *Census of Population: 1960*, PC(1)-1D, table 219; *Census of Population: 1970*, PC(1)-D1, table 245; *Census of Population: 1980*, PC80-1-D1A, table 293; *Current Population Reports*, series P-60, no. 151 (April 1986), table 32.

the same rate. A different pattern is exhibited in the 1980s when the incomes of black men in all age groups, except 65 and over, fell more rapidly than those of white men, negating some of the gains recorded in the prosperous 1960s.

Once again, we see that the progress was much less when differences in purchasing power of men are considered. The racial gap in purchasing power among men increased in 5 of the 6 age groups in the 1950s and in 4 groups in the 1960s. Since 1974 income levels have been falling and, in several age groups, the racial gap has contracted, although it was still greater in 1984 than in 1949 except for men under age 25.

The 1960s were prosperous years for black women; their relative incomes increased and the gap that separated their median income from that of white women declined. However, black women also made progress vis-à-vis white women in the 1950s and 1970s. By the early 1980s black women aged 35 to 54 enjoyed an income advantage over their white peers.

Figure 10.2 shows change on a *cohort* basis. The upper panels display the relative income of birth cohorts of blacks while the lower ones show the racial difference in median income. Cohorts are identified by the census year in which they reached ages 15 to 24. Thus, the cohort of 1970 consists of men born between 1945 and 1954 who reached ages 15 to 24 in 1970. If blacks make gains on a cohort basis, the trend line for any younger cohort should be above the trend lines for older cohorts, meaning a higher ratio of median incomes among the younger people. If a cohort of blacks gains on whites as they grow older, the trend line for that cohort will shift upward toward the upper left-hand corner in this figure. Data for this analysis come from the four most recent censuses so that we cannot obtain income information for all ages for any cohort.

A look at trends among men reveals small gains on a cohort basis. For relative earnings (the upper panel), the trend lines for recent cohorts are slightly above those for previous cohorts, which means that incoming cohorts of black and white men are more similar in income in relative terms. The pattern is more mixed with regard to racial differences in the median income of men (the lower panel). At many ages, the racial differences are larger for recent cohorts than for older ones. For example, among men who entered the labor market on the verge of the Depression, the racial difference in income when they attained ages 35 to 44 averaged $6,600. For those who entered at the start of the 1960s, the racial difference at the identical ages was $9,900. Recent cohorts of blacks are at least as far or even further behind whites than were previous cohorts in terms of absolute amounts of income and thus there is little support for the hypothesis that a cohort pattern will eliminate racial differences among men.

As the same birth cohort grows older, do racial differences get smaller or larger? Figure 10.2 provides an unambiguous answer. The trend lines for men in the upper left-hand figure are basically flat, meaning that, in relative terms, black men neither gain nor lose very much as they grow older. Their income as a percentage of that of white men stays just about constant. However, in dollar amounts, the racial difference grows much wider, at least until retirement ages where Social Security benefits reduce racial differences (see lower left-hand panel in Figure 10.2), because black and white men have very different income patterns by age. In brief, the income of white men rises rapidly as they move from their 20s to their late 40s, but income gains with age are modest for black men and almost nonexistent for women.

Figure 10.3 illustrates this by showing the pattern of income by age in 1980. These are cross-sectional data, but a similar pattern can be found if cohort data are assembled from recent censuses. The typical white man in his late 20s can look forward to two decades of rapidly escalating income, but a black man can expect no more than modest increases. As an outcome, the racial differences in actual income grow much larger as cohorts of men get older. Black men begin their careers with smaller incomes than white men and then fall further behind.

The right-hand side of Figure 10.2 presents cohort trends in income for women. There is a clear cohort pattern here, and black women who entered the labor market recently have incomes more similar to those of white women—in both relative and absolute terms—than did black women in the past. There is one other important sexual difference. Unlike the situation among men, black women catch up with their white age mates as they grow older. No matter what the yardstick of progress, black women have gained on whites.

We have examined these income trends by age to determine whether change took place on a period or cohort basis. For men, most of the change occurred on a period basis because the relative income position of black men at all ages improved in the 1960s but changed only a little in the 1950s or 1970s. If improvements took place on a cohort basis, we would expect that black men who entered the labor market in the 1970s would have incomes more similar to those of whites than did previous cohorts, but this is not the case. Rather, the 1960s may be identified as the only postwar decade in which the income of black men rose more rapidly than that of whites.

Among women, it is difficult to distinguish cohort from period change. The 1960s were a beneficial period for black women, but in all

FIGURE 10.2

Black Median Income as a Percentage of White and Racial Differences in Income by Age for Birth Cohorts

BLACK MEDIAN INCOME AS A PERCENTAGE OF WHITE BY AGE

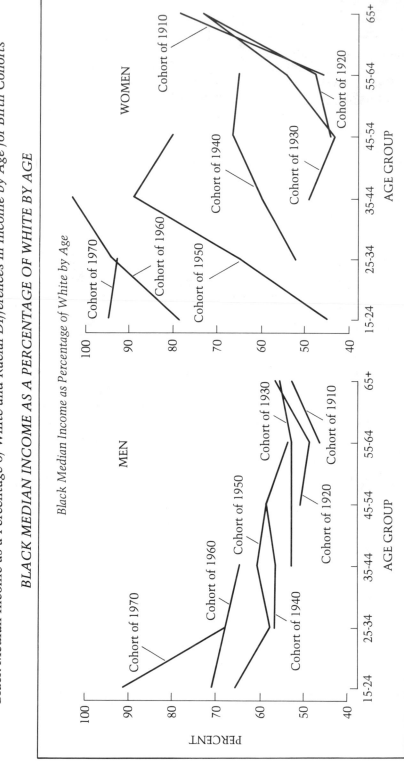

DIFFERENCES IN MEDIAN INCOMES BY AGE

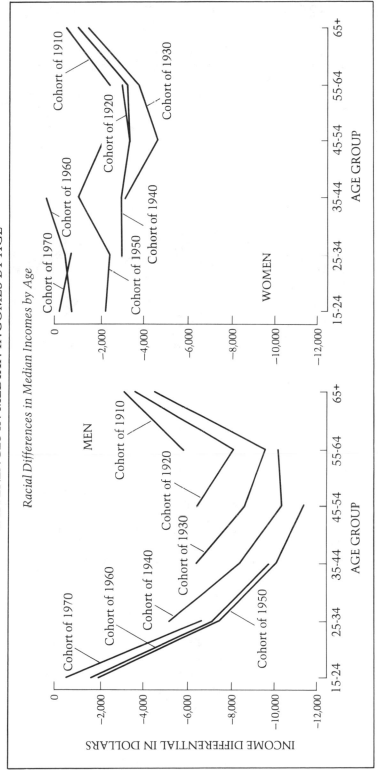

Racial Differences in Median Incomes by Age

NOTE: Data refer to persons reporting income. Differences shown in 1984 dollar amounts; cohorts identified by census year in which they reached ages 25–34.

SOURCE: See Table 10.5 and U.S. Bureau of the Census, *Census of Population and Housing: 1980*, Public Use Microdata Samples.

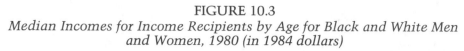

FIGURE 10.3

*Median Incomes for Income Recipients by Age for Black and White Men
and Women, 1980 (in 1984 dollars)*

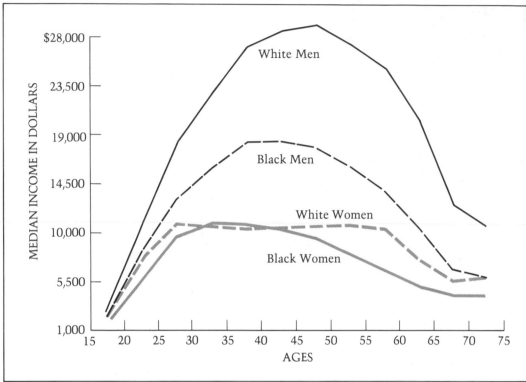

SOURCE: U.S. Bureau of the Census, *Census of Population: 1980,* PC80-1-D-1A, table 293.

decades and at most ages the incomes of black women moved closer to
those of white women.

These findings with regard to the income trends of black men are
quite pessimistic. It is difficult to imagine how any civil rights legislation or social policy could eliminate the racial gap in income among
those who are well into their careers. It is impossible to go back and
change the educational experience of blacks or retroactively eliminate
the discrimination which once slotted them into bottom-rung jobs. It is
more realistic to expect that black and white men who entered the labor force recently would share the same economic status and that the
incomes of blacks would stay close to those of whites as they aged. In
fact, this has not happened.

Summary

Following 1940 personal incomes increased among both races and both sexes leading to much higher standards of living. Most groups reported their peak incomes in the early 1970s, but the decade following the 1973–75 recession has been one of modest declines in income, especially for men.

Among men, there was some evidence of a trend towards a racial convergence of incomes between 1940 and 1970 since those of blacks rose more rapidly than those of whites. The median income of black men as a percentage of that of whites increased from 41 percent in 1939 to 58 percent in 1969. However, the racial gap in income did not close, so a focus on relative incomes leads to more favorable conclusions than does a focus on the actual racial difference. Since the mid-1970s the incomes of black men have fallen more rapidly than those of white men, leading to even larger racial differences. Until information is available for later years, we will not know whether this a temporary pause in a long-term trend toward racial convergence or the beginning of a new era of growing racial disparities.

Among women, there is a clearer pattern of narrowing racial gaps. By 1970 black women approached income parity with white women, and, by 1980, black women in many groups had higher average incomes than white women, the result, in large part, of the greater hours of employment of black women.

Blacks and whites differ greatly in their wealth holdings, with the typical white household having assets about eleven times those of the typical black household, a difference that is very much larger than the racial difference in current income. This has important consequences for the sources of incomes, because whites are about two and a half times as likely as blacks to obtain income from property or financial investments. Such sources provide 12 percent of the total income of white women and 5 percent of that of white men. For blacks, the corresponding figures are 2 percent for women and 1 percent for men. Blacks, on the other hand, are much more likely than whites to receive means-tested transfer benefits such as welfare, public assistance, and Aid to Families with Dependent Children. Such sources provide about 10 percent of the income of black women but less than 1 percent of that of white women.

The analysis of regional trends shows that the personal incomes of blacks in the South rose faster than those of blacks in other regions, but racial differences are persistently greater in the South, and there is no region in which the incomes of black men have caught up with those of white men. The Midwest is unique since black men made few gains since the 1940s.

Income levels increase with educational attainment, but the pattern of racial differences varies by sex. Black women with a high school education or more have larger incomes than do similarly educated white women. Among men, however, the income of blacks relative to that of whites hardly varies by educational attainment, and racial differences in income actually increase with education. There is no evidence that college-educated black men are now approaching income parity with college-educated white men. Remaining in school for additional years increases the income of a black man, but it will not bring his income closer to that of his white peers.

Changes in the relative incomes of black men occurred more on a period basis than on a cohort basis, with the 1960s being the decade of greatest change. Younger cohorts of black and white men are *not* closer in income than were previous cohorts when they entered the labor market. As cohorts move through the life cycle, black men fall further and further behind their white age mates. Among women, the trend toward racial parity in income occurred on both a period and a cohort basis.

Income statistics omit some important sources, so we cannot be certain whether racial differences are greater or smaller than those described in this chapter. The very large racial difference in assets suggests that if capital gains, inheritances, and gifts were included, racial differences would be larger. Whites also benefit more than blacks from perquisites as they are more likely to hold prestigious jobs. Noncash transfer programs, on the other hand, expanded rapidly in the later 1960s and 1970s and benefit proportionally more blacks than whites. If income statistics included the cash value of food stamps, school lunches, subsidized housing, and Medicaid, the income levels of blacks would be raised somewhat more than those of whites, leading, perhaps, to smaller racial differences.

APPENDIX TABLE 10.1

Comparison of Income Data Gathered in the March 1980 Current Population Survey and the April 1980 Census for the Population Aged 15 and Over (in 1980 dollars)

	Men			Women		
			Census			Census
	CPS	Census	CPS	CPS	Census	CPS
Total						
For Income Recipients:						
Mean	$14,369	$15,124	1.053	$ 6,032	$ 6,928	1.149
Median	11,845	12,192	1.029	4,354	5,263	1.209
Percent with Income	95.4	92.2	.966	88.9	76.4	.859
Per Capita Income	$13,708	$13,944	1.017	$ 5,362	$ 5,292	.987
N (in thousands)	80,218	83,825	1.045	87,980	91,483	1.040
WHITE						
For Income Recipients:						
Mean	$14,920	$15,865	1.063	$ 6,066	$ 7,030	1.159
Median	12,357	12,881	1.042	4,394	5,356	1.219
Percent with Income	96.3	93.8	.974	89.7	76.4	.852
Per Capita Income	$14,368	$14,875	1.035	$ 5,441	$ 5,371	.987
N (in thousands)	70,632	71,348	1.010	76,480	77,346	1.011
BLACK						
For Income Recipients:						
Mean	$ 9,383	$ 9,728	1.037	$ 5,678	$ 6,276	1.123
Median	7,745	7,827	1.106	4,023	4,674	1.162
Percent with Income	88.2	82.7	.938	84.5	78.7	.931
Per Capita Income	$ 8,276	$ 8,045	.972	$ 4,798	$ 4,939	1.029
N (in thousands)	8,067	8,678	1.076	9,828	10,219	1.040

SOURCE: U.S. Bureau of Census, *Current Population Reports*, series P-60, no. 129 (November 1981), table 46; *Census of Population: 1980*, PC80-1-C1, table 95.

11

THE EARNINGS
OF EMPLOYED WORKERS

I NCOME trends describe the overall economic status of blacks, but because income includes monies from rents, dividends, and transfer programs, these trends do not tell us how blacks are doing in the job market. Both the Fair Employment Practices Commission in the 1940s and the Civil Rights Act of 1964 sought to ensure that blacks and whites who perform the same jobs earn the same amount. In this chapter we examine earnings so that we can determine how blacks are faring and how their status has changed since 1960. Recall that earnings are the wages or salary a person receives for working at a job, or, if the person is self-employed, they are roughly equal to the difference between gross income and the cost of doing business. The figures from the Census Bureau which we analyze refer to pretax cash earnings and exclude those fringe benefits and perquisites some workers receive.

One of the best ways to study racial differences is to take a large sample of blacks and whites who are completing their schooling and match them with regard to test scores, aspirations, place of residence, and other factors which influence earnings. If we followed these people for two or three decades, we could ascertain whether blacks and whites got the same jobs and worked as many hours and whether those with similar jobs earned similar salaries. But it is not practical to carry out such studies since they are very expensive, and it is impossible to conveniently control for all the factors which influence earnings. As a substitute we use census data which allows us to match blacks and whites statistically on the most important determinants of earnings.

Table 11.1 presents the hourly wage rates in 1979 of blacks and whites who worked in selected occupations and had a given educational attainment.[1] These data—a small sample of those available from the census—pertain to the most stable employees since they worked full time for the entire year preceding the census. Information is shown for two age groups: those aged 25 to 34 and 55 to 64 in 1980.

Controlling for occupation, education, age, and duration of employment in 1979, black men earned less than white men. Young black lawyers, for instance, averaged about $2.00 less per hour than young white lawyers, while among men who cleaned buildings, whites enjoyed an advantage of about $1.33 per hour. In some occupations racial differences were small, at least among the young. Black men, for example, who went into teaching, engineering, or urban planning could expect to begin their careers earning just a little less than comparable white men.

Among men, age was an important factor since racial differences were much greater among those advanced in their careers than among those who recently entered the job market. This may reflect the beneficial effects of the Civil Rights Act on hiring practices, but it may also come about because the earnings of white men increase with age much more rapidly than those of black men.

Racial differences were smaller among women than among men, and in many comparisons black women earned the greater amounts. Young black women who recently graduated from law school earned $1.20 more per hour than young white women. At the other end of the educational spectrum, young black women who drove motor vehicles also earned more per hour than young white women. Table 11.1 makes 32 comparisons of the wages of black and white women. In 14 of these comparisons, blacks had higher earnings, which suggests that there was not a pervasive pattern of racial discrimination regarding the earnings of women who are full-time workers. Of course, it is possible that both black and white women suffer from sexual discrimination. In 62 of the 64 comparisons in this table, women earned less than men of comparable age and education who held similar jobs.

Investigators who observe racial differences in the pay rates of comparably educated blacks and whites frequently assume that these differences reflect the poorer quality of black schools and the fact that blacks typically score lower on standardized tests of achievement than do

---

[1]For similar tabulations for 1960 and 1970, see U.S. Bureau of the Census, *Census of Population: 1960*, PC(2)–7B; *Census of Population: 1970*, PC(2)–8B.

TABLE 11.1

Hourly Wage Rates in 1979 for Full-Time, Year-Round Workers in Selected Occupations with Specified Educations, by Race, Sex, and Age (in 1984 dollars)

	Men					
	25–34 Years			55–64 Years		
	Black	White	Differ-ence	Black	White	Differ-ence
5 YEARS OF COLLEGE OR MORE						
University Professors	$10.77	$11.09	$ – .32	$15.57	$19.13	$ – 3.56
Doctors and Dentists	13.44	16.83	– 3.39	26.83	33.09	– 6.26
Lawyers and Judges	14.49	16.41	– 1.92	21.80	31.70	– 9.90
Social Scientists and						
Urban Planners	13.56	13.78	– .22	17.84	22.78	– 4.94
All Workers	12.07	13.60	– 1.53	15.93	22.94	– 7.01
4 YEARS OF COLLEGE						
Elementary and						
Secondary Teachers	$ 8.47	$ 8.65	$ – .18	$11.03	$12.44	$ – 1.41
Engineers, Architects,						
and Surveyors	14.28	14.56	– .28	17.87	21.51	– 3.64
Mathematical and						
Computer Scientists	14.33	14.45	– .12	18.60	20.17	– 1.57
Sales Representatives in						
Finance and Business	10.75	14.53	– 3.78	9.37	20.21	– 10.84
All Workers	10.77	12.37	– 1.60	12.67	20.96	– 8.29
4 YEARS OF HIGH SCHOOL						
Financial Record Processors	$ 8.26	$ 9.42	$ – 1.16	$10.79	$11.79	$ – 1.00
Mechanics and Repairers	9.65	10.66	– 1.01	11.00	12.22	– 1.22
Motor Vehicle Operators	8.95	9.95	– 1.00	9.99	10.91	– .92
Precision Metal Workers	9.93	11.16	– 1.23	10.88	12.87	– 1.99
All Workers	8.77	10.25	– 1.48	10.07	12.78	– 2.71
1 TO 3 YEARS OF HIGH SCHOOL						
Cleaning and Building						
Service Workers	$ 6.12	$ 7.45	$ – 1.33	$ 7.38	$ 8.58	$ – 1.20
Duplicating, Mail, and Office						
Machine Operators	8.07	9.11	– 1.04	8.10	10.19	– 2.09
Equipment Cleaners, Handlers,						
and Laborers	6.68	8.42	– 1.74	8.84	10.02	– 1.18
Textile Machine Operators	6.05	7.34	– 1.29	7.17	8.10	– .93
All Workers	7.34	9.06	– 1.72	9.25	11.34	– 2.09

NOTES: For "University Professors," job title is Post-Secondary Teachers; for "Doctors and Dentists," job title is Health Diagnosing Occupations.

TABLE 11.1 *(continued)*

	Women					
	25–34 Years			55–64 Years		
	Black	White	Differ-ence	Black	White	Differ-ence
YEARS OF COLLEGE OR MORE						
University Professors	$ 9.96	$ 9.61	$+ .35	$14.15	$13.68	$+ .47
Doctors and Dentists	10.84	10.36	+ .48	23.79	25.21	− 1.42
Lawyers and Judges	14.61	13.41	+1.20	21.62	20.78	+ .84
Social Scientists and						
Urban Planners	11.98	11.66	+ .32	14.12	15.29	− 1.17
All Workers	10.28	10.08	+ .20	12.84	12.44	+ .40
YEARS OF COLLEGE						
Elementary and						
Secondary Teachers	$ 7.85	$ 7.68	$+ .17	$ 9.22	$ 9.88	$− .66
Engineers, Architects,						
and Surveyors	11.93	12.15	− .22	15.19	17.14	− 1.95
Mathematical and						
Computer Scientists	11.91	12.99	− 1.08	11.52	15.57	− 4.05
Sales Representatives in						
Finance and Business	10.43	10.25	+ .18	5.66	10.98	− 5.32
All Workers	8.94	9.11	− .17	10.41	10.34	+ .07
YEARS OF HIGH SCHOOL						
Financial Record Processors	$ 7.18	$ 6.89	$+ .29	$ 8.21	$ 7.95	$+ .26
Mechanics and Repairers	8.88	9.58	− .70	9.48	9.15	+ .33
Motor Vehicle Operators	7.12	6.69	+ .43	5.99	6.31	− .32
Precision Metal Workers	7.54	7.83	− .29	8.10	8.74	− .64
All Workers	6.84	7.06	− .22	7.08	7.81	− .73
TO 3 YEARS OF HIGH SCHOOL						
Cleaning and Building						
Service Workers	$ 4.73	$ 5.16	$− .43	$ 5.61	$ 5.71	$− .10
Duplicating, Mail, and Office						
Machine Operators	4.92	6.03	− 1.11	6.76	7.00	− .24
Equipment Cleaners, Handlers,						
and Laborers	6.68	6.23	+ .45	6.94	6.66	+ .28
Textile Machine Operators	4.98	5.21	− .23	5.27	5.71	− .44
All Workers	5.81	5.92	− .11	5.93	6.65	− .72

SOURCE: U.S. Bureau of the Census, *Census of Population: 1980*, PC80-2-8D, tables 3 and 4.

similarly educated whites,[2] but findings with regard to the earnings of black women raise doubts about this explanation because black women earn as much or more than comparably educated white women.

Measuring Racial Differences in Earnings

Investigators have often described racial differences in earnings by comparing workers who are similar on those factors that influence earnings. In the past, they were restricted to analyzing the published tabulations from censuses and surveys, a cumbersome procedure since a complete study covering the hundreds of detailed occupations would fill scores of pages. A more efficient approach uses the Public Use Microdata Samples from recent enumerations, which report the characteristics of individuals but maintain confidentiality by not disclosing names or addresses. Statistical models may be used to equate blacks and whites with regard to those factors which affect earnings, and thereby we can determine whether similar blacks and whites earn identical amounts.

These investigations serve several important purposes. First, black and white workers differ in the characteristics they bring to the labor market. Whites, on average, complete more years of schooling and are less likely than blacks to live in the South where wage rates remain low. These racial differences would produce a difference in earnings even if every employer treated whites and blacks exactly the same. These models allow us to quantitatively measure the effects of social and demographic differences on earnings and facilitate a comparison of how the present situation compares with that of the past.

Additionally, they tell us very directly how earnings are related to educational attainment, years of labor force experience, or other explanatory factors. In essence we learn the earnings payoff associated with an extra year of school or experience. If employers treated blacks and whites alike, we anticipate that an additional year of education would increment the earnings of blacks and whites by identical amounts. If we find that blacks receive much smaller financial rewards for their investments in education or their years in the labor force, we may have an indication of discrimination. For example, when employers pay blacks less than similar whites, an additional year of education will lead to smaller wage increases for blacks.

Caution is needed before equating differences of this type with racial discrimination. Our models do not control for all the variables that influence earnings and may measure others imperfectly. For instance,

[2]Hanushek (1978); Welch (1973); Sowell (1981b), p. 21.

censuses tell us how many years of schooling a person completed, but they do not tell us what a person studied or the grades he or she received.

Furthermore, the models used in this chapter describe only one type of discrimination. An employer might discriminate on the basis of race or sex in different ways.[3] He might actually pay blacks or women less than white men who do the same work. Black men, for example, who unload box cars may be called "laborers" while white men are called "equipment handlers" and have a higher pay rate. Employers might also assign layoffs and overtime in a discriminatory fashion. If blacks are the last hired and first fired, and if their opportunities, jobs, or overtime are limited, their earnings will fall below those of comparable whites even if their hourly wage rates are identical. In addition, employers may use skin color or sex to determine who gets jobs. This chapter primarily investigates whether similar blacks and whites earn the same amounts while they are working.

Since the mid-1960s, many investigators have analyzed models of this type to study racial differences in earnings. Table 11.2 presents a synopsis of those studies which used national samples. The list is not exhaustive; rather, it suggests the range of sources and the array of variables. As a dependent variable, most have used annual earnings or hourly wages, although some have used weekly wages or even annual income. The most common explanatory variables have been educational attainment, region of residence, age, or years of potential labor market experience, and a measure of how much the person worked during the period for which earnings were reported. Some investigators also included information about occupation, marital status, veteran status, or aspects of the family of origin such as father's occupation, father's education, or whether the family of origin was intact. Those who have analyzed data from the University of Michigan's Panel Study of Income Dynamics have studied the effects of on-the-job seniority, on-the-job training, and mental ability, as well as certain aspects of the employee's health. Specific publications should be consulted for detailed information about who was excluded, how the investigator coded variables, and which statistical transformation or interaction terms were used.

Four generalizations can be drawn from these analyses. First, black men have actual earnings far below those of white men, but this difference is *not* due solely to their lower educational attainment, fewer hours of work, or concentration in the South. Even those studies that included measures of mental ability or seniority with an employer reported large racial differences in earnings net of explanatory factors.

[3]For a review of economic theories of racial discrimination in employment and earnings, see Arrow (1972); Freeman (1974); Gordon (1972); Marshall and Christian (1978); Reich (1981), chap. 3; and Stiglitz (1974).

TABLE 11.2

National Studies of Racial Differences
in the Earnings or Income of Employed Blacks and Whites

Author	Date of Publication	Data Source	Sample
Siegel	1965	1960 Census	Men, Aged 25–64
Duncan	1968	1962 OCG	Native Men, Aged 25–64
Michelson	1968	1960 Census	Men, Aged 21–60
Thurow	1969	1960 Census	Men, Aged 18–64
Gwartney	1970	1940 & 1950 Census 1968 CPS	Men & Women, Aged 25 +
Welch	1973	1960 Census 1967 SEO	Urban Men, Aged 15–69
Blinder	1973	1968 PSID	All Men and White Women with Earnings
Mincer and Polachek	1974	1967 NLS	Women, Aged 30–44
Masters	1975	1960 Census 1967 SEO	Men, Aged 17–64
Haworth, Gwartney, and Haworth	1975	1960 & 1970 Census	Men, Aged 14 +
Stolzenberg	1975	1960 Census	Black and White Men in Selected Occupations
Smith & Welch	1977	1960 & 1970 Census	Men with Less Than 40 Years of Work
Garfinkel and Haveman	1977	1971 CPS 1974 PSID	Men and Women

NOTES: *Data Source:* CPS = Current Population Survey, Bureau of the Census; NLS = National Longitudinal Study of Work Experience, conducted at Ohio State University; OCG = Occupational Change in a Generation, studies conducted as supplements to CPS in 1962 and 1973; PSID = Panel Study of Income Dynamics, conducted at University of Michigan, Institute for Social Research; SEO = Survey of Economic Opportunity, conducted by Bureau of the Census in 1967; SIE = 1976 Survey of Income and Education, conducted by Bureau of the Census. *Method:* REG = regression analysis; almost without exception, these studies use ordinary least-squares models; STAND = direct or indirect standardization. *Independent Variables:* CHILD = indication of presence or number of children in respondent's household; EDUC = measure of educational attainment; ENG = measure of English language ability; EXP = various measures of duration of respondent's participation in labor force; FAM BACK = various measures of social characteristics of respondent's family of origin, including whether one- or two-parent family; FARM = respondent lived on farm as youth; FEDUC = respondent's father's educational attainment; FOCC = father's occupational achievement; FOR = foreign-born individuals;

TABLE 11.2 *(continued)*

Dependent Variable	Method	Independent Variables
Annual Earnings	STAND	EDUC, AGE, REG
Annual Earnings	REG	FEDUC, FOCC, SIBS, EDUC, OCC, MENT ABL
Annual Earnings	REG	EDUC, AGE, OCC
Median Annual Income for Groups	REG	EDUC, EXP, REG
Ratio of Nonwhite to White Annual Income	STAND	EDUC, AGE, REG
Log of Annual Earnings	REG	EDUC, AGE, WEEKS, GOV'T EMP
Log of Hourly Wage	REG	EDUC, AGE, REG, OCC, SENIORITY, UNION VET, SIBS, FEDUC, HEALTH, FAM BACK
Log of Annual Earnings	REG	EDUC, AGE, MARITAL, CHILD, WORK HIST, HEALTH, HOURS, WEEKS
Annual Earnings	REG	EDUC, AGE, REG, WEEKS, PLACE
Annual Earnings	REG	EDUC, AGE, REG, PLACE, HOURS, MARITAL, VET
Log of Hourly Wage	REG	EDUC, EXP, REG, PLACE, SELF
Log of Weekly Wage	REG	EDUC, REG, PLACE, GOV'T EMP, EXP
Log of Annual Earnings	REG	EDUC, RES, AGE, PLACE, HOURS, MARITAL, CHILD

GOV'T EMP = respondent worked for government agency or his industry was closely tied to government; HEALTH = respondent's physical condition or limitations; HOURS = number of hours respondent worked during year for which earnings were reported; LABMKT = respondent's area of residence; MARITAL = respondent's marital status; MENT ABL = respondent's performance on test of mental ability; NAT = native-born individuals; OCC = respondent's occupational achievement; PLACE place of residence other than region (e.g., city-suburb or urban-rural); REG = respondent's state, region, or division of residence; SECTOR = sector of economy in which respondent was employed; SELF respondent was self-employed; SENIORITY = work experience on given job or with particular employer; SIBS = respondent's number of siblings; UNION = respondent had unionized job; USRESID duration of residence in United States; VET = respondent's military service; VOCAT = respondent completed vocational training program; WEEKS = number of weeks respondent worked during year for which earnings were reported; WORK HIST = duration of different statuses in labor force

SOURCES: See Bibliography.

TABLE 11.2 *(continued)*
National Studies of Racial Differences
in the Earnings or Income of Employed Blacks and Whites

Author	Date of Publication	Data Source	Sample
Farley	1977	1960 & 1970 Census 1975 CPS	Men and Women, Aged 25–64
Featherman and Hauser	1978	1962 & 1973 OCG	Men, Aged 21–64
Hoffman	1979	1967–1974 PSID	Men Aged 20–39 in Labor Force in 1967
Wright	1979	1975 PSID	Men with Income
Corcoran and Duncan	1979	1976 PSID	Black and White Household Heads and Wives
Datcher	1980	1960 & 1970 Census	Women, Aged 14+
Duncan and Hoffman	1983	1967–1978 PSID	Men Aged 25–54 Who Worked 500 Hours
Farley	1984	1960 Census 1970 & 1980 CPS	Men Aged 25–64 with Earning
Hirschman and Wong	1984	1960 & 1970 Census 1976 SIE	Men Aged 25–64 with Earning
Reimers	1985	1976 SIE	Men and Women Aged 14 and Over with Earnings
Abowd and Killingsworth	1985	1976 SIE	Men and Women Aged 21 and Over with Earnings
Smith and Welch	1986	1940–1980 Censuses	Men with Earnings

NOTES: *Data Source:* CPS = Current Population Survey, Bureau of the Census; NLS = Nation. Longitudinal Study of Work Experience, conducted at Ohio State University; OCG = Occupation. Change in a Generation, studies conducted as supplements to CPS in 1962 and 1973; PSID = Pan Study of Income Dynamics, conducted at University of Michigan, Institute for Social Research; SEO Survey of Economic Opportunity, conducted by Bureau of the Census in 1967; SIE = 1976 Survey Income and Education, conducted by Bureau of the Census. *Method:* REG = regression analysis; almo without exception, these studies use ordinary least-squares models; STAND = direct or indire standardization. *Independent Variables:* CHILD = indication of presence or number of children respondent's household; EDUC = measure of educational attainment; ENG = measure of Englis language ability; EXP = various measures of duration of respondent's participation in labor force; FA BACK = various measures of social characteristics of respondent's family of origin, including wheth one- or two-parent family; FARM = respondent lived on farm as youth; FEDUC = respondent's father educational attainment; FOCC = father's occupational achievement; FOR = foreign-born individuals;

TABLE 11.2 *(continued)*

Dependent Variable	Method	Independent Variables
Annual Earnings	REG	EDUC, REG, OCC, EXP, HOURS
Annual Earnings, Log of Annual Earnings	REG	FEDUC, FOCC, SIBS, FARM, FAM BACK, EDUC, EXP, OCC, WEEKS
Log of Hourly Earnings	REG	EDUC, EXP, REG, PLACE, SECTOR, UNION, MENT ABL
Taxable Income	REG	FEDUC, FOCC, EDUC, OCC, AGE, EXP, HOURS
Log of Hourly Wage	REG	EDUC, REG, PLACE, HOURS, VOCAT, WORK HIST
Annual Earnings	REG	EDUC, AGE, REG, PLACE, HOURS
Log of Hourly Earnings	REG	EDUC, EXP, REG, PLACE
Log of Hourly Earnings	REG	EDUC, EXP, REG
Annual Earnings	REG	EDUC, AGE, REG, OCC, WEEKS, HOURS, NAT, SECTOR
Log of Hourly Earnings	REG	EDUC, EXP, FOR, ENG, HEALTH, GOV'T EMP, USRESID
Log of Hourly Wage, Log of Annual Earnings	REG	EDUC, USRESID, EXP, REG, SECTOR, LABMKT
Weekly Wage	REG	EDUC, EXP, REG, PLACE

GOV'T EMP = respondent worked for government agency or his industry was closely tied to government; HEALTH = respondent's physical condition or limitations; HOURS = number of hours respondent worked during year for which earnings were reported; LABMKT = respondent's area of residence; MARITAL = respondent's marital status; MENT ABL = respondent's performance on test of mental ability; NAT = native-born individuals; OCC = respondent's occupational achievement; PLACE = place of residence other than region (e.g., city-suburb or urban-rural); REG = respondent's state, region, or division of residence; SECTOR = sector of economy in which respondent was employed; SELF = respondent was self-employed; SENIORITY = work experience on given job or with particular employer; SIBS = respondent's number of siblings; UNION = respondent had unionized job; USRESID = duration of residence in United States; VET = respondent's military service; VOCAT = respondent completed vocational training program; WEEKS = number of weeks respondent worked during year for which earnings were reported; WORK HIST = duration of different statuses in labor force

SOURCES: See Bibliography.

Second, black men receive smaller economic returns for their investments in education and for their years of labor force experience than do white men. In addition, the reduction in earnings associated with living in the South is much greater for blacks than for whites.

Third, those who have examined changes report that racial differences in total earnings and racial differences in the value of a year of education have declined over time. This is often taken as evidence of decreasing racial discrimination in the labor market.

Fourth, racial differences in earnings themselves and in returns for investments in education are much smaller among women than among men. Several investigators report that by 1980 earnings patterns were similar for black and white women.

A Model to Describe Racial Differences in Earnings: 1960 and 1980

The model used in this study assumes that how much a person earns is influenced by his or her educational attainment, years of labor market experience, and region of residence. This investigation is restricted to the 25-to-64 age range to eliminate people who were just starting their careers and people of retirement age. The variables were measured as follows:

HOURLY WAGE RATE

For each person, we determined how much they earned by summing their wage, salary, and self-employment earnings if they had any. This was divided by the number of hours they worked during the year to estimate their hourly wage rate.[4]

---

[4] The censuses of 1960 and 1980 asked individuals how many weeks they had worked during the previous year. In 1980 individuals were asked how many hours they usually worked per week when they were employed in 1979. To estimate total hours of employment in 1979, the number of weeks of work was multiplied by the usual hours of work. Total earnings in 1979 was divided by this figure to estimate the hourly wage rate.

The 1960 enumeration asked how many hours a person worked in the week prior to the census, but did not determine usual hours of employment in 1959. To estimate hours of work, the number of weeks of work in 1959 was multiplied by the hours worked during the week before the census.

Approximately one sixth of those who worked during 1959 did not report any hours of employment in the week prior to the 1960 census since they were either unemployed or out of the labor force. Rather than delete them, we imputed their hours of employment. We first considered those who reported hours of work and regressed this number on related variables including age, education, earnings, marital status, and region separately for each of the four race-sex groups; and, for women, we also included a variable indicating

The distribution of hourly wage rates is skewed since a small number of workers earn large amounts. To avoid giving undue weight to these people, we use the logarithm of the hourly rate as the dependent variable. In statistical terms the distribution of the log of hourly wages is closer to a normal distribution than is the actual distribution of hourly wages, and thus the log transformation reduces nonlinearity in the regression equations. This has substantive consequences since it leads to smaller estimated racial and sexual differences than would an analysis based on actual hourly wages.

EDUCATIONAL ATTAINMENT

The census asks adults how many years of schooling they completed. Precollege education was distinguished from college education for two reasons. First, blacks and whites differ in the proportion of their total education which was at the college level. Using only one variable to measure attainment may bias the apparent effects of education since more of the education reported by whites is at the college level. Second, years of college education are more highly rewarded in the labor market than are years of precollege education.

The variable "Elementary and Secondary" equals years of precollege education completed by an individual and ranges from 0 to 12. The variable "College" equals the years of college or university education and ranges from 0 to 6.

YEARS OF LABOR MARKET EXPERIENCE

As individuals remain in the labor force or as they grow older, they may gain valuable experience which leads to increases in their pay. For these reasons, a variable measuring years of labor force experience is included in this analysis. It equals a respondent's current age minus his or her educational attainment minus 6. This assumes that people begin school at age 6, attend continuously, and start their careers shortly after finishing their education. Although this is fairly common among white men, blacks experience more unemployment than whites, and many women drop out of the labor force for some periods. Thus, it is a measure of the years of potential labor force experience rather then the actual number of years a person has worked. Unfortunately, the census provides no better measure of seniority or continuity of employment.

---

the presence of a child under age 6 in the household. Coefficients from these regression equations were used with the known characteristics of those people who did not report hours of work for the week before the 1960 census enumeration. See Farley (1984), p. 210.

YEARS OF LABOR MARKET EXPERIENCE SQUARED

Job skills may become obsolescent as a person grows older so earnings do not linearly increase with age, even among white men.[5] To take this into account, the square of the years of potential labor market experience is used in these models.

REGION

Wage rates for the same occupation are generally lower in the South than in other regions. If we compared national samples of blacks and whites, we would expect blacks to earn less because of their concentration in the South. To measure the effects of this regional difference, we include a dichotomous variable which distinguishes people living in the South from all others.

SOURCES OF INFORMATION

Data for this analysis are taken from Public Use Microdata Samples of the 1960 and 1980 censuses. These enumerations, conducted in April of each census year, asked about earnings and hours of employment during the previous calendar year; thus, we will frequently identify the findings as pertaining to 1960 or 1980 even though the person obtained his or her earnings in the prior year. Appendix Table 11.1 shows the average values for the variables used in this study.

SAMPLE

We wanted to use the broadest possible sample for our analysis. Since we are focusing on earnings, we necessarily eliminated all persons who reported no earnings during the year before the census. We also deleted those very few individuals (generally less than .5 percent) who claimed that they worked but had negative earnings—for example, those who traded commodities, speculated in real estate, or owned farms. A small number of people on active duty in the Armed Forces were excluded in 1960 since the census provided limited information for them. Appendix Table 11.2 reports the sample size and the reasons for deletion.

While we analyzed data for a large proportion of all adults in 1980 (90 percent of the white males, for example), a sizable proportion of

[5]Mincer (1974), chap. 5; Mincer and Polachek (1973).

black men aged 25 to 64—about 19 percent—reported they had no earnings in 1979 and thus were excluded. Do these deletions—the omission of about one black man in five and one white man in ten in 1980—bias our results? To explore this possibility, Table 11.3 arrays individuals aged 25 to 64 by their hours of employment in the previous year. It shows information about age and educational attainment and then indicates the proportions living in group quarters (institutions), reporting a disability that limits work, enrolled in school, obtaining Social Security income, and finally, living with a spouse.

People who did not work tend to be older and have completed few years of schooling. About one eighth of the black men with no hours of employment were in prisons, mental institutions, or other group quarters, and about one half of them reported disabilities which restricted their work activities. For both races, Social Security income was commonly reported by these people. A few of these individuals are aged 62 or over and claim retirement benefits, but many of them are undoubtedly reporting the supplemental income which the Social Security system provides to the disabled and blind of younger ages.

Men who did not work have very limited earnings prospects because of their age, their limited educations, their infirmities, and in some cases, their criminal backgrounds or mental problems. Presumably, if they had to support themselves, their wages would be low. This implies that our analysis of earnings describes a somewhat more successful component of the population. Furthermore, the selectivity varies by race since the sample in 1980 excluded twice as high a proportion of blacks. Restricted to earners, this analysis leads to underestimates of the true racial differences since a higher proportion of black men did not work in 1979, men who had limited earning capacities.

Selectivity with regard to employment is far less confounding for women because racial differences in hours of employment are smaller among women than among men. Furthermore, the characteristics of women who did not work are not as distinctive vis-à-vis those of employed women.

Racial Differences in Earnings and Rates of Return

The census of 1960 found that black men aged 25 to 64 had annual earnings just about 50 percent of those of white men. In the following decades earnings rose for both races, but the rate of gain was greater for blacks, and by 1980 their annual earnings reached 62 percent of those of white men.

TABLE 11.3

Characteristics of Black and White Men and Women Aged 25–64 in 1980 by Hours of Employment in 1979

Hours of Employment in 1979	Percent Distribution by Hours Worked	Average		In Group Quarters	With Work Limitation	Proportion		
		Age	Educational Attainment			Enrolled in School	With SSI	Married Spouse Present
BLACK MEN								
None	19.3%	44.3	9.4	13%	48%	4%	26%	42%
Less Than 1,000	10.6	39.2	10.8	6	17	8	5	51
1,000–1,499	6.4	39.1	11.2	3	12	8	3	56
1,500–1,999	12.8	39.5	11.4	2	7	6	1	65
2,000 or More	50.9	40.2	11.5	1	4	5	1	70
Total	100.0%	40.8	11.0	4%	15%	6%	6%	61%
WHITE MEN								
None	8.9%	50.1	10.5	7%	57%	6%	43%	64%
Less Than 1,000	6.2	41.9	12.1	2	19	11	10	63
1,000–1,499	5.1	40.2	12.4	2	14	8	4	67
1,500–1,999	10.6	40.6	12.8	1	7	5	2	76
2,000 or More	69.2	41.4	13.0	<1	5	4	1	83
Total	100.0%	42.1	12.7	1%	11%	5%	5%	79%
BLACK WOMEN								
None	34.5%	44.2	9.9	2%	34%	4%	16%	46%
Less Than 1,000	14.0	39.5	11.2	1	13	7	6	49
1,000–1,499	8.2	40.0	12.1	<1	7	8	4	53
1,500–1,999	14.3	39.4	12.2	<1	5	7	2	50
2,000 or More	29.0	39.2	12.0	<1	3	7	2	48
Total	100.0%	41.1	11.2	1%	16%	6%	8%	48%
WHITE WOMEN								
None	36.8%	45.3	11.4	1%	17%	3%	13%	83%
Less Than 1,000	15.9	40.2	12.6	<1	8	6	6	79
1,000–1,499	8.7	40.6	12.9	<1	6	6	4	74
1,500–1,999	12.4	40.9	12.9	<1	4	6	2	64
2,000 or More	26.2	40.9	12.6	<1	3	5	1	61
Total	100.0%	42.4	12.2	1%	9%	5%	7%	73%

NOTES: "In Group Quarters" includes persons living in mental institutions, homes for the aged, prisons, military barracks, college dormitories, rooming houses, and all other group quarters.

"With SSI" includes persons reporting income from the Social Security systems of any type.

Figure 11.1 presents the average earnings of blacks and whites.[6] We find, first, that the 1960s and 1970s were decades of progress for blacks since their earnings moved closer to those of whites. In every comparison, blacks and whites were more similar in earnings in 1980 than they were twenty years earlier, and investigation of the two separate decades suggests that the earnings of blacks moved closer to those of whites in both intervals.[7] We also found important sexual differences, so descriptions of racial change in earnings must distinguish men from women. In 1960 black men and women were in a similarly disadvantaged position compared with whites of the same sex, but by 1980 black women had reached earnings parity with white women.

Comparisons of annual and hourly earnings give different pictures of the racial gap. Because black men typically work fewer hours per year than white men, racial differences among men are much greater on an annual basis than on an hourly basis. The reverse is true among women because black women typically spend more time at work than do white women. The annual earnings of black women in 1980 exceeded those of white women, but on an hourly basis they lagged behind white women.

According to the model we are using, how much a person earns is influenced by his or her educational attainment, experience, and place of residence. Figure 11.2 provides information about the earnings associated with each of these factors.

Several generalizations can be drawn from information about the economic value of education shown in Figure 11.2. First, investments in education by black men are less highly rewarded than those made by white men. Let us first consider elementary and secondary schooling. In 1960 the hourly earnings of white men increased by 49 cents for each additional year.[8] The benefit to black men was less than half as much: only 19 cents per hour. By 1980 the situation had changed, and a year of

[6]The average hourly earnings shown in Figure 11.1 equal the mean of the logarithm, of hourly earnings converted first into 1959 or 1979 dollars and then inflated into 1984 dollars. Because of the log transformation, the amounts shown in Figure 11.1 are smaller than the amounts which would result from use of data for actual hourly earnings. That is, the mean of a logged earning distribution is typically smaller than the mean of the actual earnings distribution.

To estimate annual earnings, the mean logarithm of the hourly wage rate was added to the mean logarithm of hours worked during the year. This was converted first into 1959 or 1979 dollars and then into 1984 dollars.

[7]Farley (1984), pp. 65–67; Smith and Welch (1986), table 2.

[8]To determine the net change in earnings associated with an additional year of education, a year of labor market experience, or living in the South, we first estimated the hourly earnings for a worker who had the mean value on each of the independent variables. We then changed the value of one variable to reflect additional education, experience, or a southern residence. A new value for hourly earnings was estimated using appropriate coefficients. Differences between the first and second estimated amounts—in 1984 dollars—are shown in Figure 11.2.

FIGURE 11.1
*Earnings of Black and White Men and Women Aged 25–64,
Reported in Censuses of 1960 and 1980*

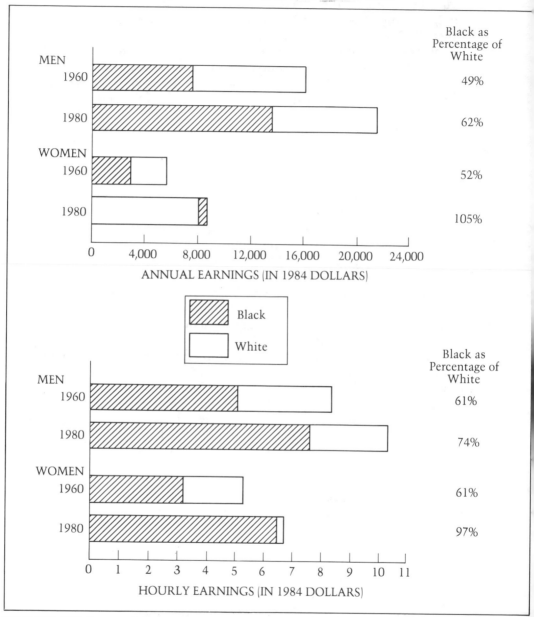

SOURCES: U.S. Bureau of the Census, *Census of Population and Housing: 1960*, Public Use Microdata Samples; *Census of Population and Housing: 1980*, Public Use Microdata Samples.

FIGURE 11.2
Rates of Return Associated with Labor Market Characteristics
(amounts per hour of employment in 1984 dollars)

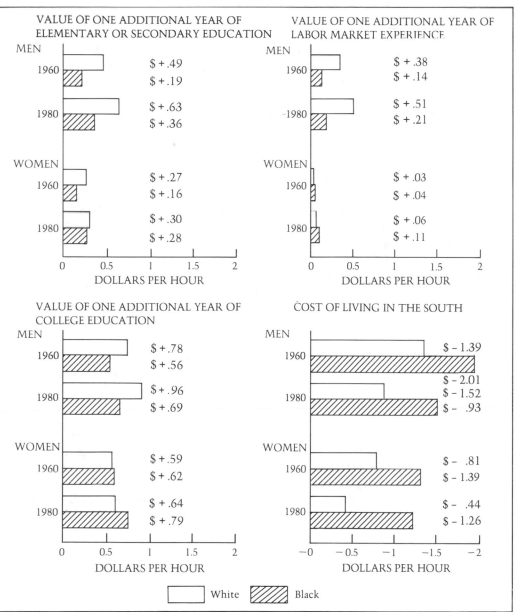

SOURCES: U.S. Bureau of the Census, *Census of Population and Housing: 1960,* Public Use Microdata Samples; *Census of Population and Housing: 1980,* Public Use Microdata Samples.

precollege education augmented the earnings of the white men by 63 cents per hour and of black men by 36 cents per hour. In 1980 a year of college added 69 cents per hour to a black man's paycheck but 96 cents per hour to a white man's paycheck.

Second, between 1960 and 1980 the economic returns associated with education rose for all groups, but the gains were greater for blacks and thus racial differences on this important indicator are declining. This may reflect improvements in the quality of education obtained by blacks as well as declines in discriminatory practices that led some employers to pay blacks less than comparably educated whites.

Third, college education is quite beneficial to black women, more than it is to white women. Even before passage of the Civil Rights Act, black women received greater returns for their investments in college education than white women; in 1960 an additional year of college added 62 cents to the average hourly earnings of black women, or 3 cents more per hour than to the earnings of white women. By 1980 their advantage grew to 15 cents per hour for each year of college.

Turning to labor market experience, we find that there are substantial racial and sexual differences. After 1960 rates of return for experience rose for all groups, but white men did not lose their very large advantage. In 1980 an additional year of experience—as measured by years elapsed since completion of school—raised the earnings of white men by 51 cents per hour, of black men by 21 cents, and of white women by 6 cents. When looking at these rates of return, we once again find that black women have an advantage over white women because an extra year of potential labor market experience raised their earnings twice as much as it did those of white women.[9]

Living in the South led to a reduction in earnings, but the cost was especially great for blacks, as Figure 11.2 shows. In 1980 black men in the South earned $1.52 less per hour, or $3,000 per year if they worked full time, than black men in the North and West who had comparable educations and years of labor market experience. Among white men, the net reduction associated with southern residence was 93 cents per hour, or $1,900 per year for the full-time employee. For both races and both sexes, the net reduction in earnings attributable to residence in the South declined after 1960, suggesting that regional differences in wage rates are waning.

Does this mean that blacks suffer from more discrimination in the

[9]There has been extensive controversy about whether the labor market experience of women should be measured by subtracting their years of schooling from their age. Men spend most of this time in the labor force, but many women withdraw from the labor force to raise families. For further information, see Corcoran (1979); Corcoran and Duncan (1979); Mincer and Polachek (1974, 1978); Polachek (1975); and Sandell and Shapiro (1978); Lamas, McNeill, and Haber (1986).

South than in other regions? The evidence in Figure 11.2 suggests that they may. If wage rates were lower in the South than in other regions for all workers regardless of their race, we would find the net reduction in earnings for living in the South the same for blacks and whites. Obviously this is not the case. Whites in the South earn considerably less than comparable whites in the North or West, but blacks who live in the South find their wages reduced by a much greater amount than whites.

Changes in Earnings and in the Cost of Being Black

The earnings of blacks have lagged behind those of whites for two reasons. First, their labor market characteristics have been less favorable since they completed fewer years of schooling and were more likely than whites to live in the South. Second, their rates of return for labor market characteristics—illustrated in Figure 11.2—have generally been inferior to those of whites. If blacks are to reach earnings parity with whites, it will require elimination of racial differences in both labor market characteristics and rates of return.

Individuals presumably have considerable control over their own labor market characteristics because they can remain in school longer or move from one region to another. Employers have control over rates of return since they set pay scales and decide who gets jobs. Between 1960 and 1980 the earnings of black men increased. Was this primarily because of an improvement in their labor market characteristics or was it because employers increased their rates of return? Figure 11.3 answers this question by showing the components of change in annual earnings during this two-decade span.[10]

The changing labor market characteristics of blacks—especially their greater educational attainment and their movement away from the South—played an important role in increasing earnings, and the net effect of changes in characteristics was to add $2,100. However, changes

[10]To estimate the net effect of changes in labor market characteristics, the 1980 characteristics of a group were inserted into the 1960 earnings equation for that group. The resulting estimate of hourly earnings was multiplied by the average hours worked in 1979 to estimate annual earnings. This was then compared with actual annual earnings from the 1960 census to determine the net effect of changing characteristics.

To ascertain the consequences of changes in rates of return, 1960 characteristics were used with the 1980 earnings equation; and an estimate of annual earnings was developed using information about the extent of employment in 1959, which was then compared with actual earnings reported in 1960 to estimate the result of changing rates of return.

The two components of change do not necessarily sum to the total change in annual earnings. There is an interaction component which is not shown in Figure 11.3.

FIGURE 11.3

Change in Annual Earnings, 1960 to 1980, and Components of Change (in 1984 dollars)

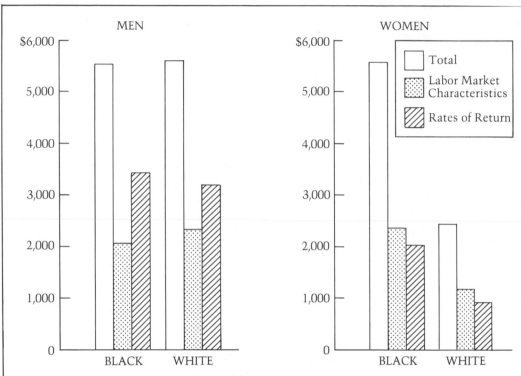

SOURCES: U.S. Bureau of the Census, *Census of Population and Housing: 1960*, Public Use Microdata Samples; *Census of Population and Housing: 1980*, Public Use Microdata Samples.

in rates of return were even more important. Employers increased the pay rates of blacks such that even if there had been no change in labor market characteristics, the average annual earnings of a black man would have risen $3,400. The higher earnings of blacks cannot be uniquely attributed to either changing labor market characteristics or higher rates of return. The characteristics of black men became more favorable, but employers also improved their pay scales.

When we turn to information about the earnings of white men, we find similar changes. The changing characteristics of white men—especially their greater educational attainment—had the independent effect of increasing their average annual earnings by about $2,400. White men also received higher rates of return, which increased their earnings as much as those of black men. As a result, the overall increase in annual earnings was about the same for men of both races: $5,600. As indicated in Figure 11.1, black men moved up in relative terms since their annual earnings rose from 49 to 62 percent of those of white men. However, they failed to catch up with white men in terms of actual earnings.

Among women, there was a dramatic racial difference: the average gain in earnings for blacks was more than double that for whites. The changing characteristics of women—particularly their greater college attendance, their increased hours of employment, and the movement of blacks to the North and West—led to much greater earnings, but higher rates of return also contributed. As Figure 11.3 indicates, the effects of both changing characteristics and changing rates of return were larger for blacks than for whites, and by 1980 the annual earnings of black women exceeded those of white women (see Figure 11.1).

The labor market characteristics of blacks are improving, and their rates of return are increasing. Are we getting to the point at which comparable black and white workers earn similar amounts? Using data from a census makes it difficult to be sure that blacks and whites are comparable on all factors that influence earnings. However, we can statistically equate workers with regard to education, years of labor force experience, and region and then determine whether ostensibly similar blacks and whites have similar earnings.

Let us assume that black men had their own rates of return—a factor over which they had little control—but had the educational attainments, years of experience, regional distribution, and hours of employment of white men. In such a case, how much would they earn?

Figure 11.4 provides an answer. If they had the characteristics of white men but their own rates of return, they would have earned 81 percent as much as white men in 1960. This is identified as expected earnings. The difference between 81 and 49—that is, 32 percent—is an estimate of the cost to black men of their shorter education, their concentration in the South, and their fewer hours on the job. In other words, their labor market characteristics cost them an amount equal to 32 percent of the average earnings of white men. There is another difference. If black men had not only the characteristics of white men, but also the rates of return of whites, they would earn just as much as whites. However, with their own rates of return, they earn only 81 percent as much as white men. This means that racial differences in pay rates cost black men an amount equal to 19 percent of the average earnings of white men.

This gap of 19 percent has been called the "cost of being black" since it indicates if a black man has the same labor market characteristics as a white man, he will earn less because rates of return are smaller

FIGURE 11.4
*Actual and Expected Annual Earnings of Blacks
as a Percentage of the Actual Earnings of Whites, 1960 and 1980*

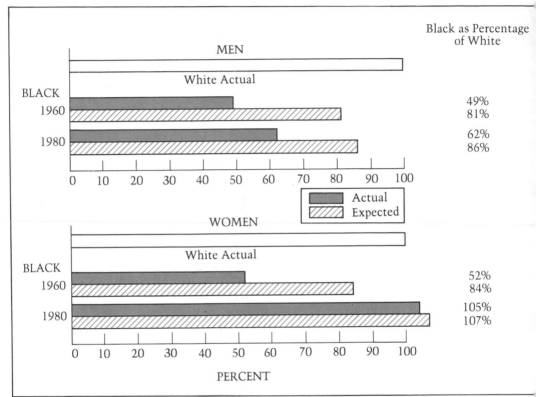

NOTE: Expected annual earnings for blacks are estimated by assuming they had their own observed rat
of return but the educational attainment, years of labor force experience, regional distribution, and hou
of employment of whites.

SOURCES: U.S. Bureau of the Census, *Census of Population and Housing: 1960*, Public Use Microda
Samples; *Census of Population and Housing: 1980*, Public Use Microdata Samples.

for blacks.[11] This is often taken as a measure of the cost of racial
discrimination in wage rates since it suggests that employers pay blacks
less than similar whites. As Rees and Shultz observe, models such as
these are valuable for studying earnings because "they are measures of
wage differences by color and sex after allowing for the effects of differences among workers in age, seniority, schooling and experience. It
therefore cannot be contended that differences between groups in these
dimensions cause the intergroup wage differentials that we observe."[12]

[11]Siegel (1965); Duncan (1968).
[12]Rees and Shultz (1970), p. 220.

We must be careful, however, in assuming that this measure offers an exact estimate of the dollar cost of racial discrimination. On the one hand, the calculation of expected earnings assumes that blacks are as extensively educated as whites and work as many hours. Quite likely, racial discrimination partly accounts for the lower educational attainment of blacks and may help explain their low hours of employment if black men are the last hired and first fired. In this sense, the estimates shown in Figure 11.4 may substantially understate the cost of racial discrimination. On the other hand, the model does not take into account all factors that influence hourly wage rates, and the inclusion of additional variables might reduce the estimated cost of discrimination. However, while studies of this topic consistently show that education, labor force experience, and place of residence are the most important determinants of earnings, investigators (see Table 11.2) who have included additional factors such as mental ability, occupation, characteristics of the family of origin, health factors, or seniority with an employer generally find that they do not substantially alter the estimated costs of racial discrimination.

Shifting from 1960 to 1980, we see that in the more recent year the actual annual earnings of black men were 62 percent of those of white men while their expected earnings were 86 percent. This implies that the labor market characteristics of black men are still a major reason for their lower earnings. If they worked as many hours as white men, stayed in school as long, and had the regional distribution of whites, their earnings would rise from 62 to 86 percent of those of white men, leading us to conclude that racial differences in characteristics cost blacks an amount equal to 24 percent of the average earnings of white men.

The findings also indicate that black men remain penalized by lower rates of return since if a black man had the characteristics of the typical white man, he would earn 14 percent less. The cost of racial differences in rates of return fell from 19 percent of the earnings of whites in 1960 to 14 percent in 1980. This is strong evidence of a declining racial difference in wage rates for male workers. Undoubtedly, the sponsors of Title VII of the Civil Rights Act hoped for a rapid convergence of the wage rates of blacks and whites. The racial difference has not been eradicated, but it was smaller in 1980 than in 1960.

In 1960 black women earned 52 percent as much as white women on an annual basis, while their expected earnings were 84 percent as much, showing that, relative to white women, they had both unfavorable labor market characteristics and lower rates of return. By 1980 the actual annual earnings of black women exceeded those of white women by 4 percent while their expected earnings were 7 percent above those of whites. Black women in 1980 completed fewer years of school than white women and were more concentrated in the South, but they spent more time at work so their overall labor market characteristics were just about as beneficial as those of white women. Their rates of return were also quite high compared with those of white women; that is, a black woman who had the characteristics of the typical white woman could expect to earn about 7 percent more, suggesting that black women no longer suffer from racial discrimination in wage rates. [13] Once again, we emphasize the distinction between racial differences and sexual differences. Women of both races received smaller benefits than men from their investments in education and for their years of labor force experience, differences which may be due to sexual differences in the quality of education and experience or to discriminatory pay policies on the part of employers.[14] Black women, however, did not receive lower rates of return than white women. Their wages may be limited once because they are women, but they are not penalized a second time because their skin color is black.

There are three possible explanations for the high expected earnings of black women. First, some employers may need to satisfy affirmative action goals so they prefer black women. If employers wish to demonstrate to a federal court or the Equal Employment Opportunity Commission that they hire many minorities and women, then recruiting and promoting black women will move them toward two goals simultaneously.

Second, employers may be more sensitive to issues of racial discrimination than sexual discrimination. Much of the justification for Title VII of the Civil Rights Act of 1964 and for the state and federal agencies which were developed to ensure equal employment opportunity originally focused on racial—not sexual—discrimination. Suppose that men and women perform similar tasks for an employer but work in separate parts of the factory, have different job titles, and earn different amounts. A white woman who recognizes the discrepancy may only raise the issue of possible sexual discrimination. A black woman may find greater receptivity for her complaints on the basis of racial discrimination.

Third, the greater earnings of black women may also result from racial differences in family structure. In 1980, 45 percent of the black women aged 25 to 64 lived with a spouse in husband-wife families compared with 74 percent of the white women.[15] In many circumstances, a

[13]For an analysis of 1960–70 changes, see Datcher (1980).
[14]For a summary of findings, see Bianchi and Spain (1986), chap. 6.
[15]U.S. Bureau of the Census, *Current Population Reports*, series P-20, no. 365 (October 1981), table 1.

wife may be under less pressure to maximize her earnings than a woman who has no spouse. There may be no racial difference in rates of return within specific marital status categories, but racial difference in marital status might produce the higher rates of return obtained by black women and their greater expected earnings.

Figure 11.5 presents results from an analysis that classified women by their marital status in 1980. Presumably, most never-married women aged 25 to 64 are responsible for their own economic status, which is also the case for a large proportion of the formerly married women—those who were separated, widowed, or divorced at the time of the census. Married-spouse-present women share their household with

FIGURE 11.5

Actual and Expected Annual Earnings of Black Women as a Percentage of the Actual Earnings of White Women in the Same Marital Status, 1980

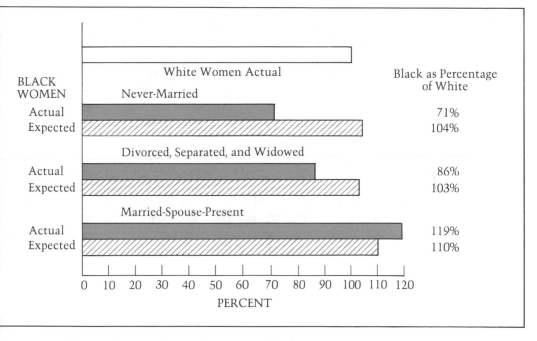

NOTE: For each marital status, the annual earnings of black women are compared with the actual annual earnings of white women of the same marital status. Expected annual earnings for blacks are estimated by assuming they had their own marital status–specific rates of return but the educational attainment, years of labor force experience, regional distribution, and hours of employment of white women of the same marital status.

SOURCE: U.S. Bureau of the Census, *Census of Population and Housing: 1980*, Public Use Microdata Samples.

a husband who, in most cases, earns much more than they do.[16] For each marital status, we show the actual and expected earnings of black women as a percentage of those of white women of the *same* marital status. The difference between the actual and expected earnings of black women reflects the fact that their labor market characteristics differ from those of white women. The difference between the expected earnings of black women and 100 percent, which represents the actual earnings of white women, results from racial differences in rates of return.

Never-married and formerly married black women earned far less than comparable white women, while the actual earnings of married black women exceeded those of married white women by almost 20 percent. Black women in all marital statuses completed fewer years of school than similar white women and were more concentrated in the South. Never-married and previously married black women also worked fewer hours than similar white women, but among women living with a husband, black wives worked longer hours than white wives: 26 hours per week versus 22 hours. This difference in labor force activity accounts for the high annual earnings of married black women.

In every marital status, black women had greater expected earnings than similar white women, and an examination of the rates of return shows that they were not the same for blacks and whites. Rather, black women, regardless of their marital status, paid a greater penalty than white women for living in the South but benefited more from their college educations and from their years of potential labor force experience. The higher rates of return for labor market experience may reflect racial differences in the quality and quantity of that experience; that is, black women spend a higher proportion of their adult years at work and are more likely to be full-time employees. They are less likely than white women to interrupt their careers and their interruptions are of shorter duration; in 1984, 52 percent of the employed black women compared with 45 percent of the white women had been on their current job for at least five years.[17] It is much more difficult to explain why college education increases the earnings of black women more than white women. Further examination of data from the census and from sources such as the Survey of Income and Program Participation may explain why the rates of return for education and experience are now higher for black women than white women.

[16]Bianchi and Spain (1983), figure 2; U.S. Bureau of the Census, *Current Population Reports*, series P-60, no. 153 (March 1986), table 10.
[17]Lamas, McNeil, and Haber (1986), table 5B; see also Corcoran (1978), table 2.2.

Regional Trends in Earnings

At the national level, the earnings of blacks increased after 1960 because of their improved characteristics and higher rates of return. Did these changes occur in all regions, or were they restricted to some locations? Michael Reich suggests that many of the economic gains recorded by blacks in the post–World War II era occurred not because of decreasing racial discrimination, but because blacks moved away from the South. This implies that an examination of earnings trends within regions will reveal a different picture than that obtained by looking at national trends.[18]

To investigate this, we classified blacks and whites by region of residence and then fitted models which said that hourly earnings were influenced by educational attainment and years of potential labor force experience. Figure 11.6 compares the actual and expected annual earnings of blacks in 1960 and 1980 with those of whites of the same sex and living in the same region. The difference between the actual and expected earnings of blacks indicates how their earnings would be changed if they had the labor market characteristics (education, labor force experience, and hours of employment) of whites rather than their own. The difference between the expected earnings of blacks and the actual earnings of whites—indicated by the 100 percent bar—reflects racial differences in rates of return. Appendix Table 11.3 shows the dollar amounts of the annual earnings plotted in Figure 11.6.

The idea that the economic gains of blacks are largely attributable to their shift away from the South is wrong. In every region, the actual earnings of black men moved closer to those of white men, and by 1980 black women in all regions outside the South earned more than comparable white women. Increases in the expected earnings of blacks relative to the earnings of whites suggests a decline in racial discrimination in every region. Stated differently, the cost of being black decreased throughout the nation and thus there would have been some racial convergence of earnings even if the regional distribution of blacks had not changed.

The idea that black gains were of the same magnitude in all regions should also be rejected. Racial differences contracted much more in the South, largely because racial differences in rates of return declined the most within this region, suggesting that more progress was made in eliminating discrimination in the South than in other regions. In 1960 black men in the Midwest and West were least handicapped by racial differences in rates of return. However, gains after 1960 were also smal-

[18]Reich (1981), pp. 46–49.

FIGURE 11.6

Actual and Expected Annual Earnings of Black Men and Women as a Percentage of the Actual Earnings of Whites, by Region, 1960 and 1980

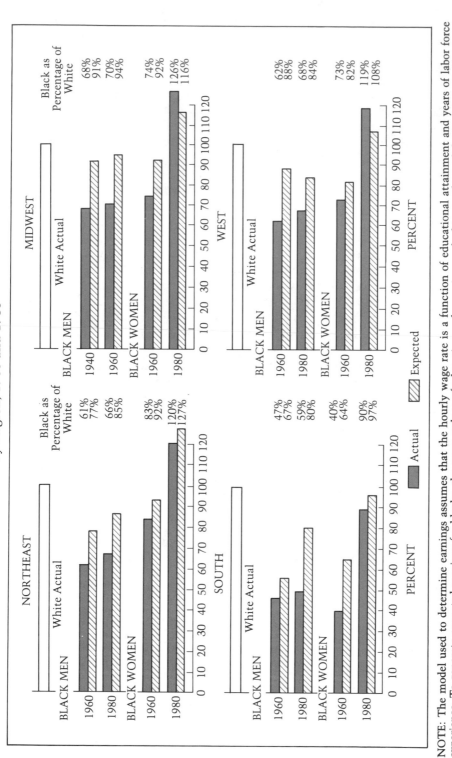

NOTE: The model used to determine earnings assumes that the hourly wage rate is a function of educational attainment and years of labor force experience. To generate expected earnings for blacks, the mean values of education and experience of white workers of the same sex, in the same region, and in the same year were inserted into the earnings equation for blacks. To estimate annual earnings, it was additionally assumed that blacks worked as many hours as whites.

The difference between the actual and expected annual earnings of blacks indicates the net effect of racial differences in educational attainment and years of experience. The difference between the expected earnings of blacks and the actual earnings of whites indicates the net effects of racial differences in rates of return for education and experience.

SOURCES: U.S. Bureau of the Census, *Census of Population and Housing: 1960. Public Use Microdata Samples; Census of Population and Housing:*

lest in those regions. Indeed, the net cost of racial differences in rates of return in the West actually rose from 12 to 16 percent of the annual earnings of white men.

Despite clear gains, racial differences in earnings among men were not eliminated by 1980 in any region, and the South remains the region in which blacks are furthest behind comparable whites. In that region, the actual earnings of black men were only 59 percent of those of white men; in other regions, 68 percent. The racial difference in rates of return cost black men in the South an amount equal to 20 percent of the average annual earnings of white men; in other regions, the net cost of being black was about 12 percent. The South was also unique with regard to women since it was the only region in which the annual earnings of black women remained below those of white women.

Earnings by Educational Attainment

Herman Miller, in his 1950 and 1960 census monographs, devoted much attention to the economic gap which separated blacks from whites. One of the striking findings which he stressed was that blacks benefited relatively little from remaining in school. Whites who completed high school earned considerably more than those who dropped out, and white college graduates earned much more than white high school graduates. For blacks, the benefits were much smaller since those who remained in school to the end of high school or college could anticipate earning just a bit more than those who dropped out. The further up the educational ladder, the greater were the economic differences between black and white men. In essence, blacks had less incentive than whites to finish their education.[19] About a decade later, Richard Freeman presented an extremely different picture. He argued that major racial changes occurred during the 1960s and that there was a "dramatic collapse in traditional discriminatory patterns in the market for highly-qualified black Americans."[20] He suggested that black college graduates were getting prestigious jobs just about as frequently as whites and that highly trained young blacks earned as much as—and sometimes more than—similar whites.

Which of these views provides the more accurate description of the situation in 1980? If the trends of the 1960s described by Freeman continued in the 1970s, we might expect black college graduates to earn as much as whites.

[19]Miller (1955), pp. 46-48; (1964), pp. 152-59; (1966), pp. 145-65.
[20]Freeman (1976), p. xx.

Figure 11.7 presents findings from an analysis of this topic. For each of three attainment groups, we show the actual annual earnings of blacks as a percentage of the earnings of whites in both 1960 and 1980. We also show the expected earnings of blacks, that is, the earnings they would receive if they had the years of labor market experience, regional distribution, and hours of employment of whites but their own rates of return. These findings are based on men and women aged 25 to 64 who had earnings in 1959 or 1979. We show findings for three educational groups: those with an elementary school education, those with four years of high school, and those with four or more years of college. Information for more attainment categories is shown in Appendix Table 11.3.

Between 1960 and 1980 the earnings of black men at all educational levels moved closer to those of similarly educated white men. However, black men, even those with college degrees, did not reach earnings parity with white men. In fact, in both census years, the ratio of black earnings to white hardly varied by educational attainment. Black men in 1980 who spent at least four years in college earned 70 percent as much as college-educated whites; black men who never entered high school earned 71 percent as much as similar white men. The actual racial difference in earnings in 1980 was much greater among college graduates, $8,800—than among those with only elementary school training—$4,200. Changes in the 1960s and 1970s did not eliminate the economic advantage enjoyed by white men in all attainment categories.

Black men are more often southern residents than white men, and at all educational levels blacks spend fewer hours at work than whites. In 1980 white college graduates, for example, worked an average of five more hours per week than black graduates. If we take these racial differences in labor market characteristics into account, will we find that highly educated blacks reached earnings parity with whites?

The data in Figure 11.7 reveal that this has not occurred.[21] When black and white male college graduates are equated with regard to characteristics, we see that blacks earned only 83 percent as much as whites, revealing that differences in rates of return cost college-educated black men an amount equal to 17 percent of the average earnings of white men (see Appendix Table 11.3). Stated differently, if we

[21]These models equate blacks and whites with regard to years of schooling but do not consider racial differences in types of training, quality of schools, test scores, or academic grades. Although scores on standardized achievement tests are correlated with characteristics of the student's family of origin and with his or her eventual attainment, several studies report that both grades while in school and scores on standardized tests have small and often insignificant net effects on the earnings of adults. See Griliches and Mason (1972); Alwin (1976); Duncan (1968), pp. 103-108; Duncan, Featherman and Duncan (1972), pp. 99-103.

FIGURE 11.7

Actual and Expected Annual Earnings of Black Men and Women as a Percentage of the Actual Earnings of Whites,
by Educational Attainment, 1960 and 1980

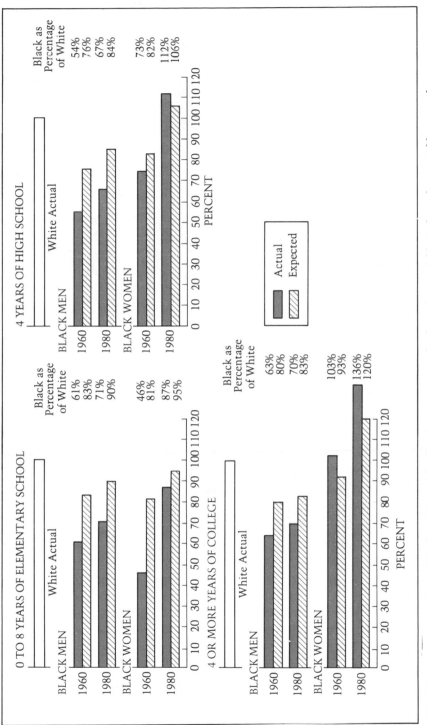

NOTE: The earnings models used the logarithm of the hourly wage rate as the dependent variable. Independent variables were the two measures of years of labor force experience and the one regional variable. To estimate expected hourly earnings for blacks, mean values of the independent variables for whites of the same sex and at the same educational attainment were inserted into the earnings equation for blacks. To estimate expected annual earnings, we also assumed that blacks worked as many hours during the year as whites.

SOURCE: U.S. Bureau of the Census, *Census of Population and Housing: 1960*, Public Use Microdata Samples; *Census of Population and Housing: 1980*, Public Use Microdata Samples.

compared black and white men in 1980 who completed at least four years of college and were similar on the other variables considered here, we would expect the white to earn $29,100 and the black to earn $23,900. To be certain, the cost to black men of racial differences in rates of return declined after 1960 at all educational levels, suggesting that progress was made in eliminating racial differences in pay rates. However, at every level, black men in 1980 earned about 10 to 15 percent less than comparable white men.

The changes in earnings among women are more consistent with the rosy views of Richard Freeman. In 1960 black women at all educational levels except college graduates earned less than white women, but in 1980 black women who remained enrolled through high school earned more than white women. In 1980 black women with a college education reported much greater earnings than white women, an excess of $4,100, largely because of their greater hours of employment.

Earnings by Age

Blacks younger than age 35 in 1980 began their occupational careers after federal laws prohibited racial discrimination in the labor market. If we compare age groups in 1960 and 1980, will we find that the greatest improvements in earnings occurred among the youngest workers? If so, there is hope that as succeeding birth cohorts enter the labor market and grow older, racial differences in earnings will diminish.

Figure 11.8 presents findings from an investigation of changes in earnings by age. Men and women have been classified into four age groups, and the actual and expected earnings of blacks are shown as a percentage of the actual earnings of similar whites in 1960 and 1980.

Racial changes may take place either on a *period* or a *cohort* basis. If, in a given span of time, the earnings of blacks of all ages moved closer to those of whites, change would occur on a *period* basis. If, in a given interval, entering cohorts of blacks and whites became more similar but the racial difference did not contract among older workers, change would occur on a *cohort* basis.

Figure 11.8 shows that the cost to black workers of having different labor market characteristics than white workers, as well as the cost of having different rates of return, declined on *both* a period and cohort basis. In 1960 black men aged 25 to 34 earned, on average, one half as much as white men. Racial differences in educational attainment, hours of work, and region of residence cost blacks an amount equal to 30 percent of the earnings of white men. The racial difference in rates of return cost them approximately 20 percent of the earnings of white men.

FIGURE 11.8

Actual and Expected Annual Earnings of Black Men and Women as a Percentage of the Actual Annual Earnings of Whites, by Age, 1960 and 1980

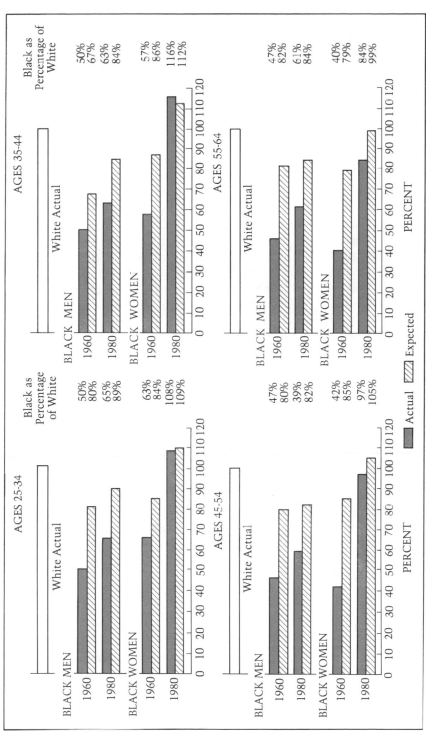

NOTE: The earnings models used the logarithm of hourly wage as the dependent variable. Independent variables were the two measures of attainment and the regional variable. To estimate expected earnings for blacks, mean values of the independent variables for whites of the same age and sex were inserted into the appropriate earnings equation for blacks. To estimate annual earnings, we also assumed that blacks worked as many hours as comparable whites.

SOURCES: U.S. Bureau of the Census, *Census of Population and Housing: 1960*, Public Use Microdata Samples; *Census of Population and Housing: 1980*, Public Use Microdata Samples.

By 1980 the cost of differences in labor market characteristics for the young group (those aged 25 to 34 in 1980) had declined to 24 percent of the average earnings of white men, while the cost of racial difference in rates of return fell to 9 percent. Cohorts of blacks and whites who entered the labor market recently are more similar in both their labor market characteristics and rates of return than those who entered prior to 1960.

A comparison of the *same* birth cohort at two time points will show whether change occurred for just those who entered the labor force or for all age groups. That is, if the economic status of the same birth cohort of blacks improved relative to whites as they grew older, we will know there was a period gain.

As cohorts of blacks aged after 1960, their earnings moved somewhat closer to those of whites, and the cost to blacks of racial differences in rates of return declined. The expected earnings of black men aged 35 to 44 in 1960 were 67 percent of those of comparable white men. Twenty years later, their expected earnings were 81 percent of those of white men, demonstrating that improvements were certainly not limited to entering cohorts. Rather, the changes in this two-decade span produced some racial covergence at all ages. Figure 11.8 reports the relative earnings of blacks. Appendix Table 11.3 shows the dollar amounts and indicates that as they grew older, black men fell behind their white age mates in terms of actual amounts of earnings even when they moved closer in relative terms. For example, black men aged 25 to 34 in 1960 earned 50 percent as much as white men while two decades later (at ages 45 to 54) they earned 59 percent as much, an apparent improvement. However, the racial gap in average earnings widened in this span from $7,600 to $10,500.[22]

When we examined regional trends or those for educational groups, we found that racial differences in earnings and in the apparent cost of discrimination declined much more among women than among men. This finding also applies to all age groups. By 1980 black women, except at the oldest ages, had both actual and expected earnings in excess of those of comparable white women.

Freeman's view of the new market for talented blacks suggests that it provides its greatest benefits to extensively educated young people.[23] We have reported that young black men (see Figure 11.8) or college-educated men (see Figure 11.7) still earn much less than comparable white men, but we have yet to analyze data for extensively educated young blacks. Perhaps we will find an absence of racial differences in this group.

[22]For a further analysis of how racial differences in earnings grow larger as cohorts age, see Ornstein (1976), pp. 173–77.
[23]Freeman (1976).

Table 11.4 provides information about the earnings of blacks and whites, classified by educational attainment, aged 25 to 34 in 1980. Looking first at the hourly wage rates of men, we find that young blacks—even those with college educations—earn less than comparable whites. Black college graduates, for example, earned 91 percent as much as white college graduates on an hourly basis. Since black men at all educational levels work 200 to 300 fewer hours per year than white men, their annual earnings are much lower than those of similarly educated whites. There are, however, suggestions of progress. When we considered all adult men, we found that the ratio of black to white earnings did not vary by educational level and that the racial difference in actual earnings was greatest among those with college degrees. That is not the case for young black men since the ratio of black to white earnings increases with attainment, and the actual racial gap in hourly earnings is smallest among college graduates. This may reflect the new market for highly trained black men; nevertheless, we find that young black college graduates earn 19 percent less than whites on an annual basis and 9 percent less on an hourly basis.

The lower panel in Table 11.4 provides information about young women. At all educational levels, young blacks earn more per hour than whites but substantially less than men of either race. Black women who complete high school work many more hours than comparable white women, meaning that their annual earnings are higher—often considerably higher—than those of white women.

Trends in Earnings: 1980–1985

Between 1960 and 1980 the racial gap in the earnings of men narrowed, and among women it disappeared. What occurred in the early 1980s? Was this also a period of economic progress for black workers, or have the trends of the recent decades been reversed?

The Current Population Survey (CPS) conducted in March 1985 asked questions about employment, labor force characteristics, and earnings similar to those appearing in recent decennial enumerations. Unfortunately, data from the CPS may not be directly compared with those from the census since they are gathered in a different manner. [24]

[24]In the CPS, trained interviewers ask numerous questions about earnings and employment, while the census obtains self reports to a few questions. The mean earnings in 1979 of black men aged 18 and over, as reported in the March CPS survey of the following year, were $10,500; in the April 1980 census, $10,100, or a difference of 4 percent. For white men there was also a difference of 4 percent in the mean earnings: $16,400 in the census, and $16,000 in the CPS. U.S. Bureau of the Census, *Census of Population: 1980*, PC80-1-D1-A, table 296; *Current Population Reports*, series P-60, no. 129 (November 1981), table 53.

TABLE 11.4

Average Hourly Wage Rate, Average Hours of Employment, and Average Annual Earnings for Blacks and Whites Aged 25–34 in 1980 (in 1984 dollars)

Educational Attainment	Average Hourly Wage Rate			Average Hours Worked in 1979		Average Annual Earnings		
	Black	White	Black as Percentage of White	Black	White	Black	White	Black as Percentage of White
MEN								
College, 4+	$10.23	$11.22	91%	1,706	1,929	$17,449	$21,640	81%
College, 1–3	8.35	9.45	88	1,595	1,899	13,320	17,938	74
High School, 4	7.50	9.00	83	1,586	1,934	11,889	17,401	68
High School, 1–3	6.30	8.04	78	1,375	1,709	8,659	13,737	63
Elementary	5.77	6.94	83	1,316	1,659	7,587	11,520	66
WOMEN								
College, 4+	$ 9.31	$ 8.64	108%	1,448	1,272	$13,482	$10,994	123%
College, 1–3	7.14	6.71	106	1,397	1,209	9,977	8,118	123
High School, 4	6.15	5.99	103	1,316	1,155	8,095	6,715	117
High School, 1–3	5.27	5.21	101	1,043	1,001	5,491	5,218	105
Elementary	4.73	4.68	101	1,023	1,094	4,842	5,121	95

SOURCE: U.S. Bureau of the Census, *Census of Population and Housing: 1980*, Public Use Microdata Samples.

To avoid the inaccuracies that would result from comparing CPS and census information, we used data from the CPS surveys carried out in March 1970, 1980, and 1985, permitting us to compare changes in the 1980s with those in the 1970s using comparable data sources. (No Public Use Microdata File was released from the March 1960 Current Population Survey.) We will describe trends for those persons aged 25 to 64 who reported earnings in the year prior to the surveys.

Figure 11.9 reports the average annual earnings of blacks and whites in constant 1984 dollars. During the 1970s the earnings of black men increased while those of white men declined slightly; thus, black earnings as a percentage of white earnings rose from 59 to 66 percent. In the first half of the 1980s, the average earnings of *both* races fell, which reflects a decrease in hours of employment and the rapid growth of the young population in the labor force—that is, people who have few years of experience and therefore earn little. Nevertheless, the decline in average earnings was sharper among blacks than among whites, and in 1985 black men reported earnings only 65 percent of those of white men.

The expected earnings of blacks are also shown in Figure 11.9. To calculate these, we assumed that blacks had the educational attainment, years of labor force experience, regional distribution, and hours of employment of whites but their own rates of return. This is identical to the strategy used with decennial census data earlier in this chapter. The actual earnings of blacks differ from their expected earnings because their labor market characteristics are not the same as those of whites. The expected earnings of blacks differ from the actual earnings of whites because of racial differences in rates of return.

Between 1970 and 1985 the characteristics of black and white workers became more alike, and the cost to black men of these differences fell from 23 to 19 percent of the average earnings of white men. The cost of racial differences in pay rates also declined in the first decade of this span, as the expected earnings of black men rose from 82 to 87 percent of those of whites. In the early 1980s this trend toward a racial convergence was reversed. The earnings penalty associated with having a black skin increased as racial differences in rates of return became larger among men.

Very different trends characterize women. For example, their earnings continued to rise in the early 1980s, albeit at a slower rate than in the previous decade. In 1970 black women typically earned less than white women, but by 1980 they had higher earnings, a difference which grew even larger in the next half decade. The higher earnings of black women come about because of declining racial differences in education and because of the greater hours black women spend on the job. The March 1985 survey, for instance, found that black women worked an

FIGURE 11.9
Actual and Expected Annual Earnings of Black Men and Women
and Actual Earnings of Whites Aged 25–64,
1970, 1980, and 1985 (in 1984 dollars)

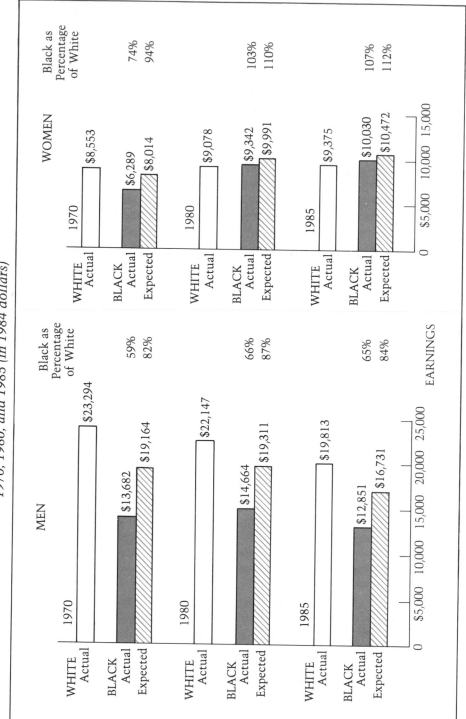

SOURCES: U.S. Bureau of the Census, Current Population Survey (March, 1970, 1980 and 1985), Public Use Microdata tape files.

average of 1,710 hours in the previous year; white women, 1,620. A look at the statistical models shows that black women continue to be more highly rewarded than white women for their investments in college education and for their years of labor market experience. According to the 1985 survey, an additional year of college added $1,460 to the annual earnings of the typical black woman and only $1,240 to those of the typical white woman.

During the early 1980s Congress passed no new civil rights laws, a more conservative ideology influenced public policy, the federal administration sought to overturn affirmative action programs, and, in several important decisions, the Supreme Court limited the remedial powers of federal courts in situations of employment discrimination.[25] Were these years in which the racial gap in earnings widened and in which the "cost of being black" increased? If we look at trends among men, we readily answer yes. The earnings of black men fell behind those of white men, the reverse of the trend of the previous four decades.[26] The cost of racial differences in rates of return—a measure which is often assumed to assess racial discrimination in the labor market—grew larger. Stated differently, if a black man in 1980 had the characteristics of the typical white man and worked as many hours, he could expect to earn $2,900 less than the white man; in 1985, $3,200 less.

If we look at the earnings of women, we reach the opposite conclusion about racial change in the first half of the 1980s. Black women averaged about 3 percent more than white women in earnings in 1980; in 1985, 7 percent more. If a black woman had the characteristics of the typical white woman, she would have earned 10 percent more than the white woman in 1980 and 12 percent more in 1985.

Summary

The preceding chapters give a mixed picture of the economic gains of blacks, and in every comparison it was necessary to distinguish trends among black men from those among black women. Turning first to employment, we found very little evidence of any improvement in the situation for blacks, despite economic booms and the civil rights legislation of the 1960s. Unemployment rates among black men persist at twice the level of white men and the proportion holding jobs has decreased much more rapidly among black men than among white men. Racial disparities in employment are becoming much larger among men

[25] *Firefighters Local Union No. 1784* v. *Stotts*, 467 U.S. 561 (1984); *Wygant* v. *Jackson Board of Education* 476 U.S. 90 L.Ed. 2d 260 (1986).
[26] Smith and Welch (1986).

at the start of their careers. Among women, the racial difference in unemployment rates hardly changed, and black women continue to be out of work much more frequently than white women. The proportion of women holding jobs has risen for both races, but the gains have been much greater among white women, and thus they are catching up with black women in terms of employment.

Trends in the occupations of workers are quite unlike those for employment itself. Blacks have gained on whites and, regardless of how they are measured, the occupational distributions of blacks and whites are now much more similar than they were in the past. This trend toward a racial convergence appears unabated by changes in economic growth, although blacks, especially black men, have a long way to go to reach occupational parity with whites.

Changes in earnings also suggest a declining significance of race. Between 1960 and 1980 there was racial convergence both because the labor market characteristics of blacks became more like those of whites; and because racial differences in the rates of return declined. These changes occurred much more rapidly among women than among men, and by the 1980s, the earnings of black women were not, in general, limited by racial discrimination.

How can we account for this mixed picture of economic gains and stagnation? Several factors need to be considered. First, macroeconomic trends played an important role both in providing opportunities for blacks at some times and in limiting their gains in others. Although it is difficult to document economic trends and specifically relate them to the status of blacks, the changes reported in these chapters are consistent with James Tobin's observation that blacks make substantial gains only when there is a severe shortage of labor.[27] The 1940s and 1960s stand out as periods of advance for black men. If the nation's economy had grown very rapidly in the 1950s, 1970s, and 1980s, perhaps racial differences in the employment and earnings of black men would be much smaller than those reported here.

We also need to take into account the changes in governmental spending and the expansion of governmental transfer payments. The racial riots of the 1960s, the trenchant reports of the Kerner Commission, and the writings of Michael Harrington, Oscar Lewis, and Harry Caudill reminded this rich nation that a substantial proportion of the population lived in extreme financial distress.[28] The federal administration, with a great deal of popular support, declared a "war on poverty," and changes were made in the allocation of financial resources. Spending for

[27]Tobin (1965).
[28]U.S. National Advisory Commission on Civil Disorders (1968); Harrington (1962); Lewis (1959, 1961, and 1965); Caudill (1964).

schools increased, and the federal government assumed responsibility for many health care costs. Undoubtedly this helps to explain the recent increases in educational attainment and the substantial declines in mortality rates among blacks. However poverty was not eliminated, and many racial differences changed only a little.

Forty years ago Myrdal observed that when it came to alleviating black deprivation, whites were often more willing to provide a dole than make the basic changes in society which would permit blacks to compete equitably with whites for educational attainment and economic security.[29] In the late 1960s and early 1970s, the food stamp program was created, housing subsidies expanded, and for much of the period AFDC payments became more available and were increased in amounts. In the short run, these programs were very beneficial to blacks, and by 1984, 25 percent of the black households received food stamps, 16 percent obtained AFDC or public assistance support, 14 percent lived in public or subsidized housing, and 51 percent of the households with children aged 5 to 18 participated in the federal school lunch program.[30] A substantial proportion of the nation's black households now benefit from the programs that were developed or expanded during the War on Poverty.

Perhaps our description of racial differences in the 1980s would be very different if, instead of expanding support programs, there had been a strong federal push to eliminate those practices that keep blacks at a disadvantage by isolating them from whites. Would the ratio of black to white unemployment rates be 2 to 1, and would black family income lag far behind that of whites if federal efforts had ensured that black and white children obtained identical educations by attending the same schools and that black families wishing to leave a deteriorating central city for the suburbs had been able to do so as readily as white families?

An explanation for the racial changes we described must also consider the changing racial views of whites.[31] The more egalitarian attitudes of whites may help account for the occupational progress of employed blacks and for the apparent decline in racial discrimination in pay rates. However, we know only a little about how employers make hiring decisions. Presumably, a white man often has to decide very quickly whether a black applicant will be a productive worker. Do em-

[29]Myrdal (1944), p. 301.

[30]For white households in 1985, the comparable figures are as follows: 6 percent received food stamps; 3 percent, AFDC or public assistance; 3 percent in public or subsidized housing; and 16 percent of the white households with a child aged 5 to 18 benefited from the school lunch program. U.S. Bureau of the Census, *Current Population Reports,* series P-60, no. 150 (November 1984), tables 5, 6, and 7; *Current Population Survey,* March 1985 (tape file).

[31]Schuman, Steeh, and Bobo (1985), chap. 3; Kluegal and Smith (1986), chap. 7.

ployers frequently assume that young black men are involved in a street culture of drugs and violence, an assumption that is not made about whites or black women?[32] Are black men viewed as competitors for the jobs of white men while women of both races are not? Are the diplomas and certifications of black men often seen as suspect or inferior because of a belief that affirmative action programs benefit unqualified blacks?

Finally, there are the changing racial attitudes of blacks.[33] Do blacks realize that there have been substantial changes in the racial situation and do they maximize their use of opportunities? A small percentage of the black population has taken advantage of the changes produced by civil rights legislation, court decisions, and the new attitudes of whites. Although young college-educated blacks lag far behind their white peers in earnings, the job market has greatly expanded for those blacks who have special skills or advanced training. Many of them compete successfully for the rewards offered in a largely white world of employment. At the same time, many other young blacks may believe that they lack skills and that their employment prospects are bleak. They may withdraw from the usual labor market and support themselves from transfer payments, illegal activities, and part-time jobs. Their status may be partly attributable to past and continuing racial discrimination, but they may also be failing to take advantage of opportunities which are now open to them.

---

[32]For observation about how employers perceive low skill black job candidates, see Liebow (1967), chap. 2, and Anderson (1980).
[33]Schuman, Steeh, and Bobo (1985), chap. 4.

APPENDIX TABLE 11.1

Mean Value of Variables and Parameters of Regression Equations for Earnings of Black and White Men and Women, 1960 and 1980

	Black Men		White Men		Black Women		White Women	
	1960 Census	1980 Census	1960 Census	1980 Census	1960 Census	1980 Census	1960 Census	1980 Census
MEANS OF VARIABLES								
Annual Earnings (1984 Dollars)	10,585	18,337	21,199	29,779	5,175	12,220	8,954	12,416
Hourly Earnings (1984 Dollars)	6.99	12.39	10.97	15.42	4.96	10.00	7.23	8.96
Years of Elementary and Secondary Schooling	7.5	10.5	9.9	11.3	8.3	11.0	10.4	11.4
Years of College	.3	.9	.7	1.7	.4	.9	.6	1.3
Percent in South	56.7	51.3	25.9	30.9	57.4	51.9	25.7	30.5
Years of Experience	27.8	22.6	26.2	22.4	26.5	21.5	25.9	21.9
Years of Experience Squared	922.3	669.4	832.5	655.1	847.3	608.3	804.2	630.0
Hours of Employment	1,780	1,849	2,076	2,106	1,283	1,600	1,431	1,543
Mean Log of Annual Earnings	7.700	9.145	8.411	9.623	6.722	8.691	7.384	8.650
Mean Log of Hourly Earnings	.363	1.759	.854	2.053	-.101	1.513	.395	1.545
PARAMETERS FROM REGRESSION EQUATIONS								
Intercept	-.000	+.982	-.092	+.788	-.407	+.869	-.277	+.792
Regression Coefficients								
Years of Elementary and Secondary Schooling	+.036	+.044	+.057	+.055	+.049	+.043	+.050	+.046
Years of College	+.105	+.079	+.089	+.082	+.175	+.115	+.107	+.090
Resident in South	-.500	-.202	-.181	-.087	-.560	-.215	-.167	-.068
Years of Experience	+.027	+.025	+.027	+.042	+.012	+.017	+.007	+.009
Years of Experience Squared	-.0004	-.0003	-.0004	-.0006	-.0002	-.0003	-.0001	-.0001
R^2	.165	.064	.110	.102	.206	.073	.078	.073
Sample Size	3,253	40,050	33,613	41,112	2,447	39,592	16,318	30,229

NOTE: Mean Log = log of current dollar amounts.

SOURCE: U.S. Bureau of the Census, *Census of Population and Housing 1960*, Public Use Microdata Samples; *Census of Population and Housing: 1980*, Public Use Microdata Samples.

APPENDIX TABLE 11.2
Sample Size and Deletions

	Men		Women	
	White	Black	White	Black
CENSUS OF 1980				
Total Number in Sample	45,462	49,943	48,525	60,97
Total	100.00%	100.00%	100.00%	100.00%
Deleted Because:				
Did Not Work in 1979	8.94	19.38	36.82	34.49
Worked but Had Negative Earnings	.47	.11	.22	.06
Worked but Had Zero Earnings	.16	.32	.66	.56
Percentage of Total Sample Included	90.43	80.19	62.30	64.91
CENSUS OF 1960				
Total Number in Sample	36,538	3,699	38,084	4,18(
Total	100.00%	100.00%	100.00%	100.00%
Deleted Because:				
Armed Forces	1.95	1.41	.03	0.0
Did Not Work in 1959	1.72	7.52	52.79	38.10
Worked but Had Negative Earnings	.13	.08	.02	.02
Worked but Had Zero Earnings	1.06	.78	2.13	1.70
Percentage of Total Sample Included	95.14	90.21	45.03	60.18

SOURCES: U.S. Bureau of the Census, *Census of Population and Housing: 1960,* Public Use Microdat Samples; *Census of Population and Housing: 1980,* Public Use Microdata Samples.

APPENDIX TABLE 11.3

Actual Average Annual Earnings of Whites and Actual and Expected Average Annual Earnings of Blacks, Classified by Region, Educational Attainment, and Age, 1960 and 1980 (in 1984 dollars)

	Men					
	1960			1980		
	White	Black		White	Black	
	Actual	Actual	Expected	Actual	Actual	Expected
REGIONS						
Northeast	$17,542	$10,686	$13,474	$21,963	$14,557	$18,746
Midwest	16,587	11,304	15,102	22,879	16,117	21,437
South	12,779	6,094	8,773	19,948	11,837	15,995
West	17,778	10,961	15,690	22,124	15,031	18,661
EDUCATIONAL ATTAINMENT						
Elementary, 0–8	$11,314	$ 6,859	$ 9,354	$14,357	$10,179	$12,881
High School, 1–3	16,017	9,322	13,167	17,344	11,563	14,954
High School, 4	18,135	9,832	13,687	20,753	13,916	17,348
College, 1–3	20,251	11,646	14,353	21,751	15,275	18,838
College, 4 +	26,320	16,556	20,953	29,429	20,582	24,341
AGE						
25–34	$15,380	$ 7,736	$12,290	$18,082	$11,822	$16,420
35–44	17,964	9,048	12,013	24,445	15,318	20,453
45–54	16,386	7,863	13,132	25,268	14,794	20,688
55–64	13,655	6,350	11,171	21,528	13,048	17,885

	Women					
	1960			1980		
	White	Black		White	Black	
	Actual	Actual	Expected	Actual	Actual	Expected
REGIONS						
Northeast	$6,414	$5,313	$5,891	$8,513	$10,162	$10,768
Midwest	5,477	4,027	5,060	7,660	9,676	8,846
South	5,338	2,134	3,421	8,186	7,366	7,940
West	5,733	4,194	4,704	8,486	10,100	9,148
EDUCATIONAL ATTAINMENT						
Elementary, 0–8	$4,519	$2,078	$3,670	$5,819	$ 5,058	$ 5,520
High School, 1–3	5,011	3,132	4,027	6,625	6,413	6,800
High School, 4	6,047	4,411	4,985	7,749	8,662	8,216
College, 1–3	6,768	5,424	5,003	8,531	10,592	9,238
College, 4 +	9,786	10,103	9,254	11,282	15,304	13,480
AGE						
25–34	$4,750	$3,107	$3,993	$7,863	$ 8,524	$ 8,556
35–44	5,665	3,207	4,886	8,066	9,347	9,039
45–54	6,550	2,726	5,531	8,809	8,571	9,258
55–64	6,308	2,505	4,982	8,302	6,887	8,263

NOTE: For the technique used to estimate expected average annual earnings of blacks, see footnotes 8, 9, and 11.

SOURCES: U.S. Bureau of the Census, *Census of Population and Housing: 1960*, Public Use Microdata Samples; *Census of Population and Housing: 1980*, Public Use Microdata Samples.

12

RACE, ANCESTRY, AND SOCIOECONOMIC STATUS: ARE WEST INDIAN BLACKS MORE SUCCESSFUL?

T HE RICH diversity that has historically characterized the black American population is clearly revealed in the 1980 census. Distinct subgroups within the black population are identifiable on the basis of regional location, urbanization, migratory patterns, socioeconomic status, and ancestry. However, the latter two categories have attracted most attention. The general hypothesis has been that foreign-born blacks (specifically those of West Indian ancestry) have achievements which distinguish them from native blacks. According to this argument, foreign-born blacks are economically more successful because of their more stable family life, greater educational attainments, higher achievement motivation, and stronger work ethic. Despite the wide dissemination of this thesis, few systematic comparisons of foreign and native blacks are to be found; thus, the validity of this contention remains in doubt.

This chapter examines the question of whether foreign-born blacks are more successful. Special attention is directed to the comparison of economic differences (that is, income, occupational, and labor force status differences) between native and immigrant blacks. In addition, we consider issues touching on ethnic diversity among blacks (for example, ancestry and languages spoken), population composition (for example, age structure and sex ratios), population distribution (for example, regional location), and social characteristics (for example, household composition and educational attainment). In each instance we are concerned with how these factors are related to and influenced by the nativity of blacks. Data on native-born persons and immigrants who are whites, Hispanics, and of other races provide additional points for comparisons.

Immigration, Social Mobility, and Assimilation: The General Case

Of the current ethnic, religious, and racial groups resident in this country, all except American Indians immigrated within the last four centuries. The immigrant status of this country's populace produced early interest in understanding the dynamics of how newly arrived groups became integrated into the established order.[1] By far, the most commonly accepted theory of assimilation is that associated with Robert Park, the University of Chicago sociologist. His model has greatly influenced the discussion and analysis of immigration in the United States. Park characterized the integration process in terms of four stages: competition, conflict, accommodation, and assimilation. The model predicts the following stages in the integration of an immigrant population.

1. *Competition:* On entering the society, immigrants compete with one another and with the native population for scarce resources such as jobs and housing.
2. *Conflict:* Inevitably, immigrants' competition with others for these resources results in conflicts which often become violent. Natives may erect barriers to keep immigrants off the payroll or out of their neighborhoods.
3. *Accommodation:* The negotiated terms of cooperation and coexistence are eventually achieved by immigrants and the native population after an extended period of conflict.
4. *Assimilation:* Eventually the immigrants are incorporated into the resident population and become indistinguishable in terms of social or economic characteristics.[2]

In its essential elements, Park's model is not only a familiar one in social science, but has popularly been accepted as a faithful description of what actually happened. It begins with the presumption of *stasis* or equilibrium. Arriving immigrants disrupt this balance, requiring the system to realign in order to achieve a new equilibrium.[3] Related in-

[1]Crevecoeur (1904), pp. 54–55.
[2]Turner (1967), p. xxxii.
[3]Park (1914).

terpretations of immigration, social patterns, and social change in urban areas were advanced by other social scientists associated with Park.[4] These studies shared the vital elements of his model and joined with it to form the "Chicago School" perspective about racial and ethnic integration.

Of the various studies influenced by Park's perspectives on organization in societies, two are of special importance for understanding immigration. Glazer and Moynihan's *Beyond the Melting Pot* presents a comparative analysis of the statuses of blacks, Puerto Ricans, Jews, Italians, and Irish in New York City.[5] This study was significant theoretically for its examination of assimilation and how this process varied across ethnic groups. The practical significance of the study derived from the 1960s upsurge in immigration from Cuba, the Caribbean, and Latin America[6] *and* from the deliberations under way in Washington concerning changes in immigration laws. Their book appeared just two years prior to enactment of the radically revised Immigration Act of 1965. Glazer and Moynihan concluded that the conception of American society as a "melting pot" was not entirely valid. The end result was not that ethnic groups lost their original identities and culture, as Park predicted. Rather, "The assimilating power of American society and culture operated on immigrant groups in different ways, to make them, it is true, something they had not been, but still something distinct and identifiable."[7]

Glazer and Moynihan acknowledged the interaction between the historical period of immigration, an ethnic group's original attributes, and the nature of American society itself in determining the extent to which newcomers were assimilated. Black Americans posed a dilemma for the "melting pot" or assimilationist model. They were among the earliest immigrants to New York City (and the United States), yet they persistently were the least assimilated group. In attempting to explain this apparent contradiction, Glazer and Moynihan compared native and foreign blacks. They concluded that native-born blacks were relatively less successful due to failings in their character. They were less frugal, less concerned with self-advancement, less industrious, more conspicuous consumers, and less well-equipped for life in urban, industrial society.[8] Careful examination reveals that Glazer and Moynihan's conclusions were largely based on impressionistic observations and questionable comparisons of statistical data.

[4]Burgess (1967); McKenzie (1967); Frazier (1937).
[5]Glazer and Moynihan (1963).
[6]Stinner, de Albuquerque, and Laporte (1982).
[7]Glazer and Moynihan (1963), pp. 13–14.
[8]Glazer and Moynihan (1963), pp. 34–37.

Oscar Handlin's conclusions in *The Newcomers: Negroes and Puerto Ricans in a Changing Metropolis* paralleled those of Glazer and Moynihan. For instance, he noted the tendency for ethnic identification to persist over three or four generations among immigrant white Europeans.[9] Similarly, he observed that many experiences were shared across immigrant populations. These included location in the same neighborhoods, comparable adjustment difficulties as rural folk entering a more complex urban society, and historic patterns of employment and occupational distribution.[10] Handlin was also adamant about the problem posed by skin color for the successful assimilation of blacks and Puerto Ricans, referring to it as "the ineradicable complication of color."[11] Like Glazer and Moynihan, Handlin attributed limited assimilation and social mobility among blacks (and, in this case, Puerto Ricans) to personal and cultural traits such as less of an orientation toward capital accumulation and saving. However, he emphasized the importance of color prejudice and the changing occupational structure of the city (that is, reduction in unskilled jobs available) as additional factors with profound consequences for the experiences of these more recent immigrants.

Immigration, Social Mobility, and Integration: Blacks as a Special Case

Ira Reid's *The Negro Immigrant* (1939) is the most comprehensive study of black immigration since the era of slavery. Because he was trained at the University of Chicago, numerous elements from the "Chicago School" perspective on immigration characterize his approach and conclusions. Most important is his analysis of the process by which native-born and foreign-born blacks accommodate one another. In this connection, Reid explores parallels in their customs, attitudes, and social backgrounds.[12] The framework for comparison is similar to those adopted by later scholars to effect cross-immigrant group analyses.[13] Unfortunately, many of the more subtle points and major contributions of this study of "inter-cultural and intraracial differentiation"[14] have been overlooked. The conclusion most often cited is: "It is estimated that in New York as high as one-third of the Negro professional population—particularly physicians, dentists and lawyers—is foreign-

[9]Handlin (1962), p. 59.
[10]Handlin (1962), p. 62.
[11]Handlin (1962), p. 62.
[12]Reid (1939), pp. 215–32.
[13]Handlin (1962); Glazer and Moynihan (1963); Sowell (1981a, 1981b).
[14]Reid (1939), p. 30.

born."[15] Although many descriptions of black immigration repeat this statistic, little consideration has been given to questions of how it was drawn (Reid does not say).

Franklin Frazier attributed differing levels of assimilation among native-born blacks to differences in socioeconomic attainments, family traditions, and "free status"—that is, their status as "free" blacks or slaves prior to emancipation.[16] As former students of Park's, both Frazier and Reid firmly believed that newly arrived migrants to cities from the South—as well as immigrants from abroad—would experience periods of disorganization and adjustment, followed eventually by assimilation. The speed with which these processes progressed would be determined by the nature of the conflict with established residents *and* the cultural capital at the disposal of the newly arrived group. In this light, the differences Reid found between the native and immigrant blacks in New York City were not unexpected. The two groups represented different waves of immigration under different circumstances—thus they possessed different cultural and capital assets.[17]

The debate over whether foreign-born blacks outperform native blacks has implications for theories of assimilation and racial discrimination. There are two contrasting models of why some ethnic or racial groups achieved greater success than others after they arrived in American cities. One model assumes that some groups brought cultural values and practices with them which ensured their success, while other groups brought values which limited their achievements. Quite frequently, the Jews and Japanese are singled out for praise.[18] It is assumed that on arrival, these groups had entrepreneurial skills, revered learning, were dedicated to deferring gratification, and maintained strong family ties. Presumably, they passed these values and practices on to subsequent generations. The Irish, on the other hand, lacked these virtues. They frequently had problems with alcohol, often resorted to fighting to solve their difficulties, uncritically accepted superstitious religious ideas, and were willing to accept menial jobs rather than starting their own enterprises.[19] According to this model of cultural determinism, the Japanese and Jews would have succeeded regardless of the conditions they faced in cities, while the Irish were destined for a more humble status.

A contrasting model recognizes that ethnic and racial groups differed in the values and experiences they brought to cities, but places greater emphasis on the economic opportunities and discrimination

---

[15]Reid (1939), p. 121.
[16]Edwards (1968).
[17]Frazier (1937). He describes the pronounced economic differences among blacks during that time.
[18]Petersen (1971); Sklare (1971).
[19]O'Dea (1959); Sowell (1981a), chap. 2.

they faced. Steinberg suggests, for example, that Jews and the Irish may have differed only a little in their dedication to intellectual pursuits.[20] Jews, however, entered cities at a time when opportunities were prospering and they could capitalize on the skills they brought with them. They found economic niches for themselves and, after achieving a level of success, educated their children. Other groups faced a lack of opportunities on arrival in cities or were the targets of such blatant discrimination that they had to accept the menial jobs no one else would take. In essence, the class position of a group shortly after arrival in cities is seen as having more to do with their subsequent economic status than the values they brought with them.

For many decades, ardent advocates of the first of these models have stressed that West Indian migrants have a culture which differentiates them from native blacks and allows them to succeed in American society. Writing in the 1920s, James Weldon Johnson observed that New York City was attracting many blacks from the South and from the British Caribbean.[21]

> Those from the British West Indies average high in intelligence and efficiency. There is practically no illiteracy among them, and many have a sound English common school education. They are characteristically sober-minded and have something of a genius for business, differing, almost totally, in these respects, from the average rural Negro of the South.

Four decades later, Glazer and Moynihan argued that "the ethos of the West Indians, in contrast to that of the Southern Negro, emphasized saving, hard work, investment, education."[22] Interestingly, these are the exact virtues which are assumed to account for the success of Jews and the Japanese in the United States.

Thomas Sowell is the most outspoken current defender of the cultural determinism perspective regarding blacks. He frequently suggests that West Indians who migrated to American cities "were much more frugal, hard-working and entrepreneurial" than native blacks.[23] Their children, he asserts, worked harder in school and scored at higher levels on tests. As a result of their cultural values, West Indian blacks were economically much more successful than native blacks. Sowell emphasizes that in 1969 family incomes of West Indian blacks were 94 percent of the national average, while the incomes of native blacks were only 62 percent.[24]

[20]Steinberg (1981), pt. 2.
[21]Johnson (1930), p. 153.
[22]Glazer and Moynihan (1963), p. 35.
[23]Sowell (1981a), p. 219.
[24]Sowell (1981b), table 1.1; (1981a), p. 220; (1978), p. 42; (1983), table 6.2; (1984), p. 77.

These data—and similar findings about the achievements of West Indians—are used by Sowell to challenge the view that color prejudice by whites is responsible for the condition of blacks in this country. Presumably, if whites consistently denied opportunities to blacks, West Indian migrants and their children would be little better off than native blacks in terms of educational attainment, occupational achievements, or income. The fact that West Indians are so successful leads Sowell to imply that the role of color prejudice has been overemphasized, while that of culture has been understated.[25]

This selective look at the literature demonstrates the need for further research. There are many unanswered questions about the immigration of foreign blacks, their social mobility, and their assimilation. What proportion of the total black population does the foreign-born compose? How do foreign-born and native blacks differ in regional distribution, socioeconomic attainments, family organization, and cultural orientation? Are foreign-born blacks really as successful as Sowell suggests? To answer these and related questions, we (1) describe the social, demographic, and economic characteristics of foreign-born blacks in the United States, and (2) compare the native and foreign-born black populations in terms of key characteristics.

To accomplish these aims, it is expedient to consider black immigrants as one component of the larger international migration flow which is rapidly altering the composition of this nation. We learn a great deal by describing those foreign-born blacks who now live in this country, but we learn even more if we compare black migrants with the three other major streams of migrants: those who come from Latin American nations, from Asia, and from the European countries that traditionally supplied the majority of migrants.

The Changing Composition of International Migration

Passage of the Hart-Celler Immigration Act of 1965 was followed by a radical change in origins of the migration flow. Before this law became effective, about 80 percent of the migrants who legally entered the United States came from European countries, while about 15 percent arrived from the Americas. Since 1968 the share of migrants coming from Europe has fallen to about 20 percent while almost 50 percent come from the Americas (Figure 12.1). In the past, Canada supplied the majority of the migrants from American countries. While there continues to be a substantial flow from Canada, this stream is now small com-

[25]Sowell (1981a), pp. 216–20; (1981b), pp. 8–10; (1983), 186–93.

FIGURE 12.1
Distribution of Immigrants by Country of Origin,
1820–1965 and 1966–1979

SOURCES: I. B. Taeuber and C. Taeuber, *People of the United States in the Twentieth Century* (Washington, DC: U.S. Government Printing Office, 1971), table III-1; U.S. Bureau of the Census, *Historical Statistics of the United States: Colonial Times to 1970* (1975), series C89-119; *Statistical Abstract of the United States: 1986* (1985), table 129.

pared with the migration from Mexico, Caribbean nations, and South America.[26]

Another major change concerned Asian migration. From the late nineteenth century through 1967, it was almost impossible for Asians to legally settle in the United States. The elimination of the national origins provisions in the 1965 law altered this, and for the last twenty years many more migrants have come from Asia than from Europe.[27]

International migration is now an important component of the nation's population increase. During the 1940s less than 10 percent of the population increase was attributable to net migration. This share rose to 11 percent in the 1950s, 16 percent in the 1960s, 20 percent in the 1970s, and 29 percent in the first half of the 1980s.[28]

It is more challenging to determine the role played by international migration in the growth of the black population since tabulations were often made for the nonwhite population only. In the 1950s no more than 3 percent of the total annual increase in nonwhite population was the consequence of legal immigration.[29] During the 1960s about 6 percent of the increase in black population came about because of immigration. Changes in admission policies and declining fertility rates among native blacks altered the situation, and during the 1970s, 11 percent of the 4 million increase in black population was attributable to net migration.[30] In the 1980s the share of black population increase due to immigration rose to 15 percent.

In the years immediately preceding 1968—the year in which the new immigration law became effective—about 20,000 blacks legally entered the United States annually and they composed about 5 percent of the net migration flow. Since then, an average of about 60,000 blacks have legally entered annually, and they now constitute about 10 percent of the volume of immigration.[31]

The Foreign-Born Population: Numbers and Countries of Birth

In 1980 the census counted 815,000 foreign-born blacks in the United States. This represents 3 percent of the total black population,

[26]Taeuber and Taeuber (1971), table III–1; U.S. Bureau of the Census, *Statistical Abstract of the United States: 1982–83* (1982), table 130. These data refer to the immigrants' country of last permanent residence, not necessarily their country of birth.

[27]U.S. Bureau of the Census, *Statistical Abstract of the United States: 1986* (1985), table 129.

[28]Taeuber and Taeuber (1971), table XI–2; U.S. Bureau of the Census, *Current Population Reports*, series P-25, no. 990, (July 1986), table A.

[29]Taeuber and Taeuber (1971), table XI–2.

or just a bit less than the population of Rhode Island or Hawaii.[32] To better describe the distinctive aspects of the foreign-born black population and to compare them with other streams of immigrants, we will use a distinctive four-way racial classification.

In the race question used in the 1980 census, respondents were asked to select one of 14 racial identities or write in an appropriate term indicating their race. Some people who wrote in a racial identity were reclassified into one of the existing 14 racial categories, while others were tabulated in a residual "other races" group. Those who wrote an ethnic or geographic origin commonly associated with blacks or whites, such as Zulu, Ibo, or Italian, were assigned to the black or white racial groups. Individuals who wrote in a Spanish origin, such as Cuban or Mexican, were categorized as "other" by race, as were people who wrote in such terms as Amerasian or Eurasian. In addition to the race question, respondents also answered a question which sought to determine whether they were of Spanish origin or descent by asking them to indicate if they were Mexican, Cuban, Puerto Rican, other Spanish, or not Spanish.

For the purposes of this chapter, we defined four mutually exclusive groups: "Blacks" include all those who gave a black racial identity, regardless of their answer to the Spanish-origin question; "Hispanics" consist of people who wrote in a Spanish response on the race question or who said they were white by race and then selected a Spanish origin or descent; "Whites" consist of people who said their race was white and their origin was not Spanish; and "Other Races" consist of those who selected American Indian, Eskimo, Aleut, one of the nine Asian and Pacific Islander races, or those who wrote in another racial identity which was not reclassified into white, black, or Hispanic.[33] Among

[30]U.S. Bureau of the Census, *Current Population Reports*, series P-25, no. 990 (July 1986), table 2.
[31]U.S. Bureau of the Census, *Current Population Reports*, series P-25, no. 990 (July 1986), tables 1 and 2.
[32]U.S. Bureau of the Census, *Census of Population: 1980*, PC80-1-A1, table 8.
[33]The distribution of responses to the racial and Spanish origin or descent questions are shown below. These reflect the reallocation made by the Census Bureau. Numbers are in thousands.

Selected Racial Identity	Number	Percent	Spanish Origin or Descent		Percent Yes
			Yes	No	
White	188,373	83.2%	8,117	180,256	4.3%
Black	26,495	11.7	391	26,104	1.5
American Indian, Eskimo, Aleut	1,422	0.6	95	1,327	6.7
Asian and Pacific Islander	3,500	1.5	166	3,334	4.7
Other Races	6,758	3.0	5,842	916	86.4
Total	226,546	100.0	14,609	211,937	6.4

SOURCE: U.S. Bureau of the Census, *Census of Population: 1980*, PC80-S1-7, table 4.

native-born people in this "Other Races" group, about 70 percent gave an Asian or Pacific Islander racial identity, while the other 30 percent were American Indians, Aleuts, or Eskimos. Among the foreign-born in the "Other Races" groups, about 97 percent were Asians.[34]

Members of these four groups were also classified by nativity. The native-born population consists of all persons born in the United States or one of its territories, such as Guam or Puerto Rico, plus persons born abroad to parents who were United States citizens. The foreign-born population refers to people born abroad to noncitizens of the United States.[35]

The native- and foreign-born populations differ greatly in their racial composition, which is illustrated in Figure 12.2. Blacks, for instance, make up 11.9 percent of the native population but only 5.5 percent of the foreign-born. A larger discrepancy involves whites: 81.3 percent of the native population is white compared with only 48.9 percent of the foreign-born. Changes in immigration laws mean that Hispanics and Asians are much overrepresented among the foreign-born.

Throughout this century, almost all black in-migration to the United States has been from the West Indies.[36] Table 12.1 reports the leading countries or areas of origin for each of the four migration streams. This table also indicates the size of the foreign-born population of each group and the proportion of total foreign-born population.

The stream of foreign-born blacks is small compared with the other foreign-born groups since there are about five times as many foreign-born Hispanics in the United States as blacks and nine times as many foreign-born whites. Jamaica is the most common country of origin for immigrant blacks; 175,000 were counted in 1980. About 84,000 blacks were born in Haiti, and just under 100,000 blacks were born elsewhere in the West Indies, including Cuba. Another 67,000 blacks were born in

---

[34]U.S. Bureau of the Census, *Census of Population: 1980*, PC80-1-D1-A, table 253.

[35]About 11 percent of both the black and white populations reporting a place of birth outside the 50 states were either born in a United States territory or were born to citizens of the United States. Data about the foreign-born population are shown below:

	Whites	Blacks	Hispanics	Other Races
Total Born Outside U.S. (000)	8,133	916	5,470	2,656
As Percentage of Total Population	4.4%	3.5%	35.7%	42.4%
Born in Puerto Rico	28	26	1,143	4
Born in Other U.S. Territories	36	2	6	28
Born Abroad of U.S. Parents	867	73	125	97
Foreign-Born	7,202	815	4,196	2,527

SOURCE: U.S. Bureau of the Census, *Census of Population and Housing: 1980*, Public Use Microdata Samples.

[36]Carpenter (1927), p. 103; Reid (1939), chap. 3.

FIGURE 12.2
Composition of the Native- and Foreign-Born Population, 1980

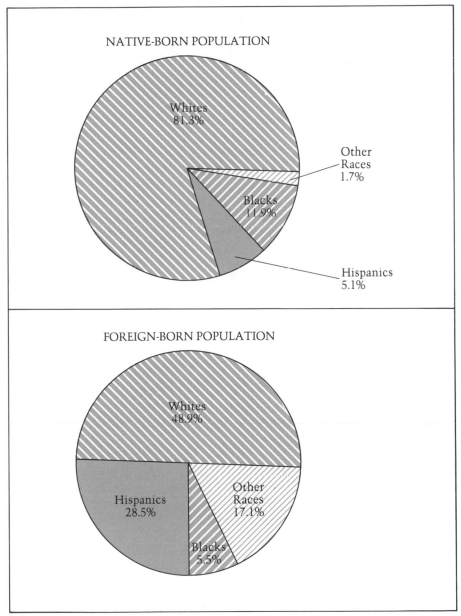

NOTE: See text for definition of these mutually exclusive groups.

SOURCE: U.S. Bureau of the Census, *Census of Population and Housing: 1980*, Public Use Microdata Samples.

TABLE 12.1

*Leading Places of Birth of the Foreign-Born Population
Classified by Race, 1980*

BLACK

Proportion of Total Population Foreign-Born	3.1%
Number Foreign-Born	814,500
Percentage of Foreign-Born from Leading Places of Birth	
Jamaica	27%
Other West Indies	17
Latin American Countries	15
Haiti	13
African Countries	10
Trinidad and Tobago	8
Cuba	2

HISPANIC

Proportion of Total Population Foreign-Born	27.4%
Number Foreign-Born	4,196,000
Percentage of Foreign-Born from Leading Places of Birth	
Mexico	54%
Latin American Countries	23
Cuba	15
Other West Indies	4

WHITE

Proportion of Total Population Foreign-Born	3.9%
Number Foreign-Born	7,202,000
Percentage of Foreign-Born from Leading Places of Birth	
Western European Nations	20%
Southern European Nations	18
Eastern European Nations	15
Canada	13
United Kingdom	10

OTHER RACES

Proportion of Total Population Foreign-Born	40.6%
Number Foreign-Born	2,527,000
Percentage of Foreign-Born from Leading Places of Birth	
South Asian Countries	49%
East Asian Countries	40%

NOTE: No allocation procedure was used to assign a country of birth to those foreign-born individuals who either gave an uncodable response to the place of birth question or did not answer it. The percentages shown in this table refer to those who reported a country of birth. The proportions of foreign-born population with no country of birth reported were: whites, 5 percent; Hispanics, 6 percent; blacks, 18 percent; and other races, 5 percent.

SOURCE: U.S. Bureau of the Census, *Census of Population and Housing: 1980,* Public Use Microdata Samples.

Central or South American nations. Quite a few of these are descendants of West Indian blacks who migrated to Central America in search of work, such as the labor movement associated with building the Panama Canal.[37] About 10 percent of the foreign-born black population, or less than 1 percent of the total black population, has an African birthplace. It is appropriate to use the term "West Indian" for black immigrants since about nine out of ten come from that area.

Looking at data for the other immigrant streams in Table 12.1, we find that more than one half of the foreign-born Hispanics come from Mexico while another one quarter were born in other Latin American countries. European birthplaces, of course, were most commonly reported by immigrating whites. The foreign-born other races population was Asian in origin with one half coming from South Asia, including the Indian subcontinent, and another two fifths from East Asia, including China, Japan, and Korea.

The Foreign-Born Population: Dates of Arrival and Citizenship

Social and economic studies of immigration stress the importance of duration of residence in a new country. Park's model, for example, assumes nearly a one-to-one correspondence between length of residence and degree of assimilation. The foreign-born groups considered here differ substantially in length of residence in the United States. Table 12.2 and Figure 12.3 indicate the date of arrival of the foreign-born. Whites, of course, have the longest residence histories since about 80 percent of those counted in the 1980 census arrived before 1970, and 40 percent before 1950. Foreign-born blacks and Hispanics are quite similar in their dates of arrival: more than 50 percent of the immigrants arrived in the 1970s and 10 percent or fewer before 1950. The other races group is distinctive because of its recent arrival in the United States; about 70 percent of these Asians came to the United States in the decade before 1980.

Becoming a citizen of a new nation indicates a commitment to remain and provides the immigrant with additional rights and employment opportunities. Descriptions of West Indian migrants living in Harlem in the 1930s noted that they were often involved in politics and, compared with southern-born blacks, they frequently made use of the courts to settle disputes.[38] Sowell reports that through the 1970s a high

---

[37]Reid (1939), pp. 67–73.
[38]Anderson (1981), pp. 299–304; Osofsky (1963), pp. 131–35.

TABLE 12.2

Distribution of Foreign-Born Population by Date of Arrival in the United States and Proportion Citizens, 1980

Dates of Arrival	Blacks		Hispanics		Whites		Other Races	
	Percent Arriving	Percent of Group Citizens	Percent Arriving	Percent of Group Citizens	Percent Arriving	Percent of Group Citizens	Percent Arriving	Percent of Group Citizens
1975–80	30%	18%	29%	11%	13%	10%	46%	8%
1974–79	25	29	24	18	8	34	23	36
1965–69	20	52	17	31	9	49	14	62
1960–64	7	67	13	52	9	60	6	65
1950–59	8	84	9	56	20	78	6	85
Before 1950	10	93	8	65	41	94	5	77
Percent Citizens		43%		30%		68%		33%
Percent Citizens Standardized for Date of Arrival		51%		34%		49%		49%

NOTE: Standardized for age using date of arrival information for the total foreign-born population in 1980.

SOURCE: U.S. Bureau of the Census, *Census of Population and Housing: 1980,* Public Use Microdata Samples.

FIGURE 12.3

Distribution of the Foreign-Born Population by Race and Date of Arrival, 1980

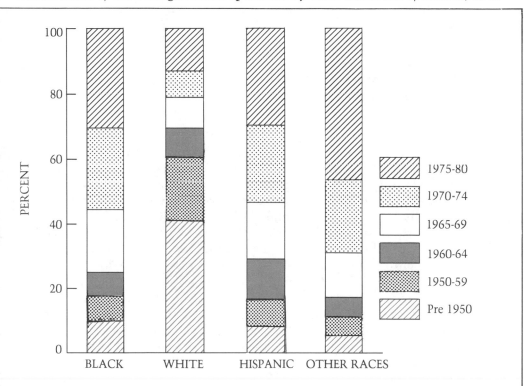

SOURCE: U.S. Bureau of the Census, *Census of Population and Housing: 1980*, Public Use Microdata Samples.

proportion of blacks elected or appointed to high offices in New York City were first- or second-generation West Indians.[39] While we cannot measure political activity, we can be certain that black immigrants often become citizens. Table 12.2 indicates the proportion of each foreign-born group who are citizens as well as citizenship by date of arrival.

At first glance, it seems that foreign-born whites are more likely to become citizens than immigrants from other races. However, citizenship depends, to a large extent, on length of residence in the United States. When this is taken into account, the proportion who are citizens is just about the same among blacks, whites, and other races, but is uniquely low among Hispanics. Indeed, if we focus on those who ar-

[39]Sowell (1981a), p. 220; Glazer and Moynihan (1963), p. 35.

rived before 1970, the only distinctive pattern is the low citizenship rate among Hispanics. Many of these immigrants are from Mexico and may anticipate returning there rather than remaining in the United States.

The Foreign-Born Population: Geographic Distribution and Languages Spoken

Geographic Distribution

Descriptions of the social processes affecting migrants often stress that the foreign-born, on arrival, concentrate themselves in a few locations. Later they, or their offspring, may spread across the nation and become distributed more like the native population. New York City has traditionally been the most common place of residence for foreign-born blacks, with Boston and New Bedford, Massachusetts, and several Florida cities following behind. This pattern was established late in the late nineteenth century and continues to the present. [40]

The Census Bureau's first thorough description of blacks, *Negro Population in the United States: 1790–1915*, reported that there were three streams of foreign-born blacks entering the United States. [41] The largest stream came from the West Indies; the second largest from Canada and Newfoundland—that is, descendants of blacks who freed themselves by following the "Underground Railroad" to Canada; the smallest from Atlantic Islands: Cape Verde and the Azores. The majority of West Indians settled in New York City, since the 1910 census reported one third of the nation's foreign-born black population lived there. Boston attracted both Canadians and West Indian blacks, and Cape Verdians often lived in the New Bedford area. [42]

The flow of blacks from Canada and the Atlantic Islands slowed after 1920 while migration from the West Indies continued, especially to New York City and its environs. [43] During the 1920s New York and New Jersey were the only states to record sizable increases in foreign-born blacks. In this decade their numbers decreased in both Mas-

[40]Carpenter (1927), p. 60; Reid (1939), pp. 85–92.

[41]U.S. Bureau of the Census, *Negro Population in the United States: 1790–1915* (1918), chap. VI.

[42]Carpenter (1927), pp. 103–4; Reid (1939), pp. 104–5. For a description of Canadian-born blacks in Boston in the late nineteenth century, see Pleck (1979). Detroit also served as a port of entry for those blacks who moved to southern Ontario before the Civil War. Zunz (1982), table 5–3.

[43]Osofsky (1963), p. 131. The laws of the 1920s apparently sought to restrict black immigration, but West Indians entered legally after 1924 by using the quotas of their colonies' mother countries.

sachusetts and Florida and by 1930, 60 percent of the 100,000 foreign-born blacks in the United States lived in or around New York.[44]

This concentration of foreign-born blacks continues today. Both the 1950 census[45] and the 1980 census found approximately half of the foreign-born black population living in proximity to New York. Table 12.3 presents the distribution of the native- and foreign-born population by states or divisions. We selected the largest states in terms of immigrant population and gave information about the remainder of the states in that geographic division. New England, for example, was the home of about 6 percent of the foreign-born blacks. As Table 12.3 indicates, there were only three areas in which the foreign-born black population was overrepresented compared with the native, and they were the areas of traditional entry: New England, New York, and Florida.

For each of the racial groups, we compared the geographic distribution of native- and foreign-born population using the index of dissimilarity. If natives and foreign-born had identical distributions across these areas, the index would equal zero. The measure would take on its maximum value of 100 if all natives lived in some locations, while all foreign-born lived in others.

The differences between the distribution of native and immigrant population are much greater for blacks than it is for Hispanics, whites, or the other races population primarily because of the concentration of foreign-born blacks in the New York and Miami areas. Looking at the other groups, we find a concentration of native Hispanics in Texas and California; indeed, half of all native-born Hispanics live in these two states. Foreign Hispanics are highly concentrated in California. Asian immigrants also have a distribution unlike that of black immigrants since they are overrepresented in Pacific states.

Table 12.4 presents data about the concentration of population in seven of the nation's largest metropolitan areas. These locations are now serving, or in the past have served, as major ports of entry for immigrants. New York City, Los Angeles, and Miami are the destinations of many newly arrived Hispanics, while Asian immigrants are highly represented in Los Angeles, New York City, and San Francisco. These concentrations are not extreme compared with the situation among blacks. New York City continues to be the leading port of entry for blacks, with more than four out of ten foreign-born blacks enumerated in that metropolis in 1980. Almost 20 percent of the black population in the New York area was born outside the United States, primarily in the

[44]U.S. Bureau of the Census, *Negroes in the United States: 1920–1934* (1935), table 10.
[45]Taeuber and Taeuber (1966), p. 109.

TABLE 12.3

Distribution of the Population by Race, Nativity, and State or Division of Residence, 1980

Place of Residence	Black Native	Black Foreign-Born	Hispanic Native	Hispanic Foreign-Born	White Native	White Foreign-Born	Other Races Native	Other Races Foreign-Born	Total Population
New England	1.7%	5.7%	2.3%	1.2%	6.2%	10.8%	2.3%	3.2%	5.4%
New York State	7.9	47.0	13.5	11.3	6.9	17.7	5.4	11.8	7.8
Other Mid-Atlantic States	7.5	6.9	4.0	5.5	8.9	11.7	3.1	6.6	8.4
Illinois	6.5	2.8	3.8	5.7	4.9	6.1	1.7	4.7	5.0
Other East North Central States	11.1	3.8	3.9	1.6	14.9	10.3	6.1	5.8	13.4
West North Central States	3.0	1.4	1.7	0.6	9.1	3.3	5.9	2.6	7.6
Florida	5.0	9.0	3.5	12.4	4.0	6.4	1.6	3.4	4.3
Other South Atlantic States	24.2	9.9	2.1	2.2	11.5	5.3	7.7	9.2	12.1
East South Central States	11.1	2.1	0.9	0.3	6.6	1.4	1.7	1.9	6.5
Texas	6.5	2.9	22.5	13.0	5.3	2.6	2.7	4.7	6.3
Other West South Central States	7.0	1.9	1.3	0.6	4.1	0.9	6.0	1.5	4.2
Mountain States	0.7	0.8	9.1	2.1	4.0	2.7	7.3	2.1	5.0
California	7.0	5.4	26.1	40.5	8.8	16.2	24.3	33.7	10.5
Other Pacific States	1.0	0.6	5.3	3.0	5.0	4.4	24.3	9.0	3.5
Total	100.0%	100.0%	100.0%	100.0%	100.0%	100.0%	100.0%	100.0%	100.0%
Indexes of Dissimilarity:									
Comparing Native- and Foreign-Born of Same Group	47		27		30		29		
Comparing Group to Native Whites	25	45	47	53	0	30	40	38	

SOURCES: U.S. Bureau of the Census, *Census of Population and Housing: 1980*, Public Use Microdata Samples; *Census of Population: 1980*, PC80-1-A1, table 10.

380

TABLE 12.4

Proportion of Population Living in Selected Metropolitan Areas for the Foreign-Born and Native, Classified by Race, 1980

	Black		Hispanic		White		Other Races		Total Population
	Foreign	Native	Foreign	Native	Foreign	Native	Foreign	Native	
New York	43.9%	6.2%	11.1%	11.7%	12.8%	2.7%	9.9%	3.5%	4.0%
Los Angeles	3.0	3.6	23.6	10.7	6.1	2.2	12.5	7.2	3.3
Chicago	2.5	5.5	5.4	3.3	5.5	2.6	4.3	1.3	3.1
Detroit	.9	3.4	.2	.6	3.8	1.8	1.1	.6	1.9
San Francisco— Oakland	.9	1.5	2.6	2.1	3.1	1.2	8.7	4.8	1.4
Washington	3.9	3.1	1.2	.5	1.6	1.1	3.4	1.6	1.4
Miami	5.0	.9	9.8	1.5	1.4	.3	1.2	.1	.7

SOURCES: U.S. Bureau of the Census, *Census of Population and Housing: 1980, Public Use Microdata Samples; Census of Population: 1980,* PC80-1-A1, table 30.

West Indies.[46] Miami, Washington, Chicago, and Los Angeles are also now serving as places of entry for West Indian blacks.

Languages Spoken

Fifty years ago when Reid wrote the definitive volume about black migrants to the United States, he found that three languages other than English were commonly spoken by this group.[47] Martiniquais, Guadeloupians, and Haitians were sufficiently numerous in New York City to support several French language social clubs and a French-language newspaper. Because of an influx from Cuba and Puerto Rico, many Spanish-speaking blacks lived in Harlem, while a smaller community of Cuban blacks produced cigars in Tampa. Reid observed that these migrants would obviously be classified as blacks by other residents of this country, but they often identified themselves as Spanish-speaking West Indians. Portuguese-speaking blacks from the Atlantic Islands worked as stevedores in New Bedford or labored in textile factories.

The situation has not changed greatly in the last five decades. Blacks who use a language other than English in the United States today typically speak either Spanish or French. The 1980 census asked several questions about the language a respondent spoke at home and his or her knowledge of English; Table 12.5 summarizes findings. Each group was classified by nativity and the proportion who reported they currently spoke a language other than English in their home is indicated. About 30 percent of foreign-born blacks and 1 percent of natives said they used a language other than English.

Among foreign-born blacks, French was most common, followed by Spanish, reflecting the immigration from Haiti, the Dominican Republic, and Central America. Among the native-born, Spanish was the typical foreign language. According to the census enumeration, about 250,000 blacks spoke Spanish in their homes and another 200,000 spoke French. A much smaller number—less than 50,000—reported speaking African languages.

Black migrants to the United States as a group are more familiar with English than any of the other migrant streams. That is, the proportion who speak only English in their homes is far greater among black immigrants than among white, Hispanic, or Asian immigrants, reflecting the distinctive origins of the majority of black migrants: English-speaking islands of the West Indies.

[46]U.S. Bureau of the Census, *Census of Population: 1980*, PC80-1-D1, table 69; PC80-1-D34, table 195.
[47]Reid (1939), chap. 4.

TABLE 12.5

Proportion Speaking a Language Other Than English
in Their Home, by Nativity and Race, 1980

	Native-Born	Foreign-Born
BLACK		
Percent Who Speak English Only at Home	97.8%	70.1%
Percent Who Currently Speak a Language		
Other Than English at Home	1.2	29.9
Most Popular Languages for Those Who		
Speak Other Than English		
Spanish	54%	32%
French	30	44
German	6	1
African Language	1	15
HISPANIC		
Percent Who Speak English Only at Home	38.1%	5.2%
Percent Who Currently Speak a Language		
Other Than English at Home	61.9	94.8
Most Popular Languages for Those Who		
Speak Other Than English		
Spanish	99%	99%
French	<1	<1
Italian	<1	<1
WHITE		
Percent Who Speak English Only at Home	96.3%	45.9%
Percent Who Currently Speak a Language		
Other Than English at Home	3.7	54.1
Most Popular Languages for Those Who		
Speak Other Than English		
Italian	14%	18%
German	17	18
Spanish	19	7
French	16	7
Eastern European Languages	7	16
OTHER RACES		
Percent Who Speak English Only at Home	76.9%	16.8%
Percent Who Currently Speak a Language		
Other Than English at Home	23.1	83.2
Most Popular Languages for Those Who		
Speak Other Than English		
Asian Languages	50%	94%
American Indian Languages	34	1
Spanish	11	2

NOTE: Respondents in the census of 1980 were asked to identify which languages other than English they spoke at home. The question applied to persons aged 3 and over. Processing errors apparently inflated the count of people who spoke languages other than English by 0.4 percent.

SOURCE: U.S. Bureau of the Census, *Census of Population and Housing: 1980*, Public Use Microdata Samples.

The Foreign-Born Population: Age, Sex, and Marital Status

Age-Sex Structure

Immigrants are often portrayed as young people—most commonly males—who leave an area of limited opportunities for a new country where they hope to find economic success. While this image may describe the stream of migrants who enter a country in a given year, it is not necessarily an accurate description of the entire foreign-born population. The immigrant population, of course, changes over time as emigration and mortality shape its demographic profile. In the United States, the foreign-born population tends to be much older than the native.[48] This is not surprising in the case of foreign-born whites since many are survivors from the large immigration streams which entered prior to the 1924 National Origins Act. It is more unexpected in the cases of foreign-born blacks.

About 40 percent of black immigrants in the United States in 1980 were in the 25-to-44 age group, which is about double the proportion among native blacks. There is also an overrepresentation of foreign blacks in the 45-to-64 age group and the median age of this population—33 years—is eight years older than the median age of the native black population and three years older than that of the country's total population. Table 12.6 presents the age distributions of the four race-nativity groups. The foreign-born black population is very different from the foreign-born white population, which is much older.

The interplay of fertility, emigration, and mortality alters the age distribution of the migrant population in less than obvious ways. For example, if a young couple moves to the United States from Haiti and starts a family, the children will show up as native-born blacks, which has the effect of keeping the age distribution youthful among the native-born. As indicated in Table 12.6, there are relatively few persons in any racial group under age 14 who were born abroad. In terms of age distribution, foreign-born blacks are similar to Hispanic and Asian immigrants.

We might assume that the foreign-born population of the United States is predominantly male since men move here to seek jobs and then later bring their families. The census of 1980, however, reports that only one of the four migration streams—that of Hispanics—has a majority male composition. Sex ratios by age are shown in Table 12.7.

The native- and foreign-born black populations differ in their sex

[48]U.S. Bureau of the Census, *Census of Population: 1980*, PC80-1-D1-A, table 255.

TABLE 12.6
Age Distribution and Median Age of the Population by Race and Nativity, 1980

Age	Black Foreign	Black Native	Hispanic Foreign	Hispanic Native	White Foreign	White Native	Other Races Foreign	Other Races Native	Total Population
0–4 Years	2.6%	9.7%	2.2%	15.1%	<1%	6.7%	3.2%	14.8%	7.2%
5–14 Years	10.5	20.2	9.7	26.1	4.0	14.9	12.6	22.3	15.5
15–24 Years	21.9	21.2	21.2	20.8	9.5	18.4	17.9	20.4	18.7
25–44 Years	39.5	25.4	41.8	23.9	25.8	28.1	45.8	24.1	27.7
45–64 Years	18.5	15.6	16.9	10.8	26.3	20.5	15.2	13.9	19.6
65–74 Years	4.0	5.1	5.1	2.2	14.6	7.1	3.0	3.2	6.9
75+ Years	3.0	2.8	3.1	1.1	18.9	4.3	2.3	1.4	4.4
Total	100%	100%	100%	100%	100%	100%	100%	100%	100%
Median Age	33	25	33	19	52	32	32	21	30
Index of Dissimilarity Comparing Foreign and Native		18		29		28		24	
Index of Dissimilarity Comparing Group with Native Whites	15	11	17	22	28	0	18	18	

SOURCES: U.S. Bureau of the Census, *Census of Population and Housing: 1980*, Public Use Microdata Samples; *Census of Population: 1980*, PC80-1-B1, table 45.

TABLE 12.7

Males per 1,000 Females for the Population by Race and Nativity, 1980

	Total Population	0–4 Years	5–14 Years	15–24 Years	25–44 Years	45–64 Years	65–74 Years	75+ Years
BLACK								
Foreign-Born	915	926	991	948	985	818	751	500
Native-Born	884	1,004	1,014	948	827	759	713	589
HISPANIC								
Foreign-Born	1,004	1,067	1,287	1,132	1,016	844	728	733
Native-Born	970	1,032	1,029	1,016	914	862	766	831
WHITE								
Foreign-Born	784	1,462	1,182	1,124	778	791	710	626
Native-Born	939	1,060	1,065	991	966	919	732	527
OTHER RACES								
Foreign-Born	822	976	882	1,018	776	630	974	1,280
Native-Born	929	1,056	945	853	945	873	1,052	613

NOTE: These data are not adjusted for census undercount.

SOURCE: U.S. Bureau of the Census, *Census of Population and Housing: 1980*, Public Use Microdata Samples.

compositions. At ages 25 to 64, for example, the ratio of men to women is considerably higher among the foreign-born. Overall, the sex ratio is 915 men per 1,000 women for foreign-born blacks and 884 for native-born. Caution is needed in drawing conclusions about this since census undercounts have traditionally lowered the sex ratio among blacks. In 1980, for example, the census reported a sex ratio of 896 for the enumerated black population. Once the figures were adjusted for undercount, the sex ratio became 948.[49] Unfortunately, we do not know whether undercount rates vary by nativity.

Marital Status and Family Patterns

Marital and family patterns of blacks in the United States are distinctive in three regards. First, blacks now marry at older ages than whites, which means that they spend more years as single individuals.[50] Second, rates of marital disruption are substantially higher among blacks than among whites.[51] As a result, the proportion of adults who are married and live with a spouse is much lower among blacks than among whites.[52] Third, after marital disruption, blacks are less likely than whites to obtain a divorce or subsequently remarry.

Are black immigrants similar to native blacks with regard to marital and family patterns, or are they closer to the native white population? Foreign-born blacks in the United States are pictured as having an interest in family life which distinguishes them from native blacks. In his history of Harlem, Osofsky argues:

> Another significant distinction between the foreign-born Negro and the American was their attitude toward family life. Slavery initially destroyed the entire concept of family for American Negroes and the slave heritage, bulwarked by economic conditions, continued into the twentieth century to make family instability a common factor in Negro life. This had not been true for most West Indians, and they arrived in America with the orthodox respect for family ties that was traditional of rural people.[53]

Although Herbert Gutman's historical investigations challenge the view that slavery destroyed the black family, his findings suggest a na-

[49]U.S. Bureau of the Census, *Current Population Reports*, series P-23, no. 115 (February 1982), table 4 and figure 9.
[50]Thornton and Rodgers (1983); Espenshade (1983).
[51]Thornton (1977); Espenshade, (1983).
[52]U.S. Bureau of the Census, *Current Population Reports*, series P-20, no. 399 (July 1985), table 1.
[53]Osofsky (1963), p. 134.

tivity difference. He reports that in New York City in 1925 the proportion of families headed by a man was somewhat greater among West Indian blacks than among the native-born.[54]

Data from the 1980 census do not support the view that native- and foreign-born blacks differ greatly in marital or family status. At the outset, we note that this information, although more comprehensive than that used in most investigations, is not ideal for testing hypotheses about marital stability or the strength of family bonds. It may provide an accurate cross-sectional description of current marital and family status, but it does not tell us how long a person has been married or separated. Another problem concerns those who are "married-spouse-absent." This is often a status people pass through on their way to becoming divorced. For the foreign-born, however, it may indicate that a spouse lives in the origin nation, perhaps a spouse who will migrate to the United States.

Figure 12.4 shows the marital status distribution of the population aged 15 and over in 1980. Since marital status varies systematically with the life cycle, these data have been standardized for age. Looking first at information about men, we observe that the proportion never married (single) was relatively great among both native- and foreign-born blacks. About three eighths of the adult black men were single compared with about three tenths of the Hispanics and whites. The proportion married-spouse-present was about 50 percent among black men and 60 percent among Hispanics and whites. Foreign-born blacks had a somewhat higher proportion married-spouse-present than native blacks, but it was lower than for either native- or foreign-born among the other three racial groups. We can be certain that the marital status distribution of foreign-born black men is quite like that of native black men and unlike that of Hispanics, whites, or other races.

Black women are unlike white women; a higher proportion never married and a much lower proportion lived with a husband. That is, only 32 percent of the adult native black women were married and living with a husband. Among native white women, 57 percent lived with their spouse. Native- and foreign-born black women did not differ greatly in marital status. The proportion married-spouse-present was larger among the foreign-born (41 percent versus 32 percent), but it was still at a much lower level than among any of the other groups. Clearly, the black population is distinguished from Hispanics, whites, and other races in terms of the small proportion of adults who live as married couples. Compared with other groups, blacks—both foreign- and native-born—remain single longer and are more likely to report separated or divorced as their marital status.

[54]Gutman (1976), p. 454.

FIGURE 12.4
Marital Status of the Population Aged 15 and Over, by Race, Sex, and Nativity, 1980

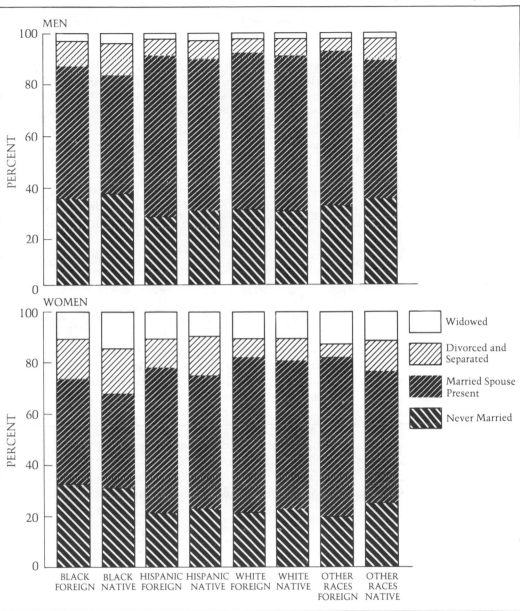

NOTE: These data have been standardized for age using the distribution of the total male and total female population aged 15 and over, respectively, as the standards.

SOURCE: U.S. Bureau of the Census, *Census of Population and Housing: 1980,* Public Use Microdata Samples.

If black migrants have values about family life which distinguish them from native blacks, we should find that the two groups differ in their types of families. According to Census Bureau definitions, a family consists of two or more related persons who share a housing unit, and a primary family is one in which a member of the family reports himself or herself as head of the household. The most common types of primary families are those that include a married couple or a woman who lives with her children but not her husband. A man who lives with relatives but not a wife might also be a family head. Persons who headed primary families were classified by race and nativity and their types of families are shown below.[55]

	Black		Hispanic		White		Other Races	
	Foreign	Native	Foreign	Native	Foreign	Native	Foreign	Native
Married Couple	64%	57%	78%	74%	85%	86%	84%	75%
Female-Headed	29	38	17	22	12	11	12	21
Male-Headed	7	5	5	4	3	3	4	4

Among blacks, there was some difference in family status by nativity since the proportion of married-couple families was greater among immigrants. The proportion headed by women was also lower for foreign-born blacks than natives. However, the family distribution of foreign-born blacks was closer to that of native blacks than it was to that of Hispanics, whites, or other races.

We also analyzed the proportion of adult men who headed married-couple primary families. Native and foreign-born blacks differed very little with regard to this measure, but the proportion of men in this status was distinctively lower among blacks than among any of the other groups.[56]

At some point in the past, West Indian migrants may have had a marital status and family distribution that distinguished them from

[55]Families are classified according to the race-nativity of the person designated as householder. If a foreign-born Asian married a native Hispanic who was designated as head of a primary family, the family was classified in the native Hispanic category.

[56]The age-standardized proportions of men aged 16 and over who headed husband-wife primary families in 1980 are indicated below:

	Foreign	Native
Black	42.4%	42.2%
Hispanic	55.8	56.7
White	60.1	56.9
Other Races	50.3	55.6

blacks born in the United States.[57] This is no longer the case since native- and foreign-born blacks are quite similar on all the indexes we considered, but are quite different from the other racial-ethnic groups.

The Foreign-Born Population: Educational Attainment, Employment Status, and Occupations

Educational Attainment

One of the most consistent findings in the field of migration concerns selectivity. Many investigators observe that propensity to migrate is directly proportional to educational attainment. When describing New England factory towns several decades ago, E. A. Ross likened them to fished-out mill ponds in which only carp and suckers were left since migration drained off the most promising citizens.[58] Where long distances are involved, the educational selectivity of the migration stream is particularly great.

With regard to international migration, there is an additional reason to suspect educational selectivity. The current admission laws give preference to people who have skills that are in short supply in the United States. The table below reports the average years of educational attainment for men and women aged 25 and over in each race-nativity group. Figure 12.5 shows educational distributions. Census data do not permit us to determine how much of the attainment of the foreign-born was obtained in the United States. These data have been standardized for age.

	Men		Women	
	Foreign	Native	Foreign	Native
Black	11.5	10.4	10.3	10.7
Hispanic	9.3	10.1	9.0	9.7
White	12.2	12.3	11.0	12.0
Other Races	13.0	11.6	11.3	11.3

A comparison of native- and foreign-born black men reveals that immigrants are more extensively educated. The proportion with a college degree—16 percent among the foreign-born—is double the propor-

---

[57]Although many claims have been made about the stability of West Indian families, Pleck's analysis of Boston blacks in the late nineteenth century reports that blacks born in Canada were more likely to be in female-headed families than blacks born in Massachusetts or the South. Pleck (1979), table VI-2.

[58]Cited in Lee (1964), p. 130.

FIGURE 12.5

Educational Attainment for the Population Aged 25 and Over, Standardized for Age and Classified by Race and Nativity, 1980

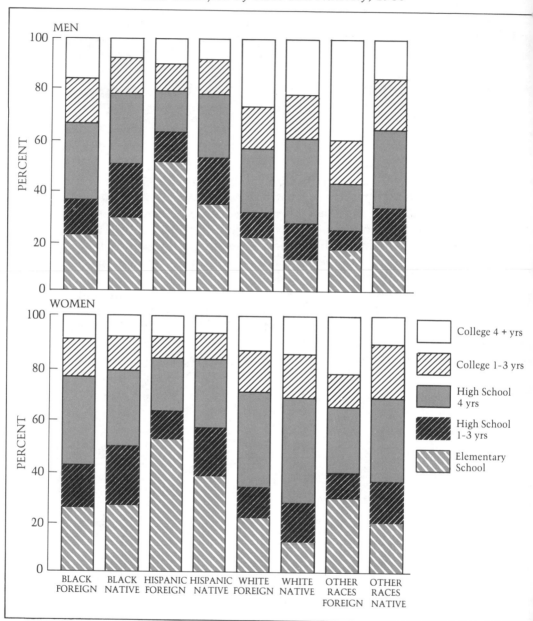

NOTE: These data have been standardized for age using the distribution of the total population of each sex as a standard.

SOURCE: U.S. Bureau of the Census, *Census of Population and Housing: 1980*, Public Use Microdata Samples.

tion for blacks born in the United States, and their average attainment exceeds that of native black men by more than a year. Nevertheless, when foreign-born black men are compared with whites—either native or foreign—we find that they lag behind in attainment. Among women, the findings are quite different since blacks born outside the United States spent fewer years in school, on average, than native black women.

Recent Asian migrants have distinctive attainment patterns; about 40 percent of the men and 25 percent of the women in this group completed college. Their attainment greatly exceeds that of both native and foreign whites. Hispanics are distinguished by their low attainment levels. Those born in the United States completed fewer years of school, on average, than native blacks, and Hispanic immigrants reported even lower attainments. This lack of selectivity may be explained by their geographic origins since the cost of getting to the United States is much less for persons from Mexico—and one half of the Hispanic migration stream comes from there—than it is for immigrants coming from Asia, Europe, or the West Indies.

Employment Status

Blacks in the United States have distinctive labor force characteristics in several regards: (1) since the end of World War II blacks have had high unemployment rates compared with those of whites; (2) an unusually large proportion of black men are nonparticipants in the labor force; and (3) black women have traditionally had higher rates of labor force participation and employment than white women. Are foreign-born blacks similar to native blacks with regard to these measures of economic activity? If foreign blacks coming to the United States bring a dedication to the work ethic and face little discrimination, as the writings of Sowell suggest, we should find that black immigrants are quite different from native blacks.

Figure 12.6 shows the labor force status of men and women aged 16 and over, classified by race and nativity. Data are shown for three labor force statuses: employed at the time of the 1980 census, unemployed, or not in the labor force. The data have been standardized for age to take differences in age structure into account. As indicated in Chapter 8, the "not in labor force" category is a heterogeneous one including retirees, full-time students, full-time homemakers, and some people who have given up the search for a job.

Foreign-born black men were quite similar to natives in their labor force status and quite unlike native white men. A high proportion of black men—native- or foreign-born—were out of the labor force or unemployed, and, compared with all other race-nativity groups, relatively few black men held jobs when the census was conducted.

Once again, we find an important sexual difference. Foreign-born black women had high rates of employment, and, compared with all other groups of women, they had low rates of nonparticipation in the labor force.

The table below presents the conventional unemployment rate for these groups as of April 1980; that is, it indicates the proportion of the labor force who were out of work and searching for a job. These rates are based on the age-standardized data shown in Figure 12.6.

	Men	Women
Black		
Foreign	9.0%	9.3%
Native	11.0	10.6
Hispanic		
Foreign	7.4	11.5
Native	7.8	8.4
White		
Foreign	5.3	5.5
Native	5.7	5.7
Other Races		
Foreign	4.4	8.0
Native	10.1	8.0

There is a racial difference in unemployment since whites generally have the lowest rates. Among men blacks were most likely to be unemployed, and among women blacks and Hispanics were most likely to be out of work. There also appears to be a nativity effect since the foreign-born are somewhat less likely to be unemployed than natives. This should not be surprising since many legal immigrants need to demonstrate their employability. Nevertheless, the unemployment rates for foreign-born blacks were only about 2 points lower than those of blacks born in the United States, and thus they were much higher than those of native whites. Asian male immigrants (the other races category) had uniquely low unemployment rates. This may, of course, reflect the selectivity of migration since many of these men have the educational attainments which qualify them for professional and administrative jobs.

Occupations and Entrepreneurial Activity

Many who comment about the occupations of West Indian blacks in the United States note the frequency with which they pursue profes-

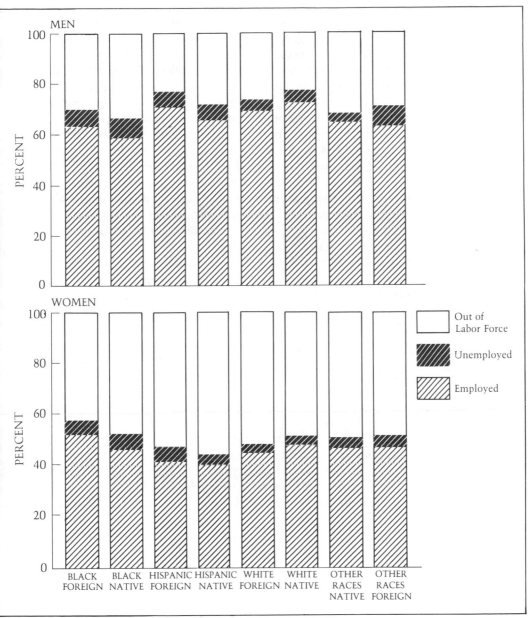

FIGURE 12.6

Labor Force Status of the Population Aged 16 and Over, Standardized for Age and Classified by Race and Nativity, 1980

SOURCE: U.S. Bureau of the Census, *Census of Population and Housing: 1980*, Public Use Microdata Samples.

sional careers, engage in the arts, or manage their own businesses.[59] A list of prominent West Indians in the pre-Depression period includes Marcus Garvey, Claude McKay, and Eric Walrand.[60] This approach has not changed, although the recent lists include such eminent blacks as W. Arthur Lewis, Sidney Poitier, Stokely Carmichael and Malcolm X.[61] Few of those who suggest that black migrants are more successful than native-born blacks have scrutinized comparative data. Gutman, on the other hand, examined the jobs held by black men in New York City in the mid-1920s and found that the foreign-born were no more than one step ahead of natives. His basic conclusion was that black men, regardless of their birthplace, were restricted to menial and low-paying jobs. More than 70 percent of both the native and West Indian blacks were unskilled laborers or service workers.[62] Reid observed that while some black immigrants achieved occupational success, many of these men worked as waiters, elevator operators, or factory hands, while West Indian women often took jobs in garment factories.[63]

Do black immigrants work at higher-status jobs than blacks born in the United States? Are they overrepresented among the ranks of the self-employed? The census of 1980 allows us to answer these questions with a degree of certainty. Table 12.8 shows the occupational distributions of employed men and women aged 16 and over in each of eight race-nativity groups. These data have been standardized for age. Information is shown for the 13 broad occupational categories which were used for the first time in 1980.

Looking first at data for men, we find that black immigrants had a slight occupational advantage over the native-born. In a 1984 publication, Sowell asserted, "West Indian representation in professional occupations is double that of blacks, and slightly higher than that of the U.S. population as a whole."[64] This claim is quite exaggerated since in 1980, 14 percent of the foreign-born blacks compared with 11 percent of the native blacks held professional or executive jobs. Twelve percent of the native men worked in the lowly-paid blue collar category of handlers, cleaners, and laborers compared with 7 percent of the foreign-born.

To summarize the status of occupational distributions, socioeconomic indexes were calculated for each race-nativity group using a procedure described in chapter 9.[65] If all members of a group worked as professionals, their average socioeconomic index would be 70; if they

[59]Sowell (1981a), p. 220.
[60]Osofsky (1963), p. 133.
[61]Anderson (1981), p. 302.
[62]Gutman (1976), table A-42.
[63]Reid (1939), p. 120.
[64]Sowell (1984), p. 77; (1978), table 1.
[65]Stevens and Cho (1985), table 1.

were all handlers, cleaners, and laborers, their average would be 19 points. Returning to Table 12.8, we see that the average socioeconomic index for foreign-born black men was 30 points or slightly greater than that of the native black men.

Table 12.8 also presents information about the jobs of employed women. Black immigrants have no occupational advantage over those women born in the United States. In both groups, about one woman in four worked in an office job—that is, in an administrative support position—and about one in ten worked as a domestic. This is more than double the proportion for any other racial group.

Indexes of dissimilarity are shown comparing the foreign and native workers of each group, and another set of indexes compares each group to native whites. These measures imply that foreign and native blacks have occupational distributions which are quite similar to each other, but they differ substantially from those of native whites.

When foreign-born blacks are compared with the other immigrant streams, we see that their occupational distribution resembles that of Hispanics. Both groups are quite highly concentrated in the blue collar occupations. Immigrants who reported white or Asian as their racial identity were often found in the higher-paying white collar jobs, undoubtedly reflecting the educational advantage of these groups.

Decennial censuses are not the ideal source for obtaining information about racial differences in entrepreneurial activities, but they do include a series of questions about income from various sources. A self-employment rate may be computed by determining the percentage of a group who reported income—either positive or negative—from a non-farm, unincorporated business, professional enterprise, or partnership. The proportions of persons aged 16 and over reporting self-employment income for the eight race-nativity groups are shown here.

	Men		Women	
	Foreign	Native	Foreign	Native
Black	3.4%	2.9%	1.1%	.9%
Hispanic	5.3	3.8	1.4	1.1
White	9.7	8.4	2.3	2.5
Other Races	8.3	6.4	2.5	2.8

Foreign-born blacks have a higher self-employment rate than natives: 34 per 1,000 black male immigrants reported income from this source compared with 29 per 1,000 among those born in the United States. However, the self-employment rate for blacks—be they West Indians or native-born—was below that of Hispanics and substantially below that of whites or the other races group. In his study of black Manhattan in the 1920s, Gutman concluded that neither West Indians

TABLE 12.8
Occupational Distribution of Employed Men and Women by Race, Sex, and Nativity, 1980

	Black		Hispanic		White		Other Races	
	Foreign	Native	Foreign	Native	Foreign	Native	Foreign	Native
MEN								
Executive, Administrative, Managerial	7.6%	5.3%	5.6%	10.1%	12.7%	12.9%	9.7%	10.6%
Professional Specialty	6.8	5.7	5.9	5.2	14.8	11.4	19.3	11.3
Technical	3.3	1.7	1.1	1.5	3.5	2.8	6.0	4.2
Sales	4.6	3.9	5.6	5.7	9.4	10.7	8.9	10.1
Administrative Support	12.3	8.7	3.9	8.5	5.6	6.6	8.9	7.6
Private Household	.7	.3	0	.1	0	0	0	0
Protective Service	4.4	3.2	.2	2.6	1.1	2.6	2.0	2.5
Other Service	15.8	15.6	16.2	11.5	10.1	6.3	19.9	8.7
Farm, Fisheries	2.4	4.3	11.4	4.4	1.9	5.4	2.3	6.6
Precision Production	15.0	14.8	18.9	20.6	22.5	20.2	9.3	17.1
Operators	12.7	14.0	15.5	11.4	8.4	8.7	7.1	6.5
Transportation Occupations	7.8	10.8	5.2	9.4	3.8	6.6	3.0	6.8
Handlers, Cleaners, Laborers	6.6	11.7	10.7	9.0	6.2	5.8	3.6	8.0
Total	100.0%	100.0%	100.0%	100.0%	100.0%	100.0%	100.0%	100.0%
Socioeconomic Index	30	28	28	30	36	35	38	30
Index of Dissimilarity Comparing Foreign and Native	11		18		11		23	
Index of Dissimilarity Comparing Group to Native Whites	24	28	28	22	11	0	27	8

WOMEN

Executive, Administrative, Managerial	2.6%	4.3%	5.0%	6.2%	7.0%	7.7%	11.2%	10.3%
Professional Specialty	9.0	10.5	5.1	8.0	12.4	14.4	9.5	14.7
Technical	2.8	3.0	1.6	4.5	3.5	2.8	4.7	2.7
Sales	6.0	5.9	8.8	10.2	14.2	12.8	7.7	8.0
Administrative Support	24.1	22.2	16.6	28.2	25.8	31.7	14.9	33.1
Private Household	9.6	10.5	4.7	4.2	2.2	1.3	5.7	.6
Protective Service	.2	.5	.1	.1	.4	.4	.1	0
Other Service	32.2	25.5	19.3	17.8	16.7	15.8	19.4	20.9
Farm, Fisheries	.4	.7	3.6	1.5	.6	1.1	2.7	1.0
Precision Production	2.1	2.2	4.5	2.8	3.4	2.3	2.9	2.5
Operators	8.5	11.2	25.8	13.4	11.0	7.1	18.4	4.7
Transportation Occupations	.4	.8	.6	.4	.9	.8	.6	.5
Handlers, Cleaners, Laborers	2.1	2.7	4.3	2.7	1.9	1.8	2.2	1.0
Total	100.0%	100.0%	100.0%	100.0%	100.0%	100.0%	100.0%	100.0%
Socioeconomic Index	27	29	28	32	35	36	32	37
Index of Dissimilarity Comparing Foreign and Native	9		20		9		25	
Index of Dissimilarity Comparing Group to Native Whites	26	26	15	15	9	0	27	10

NOTE: Data refer to the population aged 16 and over and have been standardized for age.

SOURCE: U.S. Bureau of the Census, *Census of Population and Housing: 1980*, Public Use Microdata Samples.

nor native blacks had gained any significant foothold in business of any kind.[66] This conclusion may still be appropriate.

The Earnings of Employed Workers: Does Nativity Make a Difference?

In his efforts to suggest that culture rather than discrimination accounts for the current status of blacks, Sowell states:

> West Indians living in the United States are a group physically indistinguishable from black Americans, but with a cultural background that is quite different. If current employer racial discrimination is the primary determinant of below-average black income, West Indians' incomes would be similarly affected.[67]

He then points out that West Indians in the United States had family incomes that were 94 percent of the national average, while blacks' family incomes were only 62 percent. If black migrants possess the cultural characteristics Sowell attributes to them—industriousness, frugality, independence, responsibility, and a dedication to learning—employers might pay them more than native-born blacks who, presumably, lack such desirable traits. It is possible that they would earn as much as white workers. On the other hand, if employers make assumptions about the employability and productivity of all blacks and fail to distinguish West Indians, we should find that they earn about as much as blacks born in the United States.

In Chapter 11, racial differences in economic status were examined by relating individuals' hourly earnings to their educational attainment, years of labor force experience, and place of residence. We found that not only did black men earn less than whites, but that if racial differences in the factors that influence earnings were taken into account, black men were at a substantial disadvantage. A similar procedure can be used to determine whether the earnings of foreign-born black men compare favorably with those of either native black men or white men.

Using data from the decennial census for men aged 25 to 64, hourly earnings in 1979 were related to years of elementary and secondary education; years of college training, if any; labor market experience; occupational category; and a regional variable. Since a large proportion of foreign-born black men live in or around New York, we identified that area. We also distinguished those who lived in the South, the region where wage rates remain low. As we noted, census data do not permit

[66]Gutman (1976), p. 453.
[67]Sowell (1984), p. 77.

us to determine whether foreign-born men obtained their education in the United States or elsewhere.

Table 12.9 provides a summary of findings concerning the effects of race and nativity on earnings. For each group, we report the estimated earnings in 1979 of a 40-year-old man who worked full time for the entire year. We present information for a professional worker with five years of college; an administrative support worker with two years of college; and a machine operator with two years of high school. Data are shown for those living in the New York area, the South, and the remainder of the nation. These amounts were determined using coefficients from the regression equations shown in Appendix Table 12.1.[68]

Controlling for educational attainment, occupation, and place of residence, black men earned considerably less than white men. This racial difference was not limited to native blacks. In most comparisons, black immigrants earned as little or even a bit less than black men born in the United States. Consider men with five years of college working in professional jobs in the New York area. The census data report that a native-born black man aged 40 averaged about $19,000; a foreign-born black man, $18,000; and a native-born white man, $23,000. In the South, a native white with two years of high school who operated a machine could expect to earn about $15,000 in 1979; a native-born black, about $4,000 less; and a black migrant from the West Indies, an additional $1,000 less. The coefficients in Appendix Table 12.1 may be used to estimate the earnings of men of other ages, educational attainments, or occupations.

Data in Table 12.9 allow us to compare black immigrants with three other streams. Foreign-born blacks earned much less than ostensibly similar white immigrants, but in most cases their earnings were similar to those of Hispanic immigrants. Asian immigrants who had a college education and who held professional or executive jobs reported unusually large earnings. Indeed, their average earnings exceeded those of comparable native white men. Asian immigrants with fewer years of educational attainment or less prestigious jobs reported earnings below those of similar white men; they were close to those of Hispanics and black immigrants.

If we had found that West Indian migrants consistently earned more than native blacks, the evidence would have supported Sowell's hypothesis that immigrants have a culture differentiating them from native blacks. If we additionally had found that foreign-born blacks earned as much as comparable whites, the evidence would have chal-

---

[68]For an analysis of the earnings of foreign-born blacks in 1976, see Chiswick (1979), pp. 379–81.

TABLE 12.9

Estimated Annual Earnings in 1979 for Men Aged 40 Who Worked Full Time for Entire Year, Specific for Occupation, Education, Place of Residence, and Race-Nativity (in 1979 dollars)

	Professional Worker with Five Years of College			Administrative Support Worker with Two Years of College			Machine Operator with Two Years of High School		
	New York	South	Elsewhere	New York	South	Elsewhere	New York	South	Elsewhere
Estimated Earnings in 1979									
WHITE									
Native-Born	$22,700	$20,480	$22,300	$17,370	$15,700	$17,080	$16,610	$14,980	$16,340
Foreign-Born	21,880	21,730	22,360	17,310	17,200	17,170	14,500	14,410	14,820
BLACK									
Native-Born	18,720	15,470	19,750	13,510	11,500	14,240	12,490	10,630	13,170
Foreign-Born	17,600	15,950	18,320	13,980	12,650	14,530	10,550	9,550	10,970
HISPANIC									
Native-Born	19,450	19,090	20,070	14,800	14,520	15,270	12,990	12,740	13,410
Foreign-Born	19,850	15,970	19,900	13,870	11,150	13,900	11,330	9,110	11,360
OTHER RACES									
Native-Born	18,520	19,040	20,170	15,860	16,310	17,280	10,740	11,050	11,700
Foreign-Born	22,590	23,260	23,270	12,780	13,160	13,160	9,820	10,110	10,120
Earnings as Percentage of Those of Native White Men									
WHITE									
Foreign-Born	96%	106%	100%	100%	110%	101%	87%	96%	91%
BLACK									
Native-Born	82	76	89	78	73	84	75	71	81
Foreign-Born	78	78	82	80	81	85	64	64	67
HISPANIC									
Native-Born	86	93	90	85	93	84	78	85	83
Foreign-Born	87	78	89	80	71	81	68	61	70
OTHER RACES									
Native-Born	82	93	90	91	104	101	65	74	72
Foreign-Born	100	103	104	74	84	77	59	68	62

SOURCE: U.S. Bureau of the Census, *Census of Population and Housing: 1980*, Public Use Microdata Samples.

lenged the hypothesis that racial discrimination accounts for the current economic status of blacks. However, data from the 1980 census are unambiguous in reporting that foreign-born blacks earned just about as much as native blacks and that both groups earned about 20 percent less than comparable whites. This chapter does not present a thorough analysis of the cost of racial discrimination, but findings are consistent with the hypothesis that blacks—be they born in New York, Birmingham, Kingston, or Port au Prince—earn less than white men of similar age and educational attainment, a difference that is commonly assumed to be the outcome of racial discrimination.

Family Income

To complete this analysis of nativity differences in the status of blacks, family income was examined. Once again, primary families were classified by the race-nativity of the person who identified himself or herself as head of the household. Since income levels differ greatly by type of family, data are shown in Table 12.10 for married-couple families and for those headed by women not living with a husband. Family income is a crucial indicator of a group's economic status, but caution is needed when comparing the incomes of the different races. Family income is influenced by the employment of spouses, by the presence and employment of children or other relatives, and by the receipt of transfer payments, be they from annuities or the welfare system.

Families headed by West Indian migrants in 1980 reported larger incomes than those headed by natives. Among married couples, the immigrants' advantage was about 4 percent; among female-headed families, almost 20 percent. Despite this advantage, foreign-born blacks lagged far behind whites in family income. At some point in the past, West Indian families may have had incomes that were 94 percent of the national average. A more accurate generalization for 1980 is that the income of West Indian families is about 80 percent of the national average; and of native black families, about 70 percent.

Families headed by foreign-born blacks were economically similar in income to those headed by Hispanic migrants. Both of these groups had incomes quite far below those of families headed by whites or by persons of other races.

Summary and Conclusions

Those who have described black immigrants to the United States since Emancipation present two contrasting pictures. Pleck's study of

TABLE 12.10

Average Income of Families in 1979 by Race, Type, and Nativity (in 1979 dollars)

	Husband-Wife Families	Female-Headed Families
Average (Mean) Family Income in 1979		
BLACK		
Foreign	$20,990	$12,300
Native	20,100	10,260
HISPANIC		
Foreign	19,430	11,480
Native	20,440	10,180
WHITE		
Foreign	25,860	15,400
Native	25,700	14,800
OTHER RACES		
Foreign	25,420	16,110
Native	24,200	13,380
Income as a Percentage of That of Native Whites		
BLACK		
Foreign	82%	83%
Native	78	69
HISPANIC		
Foreign	76	78
Native	80	69
WHITE		
Foreign	101	104
OTHER RACES		
Foreign	99	109
Native	94	90

SOURCE: U.S. Bureau of the Census, *Census of Population and Housing: 1980*, Public Use Microdata Samples.

Boston in the late nineteenth century, Gutman's analysis of Manhattan in the 1920s, and Reid's overview of immigrant blacks in the first third of this century report that black immigrants were similar in social and economic status to those native blacks. On the other hand, the observations of Osofsky, Johnson, and Anderson about West Indian blacks in New York between World Wars I and II and those of Glazer, Moynihan, and Sowell concerning this group in the 1960s and 1970s are strikingly different, and immigrants are seen as distinguished from native blacks in their dedication to education, occupational achievement, and stable families. Sowell goes far beyond other authors in arguing that this distinctiveness reflects a West Indian culture which is much more compatible with achievement in the United States than is the culture of native blacks.

We have examined the demographic characteristics of black immigrants and compared them with whites and with three other streams of migrants. In recent decades, an increasing number of foreign-born blacks have moved to the United States, almost all of them from the West Indies. Immigration is now a more important component of black population growth than at points since the end of slavery, and about 3 percent of the black population was born outside this country. The geographic distribution of the foreign-born differs substantially from that of native blacks since immigrants settle in traditional ports of entry, especially New York and Miami. As a group, foreign-born blacks are older than those born in the United States, but similar in both age and dates of arrival to Hispanic and Asian migrants.

Census data certainly do not provide information about dedication to marital stability, but with regard to family structure and household headship, West Indian migrants are similar to native blacks and quite different from the other racial groups. The educational attainment of foreign-born black men exceeds that of natives, but is less than that of white men born in the United States or elsewhere. Black women who migrate to this country do not have an educational advantage over those born here.

In 1980 unemployment rates were lower for foreign-born blacks than for natives, but were far in excess of those of whites, Hispanics, or other races. Black male immigrants were more likely to be holding professional or executive jobs than natives, but the differences in occupational distributions were quite small. The 1980 census data also imply that often-stated claims about the entrepreneurial accomplishments of West Indians are greatly exaggerated. Similar to native blacks, black immigrants have a low rate of self-employment. With regard to the earnings of employed men, our estimates place those of foreign-born blacks close to those of blacks born in the States, which are about 20 percent below the earnings of comparable white men.

APPENDIX TABLE 12.1

Regression of Log of Hourly Wage Rate on Explanatory Variables for Men Aged 25–64 Employed in 1979, Classified by Race and Nativity

	Black		Hispanic		White		Other Races	
	Foreign	Native	Foreign	Native	Foreign	Native	Foreign	Native
Mean LN of Hourly Wage	1.7752	1.7579	1.7028	1.8603	2.1405	2.0590	1.9306	1.9186
REGRESSION COEFFICIENTS								
Education								
Years of Elementary and Secondary Schooling	.011*	.042	.025	.046	.022	.049	.029*	.077
Years of College	.048	.067	.055	.070	.067	.068	.051	.061
Experience								
Years of Labor Force Experience	.032	.022	.019*	.035	.052	.039	.051	.020*
Years of Experience Squared	−.000	−.000	−.000*	−.000	−.001	−.001	−.001	−.000*
Residence								
Lives in South	−.138	−.214	−.220	−.050*	−.021*	−.086	−.000*	−.057*
Lives in New York Area	−.039*	−.053	−.002*	−.031*	−.028	.016*	−.029	−.086*
Occupation								
Executive, Managerial	.403	.407	.747	.264	.324	.591	.524	.576
Professional Specialty	.465	.420	.597	.195*	.278*	.501	.689	.568
Technical	.529	.430	.199*	.188*	.037*	.505	.520	.638
Sales	.193*	.231	.426	−.149*	.146*	.429	.110*	.280*
Administrative Support	.287	.345	.350	.031*	.101*	.383	.130*	.537
Protective Service	.183*	.298	.361	.168*	−.159*	.337	.010*	.078*
Other Service	.043	.064	.168*	−.248	−.128*	.070	−.093*	.269*
Precision Production	.338	.397	.459	.212	.190*	.480	.283*	.518
Operators	.081*	.326	.271	.067	.040*	.401	−.011*	.364
Intercept	1.017	.755	.895	.787	.968	.486	.866	.222
R²	.088	.085	.100	.101	.134	.134	.250	.149
Sample Size	1,929	38,121	1,055	1,554	1,462	38,014	582	580

OTHER VARIABLES

Mean Hourly Wage	8.19	8.69	8.37	9.06	12.01	10.79	11.29	9.49
Mean Annual Earnings	12,310	12,844	14,303	15,086	23,526	20,910	18,820	16,462
Mean Hours of Work	1,784	1,852	1,897	1,920	2,070	2,115	1,980	1,976

*Regression coefficient *is not* significantly different from zero at .05 level.

NOTES: Regression models used the natural log of the hourly wage rate as that dependent variable. Analysis refers to men who reported an employment in 1979 and who had earnings of one dollar or more.

For the dichotomous geographic variables, the omitted category refers to men who did not live in the south or the New York area. For this analysis, the New York area consists of the states of Connecticut, New Jersey and New York.

For the dichotomous occupation variables, the omitted category refers to men employed as household servants, in farming, fishing, or forestry, or with occupations not reported.

To estimate the earnings of workers with specific characteristics, appropriate values for independent variables should be associated with the regression coefficients and intercept shown on this page. Consider a native-born black man aged 40 who lived in New York City, had a college education, and worked as an executive. He had a value of 12 on the elementary and secondary education variable, 4 on the years of college variable, 18 on the years of labor market experience variable since he was 40 years old, and 324 for years of experience squared. He was coded as a resident of New York and an executive worker. The regression equations (using more digits than shown in this table) estimate that the natural log of his hourly wage was 2.19567, which implies an hourly wage in 1979 of $8.99. If he worked 40 hours for each of 50 weeks, he earned $17,980.

SOURCE: U.S. Bureau of the Census, *Census of Population and Housing: 1980,* Public Use Microdata Samples.

13

A WORLD WITH NO COLOR LINE: RACE AND CLASS IN TWENTY-FIRST-CENTURY AMERICA

I found new friends and lived in a wider world than ever before—a world with no color line.[1]

WITH these words, W. E. B. DuBois concluded his autobiography. His was a remarkable life, distinguished by prolific contributions to the scholarship on race and culture in America. Fortunately, he lived long enough to see the profound transitions black people experienced during the 95 years from his birth to his death. He was born February 23, 1868, a mere five years after the Emancipation Proclamation abolished this "peculiar institution." He died August 27, 1963, a day before the Great March on Washington for black civil rights, led by the Reverend Dr. Martin Luther King, Jr.

DuBois' concept of the place of race in the lives of black Americans changed over his lifetime. Some argue that the evolution in DuBois' thinking paralleled changes in how this society approached race. During the early years DuBois saw race as the supreme factor and predicted that the twentieth century would be dominated by "the problem of the color line." In his later years DuBois revised his stance to place more emphasis on economic factors, and he declared that the dominant problem of the twenty-first century would be the search for a new economic order in the world.

> If tomorrow Russia disappeared from the face of the earth, the basic problem facing the modern world would remain; and that is: Why is it,

[1]DuBois (1968), p. 395.

with the earth's abundance and our mastery of natural forces, and miraculous technique; with our commerce belting the earth; and goods and services pouring from our stores, factories, ships and warehouses; why is it that nevertheless most human beings are starving to death, dying from preventable disease, and too ignorant to know what is the matter, while a small minority are so rich that they cannot spend their income?[2]

The years of DuBois' life coincided with a historic period in American race relations. In 1875 the future for Africans in America was dismal; blacks were forced into rural serfdom as an alternative to slavery. The years following World War I did not hold much promise; cities exploded with antiblack violence designed to protect white advantages in housing, employment, and education in the face of the massive in-migration of blacks. Certainly the summer of 1965 offered little hope for a peaceful, prosperous future; black economic disadvantages stubbornly persisted and over 100 of the nation's cities were in flames from urban rebellions. Although today some predict a bright future for race relations in the United States, others are less optimistic.

Scholars agree that race relations in the United States have changed in significant ways. Slavery was abolished; blacks moved out of the rural South to the urban North; and the laws supporting racial segregation were overturned. Beyond these points of fundamental agreement, however, there is much debate. Scholars disagree on the extent to which race relations have changed, the causes of these changes, or their consequences. Have blacks overcome? Do they enjoy full equality in the United States? These questions provide the basis for debates over the declining significance of race, the proper role for government policy in affirmative action, and the explanation of contradictory economic trends among blacks. Without systematic data, efforts to resolve these debates will prove futile.

This book provides a statistical comparison of blacks and whites based on the 1980 census. In conducting this inquiry we were inextricably drawn into comparisons of relative economic status. Economic differences were highlighted by each comparison of black and white characteristics, whether the topic of primary concern was family structure, educational attainment, or health status. In a larger sense, therefore, the book came to be about the nexus of race and class in the United States. Of specific interest was the question, "What has been the economic progress of blacks relative to whites?" Similarly, we found ourselves asking, "What is the relative quality of life experienced by blacks and whites in contemporary America?"

---

[2]DuBois (1968), p. 376.

The Comparative Status of Black and White Americans

Black life in this country has always been characterized by an existence on the periphery. Black Americans were denied the most fundamental human rights; indeed, the Supreme Court's Dred Scott decision proclaimed that blacks possessed "no rights which whites were bound to respect." The result has been a pattern of systematic underdevelopment. Although the most oppressive aspects of black subjugation have been eliminated, the legacy remains. The legacy is evident in the persistent racial inequities in educational achievement, family structure, occupational attainment, patterns of unemployment, income levels, residential segregation, and the distribution of wealth. Continuing discrimination, together with accumulated disadvantages, sustains black deprivation.

After a wide-ranging study of the urban rebellions during the 1960s, the Kerner Commission concluded that "our nation is moving toward two societies, one black, one white—separate and unequal."[3] Our findings support this conclusion to a large extent. Detailed examination of empirical data shows that for the *majority* of black Americans life continues to be experienced as "separate and unequal."

The basic findings from this study of race differences in social and economic status are summarized below. These findings reveal areas of considerable gain as well as areas of considerable loss. Black Americans continue to be substantially disadvantaged relative to whites in this country.

Since 1940 death rates have fallen for both races. The declines have been slightly greater for blacks, but there is still a major racial discrepancy in mortality since whites live about six years longer than blacks. Racial differences in mortality are much smaller among women than among men; the life span of black women may equal that of white women early in the next century. However, at current rates such a convergence will not occur among men until well after 2050, if at all.

Birthrates rose rapidly among both races between 1940 and the late 1950s. Since that time births have fallen at a similar rate for both races. Nevertheless, the fertility level of blacks remains about 30 percent higher than that of whites.

The races are becoming increasingly distinct as regards the timing of childbearing and the marital status of mothers. White women become mothers at later ages, while black women concentrate their childbearing in their teens and early 20s. The shift toward a higher proportion of births "out of wedlock" is evident among both races but is much

[3]U.S. National Advisory Commission on Civil Disorders (1968), p. 1.

more pronounced among blacks. About 60 percent of black children compared with 12 percent of white children are delivered to unmarried women.

On all indicators of marital and family status, we find that both races are moving away from a traditional pattern of early marriage, fertility within marriage, and the rearing of children by both parents. The shifts have been much greater among blacks, and racial differences in the proportion of young women who are married, the proportion of adult women who are wives, and the proportion of children living with both parents have grown larger. A minority of black children now live with a father.

In terms of school enrollment and years of schooling completed, racial differences have narrowed substantially since 1940. The racial gap in attainment has certainly not been eliminated but a continuation of recent trends offers hope that young blacks will soon complete as many years of schooling as whites. In the 1970s racial differences on standardized tests of academic achievement narrowed slightly, but black youth continue to score lower on these tests than white youth.

Our study recounts dramatic changes in the spatial distribution of blacks. During the 1940s, 1950s, and 1960s there was a substantial flow of black migrants from the South to the North and West. The migrants tended to be less extensively educated than native-born blacks; as a result, they held less prestigious jobs and had lower earnings.

In the 1970s we find evidence of a modest stream of black migrants from the North into the South. They tended to be extensively educated and thus held better jobs and earned more than native-born blacks in either the North or South. Indeed, these black migrants to the South compared favorably with the native white population of the South.

In the 1970s—for the first time—there was a substantial migration of blacks from central cities into the suburban ring. In many urban areas the black population is declining in the central city, but growing rapidly in the suburbs. However, levels of black-white residential segregation declined little in the 1970s. Despite black economic advances, more liberal white attitudes, and federal fair housing laws, blacks and whites remain segregated.

Residential segregation continues to affect racial patterns in schools. In many small and medium-sized metropolitan areas public schools have been racially integrated. But in the large metropolitan areas—where the majority of blacks live—little progress has been made toward integrating elementary and secondary schools. In many large cities black and white students are as segregated now as they were before the *Brown* ruling.

With regard to the labor force participation of men, rates are down for both races but the decline has been much greater among blacks. For the first time in the nation's history a substantial percentage of adult black men are neither working nor looking for a job. In 1985, 13 percent of black men aged 34 to 44 were not participating in the labor force; for white men, the figure was 5 percent.[4]

Labor force participation rates have risen for both black and white women, but more rapidly among whites in recent times. In the past the proportion of women employed was greater for black women than for whites. This has changed and white women have "caught up" with blacks with regard to labor force participation and employment. Black women, however, have persistently higher unemployment rates.

Since the mid-1950s, black male unemployment rates have been double those of white men, and there is no evidence suggesting convergence on this important indicator of status. Both economic expansion and civil rights laws have failed to lower the two to one ratio. In terms of employment, there is no evidence of any improvement in the status of black men.

If we look at the occupations of employed men, we find a very clear and seemingly uninterrupted trend toward a racial convergence. Black men are clearly moving into the more attractive and higher-paying occupations once reserved for whites. Despite the obvious and substantial progress, however, racial differences in the occupational distribution of men remain large. When the occupations of employed women are considered, we find that as recently as 1960 black women were concentrated in domestic service. There has been a major upgrading of the jobs held by black women, and their occupational distributions are also converging with those of white women. In 1980 black males and females each had occupational distributions that were similar to those achieved by whites in 1950, a full generation earlier.

The salaries of employed black men have risen more rapidly than those of white men, although blacks continue to earn less than whites. Models that seek to measure racial discrimination in pay rates suggest that there has been a substantial decline since 1940. Nevertheless, black men in 1985 could expect to earn about 16 percent per hour less than ostensibly similar white men, a difference some would accept as a rough estimate of the cost of racial discrimination in pay rates. Black women, on the other hand, have attained earnings parity with white women. It may be disadvantageous to be a woman in today's labor market, but black women are not at any more of a disadvantage than white women.

Since the turn of this century, commentators have pointed to the

---

[4]U.S. Bureau of Labor Statistics, *Employment and Earnings*, vol. 33, no. 1 (January 1986), table 4.

economic success of black West Indian migrants in the United States and taken this as evidence of an absence of racial discrimination. They believe that the achievements of the West Indians demonstrate what blacks may accomplish. Our analysis of this topic reveals that foreign-born blacks in 1980 were just one step ahead of native-born blacks on measures of educational attainment, occupational achievement, and earnings. In particular, the economic characteristics and achievements of foreign-born blacks resemble those of native-born blacks much more than they resemble those of whites. Most claims concerning the achievements of West Indian blacks in the United States are greatly exaggerated.

A Research Agenda for the Future

The comprehensive study of racial stratification requires the development of more sophisticated models. Such models need to incorporate a focus on societal-level factors (for example, economic relations, social history), middle-range or institutional-level factors (for example, labor force participation, family structure), and individual-level factors (for example, attitudes, behavioral patterns). The complex of variables and relationships implicated in racial stratification must be systematically outlined. Ideally, any empirical investigation of these relationships should employ a multi-method approach, collecting data from government statistics, social surveys, and field studies.

In recommending the systematic study of racial stratification, we also suggest an agenda for research. This list is not as ambitious as the 100-year research plan for the study of black life and culture outlined by W. E. B. DuBois during his years at Atlanta University.[5] The research topics we suggest are neither exhaustive nor detailed. They indicate avenues for inquiry into the connections between race and economics in American life.

Macro-Level Research Issues
Racial Division of Labor in the Changing U.S. Economy

There is need for systematic, empirical study of transitions in the economic base of the United States as related to patterns of racial stratification. Specifically, the nation's economy has shifted from a focus on labor-intensive, agricultural activities to capital-intensive, service/financial activities. We need research that details the changing

[5]DuBois (1968), pp. 205–35.

patterns of black participation in the American economy and the implications of these changes.

North-South Economic Relations and Racial Stratification in America

The world's industrialized, developed northern hemisphere has tended to dominate recent economic relationships with the preindustrial, developing southern hemisphere. Coincidentally or not, the North-South split roughly corresponds to a white-nonwhite distinction. Research should investigate the contribution that the United States makes to the operation of the international system of racial stratification. What implications does this stratification system hold for domestic racial issues?

The Ideology of Racial Stratification

Around the end of the nineteenth century in the United States, and earlier in Europe, a systematic doctrine justifying racial stratification crystallized and was institutionalized. The process by which these ideologies were articulated, propagated, and accepted into the society's mores should be investigated.

The racial hierarchy theories espoused by the Social Darwinists nine decades ago no longer have scholarly defenders, but their legacy persists. Our society continues to accept stereotypes that stress black limitations in ability and skills. As DuBois observed, blacks in the United States view themselves, in some degree, through the stereotypes that whites believe.[6] Quite likely, many young blacks still receive, and accept, the messages from the schools and from the media which say that they may aspire to become musicians or athletes, but that they should not hope to become mathematicians or physicians.

Middle-Range Research Issues
Racial Stratification and Health in the United States

The system of health care is changing rapidly; therefore, it is important to examine how the delivery of medical services will be influenced by racial stratification. For instance, how will the persistent

---

[6]DuBois (1961), p. 17.

racial gap in morbidity and mortality be affected by the current cost containment movement in health care or by the increasing privatization of health care delivery?

Racial Stratification and Education

Educational attainment is one of the primary determinants of economic and occupational mobility in America. Research that examines the relationship of racial stratification to educational options, experiences, and outcomes is required. A vitally important question asks whether educational opportunities, the quality of schooling, and returns for investments in schooling are comparable for blacks and whites. More systematic racial comparisons of educational progress are required.

Race and Family Organization in America

The organization of black family life has changed dramatically over the past 25 years, and a continuation of current trends will lead to even larger racial differences in family structure. Research is needed to clarify the sources and implications of changes in black family structure. For example, how is the trend toward female-headed households related to trends in black male unemployment, female employment, and black marginality in the labor market? Do changes in black family organization presage similar changes for whites and other race/ethnic groups?

Occupational Mobility and Race in America

We described two important, but contradictory, trends in the working lives of blacks. A group of highly educated blacks have found unprecedented opportunities for occupational mobility, but at the same time a sizable proportion of the black population is mired in crisis-level unemployment. Research needs to be conducted into the pattern of employment opportunities available to blacks. What are the implications of chronic exclusion from the labor force for the work attitudes of young blacks (many of whom have *never* been gainfully employed) as they enter adulthood? What combination of factors explains the chronic unemployment of black males since World War II and the faster expansion of white female employment relative to that of black females?

The "Black Underclass" and American Society

The "black underclass" resembles the rural, southern peasantry—white and black—of the late nineteenth century and early twentieth century. Research is needed to describe the characteristics of this population. How are they connected to the larger society? What proportion are able to break the vicious cycle of poverty? How do they do so? Is the "black underclass" only the most visible segment of a larger multiracial "underclass" in America?

Racial Segregation of Residence in America

Residential segregation has persisted stubbornly in this country. Research should focus on the complex of attitudes and economic relationships through which this racial isolation is maintained. It is important to assess the difference that racial segregation makes in opportunities for schooling, interracial contact, employment, and quality of life issues.

Black Community Institutions

Research is required to determine the organization and operation of indigenous institutions in the black community. To what extent are these institutions engaged in "self-help" activities? How was the place of these institutions in black community life redefined because of desegregation? What has been the effect of the changing role of black community institutions on community cohesion? Are low-income, urban black communities in fact disorganized—that is, lacking in institutional organization?

Micro-Level Issues
Racial and Economic Attitudes in America

Racial attitudes in this society have changed considerably but, given the persistent economic deprivation of blacks, it is impossible to separate racial attitudes from economic attitudes. It becomes critical, therefore, to examine attitudes in the United States concerning affirmative action, social welfare policies, and the distribution of wealth. What strategies, if any, are considered appropriate approaches to the redress of historic patterns of discrimination? Is there a national consensus on what represents reasonable or tolerable levels of deprivation? To what extent are economically deprived people considered to be responsible for their own status?

Black Self-Concept, Motivation, and Cultural Values

Life as a discriminated minority affects the self-concept, motivation, attitudes, and cultural values of black Americans. Yet for all of the theorizing about these issues, systematic studies of black self-esteem, achievement, motivation, attitudes, and values are limited. More research along these lines is indicated if we are to clarify the human response to oppression and discrimination.

Race and Social Policy in America

Confronted with these research findings, the perceptive policy analyst questions their meaning for social policy. William Wilson succinctly summarized the two major competing ideological perspectives from the area of race and social policy in American scholarship:

> Liberals have traditionally emphasized how the plight of disadvantaged groups can be related to the problems of the broader society, including problems of discrimination and social class subordination. . . . Conservatives, in contrast, have traditionally stressed the importance of different group values and competitive resources in accounting for the experiences of the disadvantaged.[7]

He then goes on to relate these different ideological positions to predictable differences in the social policy recommendations advanced by the competing groups.

The different schools of thought result in profoundly different proposals for the resolution of economic and social inequities by race in the society. Writing from the "conservative" position, Charles Murray recommends:

> The proposed program . . . consists of scrapping the entire federal welfare and income-support structure for working-aged persons, including AFDC, Medicaid, Food Stamps, Unemployment Insurance, Workman's Compensation, subsidized housing, disability insurance and the rest. It would leave the working-aged person with no recourse whatsoever except the job market, family members, friends and public or private locally funded services.[8]

[7]Wilson (1985).
[8]Murray (1984), pp. 227–228.

Writing from the opposing "liberal" view, Robert Hill concludes:

> Our study has indicated that the popular generalization "government programs have not worked for most poor blacks" needs to be significantly modified. . . . [M]ost government programs for the poor have only reached a fraction of those who needed them. But the record shows that those low-income blacks who were reached did benefit from most of them. . . . A major priority should be given to providing meaningful employment opportunities to female-headed families before they go on welfare or are in dire economic straits. Secondly, all government policies . . . should be periodically evaluated and modified to ensure that they are complementing and reinforcing family strengths.[9]

This dialogue leads us to acknowledge the conflicting interpretations, policy prescriptions, and positions for which our findings potentially provide support. On the one hand, those who are so inclined might look into these findings and see evidence of black failing. As the argument goes, black Americans have not achieved economic parity despite constitutional guarantees banning discrimination and greater access to opportunities in education, the labor market, and housing because of countercultural—that is, "culture of poverty"—values, which are self-limiting. For reasons of laziness or pride, black men refuse to accept the jobs that are available, and black women are content to rely on AFDC to support themselves and their illegitimate children. The very fabric of the black community has come apart, and black life has become socially disorganized. This line of argument concludes by suggesting that until black Americans adopt the values of the majority regarding hard work, deferred gratification, frugality, and self-respect, they will continue to lag behind. Black Americans have the same chance as other immigrant groups to prosper; why must they rely on special government favors? To succeed, black Americans need only take advantage of the many opportunities which our great country offers.

This argument effectively summarizes the conclusions that many endorse. A contrasting view points to the absence of empirical support for the notion that blacks are lacking in the core values of American culture. Further, this argument goes, the legacy of this restrictive system persists today under the guise of *de facto* residential segregation and racial discrimination in education and employment, with the result that black efforts at self-improvement and community development continue to be systematically undermined. The fact that black Americans were able to manage any progress whatsoever in the face of such

[9]Hill (1981).

awesome odds is testament to their determination. It is likely that elements of the black American experience, culture, and character combine with racial discrimination and stratification to effectively account for persistent racial inequalities in American society. However, to suggest that black Americans bear major responsibility for their own plight is to ignore history and denies the reality of race discrimination.

Charles Johnson and Robert Park, writing about the bloody racial riot in Chicago in 1919, observed:

> It is important for our white citizens always to remember that the Negroes alone of all our immigrants came to America against their will by the special compelling invitation of the whites; that the institution of slavery was introduced, expanded, and maintained in the United States by the white people and for their own benefit; and that they likewise created the conditions that followed emancipation. Our Negro problem, therefore, is not of the Negro's making. No group in our population is less responsible for its existence. But every group is responsible for its continuance.[10]

History shows that when opportunities for black advancement became available, black people moved swiftly to take advantage, often at great personal risk and sacrifice. The historical record also reveals this society's discomfort with sustained black progress, particularly when this progress results in any diminution in the economic status of whites. Thus, the pattern has traditionally been one where "windows of opportunity" were promptly closed just as the gains began to produce significant reduction in black-white economic disparities.

The challenge that confronts this country involves the task of arriving at conditions that allow blacks and whites to live equitably. Clearly we are closer to such equality now than we were 30 years ago, yet vast differences persist in the quality of life experienced by Americans due to racial and economic stratification. In a society dedicated to the norm of equality, significant deprivation in the midst of prosperity represents a constant potential for explosion, since a failed promise is like a time bomb ticking.

[10]Chicago Commission on Race Relations (1922), p. xxiii.

APPENDIX:
THE QUALITY OF CENSUS DATA

T HE CENSUS of 1840 was one of the most controversial. This enumeration was the first of six requiring that census takers report the number of insane and idiotic individuals. In 1842 a physician practicing in Louisville, Edward Jarvis, observed that freedom apparently disturbed the mental condition of blacks.[1] In free states the rate of insanity for blacks was six times that for whites; in the slave states the rate for blacks was only 60 percent that for whites. The census reported that 0.6 percent of the blacks in the North were insane or idiotic; in slave states, only .06 percent.

Two years later Jarvis published a retraction.[2] After examining data for specific northern cities, he found that in some of them almost all blacks were reported insane; in other locations the count of impaired blacks exceeded the total count of blacks. He concluded that census data about the health status of blacks in free states were fraudulent.

Former President John Quincy Adams, then serving in the House, insisted that the State Department, which was responsible for the enumeration, correct its errors. Instead, Secretary of State John C. Calhoun defended the count and argued that in a state of freedom Negroes ". . . invariably sunk into vice and pauperism, accompanied by the bodily and mental afflictions incident thereto—deafness, blindness, and insanity. . . ."[3] Coming under greater criticism, Calhoun

[1]Jarvis (1842).
[2]Jarvis (1844).
[3]Cited in Stanton (1960), p. 61.

eventually asked William Weaver, who superintended the census, to investigate its quality. Weaver found the census to be accurate and Calhoun concluded that his opponents were defaming the census because they would not admit that Negroes could survive in America only if they were held in bondage.[4] Patricia Cohen's investigation of manuscripts from the 1840 enumeration suggests that both protagonists were in error since unintentional mistakes and deficient editing produced the surprising results. Census marshals could easily be confused and transpose columns on the multiple-page enumeration forms. They often listed senile whites in a column which should have been reserved for insane blacks. If an area contained many elderly whites but few blacks, the insanity rate for blacks was greatly inflated.[5]

Although the specific issues are very different, the accuracy of census data for blacks remains a controversial issue. In recent decades most attention has been devoted to the undercount of blacks since this may diminish their political representation and reduce their share of benefits allocated by governmental programs.[6] In this appendix we first discuss net census undercount and then move on to a description of the quality of data. A final section provides information about the sample used in this analysis.

Net Census Undercount

Each of the censuses has failed to count some percentage of the population, but the undercount rate differs by age and sex and varies from one enumeration to the next. Since the late nineteenth century, demographers have believed that the count of blacks is less complete than that of whites. For example, Figure 2.2 shows that between 1870 and 1880 the black population grew at about 3 percent annually, a much higher rate than in the preceding or following decade. This does not represent a spurt in fertility or an influx of immigrants from the Caribbean. Rather, the 1870 enumeration in the South was seriously deficient because it was conducted by federal appointees during the disrupted Reconstruction era. In the early twentieth century Census Bureau demographers estimated that the count of blacks in 1870 should have been 5.4 million instead of the 4.9 million reported.[7]

---

[4]Stanton (1960), pp. 58–63.
[5]Cohen (1982), pp. 201–204.
[6]Siegel (1968); Parsons (1972), chaps. 1 and 2; National Academy of Sciences (1978), pp. 9–14; Slater (1980); Bryce (1980); Carlucci (1980).
[7]U.S. Bureau of the Census, *Negro Population in the United States: 1790–1915* (1918), p. 27; for further information, see Walker (1890).

Low decennial growth rates of the population have led analysts to conclude that other enumerations missed many blacks. For example, the census of 1890 may have been deficient, and one correction offered by the Census Bureau raised that count of blacks from 7.5 million to 7.8 million.[8] The 1920 enumeration was conducted in January, while those of 1910 and 1930 were carried out in April. Winter weather impeded the 1920 count in rural areas, and Census Bureau publications imply that the count of blacks, 10.5 million, was at least 150,000 too low.[9] The statistician T.J. Woofter suggested the true undercount was closer to 450,000 blacks.[10]

These estimates of undercount were based on reports of the difficulties of census enumeration or on assumptions that growth rates in adjoining decades should be similar, because no other statistical system provided information about population size or demographic rates. The registration of vital events spread throughout the nation in the early decades of this century, and by 1933 all states had joined a national registration area.[11] The number of births registered in a given year may be "survived" to the date of a census by subtracting the deaths recorded to that birth cohort. The number at a given age estimated from the vital registration system may then be compared to the census count to determine the completeness of enumeration. Estimates of undercount calculated in this fashion are available for recent dates and suggest that earlier estimates were probably too low.

Table A.1 presents net census undercount rates by age, sex, and race for 1950–1980. Blacks have been less completely enumerated than whites but, in the last three decades, the Census Bureau has substantially improved its count. In 1950, for example, apparently 13 percent of the nonwhite population and 2 percent of the white were omitted. This implies that the "true" black population in 1950 was about 17 million rather than the 15 million shown in census reports. Twenty years later, estimates of census undercount were 8 percent for blacks and about 1.5 percent for whites. Preliminary estimates for 1980 suggest a further improvement to an undercount rate of 5 percent for blacks and a small overcount for whites. The actual number of missed blacks, which may have approached 2 million in 1950, fell to 1.3 million in 1980.

[8]U.S. Bureau of the Census, *Negro Population in the United States: 1790–1915* (1918), p. 27.

[9]U.S. Bureau of the Census, *Fifteenth Census of the United States,* (1932) vol. III, pt. 1, p. 7; *Negroes in the United States: 1920–32* (1935), table 1. Congressional representatives from rural states contended that the 1920 count—the first to show that a majority lived in urban areas—was exceptionally deficient in farming areas. The census results were not used to reapportion the House of Representatives.

[10]Woofter (1931); Miller (1922).

[11]Vance (1959), pp. 287–88; Linder and Grove (1947), pp. 95–101.

Net Census Undercount by Age, Sex, and Race, 1950–1980

Age	Males				Females			
	1950	1960	1970	1980	1950	1960	1970	1980
BLACK OR NONWHITE								
Total	15%	9.7%	10.1%	8.0%	11%	6.3%	5.3%	2.7%
0–4	11	6.6	10.3	9.2	10	5.1	9.3	8.7
5–9	12	5.1	7.4	6.1	10	4.2	6.9	5.6
10–14	7	5.0	3.8	1.4	7	3.9	3.0	1.2
15–19	18	12.3	4.4	.2	12	9.6	3.3	− .5
20–24	19	18.4	12.7	8.1	8	9.5	5.8	2.5
25–34	20	18.5	19.0	12.9	8	6.5	6.5	2.7
35–44	16	11.5	17.3	17.0	10	3.8	3.5	4.1
45–54	12	11.0	11.8	13.6	14	9.0	4.9	2.9
55–64	21	8.5	8.8	6.3	33	11.6	6.6	.6
65+	12	− 5.8	.4	− 1.7	5	2.8	1.9	− 1.2
WHITE OR NONBLACK								
Total	2.7	2.4	2.1	.6	2.3	1.6	.9	− .9
0–4	4.5	1.9	2.4	.1	3.8	1.1	2.0	− .1
5–9	3.1	2.4	2.6	.5	2.5	1.5	2.4	.4
10–14	1.1	2.5	.8	− 1.0	1.1	1.5	.6	− .9
15–19	2.4	3.8	.8	− .8	1.5	2.4	.1	− 1.1
20–24	3.7	4.3	1.4	0.0	1.8	2.4	.3	− 1.3
25–34	3.6	3.6	3.6	1.5	1.3	1.0	1.3	− .7
35–44	1.9	2.2	3.1	2.8	.8	− .2	− .2	.4
45–54	1.4	2.5	2.3	1.9	2.4	2.4	− .4	− 1.4
55–64	3.2	.5	2.0	.1	6.7	1.7	1.3	− 2.4
65+	2.0	0.0	2.3	− 1.2	1.8	3.5	1.8	− 1.4

NOTES: Figures indicate the percentage by which the enumerated population differs from the estimated "true" population; negative percentages indicate that the enumerated population exceeded the estimated "true" population. Figures for 1950 refer to nonwhites and whites, those for 1960 refer to blacks and whites, and those for 1970 and 1980 refer to blacks and nonblacks.

SOURCES: A. J. Coale, "The Population of the United States in 1950 Classified by Age, Sex, and Color—A Revision of Census Figures," *Journal of the American Statistical Association* 50 (March 1955), table 7; J. S. Siegel, "Estimates of the Coverage of the Population by Sex, Race, and Age in the 1970 Census," *Demography* 11 (February 1974), table 6; J. S. Passel and J. G. Robinson, "Revised Estimates of the Coverage of the Population in the 1980 Census Based on Demographic Analysis: A Report on Work in Progress," *1984 Proceedings of the Social Statistics Section*, American Statistical Association, table 3.

Table A.1 shows that undercount rates are much larger for men than for women and vary by age. Black men aged 20 to 54 have been difficult to enumerate, and despite several decades of improvements about one man in eight in this age range was missed in the 1980 enumeration.

In several chapters of this monograph we alert readers to problems of census coverage. However, our analyses were carried out with the data reported by the Census Bureau. This decision was made for three reasons. First, much of our study describes characteristics of blacks and whites such as marital status, educational attainment, income, or earnings. Estimates of undercount have been developed for the national population classified by age and sex, but there are no undercount rates for the population classified by social and economic characteristics and there is little information about regional differences. Were black men working as machine operators undercounted to a greater or lesser degree than those who held administrative support jobs? Were black men born in the South more completely or less completely counted than those born in the North? Was the undercount rate higher in Georgia or California?[12] These questions cannot be answered with the available estimates of census undercount.

Second, there is no one set of undercount rates. Each set of estimates depends on assumptions about patterns of errors in the census, the omission rate of births and deaths from the registration system, and net international migration. In many situations, there are two or more plausible but different corrections of census data.[13]

Third, at the time this volume was written the Census Bureau had not completed its analysis of net census undercount in 1980, and the only rates available to use were one set which was described as preliminary and another set labeled "work in progress."[14]

Does census undercount affect our description of the black population or our analysis of racial differences? There may be some circumstances in which it does, but in many others it may not. An example of a situation in which undercount seriously distorts our knowledge involves the sex ratio at young adult ages. There has been speculation that a shortage of black men helps to explain why a high proportion of black women remain single through much of the childbearing period.[15] The census counts in 1980 suggest that there were only 88 black men aged 20 to 34 for every 100 black women in this age group. However, if the data are corrected for net census undercount, we find that there were 97 black men aged 20 to 34 for every 100 women in this age group.

[12]For attempts to estimate undercount by states, see Siegel et al. (1977); Hill (1980); National Academy of Sciences (1978): app. A.

[13]For a comparison of different estimates for 1970 and 1980, see Siegel (1974), table 6.; Passel, Siegel, and Robinson (1982), table 4; Passel and Robinson (1984); table 3. For a comparison of different estimates in 1940 and 1950 see Coale (1955), table 2; Land, Hough, and McMillen (1984), tables 9 and 10.

[14]Passel, Siegel, and Robinson (1982), table 4; Passel and Robinson (1984), table 3.

[15]Guttentag and Secord (1983); Cox (1940); Jackson (1971); Spanier and Glick (1980); Wilson and Neckerman (1986), pp. 254–58.

424

This is hardly different from the balanced ratio observed among whites at these ages: 100 men per 100 women.[16]

Most racial comparisons are not seriously distorted because of net census undercount since almost all blacks and whites were counted, and the characteristics of the omitted population would have to be extremely different from those of the enumerated to distort basic comparisons. Let us examine several extreme examples. The census of 1980 reported that 19.8 percent of the white men aged 25 to 29 and 25.4 percent of the black men were unemployed at some point in 1979.[17] In this age group the effects of undercount are greatest. The black unemployment rate was 28 percent greater than the white rate and the racial difference was 5.6 percentage points. Suppose that men missed in the census were 50 percent more likely to have been unemployed in 1979 than those who were counted. Such an assumption leads to the conclusion that 20.0 percent of the white men and 27.1 percent of the black men were out of work at some point in 1979. The black rate, under this extreme assumption, was 35 percent above the white rate and the racial difference was 7.1 points.

White men at ages 30 to 34 had mean incomes in 1979 of $17,840, while black men averaged $11,340.[18] The racial gap was $6,500, and blacks averaged only 64 percent as much as whites. Using the undercount rates in Table A.1, let us make the extreme assumption that men missed by the census had average incomes only three fifths as large as those of the enumerated population. This would lead us to conclude that the average income for white men aged 30 to 34 was $17,750; black men, $10,750. This implies that black men received only 61 percent as much as white men and that the actual racial difference was $7,000.

These examples show that census undercount influences our knowledge of racial differences, and if the omitted black population consists of people who have high unemployment rates, little income, and weak family ties, then their omission minimizes racial differences. But we have no convincing evidence that allows us to conclude that these are the characteristics of the unenumerated. The extreme assumptions we made about men at the ages of peak census undercount lead us to be cognizant of the effects of underenumeration, but they also allow us to have confidence in the racial differences described through the use of reported data, since the magnitude of racial differences was basically similar before and after the corrections.

[16]U.S. Bureau of the Census, *Census of Population: 1980*, PC80–1-B1, table 45; Passel and Robinson (1984), table 3.

[17]U.S. Bureau of the Census, *Census of Population: 1980*, PC80-1-D1-A, table 273.

[18]U.S. Bureau of the Census, *Census of Population: 1980*, PC80-1-D1-A, table 293.

Census Procedures and the Accuracy of Data
Substitutions

Since 1960, censuses have been conducted primarily through self-enumeration. In 1980, for example, a questionnaire was mailed to every household shortly before April 1. Approximately 95 percent of the households were told to complete the questionnaire and mail it to a local processing office. About 5 percent of the population—people in remote locations—were told to hold their questionnaires for an enumerator who would visit them.

Eighty-three percent of those instructed to return a questionnaire by mail did so. Those who did not were visited by an enumerator up to four times. Those who sent in a form with incomplete or inconsistent information were contacted by telephone or visited by an enumerator if the errors exceeded a specified minimum.[19]

If, after repeated visits, enumerators could not contact the occupants of a housing unit, they asked neighbors and landlords about the residents. If they determined that a unit was occupied but could obtain no information from its residents, a substitution procedure was used. In other circumstances, information was secured about some residents of a household but no data were available for other persons known to be members of the household. In such cases, the substitution procedure was also used. Finally, in a few circumstances data were appropriately gathered but subsequently lost because of mechanical failure such as an erasure by a computer. Substitution was also used in these cases.

Census Bureau officials describe substitution as "the imputation of data for a person or housing unit known to be present but for which there is no information on the questionnaire. A previously processed person or unit is drawn from the file under certain criteria, and the full set of characteristics for the person or unit is duplicated."[20]

In 1980 substitution led to the imputation of 3,342,000 individuals in 1,348,000 households. This represents 1.5 percent of the estimated total population and 1.7 percent of the households. Approximately 700,000 persons were imputed to housing units known to be occupied but for which no other information was obtained. Almost 2.5 million persons were imputed to occupied housing units about which enumera-

---

[19]U.S. Bureau of the Census, *Census of Population: 1980*, PC80-1-B1, app. C; Citro and Cohen (1985), chap. 3.

[20]U.S. Bureau of the Census, *Census of Population and Housing: 1980*, Public Use Microdata Samples, Technical Documentation K-48.

TABLE A.2

Racial, Age, and Marital Status Characteristics of the Enumerated, Substituted, Allocated, and Total Population in 1980 (complete count data)

	Enumerated and Reported	Substituted	Allocated	Total
RACE				
Total Number (in thousands)	219,879	3,354	3,312	226,546
Total	100.0%	100.0%	100.0%	100.0%
White	83.5	72.6	75.5	83.0
Black	11.5	20.6	13.1	11.6
Native American	.6	.8	1.3	.9
Asian/Pacific Islander	1.5	1.5	2.2	1.5
Other	2.9	4.5	7.9	3.0
AGE				
Total Number (in thousands)	216,621	3,368	6,557	226,546
Total	100.0%	100.0%	100.0%	100.0%
Under 15 Years	22.8	22.3	18.4	22.6
15–19 Years	9.4	8.7	6.2	9.3
20–24 Years	9.4	9.8	11.1	9.4
25–34 Years	16.3	17.2	17.9	16.4
35–44 Years	11.3	11.1	10.6	11.3
45–54 Years	10.1	9.9	9.7	10.1
55–64 Years	9.6	9.6	10.4	9.6
65+ Years	11.1	11.4	15.7	11.3
Median Age	30.2	30.3	33.1	30.3
MARITAL STATUS FOR PERSONS AGED 15 AND OVER				
Total Number (in thousands)	170,360	2,621	2,274	175,255
Total	100.0%	100.0%	100.0%	100.0%
Single	26.2	27.0	35.8	26.3
Married	57.5	52.4	42.0	57.4
Separated	2.3	3.6	4.1	2.3
Widowed	7.7	8.9	9.0	7.7
Divorced	6.3	8.1	9.1	6.3

SOURCE: U.S. Bureau of the Census, *Census of Population: 1980*, PC80-1-B1, table B-1.

tors secured some minimal information. Just under 250,000 persons were imputed because of mechanical failures.[21]

Table A.2 shows the distribution of the enumerated population by race, age, and marital status and also indicates the distribution of the substituted population. There is a racial difference since blacks made up 21 percent of the substituted population compared with 12 percent of those who were both enumerated and reported their race. The substitution procedure raised the black population from 25.4 million to 26.1 million. Stated differently, 2.6 percent of the total black population count results from imputations. If this had not been done, the undercount rate for blacks in 1980 would have been about 8 percent rather than the 5 percent estimated by Passel and Robinson.[22] Approximately 1.3 percent of the white population count resulted from this substitution procedure.

Table A.2 shows that the substituted population had an age distribution similar to the enumerated, but their marital status was somewhat different. Among the substitutes, relatively few were married but a high proportion were widowed or divorced.

Allocation

Some enumerated persons failed to answer one or more questions or gave inconsistent answers such as a man who might report his age as 17, claim to be enrolled in twelfth grade, but also report that he was a World War I veteran. If errors or omissions exceeded a minimum, the respondent was contacted by phone or was visited by an enumerator. Since it was not possible to eliminate all nonresponse or resolve all inconsistencies, the questionnaires were edited. This involved first checking for completeness and consistency and then using a computerized allocation procedure described as follows:

> The process by which a characteristic (for example, age, race or rent) is assigned to a person or housing unit in the absence of an acceptable entry on the census or survey questionnaire. The general procedure for inserting omitted entries or changing unacceptable entries is to assign an entry for a person that is consistent with other entries for that person or entries for other persons with similar characteristics.[23]

[21]Citro and Cohen (1985), p. 96. Substitution is a procedure that reduces net census undercount. For a discussion of the extent of undercount, its causes, and possible corrections, see Heer (1968); Parsons (1972); National Academy of Sciences (1978); U.S. Bureau of the Census, *Proceedings of the 1980 Conference on Census Undercount* (1980).
[22]Passel and Robinson (1984).
[23]U.S. Bureau of the Census, *Census of Population and Housing: 1980*, Public Use Microdata Samples, Technical Documentation K-4.

The allocation procedure facilitates the analysis of census data by eliminating obvious inconsistencies and ensures that responses are available for most questions.[24]

At this point, we must distinguish complete count data from sample data. All households in 1980 received a short questionnaire which asked about the age, sex, race, Spanish-origin, marital status, and relationship to head of household of every occupant. One sixth of the households in places of 2,500 or more population and one half of those in smaller places received a long-form questionnaire.[25] This included the "complete count" or short-form questionnaire, but also contained an array of questions about the social, economic, and demographic characteristics of household members. Of all households, 80.6 percent received the "complete count" questionnaire; 19.4 percent, the long form.

Of those completing the short form, 10.5 percent had one or more items allocated. Allocation rates for basic demographic items asked of everybody are listed below:[26]

Relationship to Head of Household	2.1%
Sex	.8
Age	2.9
Race	1.5
Spanish Origin	4.2
Marital Status (Age 15 and Over)	1.3

Table A.2 also presents information about the race, age, and marital status distribution of the allocated population. Approximately 3.3 million persons had their racial identity imputed. Thirteen percent of those were classified as black compared with 12 percent of the enumerated population, meaning that allocation augmented the black population by 430,000. Approximately 1.6 percent of the total black population and 1.3 percent of the white population had their race assigned to them.

A look at Table A.2 also reveals that the marital status distribution of the allocated population differed from that of the enumerated population since many of those who did not report marital status were classified as single, widowed, or divorced. The age distribution of the allocated population was considerably older than that of the enumerated or substituted population.

[24]Allocation was not used to eliminate nonresponses to certain questions, including ancestry of all respondents and country of birth of persons who reported they were foreign-born.

[25]Citro and Cohen (1985), p. 91.

[26]U.S. Bureau of the Census, *Census of Population: 1980*, PC80-1-B1, table B-4. These allocation rates refer to the "complete count" items which were asked of all respondents.

Allocation rates were higher for the sample items which appeared only in the long-form questionnaire. Listed below are the proportions of individuals in the sample who had specific characteristics imputed by allocation:[27]

Place of Birth	4.9%
Language Spoken at Home	14.8
Place of Residence in 1975	14.6
Children Ever Born (Women Aged 15 to 44)	6.1
Years of School (Persons Aged 25 and Over)	9.7
Labor Force Status (Persons Aged 16 and Over)	3.9
Occupation (Employed Persons)	6.7
Disability (Persons Aged 16 and Over)	6.4
Personal Income—Men (Aged 15 and Over)	11.8
Personal Income—Women (Aged 15 and Over)	11.2

Is the allocated population quite similar to the enumerated or very different? Does allocation confound or improve our understanding of social and economic characteristics? Table A.3 considers four characteristics—labor force status, educational attainment, occupation, and personal income—and shows distributions for the enumerated population, the allocated population, and the total population. These data are for both sexes and all races.

Persons with allocated characteristics in 1980 differed from those with reported characteristics. Their unemployment rate was about 40 percent higher; their median income was 18 percent lower; they were less extensively educated and a smaller share of them held jobs in the prestigious categories at the top of the occupational rank. Nevertheless, the impact of allocations on the overall population distributions was not great, primarily because characteristics were being allocated for a small percentage of the total population. For example, the unemployment rate was estimated at 6.3 percent before allocation; 6.5 percent after.

The public use microdata samples from the census of 1980 indicate whether each of 48 characteristics was allocated, along with the reason for allocation—that is, omitted response or inconsistent response. Using these data, it is possible to compare blacks and whites prior to allocation and to determine how the relationship of variables such as that of earnings to educational attainment is influenced by specific types of allocations. At the outset of this research project, we intended to conduct such an investigation but budgetary and time constraints restricted our efforts.

[27]U.S. Bureau of the Census, *Census of Population: 1980*, PC80-1-C1, table C-2.

Accuracy of Reported Information

Individuals may misreport their age or their marital status, they may exaggerate their educational attainment, and they frequently underreport their income. The Census Bureau has used a variety of strategies to explore issues of data quality.[28] These include post-enumeration reinterviews with samples of the enumerated population, a comparison of characteristics reported by the same individuals in the census and the contemporaneous Current Population Survey, and a comparison of characteristics reported on the census with those reported in other administrative records such as income tax returns.[29] Quite often the focus has been on the consistency of reporting, although some studies investigated accuracy. Very few of these studies have specifically examined racial differences in the quality of demographic data.[30] When this volume was written, findings from the 1980 post-enumeration program and from the evaluation and research program had not yet been published. Rather than summarizing the sparse literature about racial differences in the accuracy of characteristics reported in the census, these issues are discussed within substantive chapters. The income chapter, for example, describes the substantial underreporting of income.

Characteristics of the Sample

Many of the statistics published in 1980 census documents are based on tabulations from the long-form questionnaire which was sent to 19.4 percent of the nation's households. The Census Bureau selected a subsample containing information about 5 percent of the nation's households and made this publicly available. This data source was used as a starting point for our analysis of racial differences.

Budgetary constraints prevented us from tabulating and analyzing data for the entire 5 percent sample—approximately 11.5 million people—contained on 26 reels of computer tape. We selected a smaller sample which could be analyzed more economically. To make certain that it was sufficiently large to allow us to study differentiation within small proportions of the black population, such as those born abroad, we used a larger sampling ratio for blacks than for nonblacks. We also

[28]U.S. Bureau of the Census, *The Post-Enumeration Survey: 1950* (1960); *Evaluation and Research Program of the U.S. Censuses of Population and Housing, 1960*, series ER (published 1964 through 1970); *1970 Census of Population and Housing: Evaluation and Research Program*, series PHC(E) (published 1973 through 1975).

[29]Citro and Cohen (1985), chap. 4.

[30]Farley (1968).

TABLE A.3

Labor Force, Educational, Occupational, and
Income Characteristics of the Population Before
and After Allocations, 1980 (sample data)

	Before Allocation	Allocated Population	After Allocation
LABOR FORCE STATUS OF POPULATION AGED 16 AND OVER			
Total Number			
(in thousands)	164,490	6,725	171,214
Total	100.0%	100.0%	100.0%
In Labor Force	62.3	53.6	62.0
Civilian Employed	57.4	48.2	57.0
Armed Forces	1.0	.7	1.0
Unemployed	3.9	4.7	4.0
Not in Labor Force	37.7	46.4	38.0
Percentage of Labor Force			
Unemployed	6.3%	8.8%	6.5%
YEARS OF SCHOOL COMPLETED BY POPULATION AGED 25 AND OVER			
Total Number			
(in thousands)	119,981	12,855	132,836
Total	100.0%	100.0%	100.0%
Elementary			
0–4 Years	3.5	4.4	3.6
5–7 Years	6.4	8.8	6.7
8 Years	7.9	9.0	8.0
High School			
1–3 Years	14.9	18.3	15.3
4 Years	35.5	26.8	34.5
College			
1–3 Years	15.3	18.8	15.7
4 Years	8.9	6.1	8.6
5+ Years	7.6	7.8	7.6
Percentage High School			
Graduate	67.2%	59.5%	66.5%

desired a household rather than an individual sample since, at some points in the analysis, we wished to compare the characteristics of spouses or simultaneously look at children and their parents.

To facilitate the selection of subsamples, each record on the 1980 public use sample contains a two-digit subsample number. Selecting records with a given two-digit number and evenly spaced increments of that number permits a user to obtain samples smaller than 5 percent.[31]

[31]U.S. Bureau of the Census, *Census of Population and Housing: 1980,* Public Use Microdata Samples, Technical Documentation, p. 42.

	Before Allocation	Allocated Population	After Allocation
OCCUPATIONS OF EMPLOYED PERSONS AGED 16 AND OVER			
Total Number (in thousands)	91,090	6,549	97,639
Total	100.0%	100.0%	100.0%
Executive, Administrative	10.6	7.9	10.4
Professional Specialty	12.5	8.6	12.3
Technical	3.1	2.6	3.1
Sales	10.0	9.6	10.0
Administrative Support	17.2	16.3	17.2
Private Household	.6	1.2	.6
Protective Service	1.5	1.6	1.5
Other Service	10.6	13.8	10.8
Farming, Fishing, Forestry	2.9	3.3	2.9
Precision Production, Crafts	12.9	12.8	12.9
Machine Operators	9.2	11.4	9.3
Transport Occupations	4.5	4.8	4.5
Handlers, Cleaners, Laborers	4.4	6.1	4.5
INCOME OF PERSONS AGED 15 AND OVER			
Total Number (in thousands)	155,118	20,190	175,308
Total	100.0%	100.0%	100.0%
No Income	16.1	14.9	16.1
Less Than $1,999	11.7	13.2	11.8
$2,000 –2,999	6.2	7.3	6.4
$3,000 –4,999	10.7	12.3	10.8
$5,000 –7,999	12.4	14.1	12.6
$8,000 –9,999	7.1	7.3	7.1
$10,000–11,999	6.2	5.8	6.2
$12,000–14,999	7.4	5.9	7.2
$15,000–24,999	14.7	12.5	14.4
$25,000–49,999	6.3	5.4	6.2
$50,000 or More	1.2	1.3	1.2
Median for Those With Income	$8,250	$6,760	$8,090

SOURCE: U.S. Bureau of the Census, *Census of Population: 1980*, PC80-1-C1, table C-2.

We selected a 1 percent sample of blacks and a 0.1 percent sample of nonblacks. We might have considered the race of the household head and taken entire households in accord with the sampling ratios. This procedure would lead us to select blacks who lived in households headed by nonblacks at the nonblack sampling ratio, which would have produced an underrepresentation of such blacks. The proportion of blacks living in households containing nonblacks is much greater than the proportion of nonblacks living in households which contain a black. In 1980, for example, 1.9 percent of black married-spouse-present wives had nonblack husbands. Only 0.3 percent of nonblack married-spouse-present wives had black husbands.[32] To avert this underrepresentation of blacks, we scanned the race of all occupants of a household. If a black was present, we selected the household using the black sampling ratio (one in one hundred). If no black was present, the nonblack sampling ratio was used (one in one thousand). This led to a sample of 266,840 blacks and 206,691 nonblacks. Comparing this sample with complete count data, we have a 1.0071 percent sample of blacks and a 0.1032 sample of nonblacks. Distributions by race of complete count census data, the Census Bureau's 19.4 percent sample, and the sample used in this study are shown below:[33]

	Complete Count	Census Sample	Monograph Sample
White	83.2%	83.4%	83.4%
Black	11.7	11.7	11.4
Native Americans	.6	.7	.8
Asian/Pacific Islanders	1.5	1.6	1.6
Other Races	3.0	2.6	2.8
Total	100.0%	100.0%	100.0%

Although providing a full 1 percent sample of blacks, our sample is slightly overrepresentative of the nonblack population because nonblack persons living in households that contain a black resident were selected at the black sampling ratio.

We wished to be certain that the characteristics of the 474,000 persons in our sample were basically similar to those reported by the Census Bureau for the complete count or for their 19.4 percent sample. Since most of the analysis describes blacks and whites, we will focus on these racial groups. Table A.4 presents regional, age, and marital status

[32]U.S. Bureau of the Census, *Current Population Reports*, series P-20, no. 366, table 16.

[33]U.S. Bureau of the Census, *Census of Population: 1980*, PC8-80-1-B1, table 46; PC80-1-D1-A, table 253; *Census of Population and Housing: 1980*, Public Use Microdata Samples (Tape File).

Comparison of the Regional, Age, and Marital Status Distributions
of the Black and White Populations, 1980, from Complete Count Data,
the Census Bureau's Sample Data, and the Sample Drawn for This Monograph

	Whites			Blacks		
	Complete Count	Census Sample	Monograph Sample	Complete Count	Census Sample	Monograph Sample
DISTRIBUTION BY REGIONS						
New England	6.2%	6.2%	6.2%	1.8%	1.8%	1.8%
Mid Atlantic	16.3	16.3	16.3	16.5	16.5	16.5
East North Central	19.2	19.1	19.0	17.2	17.2	17.2
West North Central	8.5	8.5	8.5	3.0	3.0	3.0
South Atlantic	15.2	15.2	15.1	28.9	28.9	28.9
East South Central	6.2	6.2	6.1	10.8	10.8	10.8
West South Central	9.9	9.9	9.9	13.3	13.3	13.2
Mountain	5.3	5.3	5.3	1.0	1.0	1.0
Pacific	13.2	13.3	13.6	7.5	7.5	7.6
Total	100.0	100.0	100.0	100.0	100.0	100.0
DISTRIBUTION BY AGE						
0–14 Years	21.3	21.4	21.6	28.6	28.7	29.4
15–24 Years	18.2	18.2	18.2	21.6	21.5	21.3
25–44 Years	27.7	27.7	27.9	26.1	26.2	25.9
45–64 Years	20.6	20.6	20.4	15.8	15.8	15.7
65+ Years	12.2	12.1	11.9	7.9	7.8	7.7
Total	100.0	100.0	100.0	100.0	100.0	100.0
DISTRIBUTION BY MARITAL STATUS						
Single	21.2	21.2	21.4	34.3	34.1	33.8
Married	57.4	57.8	57.5	35.0	35.5	35.9
Separated	1.8	1.7	1.9	8.8	8.6	8.7
Widowed	12.6	12.5	12.3	12.8	12.7	12.6
Divorced	7.0	6.8	6.9	9.1	9.1	9.0
Total	100.0	100.0	100.0	100.0	100.0	100.0

SOURCES: U.S. Bureau of the Census, *Census of Population: 1980*, PC80-B1, tables 45, 46, and 62;
PC80-1-C1, tables 120, 232, and 264; *Census of Population and Housing: 1980*, Public Use Microdata
Samples (Tape File).

distributions from three sources. The complete count data include the
responses of those people who only filled out the short enumeration
form as well as those in the Census Bureau 19.4 percent sample who
completed both the short form and the numerous additional questions.
The complete count and Census Bureau sample distributions differ primarily because of sampling. However, the sample data were also subject
to a much more thorough editing procedure before they were released.

In Table A.5 data are presented about three key economic indica-

TABLE A.5

Comparison of Years of Schooling, Occupation, and Household Income, 1980,
from the Census Bureau's Sample Data and the Data Drawn for This Monograph

	White				Black			
	MEN		WOMEN		MEN		WOMEN	
	Census Sample	Monograph Sample	Census Sample	Monograph Sample	Census Sample	Monograph Sample	Census Sample	Monograph Sample
YEARS OF SCHOOLING FOR POPULATION AGED 25 AND OVER								
Elementary								
0–4 Years	2.8%	2.8%	2.5%	2.6%	10.0%	10.3%	6.8%	6.8%
5–7 Years	6.0	5.9	5.6	5.5	12.0	11.9	11.6	11.4
8 Years	8.1	8.0	8.4	8.2	6.7	6.8	7.3	7.3
High School								
1–3 Years	13.6	13.4	15.5	15.1	20.5	20.6	22.9	23.0
4 Years	31.8	32.0	39.0	39.2	28.4	27.8	29.9	30.2
College								
1–3 Years	16.4	16.3	15.6	16.0	14.0	14.1	13.2	13.1
4 Years	10.5	10.4	7.9	7.9	4.3	4.4	4.5	4.4
5+ Years	10.8	11.2	5.5	5.5	4.1	4.1	3.8	3.8
Total	100.0	100.0	100.0	100.0	100.0	100.0	100.0	100.0

TABLE A.5 *(continued)*

	White		Black	
	Census Sample	Monograph Sample	Census Sample	Monograph Sample
OCCUPATIONS OF EMPLOYED PERSONS AGED 16 AND OVER				
Executive, Administrative	11.1%	10.9%	5.2%	5.2%
Professional Specialty	12.8	13.1	8.9	8.9
Technical	3.1	3.0	2.7	2.6
Sales	10.7	10.9	5.0	5.0
Administrative Support	17.2	17.5	17.5	17.5
Private Household	.4	.4	2.6	2.7
Protective Service	1.5	1.5	1.9	1.9
Other Service	9.8	9.8	18.5	18.7
Farming, Forestry, Fishing	2.9	2.9	2.0	2.0
Precision Production, Crafts	13.4	13.1	9.0	8.9
Machine Operators	8.6	8.7	13.5	13.6
Transport Occupations	4.4	4.2	6.0	6.0
Handlers, Cleaners, Laborers	4.1	4.0	7.2	7.0
Total	100.0	100.0	100.0	100.0
HOUSEHOLD INCOMES				
Less Than $5,000	11.6%	11.7%	25.6%	25.6%
$5,000 –7,499	7.6	7.7	11.2	11.3
$7,500 –9,999	7.6	7.6	9.8	9.8
$10,000–14,999	15.2	15.1	16.3	16.3
$15,000–19,999	14.3	14.2	12.4	12.1
$20,000–24,999	12.9	13.0	8.9	9.1
$25,000–34,999	16.6	16.4	9.8	9.8
$35,000–49,999	9.2	9.3	4.5	4.5
$50,000 or More	5.0	5.0	1.5	1.5
Total	100.0	100.0	100.0	100.0
Median	$17,800	$17,780	$11,040	$11,010

SOURCES: U.S. Bureau of the Census, *Census of Population: 1980*, PC80-1-C1, tables 123, 125, and 128; *Census of Population and Housing: 1980*, Public Use Microdata Samples (Tape File).

tors: years of schooling completed by the population aged 25 and over, the occupations of employed persons, and household income. For these variables, the Census Bureau's sample is compared with the one drawn for this analysis.

The distributions in Tables A.4 and A.5 allow us to have confidence in our sample. With regard to each of the six variables, the sample drawn for this project closely resembles the Census Bureau's sample.

Bibliography

Abowd, John M., and Mark R. Killingsworth "Employment, Wages, and Earnings of Hispanics in the Federal and Non-federal Sectors: Methodological Issues and Their Empirical Consequences." In George J. Borjas and Marta Tienda, eds. *Hispanics in the U.S. Economy.* New York: Academic Press, 1985.

Abraham, Sidney; Margaret D. Carroll; Clifford L. Johnson; and Connie M. Villa Dresser "Caloric and Selected Nutrient Values for Persons 1–74 Years of Age." U.S. National Center for Health Statistics, *Vital and Health Statistics*, series 11, no. 209 (June 1979).

Abraham, Sidney; Clifford L. Johnson; and Matthew F. Najjar "Height and Weight of Adults 18–74 Years of Age in the United States." U.S. National Center for Health Statistics, *Advancedata*, no. 3 (November 19, 1976).

———— "Dietary Intake of Persons 1–74 Years of Age in the United States." U.S. National Center for Health Statistics, *Advancedata*, no. 6 (March 30, 1977).

———— "Weight and Height of Adults 18–74 Years of Age." U.S. National Center for Health Statistics, *Vital and Health Statistics*, series 11, no. 211 (May 1979).

Advisory Committee for the Department of Housing and Urban Development *Freedom of Choice in Housing.* Washington, DC: National Academy of Sciences, 1972.

Allen, Walter R. "Black Family Research in the United States: A Review, Assessment and Extension." *Journal of Comparative Family Studies* 9 (Autumn 1978):167–89.

———— "Class, Culture and Family Organization: The Effects of Class and Race on Family Structure in Urban America." *Journal of Comparative Family Studies* 10 (Autumn 1979):301–13.

Alwin, Duane F. "Socioeconomic Background, Colleges, and Post-Collegiate Achievements." In William H. Sewell, Robert M. Hauser, and David L. Featherman, eds. *Schooling and Achievement in American Society.* New York: Academic Press, 1976.

Anderson, Elijah *A Place on the Corner.* Chicago: University of Chicago Press, 1978.

———— "Some Observations of Black Youth Employment." In Bernard E. Anderson and Isabel V. Sawhill, eds. *Youth Employment and Public Policy.* Englewood Cliffs, NJ: Prentice-Hall, 1980.

Anderson, James "The Schooling and Achievement of Black Children: Before and After *Brown* v. Topeka, 1900–1980." In D. Brotz and M. Maehr, eds. *Advances in Motivation and Achievement.* Greenwich, CT: JAI Press, 1984.

Anderson, Jervis *This Was Harlem: 1900–1950.* New York: Farrar, Straus, & Giroux, 1981.

Anderson, John E. "Planning Status of Marital Births, 1975–76." *Family Planning Perspectives* 13 (1981):62–70.

Anderson, Kristin, and Walter R. Allen "Correlates of Extended Household Structure." *Phylon* 45 (June 1984):144–57.

Anderson, Martin *Welfare: The Political Economy of Welfare Reform in the United States.* Stanford, CA: Stanford University Press, 1978.

Andrews, Frank; James Morgan; John Sonquist; and Laura Klem *Multiple Classification Analysis.* Ann Arbor: Institute for Social Research, University of Michigan, 1973.

Angel, Ronald, and Marta Tienda "Determinants of Extended Household Structure: Cultural Pattern or Economic Need?" *American Journal of Sociology* 87 (May 1982):1360–83.

Armstrong, Roger J. "A Study of Infant Mortality from Linked Records by Birth Weight, Period of Gestation, and Other Variables, United States, 1960 Live-Birth Cohort." U.S. National Center for Health Statistics, *National Vital Statistics System*, series 20, no. 10 (May 1972).

Arrow, Kenneth "Models of Job Discrimination." In A. H. Pascal, ed. *Racial Discrimination in Economic Life.* Lexington, MA: Heath, 1972.

Athearn, Robert G. *In Search of Caanan.* Lawrence: Regents Press of Kansas, 1978.

Bachrach, Christine A., and William D. Mosher "Use of Contraception in the United States, 1982." U.S. National Center for Health Statistics, *Advancedata*, no. 102 (December 4, 1984).

Ball-Rokeach, S., and J. F. Short "Collective Violence: The Redress of Grievance and Public Policy." In Lynn Curtis, ed. *American Violence and Public Policy.* New Haven, CT: Yale University Press, 1985.

Bancroft, Gertrude *The American Labor Force.* New York: Wiley, 1958.

Banfield, Edward C. *The Unheavenly City.* Boston: Little, Brown, 1968.

Batchelder, Alan "Declines in the Relative Income of Negro Men." *Quarterly Journal of Economics* 78 (November 1964):525–48.

Beale, Calvin L. "The Negro in American Agriculture." In John P. Davis, ed. *The American Negro Reference Book.* New York: Prentice-Hall, 1966.

Bean, Frank D., and John P. Marcum "Differential Fertility and the Minority Group Status Hypothesis: An Assessment and Review." In Frank D. Bean and W. Parker Frisbie, eds. *The Demography of Racial and Ethnic Groups.* New York: Academic Press, 1978.

Bean, Frank D., and Charles H. Wood "Ethnic Variation in the Relationship Between Income and Fertility." *Demography* 11 (November 1974):629–40.

Becker, Gary S. *The Economics of Discrimination.* Chicago: University of Chicago Press, 1957.

———. *Human Capital.* New York: Columbia University Press, 1964.

Beebe, Gilbert W. *Contraception and Fertility in the Southern Appalachians.* Baltimore: Williams & Wilkins, 1942.

Beller, Andrea H. "Trends in Occupational Segregation by Sex and Race, 1960–1981." In Barbara F. Reskin, ed. *Sex Segregation in the Workplace.* Washington, DC: National Academy Press, 1984.

Bennett, Lerone, Jr. *Before the Mayflower: A History of the Negro in America: 1619–1964.* New York: Penguin Books, 1984.

Bernard, William S. "A History of U.S. Immigration Policy." In *Immigration: Dimensions of Ethnicity, Selections from the Harvard Encyclopedia of American Ethnic Groups.* Cambridge, MA: Belknap Press of Harvard University Press, 1982.

Bernstein, Blanche, and William Meezan *The Impact of Welfare on Family Stability.* New York: New School for Social Research, Center for New York City Affairs, 1975.

Bianchi, Suzanne M. *Household Composition and Racial Inequality.* New Brunswick, NJ: Rutgers University Press, 1981.

_____, and Reynolds Farley "Racial Differences in Family Living Arrangements and Economic Well-Being: An Analysis of Recent Trends." *Journal of Marriage and the Family* 41 (August 1979):537–51.

_____; and Daphne Spain Racial Inequalities in Housing: An Examination of Recent Trends." *Demography* 19 (February 1982):37–51.

Bianchi, Suzanne, and Nancy Rytina "The Decline in Occupational Sex Segregation During the 1970s: Census and CPS Comparisons." *Demography* 23 (February 1986):79–86.

Bianchi, Suzanne, and Daphne Spain *Wives Who Earn More Than Their Husbands.* Washington, DC: U.S. Bureau of the Census, Special Demographic Analyses CDS-80-9 (November 1983).

_____ *American Women in Transition.* New York: Russell Sage Foundation, 1986.

Billingsley, Andrew *Black Families in White America.* Englewood Cliffs, NJ: Prentice-Hall, 1968.

Blake, Judith "Abortion and Public Opinion: the 1960–70 Decade." *Science* 171 (February 12, 1971):540–49.

Blalock, Hubert M., Jr. *Toward a Theory of Minority-Group Relations.* New York: Wiley, 1967.

Blassingame, John W. *Black New Orleans: 1860–1880.* Chicago: University of Chicago Press, 1973.

Blau, Francine D., and Wallace E. Hendricks "Occupational Segregation by Sex: Trends and Prospects." *Journal of Human Resources* 14 (Spring 1979):197–210.

Blau, Peter M., and Otis Dudley Duncan *The American Occupational Structure.* New York: Wiley, 1967.

Blinder, Derrick A., Jr. "Wage Discrimination: Reduced Form and Structural Estimates." *Journal of Human Resources* 3 (Fall 1973):436–55.

Bonacich, Edna "A Theory of Ethnic Antagonism: The Split Labor Market." *American Sociological Review* 37 (October 1972):547–59.

_____ "Advanced Capitalism and Black/White Relations in the United States: A Split Labor Market Interpretation." *American Sociological Review* 41 (February 1976):34–51.

Borchert, James *Alley Life in Washington.* Urbana: University of Illinois Press, 1980.

Boskin, Joseph *Urban Racial Violence in the Twentieth Century.* Beverly Hills, CA: Glencoe Press, 1969.

Bowen, William G., and T. Aldrich Finegan *The Economics of Labor Force Participation.* Princeton, NJ: Princeton University Press, 1969.

Bridenbaugh, Carl *Cities in the Wilderness.* New York: Capricorn Books, 1964 [1938].

Brookover, Wilbur B. *School Social Systems and Student Achievement: Schools Can Make a Difference.* New York: Praeger, 1979.

Bryce, Herrington J. "The Impact of the Undercount on State and Local Government Transfers." In *Proceedings of the 1980 Conference on Census Undercount.* Washington, DC: U.S. Government Printing Office, 1980.

Bumpass, Larry L.; Ronald Rindfuss; and Richard B. Janosik "Age and Marital

Status at First Birth and the Pace of Subsequent Fertility." *Demography* 15 (1978):75–86.

Burgess, Ernest W. "The Growth of the City: An Introduction to a Research Project." In Robert E. Park, Ernest W. Burgess, and R. D. McKenzie, eds. *The City.* Chicago: University of Chicago Press, 1967 [1925].

Burnham, Drusilla "Induced Termination of Pregnancy: Reporting States, 1977 and 1978." U.S. National Center for Health Statistics, *Monthly Vital Statistics Report,* vol. 30, no. 6, supplement (September 28, 1981).

———— "Induced Termination of Pregnancies: Reporting States, 1980." U.S. National Center for Health Statistics, *Monthly Vital Statistics Report,* vol. 32, no. 8, supplement (December 20, 1983).

Burstein, Paul "EEO Legislation and the Income of Women and Nonwhites." *American Sociological Review* 44 (June 1979):367–91.

———— *Discrimination, Jobs, and Politics.* Chicago: University of Chicago Press, 1985.

Burton, Nancy W., and Lyle V. Jones "Recent Trends in Achievement Levels of Black and White Youth." *Educational Researcher* 11 (1982):10–14.

Butler, Richard, and James J. Heckman "The Government's Impact on the Labor Market Status of Black Americans: A Critical Review." In Leonard Hausman et al., eds. *Equal Rights and Industrial Relations.* Madison, WI: Industrial Relations Research Assn., 1977.

Butz, William P., and Michael P. Ward "The Emergence of Countercyclical U.S. Fertility." *American Economic Review* 69 (June 1979):318–28.

Cafferty, P. S. J.; Pastora San Juan; Barry R. Chiswick; Andrew M. Greeley; and Teresa A. Sullivan *The Dilemma of American Immigration.* New Brunswick NJ: Transaction Books, 1983.

Cagan, Phillip, and Geoffrey H. Moore "Some Proposals to Improve the Consumer Price Index." *Monthly Labor Review* 104 (September 1981):20–25.

Cain, Glen *Married Women in the Labor Force.* Chicago: University of Chicago Press, 1966.

Callahan, David W. "Defining the Rate of Underlying Inflation." *Monthly Labor Review* 104 (September 1981):16–19.

Campbell, Arthur "The Role of Family Planning in the Reduction of Poverty." *Journal of Marriage and the Family* 30 (1968):236–45.

Carlson, Edwood D. "Social Determinants of Low Birth Weight in a High-Risk Population." *Demography* 21 (May 1984):207–16.

Carlucci, Carl P. "The Impact of Adjustment to the 1980 Census on Congressional and Legislative Reapportionment." In *Proceedings of the 1980 Conference on Census Undercount.* Washington, DC: U.S. Government Printing Office, 1980.

Carpenter, Niles *Immigrants and Their Children.* Census Monograph no. 7. Washington, DC: U.S. Government Printing Office, 1927.

Cassedy, James H. *Demography in Early America: Beginnings of the Statistical Mind: 1600–1800.* Cambridge, MA: Harvard University Press, 1969.

Caudill, Henry *Night Comes to the Cumberlands.* Boston: Little, Brown, 1964.

Cebula, Richard J. "Interstate Migration and the Tiebout Hypothesis: An Analysis According to Race, Sex and Age." *Journal of the American Statistical Association* 69 (December 1974):876–89.

Chase, Helen C. *A Study of Infant Mortality from Linked Records Comparison of Neonatal Mortality from Two Cohort Studies.* U.S. National Center for Health Statistics. *Vital and Health Statistics* , series 20, no. 13 (June 1972).

Chaudacuff, Howard P. *Mobile Americans: Residential and Social Mobility in Omaha: 1880–1920.* New York: Oxford University Press, 1972.

Cherlin, Andrew J. *Marriage, Divorce, Remarriage.* Cambridge, MA: Harvard University Press, 1981.

Chicago Commission on Race Relations *The Negro in Chicago: A Study of Race Relations and a Race Riot.* Chicago: University of Chicago Press, 1922.

Chilman, Catherine S. *Adolescent Pregnancy and Childbearing: Findings from Research.* Washington, DC: U.S. Department of Health and Human Services (December 1980).

Chiswick, Barry R. "The Economic Progress of Immigrants: Some Apparently Universal Patterns." In William Fellner, ed. *Contemporary Economic Problems.* Washington, DC: American Enterprise Institute, 1979.

Citro, Constance F., and Michael L. Cohen, eds. *The Bicentennial Census: New Questions for Methodology in 1990.* Washington, DC: National Academy Press, 1985.

Clague, Alice J., and Stephanie J. Ventura "Trends in Illegitimacy, United States: 1940–1965." U.S. National Center for Health Statistics, *Vital and Health Statistics,* series 21, no. 15 (February 1968).

Clark, Taliaferro *The Control of Syphilis in Southern Rural Areas.* Chicago: Julius Rosenwald Fund, 1932.

Coale, Ansley J. "The Population of the United States in 1950 Classified by Age, Sex and Color—A Revision of Census Figures." *Journal of the American Statistical Association* 50 (March 1955):16–54.

——— "Demographic Transition." *Proceedings of the International Population Conference, Liege, 1973* 1 (1974):53–72.

———, **and Paul Demeny** *Regional Model Life Tables and Stable Populations.* Princeton, NJ: Princeton University Press, 1966.

Coale, Ansley J., and Norfleet W. Rives, Jr. "A Statistical Reconstruction of the Black Population of the United States, 1880–1970: Estimates of True Numbers by Age and Sex, Birth Rates and Total Fertility." *Population Index* 39 (January 1973):3–35.

Coale, Ansley J., and Melvin Zelnik *New Estimates of Fertility and Population in the United States.* Princeton, NJ: Princeton University Press, 1963.

Cogan, John F. "The Decline in Black Teenage Employment: 1950–70." *American Economic Review* 72 (September 1982):621–39.

Cohen, Patricia Cline *A Calculating People.* Chicago: University of Chicago Press, 1982.

Conk, Margo Anderson *The United States Census and Labor Force Change: A History of Occupational Statistics, 1870–1940.* Ann Arbor: UMI Research Press, 1978.

Conot, Robert *American Odyssey.* New York: Bantam Books, 1974.

Conrad, Alfred H., and John R. Meyer *The Economics of Slavery.* Chicago: Aldine, 1964.

Corcoran, Mary "Work Experience, Work Interruption, and Wages." In Greg J. Duncan and James N. Morgan, eds. *Five Thousand American Families: Patterns of Economic Progress.* Ann Arbor: Institute for Social Research, University of Michigan, 1978.

——— "Work Experience, Labor Force Withdrawals, and Women's Earnings: Empirical Results Using the 1976 Panel Study of Income Dynamics." In Cynthia B. Lloyd, E. Edwards, and C. L. Gilroy, eds. *Women in the Labor Market.* New York: Columbia University Press, 1979.

_____, **and Greg J. Duncan** "Work History, Labor Force Attachment and Earnings: Differences Between the Races and Sexes." *Journal of Human Resources* 14 (Winter 1979):3–20.

Courant, Paul N. "Racial Prejudice in a Search Model of the Urban Housing Market." *Journal of Urban Economics* 5 (1973):329–45.

Cox, Oliver C. "Sex Ratio and Marital Status Among Blacks." *American Sociological Review* 5 (1940):937– 47.

_____ "Race and Caste: A Distinction." *American Journal of Sociology* 50 (1945):360–68.

Crain, Robert L., and Carol Sachs Weisman *Discrimination, Personality and Achievement.* New York: Seminar Press, 1972.

Crèvecoeur, J. Hector *Letters from an American Farmer.* New York: Fox, Duffield, 1904.

Crimmins, Eileen M. "The Changing Pattern of American Mortality Decline, 1940–77, and Its Implications for the Future." *Population and Development Review* 7 (June 1981):229–54.

Curry, Leonard P. *The Free Black in Urban America: 1800–1850.* Chicago: University of Chicago Press, 1981.

Curtin, Philip D. *The Atlantic Slave Trade: A Census.* Madison: University of Wisconsin Press, 1969.

Cutright, Philip "AFDC, Family Allowances and Illegitimacy." *Family Planning Perspectives* 2 (October 1970):4–9.

_____, **and Edward Shorter** "The Effects of Health on the Completed Fertility of Nonwhite and White U.S. Women Born Between 1867 and 1935." *Journal of Social History* 13 (1979):191–217.

Danchik, Kathleen M., and Joel C. Kleinman "Hypertension." In *Health, United States, 1980.* Washington, DC: U.S. Public Health Service, 1980.

Daniel, Pete *The Shadow of Slavery: Peonage in the South: 1901–1969.* Urbana: University of Illinois Press, 1972.

_____ *Breaking the Land: The Transformation of Cotton, Tobacco, and Rice Cultures Since 1880.* Urbana: University of Illinois Press, 1985.

Datcher, Linda "Technical Appendix: Relative Economic Status of Women: 1960–1970." In Phyllis A. Wallace, ed. *Black Women in the Labor Force.* Cambridge, MA: MIT Press, 1980.

Dauer, Carl C.; Robert F. Korns; and Leonard M. Schuman *Infectious Diseases.* Cambridge MA: Harvard University Press, 1968.

Davie, Maurice R. *Negroes in American Society.* New York: McGraw-Hill, 1949.

Davis, James A. "Hierarchical Models for Significance Tests in Multivariate Contingency Tables: An Exegesis of Goodman's Recent Papers." In Leo A. Goodman, ed. *Analyzing Qualitative/Categorical Data.* Cambridge, MA: Abt Books, 1978.

Demeny, Paul, and Paul Gingrich "A Reconsideration of Negro-White Mortality Differences in the United States." *Demography* 4 (1967):820–37.

Devaney, Barbara "An Analysis of Variations in U.S. Fertility and Female Labor Force Participation Trends." *Demography* 20 (May 1983):147–63.

Dickerson, Dennis C. *Out of the Crucible: Black Steelworkers in Western Pennsylvania, 1975–1980.* Albany, NY: State University of New York Press, 1986.

Dimond, Paul R. *Beyond Busing: Inside the Challenge to Urban Segregation.* Ann Arbor: University of Michigan Press, 1985.

Drury, Thomas F. "Access to Ambulatory Health Care: United States, 1974."

U.S. National Center for Health Statistics, *Advancedata*, no. 17 (February 23, 1978).

Dryfoos, G. "The United States National Family Planning Program, 1968–74." *Studies in Family Planning* 1 (March 1976):80–92.

DuBois, W. E. B. *The Philadelphia Negro*. Philadelphia: University of Pennsylvania Press, 1899.

–––––– "The Freedman's Bureau." *Atlantic Monthly* 87 (1901):354–65.

–––––– *Black Reconstruction in America: 1860–1880*. Cleveland: World, 1935.

–––––– "Prospect of a World Without Race Conflict." *American Journal of Sociology* 49 (March 1944):450–56.

–––––– *The Souls of Black Folk*. New York: Fawcett World Library, 1961 [1953].

–––––– *The Suppression of the African Slave Trade*. Baton Rouge: Louisiana State University Press, 1965 [1896].

–––––– *The Autobiography of W. E. B. DuBois*. New York: International Publishers, 1968.

–––––– *The Negro American Family*. Cambridge, MA: MIT Press, 1970 [1909].

Duncan, Beverly, and Otis Dudley Duncan "Family Stability and Occupational Success." *Social Problems* 16 (1969):273–85.

Duncan, Greg J., and Saul D. Hoffman "A New Look at the Causes of Improved Economic Status of Black Workers." *Journal of Human Resources* 17 (1983):268–82.

Duncan, Otis Dudley "A Socioeconomic Index for All Occupations." In A. J. Reiss, ed. *Occupations and Social Status*. New York: Free Press, 1961.

–––––– "Inheritance of Poverty or Inheritance of Race?" In Patrick Moynihan, ed. *Understanding Poverty*. New York: Basic Books, 1968.

–––––– "Ability and Achievement." *Eugenics Quarterly* 11 (March 1968):1–11.

––––––; David L. Featherman; and Beverly Duncan *Socioeconomic Background and Achievement*. New York: Seminar Press, 1972.

Easterlin, Richard A. *The American Baby Boom in Historical Perspective*. New York: National Bureau of Economic Research, 1962.

–––––– *Birth and Fortune*. New York: Basic Books, 1980.

Eblan, Jack E. "Growth of the Black Population in Ante Bellum America, 1820–1860." *Population Studies* 26 (July 1972):273–89.

–––––– "New Estimate of the Vital Rates of the United States Black Population." *Demography* 11 (May 1974):301–20.

Eckler, A. Ross *The Bureau of the Census*. New York: Praeger, 1972.

Edwards, G. Franklin *E. Franklin Frazier on Race Relations*. Chicago: University of Chicago Press, 1968.

Elder, Glen H., Jr. *Children of the Great Depression*. Chicago: University of Chicago Press, 1974.

–––––– "Household, Kinship and the Life Course: Perspectives on Black Families and Children." In G. Brookins and W. Allen, eds. *Beginnings: the Social and Affective Development of Black Children*. Hillside, NJ: Erlbaum, 1985.

Eldridge, Hope T., and Dorothy Swaine Thomas *Population Redistribution and Economic Growth, United States: 1870–1950*. Philadelphia: American Philosophical Society, 1964.

Ellwood, David T., and Mary Jo Bane "The Impact of AFDC on Family Structure and Living Arrangements." *Journal of Labor Economics* 7 (1984): 137–207.

Engerman, Stanley "Changes in Black Fertility, 1880–1940." In Tamara K. Hareven and Maris A Vinovskis, eds. *Family and Population in Nineteenth Century America*. Princeton, NJ: Princeton University Press, 1978.

445

Engram, Eleanor *Science, Myth, Reality: The Black Family in One Half Century of Research.* Westport, CT: Greenwood Press, 1982.

Ericksen, Eugene P. "Can Regression Be Used to Estimate Local Undercount Adjustments?" In *Proceedings of the Conference on Census Undercount.* Washington, DC: U.S. Government Printing Office, 1980.

Espenshade, Thomas J. "Black-White Differences in Marriage, Separation, Divorce and Remarriage." Paper presented at the annual meetings of the Population Association of America, Pittsburgh, April 1983.

_____ "Economic Impacts of Immigration." Paper presented at the annual meetings of the American Statistical Association, Las Vegas, August 6, 1985.

Evans, M. D. R. "American Fertility Patterns: A Comparison of White and Nonwhite Cohorts Born 1903–56. *Population and Development Review* 12 (June 1986):267–294.

Ezzard, N. V.; W. Cates, Jr.; D. G. Kramer; and C. Tietze "Race-Specific Patterns of Abortion Use by American Teenagers." *American Journal of Public Health* 82 (August 1982):809–14.

Farley, Reynolds "The Demographic Rates and Social Institutions of the Nineteenth-Century Negro Population: A Stable Population Analysis." *Demography* 2 (1965):386–98.

_____ "The Quality of Demographic Data on Nonwhites." *Demography* 5 (1968):1–10.

_____ *Growth of the Black Population: A Study of Demographic Trends.* Chicago: Aldine, 1970.

_____ "Family Types and Family Headship: A Comparison of Trends Among Blacks and Whites." *Journal of Human Resources* 6 (Summer 1971):275–96.

_____ "Residential Segregation in Urbanized Areas of the United States: 1970: An Analysis of Social Class and Racial Differences." *Demography* 14 (November 1977a):497–518.

_____ "Trends in Racial Inequalities: Have the Gains of the 1960s Disappeared in the 1970s?" *American Sociological Review* 42 (1977b):189–208.

_____ *Blacks and Whites: Narrowing the Gap?* Cambridge, MA: Harvard University Press, 1984.

_____; **Suzanne Bianchi; and Diane Colasanto** Barriers to the Racial Integration of Neighborhoods: The Detroit Case." *Annals of the American Academy of Political and Social Science* 444 (January 1980):97–113.

Farley, Reynolds; Suzanne M. Bianchi; and Daphne Spain Racial Inequalities in Housing: An Examination of Recent Trends." *Demography* 19 (February 1982):37–51.

Farley, Reynolds; Howard Schuman; Suzanne Bianchi; Diane Colasanto; and Shirley Hatchett "Chocolate City, Vanilla Suburbs: Will the Trend Toward Racially Separate Communities Continue?" *Social Science Research* 7 (December 1978):319–44.

Featherman, David L., and Robert M. Hauser "Prestige or Socioeconomic Scales in the Study of Occupational Achievement?" *Sociological Methods and Research* 4 (1976):403–22.

_____ *Opportunity and Change.* New York: Academic Press, 1978.

Featherman, David L., and Gillian Stevens "A Revised Socioeconomic Index of Occupational Status: Applications in Analysis of Sex Differences in Attainment." In Mary G. Powers, ed. *Measures of Socioeconomic Status: Current Issues.* Boulder, CO: Westview Press, 1982.

Feinberg, Stephen *The Analysis of Cross-Classified Categorical Data.* Cambridge, MA: MIT Press, 1977.

446

Feinberg, William E. "At a Snail's Pace: Time to Equality in Simple Models of Affirmative Action Programs." *American Journal of Sociology* 90 (July 1984):168–81.

Fields, Barbara Jeanne *Slavery and Freedom on the Middle Ground.* New Haven: Yale University Press, 1985.

Fingerhut, Lois A. "Changes in Mortality Among the Elderly." U.S. National Center for Health Statistics, *Vital and Health Statistics*, Analytic Studies, series 3, no. 22 (1982).

———, **and Harry M. Rosenberg** "Mortality Among the Elderly." In *Health, United States, 1981.* Washington, DC: U.S. Public Health Service, 1981.

Fligstein, Neil *Going North: Migration of Blacks and Whites from the South, 1900–1950.* New York: Academic Press, 1981.

Fogel, Robert William, and Stanley L. Engerman *Time on the Cross: The Economics of American Negro Slavery.* Boston: Little, Brown, 1974.

——— *Time on the Cross: Evidence and Methods* (Supplement). Boston: Little, Brown, 1974.

Foner, Philip S. *Organized Labor and the Black Worker: 1619–1981.* New York: Praeger, 1981 [1974].

Forrest, Jacqueline Dorroch, and Stanley K. Henshaw "What U.S. Women Think and Do About Contraception." *Family Planning Perspectives* 15 (July-August 1983):157–66.

Foundray, Elbertie *United States Abridged Life Tables: 1919–20.* Washington, DC: U.S. Government Printing Office, 1923.

Francese, Peter K. "The 1980 Census: The Counting of America." *Population Bulletin* 34 (September 1979):3–39.

Franklin, John Hope *Reconstruction After the Civil War.* Chicago: University of Chicago Press, 1961.

——— *From Slavery to Freedom: A History of Negro Americans.* 3rd ed. New York: Knopf, 1980.

Frazier, E. Franklin *The Free Negro Family.* Nashville: Fisk University Press, 1932.

——— "Negro Harlem: An Ecological Study." *American Journal of Sociology* 43 (July 1937):72–88.

——— *The Negro in the United States.* New York: Macmillan, 1949.

——— *Race and Culture Contacts in the Modern World.* Boston: Beacon Press, 1957.

——— *The Negro Family in the United States.* Chicago: University of Chicago Press, 1966 [1939].

Frederickson, George M. *White Supremacy: A Comparative Study in American and South African History.* New York: Oxford University Press, 1981.

Freedman, Ronald; David Goldberg; and Harry Sharp "Ideals About Family Size in the Detroit Metropolitan Area: 1954." *Milbank Memorial Fund Quarterly* 33 (April 1955):185–97.

Freeman, Richard B. "Decline of Labor Market Discrimination and Economic Analyis." *American Economic Review* 63 (May 1973):280–86.

——— "Alternative Theories of Labor-Market Discrimination: Individual and Collective Behavior." In George M. von Furstenberg, Ann R. Horowitz, and Bennett Harrison, eds. *Employment and Income, Patterns of Racial Discrimination*, vol. 2. Lexington, MA: Heath, 1974.

——— *Black Elite: The New Market for Highly Educated Black Americans.* New York: McGraw-Hill, 1976.

——— "Why Is There a Youth Labor Market Problem?" In Bernard E. Anderson and Isabel V. Sawhill, eds. *Youth Employment and Public Policy*. Englewood Cliffs, NJ: Prentice-Hall, 1979.

———, **and David A. Wise** *The Youth Labor Market Problem: Its Nature, Causes and Consequences*. Chicago: University of Chicago Press, 1982.

Fulwood, Robinson; Sidney Abraham; and Clifford Johnson "Height and Weight of Adults 18–74 Years by Socioeconomic and Geographic Variables." U.S. National Center for Health Statistics, *Vital and Health Statistics*, series 11, no. 224 (August 1981).

Furstenberg, Frank J., Jr. *Unplanned Parenthood*. New York: Free Press, 1976.

———; **T. Hershberg; and J. Modell** "The Origins of the Female-Headed Black Family: The Impact of the Urban Experience." *Journal of Interdisciplinary History* 6 (Autumn 1985):211–33.

Furstenberg, Frank J., Jr.; Richard Lincoln; and Jane Menken, eds. *Teenage Sexuality, Pregnancy, and Childbearing*. Philadelphia: University of Pennsylvania Press, 1981.

Garfinkel, Irwin, and Robert H. Haveman *Earnings Capacity, Poverty, and Inequality*. New York: Academic Press, 1977.

Genovese, Eugene D. *The Political Economy of Slavery*. New York: Vintage Books, 1965.

Gilder, George *Wealth and Poverty*. New York: Basic Books, 1981.

Glazer, Nathan, and Daniel Patrick Moynihan *Beyond the Melting Pot: The Negroes, Puerto Ricans, Jews, Italians, and Irish of New York City*. Cambridge, MA: MIT Press, 1963.

Glick, Paul C. "A Demographic Picture of Black Families." In Harriette P. McAdoo, ed. *Black Families*. Beverly Hills, CA: Sage, 1981.

Godley, Frank, and Ronald W. Wilson "Health Status of Minority Groups." In *Health, United States, 1979*. Washington, DC: U.S. Public Health Service, 1980.

Gold, Rachel Benson, and Barry Nestor "Public Funding of Contraceptive, Sterilization and Abortion Services, 1983." *Family Planning Perspectives* 17 (January-February 1985):25–30.

Goldenberg, Robert L.; Joan L. Humphrey; Christiane B. Hale; Beverly W. Boyd; and John B. Wayne "Neonatal Deaths in Alabama, 1970–1980: An Analysis of Birth Weight—and Race Specific Neonatal Mortality Rates." *American Journal of Obstetrics and Gynecology* 145 (March 1, 1983):545–52.

Goldscheider, Calvin, and Peter R. Uhlerberg "Minority Group Status and Fertility." *American Journal of Sociology* 74 (January 1969):361–72.

Gordon, David M. *Theories of Poverty and Unemployment*. Lexington, MA: Heath, 1972.

Grabill, Wilson H.; Clyde V. Kiser; and Pascal Whelpton *The Fertility of American Women*. New York: Wiley, 1958.

Grandberg, Donald, and Beth Wellman "Abortion Attitudes, 1965–1980: Trends and Determinants." *Family Planning Perspectives* 12 (September 1980):250–61.

Green, Constance McLaughlin *The Secret City: A History of Race Relations in the Nation's Capital*. Princeton, NJ: Princeton University Press, 1967.

Green, Lorenzo Johnston *The Negro in Colonial New England*. New York: Atheneum, 1968 [1942].

Greville, Thomas N. E. "United States Life Tables and Actuarial Tables, 1939–1941." U.S. Bureau of the Census, *Sixteenth Census of the United States: 1940*. Washington, DC: U.S. Government Printing Office, 1946.

Griliches, Zvi, and William M. Mason "Education, Income, and Ability." In

Arthur S. Goldberger and Otis Dudley Duncan, eds. *Structural Equation Models in the Social Sciences*. New York: Seminar Press, 1972.

Grossman, Michael, and Steven Jacobowitz "Variations in Infant Mortality Rates Among Counties of the United States: The Roles of Public Policies and Programs." *Demography* 18 (November 1981):695–713.

Grove, Robert D. *Studies in the Completeness of Birth Registration*. U.S. Public Health Service, *Special Report Series* 17 (April 20, 1943):224–30.

_____, **and Alice M. Hetzel** *Vital Statistics Rates in the United States: 1940–1960*. Washington, DC: U.S. Government Printing Office, 1968.

Guest, Avery M., and James A. Weed "Ethnic Residential Segregation: Patterns of Change." *American Journal of Sociology* 81 (March 1976):1088–111.

Gurock, Jeffery S. *When Harlem Was Jewish: 1870–1930*. New York: Columbia University Press, 1979.

Gutman, Herbert G. "Persistent Myths About the Afro-American Family." *Journal of Interdisciplinary History* 6 (Autumn 1975):181–210.

_____ *The Black Family in Slavery and Freedom, 1750–1925*. New York: Pantheon Books, 1976.

Guttentag, Marcia, and Paul F. Secord *Too Many Women? The Sex Ratio Question*. Beverly Hills, CA: Sage, 1983.

Gwartney, James "Changes in the Non-White/White Income Ratio—1939–67." *American Economic Review* 60 (December 1970):872–83.

Halacy, Dan *Census: 190 Years of Counting America*. New York: Elsevier/Nelson, 1980.

Hamill, Peter V.; Francis E. Johnston; and Stanley Lemeshaw "Height and Weight of Children: Socioeconomic Status." U.S. National Center for Health Statistics, *National Health Survey*, series 11, no. 119 (October 1972).

Handlin, Oscar *The Newcomers: Negroes and Puerto Ricans in a Changing Metropolis*. Garden City, NY: Anchor Books, 1962.

Hannerz, Ulf *Soulside: Inquiries into Ghetto Culture and Community*. New York: Columbia University Press, 1969.

Hanushek, Eric "Ethnic Income Variations: Magnitudes and Explanations." In Thomas Sowell, ed. *Essays and Data on American Ethnic Groups*. Washington, DC: Urban Institute, 1978.

Harding, Vincent *There Is a River*. New York: Vintage Books, 1983 [1981].

Hare, Bruck, and D. Levine *Toward Desegregated Schools*. Washington, DC: U.S. Department of Education, 1984.

Harrington, Michael *The Other America*. New York: Macmillan, 1962.

Harris, William H. *The Harder We Run: Black Workers Since the Civil War*. New York: Oxford University Press, 1982.

Hartley, Shirley Foster *Illegitimacy*. Berkeley: University of California Press, 1975.

Hauser, Philip "Demographic Factors in the Integration of the Negro." *Daedalus* 94 (Fall 1965):847–77.

Hauser, Robert M., and David L. Featherman *The Process of Stratification: Trends and Analyses*. New York: Academic Press, 1977.

Haveman, Robert H., ed. *A Decade of Federal Antipoverty Programs*. New York: Academic Press, 1977.

Hawley, Amos H, and Vincent P. Rock, eds. *Segregation in Residential Areas*. Washington, DC: National Academy of Sciences, 1973.

Haworth, J.; James Gwartney; and C. Haworth "Earnings, Productivity and Changes in Employment Discrimination During the 1960s." *American Economic Review* 65 (March 1975):158–68.

Hazen, H. H. "Syphilis in the American Negro." *American Journal of Syphilis, Gonorrhea and Venereal Diseases* 20 (September 1936):530–61.

Heer, David M., ed. *Social Statistics and the City.* Cambridge: Joint Center for Urban Studies of the Massachusetts Institute of Technology and Harvard University, 1968.

Heiss, Jerold *The Case of the Black Family: A Sociological Inquiry.* New York: Columbia University Press, 1975.

Helper, Rose *Racial Policies and Practices of Real Estate Brokers.* Minneapolis: University of Minnesota Press, 1969.

Hendershot, Gerry E., and Paul J. Placek, eds. *Predicting Fertility.* Lexington, MA.: Heath, 1981.

Henshaw, Stanley K.; Nancy J. Binkin; Ellen Blaine; and Jack C. Smith "A Portrait of American Women Who Obtain Abortions." *Family Planning Perspectives* 17 (March-April 1985):90–96.

Henshaw, Stanley K., and Kevin O'Reilly "Characteristics of Abortion Patients in the United States, 1979 and 1980." *Family Planning Perspectives* 15 (January-February 1983):5–15.

Hermalin, Albert I., and Reynolds Farley "The Potential for Residential Integration in Cities and Suburbs: Implications for the Busing Controversy." *American Sociological Review* 38 (October 1973):595–619.

Hiestand, Dale L. *Economic Growth and Employment Opportunities for Minorities.* New York: Columbia University Press, 1964.

Higgs, Robert *Competition and Coercion: Blacks in the American Economy: 1865–1914.* Chicago: University of Chicago Press, 1977.

Hill, Robert B. *The Strengths of Black Families.* New York: Emerson Hall, 1971.

—— "The Synthetic Method: Its Feasibility for Deriving the Census Undercount for States and Local Areas." In *Proceedings of the Conference on Census Undercount.* Washington, DC: U.S. Government Printing Office, 1980.

—— "The Economic Status of Black Americans." In James D. Williams, ed. *The State of Black America: 1981.* Washington, DC: National Urban League, 1981.

Hirsch, Arnold R. *Making the Second Ghetto: Race and Housing in Chicago: 1940–1960.* New York: Cambridge University Press, 1983.

Hirschman, Charles, and Morrison G. Wong "Socioeconomic Gains of Asian Americans, Blacks and Hispanics: 1960–1976." *American Journal of Sociology* 90 (November 1984):584–607.

Hochschild, Jennifer L. *The New American Dilemma: Liberal Democracy and School Desegregation.* New Haven: Yale University Press, 1984.

Hofferth, Sandra L., and Kristin A. Moore "Early Childbearing and Later Economic Well-Being." *American Sociological Review* 44 (October 1979):784–815.

Hoffman, Frederick L. "Race Traits and Tendencies of the American Negro." *Publications of the American Economic Association* 11, nos. 1–3 (August 1896).

Hoffman, Saul D. "Black-White Life Cycle Earnings Differences and the Vintage Hypothesis: A Longitudinal Analysis." *American Economic Review* 69 (December 1979):855–67.

Hogan, Dennis P., and David L. Featherman "Racial Stratification and Socioeconomic Change in the American North and South." *American Journal of Sociology* 83 (July 1977):100–26.

Hogan, Dennis P., and Evelyn M. Kitagawa "Family Factors in the Fertility of

Black Adolescents." Paper presented at the annual meeting of the Population Association of America, Pittsburgh, 1983.

———. "The Impact of Social Status, Family Structure and Neighborhood on the Fertility of Black Adolescents." *American Journal of Sociology* 90 (January 1985):825–55.

Hout, Michael "Opportunity and the Minority Middle Class: A Comparison of Blacks in the United States and Catholics in Northern Ireland." *American Sociological Review* 51 (April 1986):214–23.

Hurn, Christopher J. *The Limits and Possibilities of Schooling.* 2nd ed. Boston: Allyn & Bacon, 1985.

Hyman, Herbert, and John Reed "Black Matriarchy Reconsidered: Evidence from Secondary Analysis of Sample Surveys." *Public Opinion Quarterly* 33 (Fall 1969):346–54.

Jackman, Mary, and Robert Jackman "Racial Inequalities in Homeownership." *Social Forces* 58 (1980):1221–34.

Jackson, Jacquelyne J. "But Where Are the Men?" *Black Scholar* (December 1971):30–41.

Jacobson, Paul H. "An Estimate of the Expectation of Life in the United States in 1850." *Milbank Memorial Fund Quarterly* 35 (April 1957):197–201.

Janowitz, Morris, ed. *W. I. Thomas on Social Organization and Social Personality.* Chicago: University of Chicago Press, 1966.

Jarvis, Edward "Statistics of Insanity in the United States." *Boston Medical and Surgical Journal* 27 (1842):116–21; 281–82.

———. "Insanity Among the Colored Population of the Free States." *American Journal of the Medical Sciences* 7 (1844):71–83.

Jaynes, Gerald David *Branches Without Roots: Genesis of the Black Working Class in the American South: 1962–1982.* New York: Oxford University Press, 1986.

Jencks, Christopher *Inequality: An Assessment of Family and Schooling in America.* New York: Harper & Row, 1972.

———. *Who Gets Ahead? The Determinants of Economic Success in America.* New York: Basic Books, 1979.

Jiobu, Robert M., and Harvey H. Marshall, Jr. "Urban Structure and the Differentiation between Blacks and Whites." *American Sociological Review* 36 (August 1971):638–49.

Johnson, Charles S. *Shadow of the Plantation.* Chicago: University of Chicago Press, 1934.

———. *Patterns of Negro Segregation.* New York: Harper, 1943.

Johnson, Daniel M., and Rex R. Campbell *Black Migration in America.* Durham, NC: Duke University Press, 1981.

Johnson, James Weldon *Black Manhattan.* New York: Atheneum, 1983.

Johnson, Nan E. "Minority-Group Status and the Fertility of Black Americans, 1970: A New Look." *American Journal of Sociology* 84 (May 1979): 1386–1400.

———, and Ryoko Nishida "Minority-Group Status and Fertility." *American Journal of Sociology* 86 (November 1980):496–511.

Jones, Elise J., and Charles F. Westoff "Attitudes Toward Abortion in the United States in 1970 and the Trend Since 1965." In Charles F. Westoff and Robert Parke, Jr., eds. *Demographic and Social Aspects of Population Growth.* Commission on Population Growth and the American Future, vol. 1. Washington, DC: U.S. Government Printing Office, 1972.

Jones, James H. *Bad Blood.* New York: Free Press, 1981.

Jordan, Winthrop D. *White Over Black*. Chapel Hill: University of North Carolina Press, 1968.

Kain, John F. "Housing Segregation, Negro Employment and Metropolitan Decentralization." *Quarterly Journal of Economics* 82 (May 1968):175–97.

————, **and J. M. Quigley** *Housing Markets and Racial Discrimination: A Micro-Economic Analysis*. New York: National Bureau of Economic Research, 1975.

Kampmeier, Rudolph H. *Essentials of Syphilogy*. Philadelphia: Lippincott, 1943.

Kantrowitz, Nathan *Ethnic and Racial Segregation in the New York Metropolis*. New York: Praeger, 1972.

Katzman, David M. *Before the Ghetto*. Urbana: University of Illinois Press, 1973.

Keeley, Charles B. "Immigration: Considerations on Trends, Prospects and Policy." In Charles F. Westoff and Robert Parke, Jr., eds. *Demographic Aspects of Population Growth*. Commission on Population Growth and the American Future, vol. 1. Washington, DC: U.S. Government Printing Office, 1972.

Keyfitz, Nathan, and Wilhelm Flieger *World Population: An Analysis of Vital Data*. Chicago: University of Chicago Press, 1968.

Killingsworth, Charles C. *Jobs and Income for Negroes*. Ann Arbor: Institute of Labor and Industrial Relations, University of Michigan, 1968.

King, A. Thomas *Discrimination in Mortgage Lending: A Study of Three Cities*. Monograph Series in Finance and Economics. New York: New York University, 1981.

————, **and P. Mieszkowski** "Racial Discrimination, Segregation and the Price of Housing." *Journal of Political Economy* 81 (May-June 1973):590–606.

Kiple, K., and V. King *Another Dimension of the Black Diaspora: Diet, Disease and Racism*. New York: Cambridge University Press, 1981.

Kirby, John B. *Black Americans in the Roosevelt Era*. Knoxville: University of Tennessee Press, 1980.

Kiser, Clyde Vernon *Sea Island to City*. New York: Columbia University Press, 1932.

———— "Fertility Trends and Differentials Among Nonwhites in the United States." *Milbank Memorial Fund Quarterly* 46 (April 1958):149–97.

Kitagawa, Evelyn M., and Philip M. Hauser *Differential Mortality in the United States*. Cambridge, MA: Harvard University Press, 1973.

Klebba, A. Joan "Mortality Trends in the United States: 1954–1963." U.S. National Center for Health Statistics, *Vital and Health Statistics*, series 20, no. 2 (1966).

————, **and A. B. Dolman** "Comparability of Mortality Statistics for the Seventh and Eighth Revisions of the International Classification of Diseases." U.S. National Center for Health Statistics, *Vital and Health Statistics*, Series 2, no. 66 (1975).

Klebba, A. Joan; Jeffery D. Mauer; and Evelyn J. Glass "Mortality Trends for Leading Causes of Death: United States, 1950–69." U.S. National Center for Health Statistics, *Vital and Health Statistics*, series 20, no. 16 (1974).

Kleinman, Joel C. "Trends and Variations in Birth Weight." In *Health, United States, 1981*. Washington, DC: U.S. Public Health Service, 1981.

————; **Lois A. Fingerhut; and Jacob J. Feldman** Trends in Mortality." In *Health, United States, 1980*. Washington, DC: U.S. Public Health Service, 1980.

Kluegel, James R., and Eliot R. Smith *Beliefs About Inequality: Americans' Views of What Is and What Ought to Be*. New York: Aldine de Gruyter, 1986.

452

Kusmer, Kenneth L. *A Ghetto Takes Shape: Black Cleveland, 1870–1930.* Urbana: University of Illinois Press, 1976.

Lake, Robert W. *The New Suburbanites: Race and Housing in the Suburbs.* New Brunswick, NJ: Center for Urban Policy Research, Rutgers University, 1981.

Lamb, Charles M. "Equal Housing Opportunities." In Charles S. Bullock III and Charles M. Lamb, eds. *Implementation of Civil Rights Policy.* Monterey, CA: Brooks/Cole, 1984.

Lammamier, Paul J. "The Urban Black Family of the Nineteenth Century: A Study of Black Family Structure in the Ohio Valley, 1850–1880." *Journal of Marriage and the Family* 35 (August 1973):440–56.

Land, Kenneth C.; George C. Hough, Jr.; and Marilyn M. McMillen "New Mid-Year Age-Sex-Color-Specific Estimates of the U.S. Population for the 1940s and 1950s: Including a Revision of Coverage Estimates for the 1940 and 1950 Censuses." *Demography* 21 (November 1984):623–46.

Lane, Roger *Roots of Violence in Black Philadelphia: 1860–1900.* Cambridge, MA: Harvard University Press, 1986.

Lantz, Herman, and Lewellyn Hendrix "Black Fertility and the Black Family in the Nineteenth Century: A Re-examination of the Past." *Journal of Family History* 3 (1978):251–61.

Lasch, Christopher *The Culture of Narcissism.* New York: Warren Books, 1979.

Lee, Everett S. "Internal Migration and Population Redistribution in the United States." In *Population: The Vital Revolution.* New York: Doubleday, 1964.

_____, **and Anne S. Lee** "The Differential Fertility of the American Negro." *American Sociological Review* 17 (August 1952):437–47.

_____ "The Future Fertility of the American Negro." *Social Forces* 3 (1959):228–31.

Lee, Ronald "Demographic Forecasting and the Easterlin Hypothesis." *Population and Development Review* 2 (September-December 1976):459–68.

Leibowitz, Arleen; Marvin Eisen; and Winston Chow "Decision Making in Teenage Pregnancy: An Analysis of Choice." Paper presented at the annual meeting of the Population Association of America, Denver, April 1980.

Levitan, Sar A. *Programs in Aid of the Poor for the 1970s.* Baltimore: Johns Hopkins University Press, 1973.

_____ *Programs in Aid of the Poor for the 1980s.* 4th ed. Baltimore: Johns Hopkins University Press, 1980.

Lewis, Oscar *Five Families.* New York: Basic Books, 1959.

_____ *The Children of Sanchez.* New York: Random House, 1961.

_____ *La Vida: A Puerto Rican Family in the Culture of Poverty—San Juan and New York.* New York: Random House, 1965.

Lichter, Daniel T.; Glenn V. Fugitt; and Tim B. Heaton "Racial Differences in Nonmetropolitan Population Distribution." *Social Forces* 64 (December 1985):499–506.

Lieberson, Stanley *Ethnic Patterns in American Cities.* New York: Free Press, 1963.

_____ *A Piece of the Pie: Black and White Immigrants Since 1880.* Berkeley: University of California Press, 1980.

_____, **and Glenn V. Fuguitt** "Negro-White Occupational Differences in the Absence of Discrimination." *American Journal of Sociology* 73 (September 1967):188–200.

Lieberson, Stanley, and Christy A. Wilkinson "A Comparison Between North-

453

ern and Southern Blacks Residing in the North." *Demography* 13 (May 1976):199–224.

Liebow, Elliot *Tally's Corner*. Boston: Little, Brown, 1967.

Lilienfeld, Abraham M.; Morton L. Levin; and Irving Kessler *Cancer in the United States*. Cambridge, MA: Harvard University Press, 1972.

Lillard, Lee; James P. Smith; and Finis Welch "What Do We Really Know About Wages? The Importance of Nonreporting and Census Imputation." *Journal of Political Economy* 94 (June 1986):489–506.

Linder, Forrest E., and Robert D. Grove *Vital Statistics Rates in the United States: 1900–1940*. Washington, DC: U.S. Government Printing Office, 1947.

Litwack, Leon F. *North of Slavery*. Chicago: University of Chicago Press, 1961.

_____ *Been in the Storm So Long: The Aftermath of Slavery*. New York: Knopf, 1979.

Logan, John R., and Mark Schneider "Racial Segregation and Racial Change in American Suburbs: 1970–1980." *American Journal of Sociology* 89 (January 1984):874–88.

Long, Larry "Poverty Status and Receipt of Welfare Among Migrants and Non-migrants in Large Cities." *American Sociological Review* 39 (February 1974):46–57.

_____, **and Diane DeAre** "The Suburbanization of Blacks." *American Demographics* 3 (September 1981):16–21.

Long, Larry, and Lynne R. Heltman "Migration and Income Differences Between Black and White Men in the North." *American Journal of Sociology* 80 (May 1975):1391–1409.

Lowell, Anthony M.; Lydia B. Edwards; and Carroll E. Palmer) *Tuberculosis*. Cambridge, MA: Harvard University Press, 1969.

MacMahon, Brian; Mary Grace Kovar; and Jacob J. Feldman *Infant Mortality Rates: Socioeconomic Factors*. U.S. National Center for Health Statistics, *National Vital Statistics System*, series 22, no. 14 (March 1972).

_____ *Infant Mortality Rates: Relationships with Mothers' Reproductive History: United States*. U.S. National Center for Health Statistics, *Vital and Health Statistics*, series 22, no. 15 (April 1973).

Mandle, Jay R. *The Roots of Black Poverty: The Southern Plantation Economy After the Civil War*. Durham, NC: Duke University Press, 1978.

Mannix, Daniel P. *Black Cargoes: A History of the Atlantic Slave Trade*. New York: Viking Press, 1962.

Manton, Kenneth G.; Susan S. Poss; and Steve Wing "The Black/White Mortality Crossover: Investigations from the Perspective of the Components of Aging." *Gerontologist* 19 (1979):291–300.

Manton, Kenneth G., and Eric Stallard "Methods for Evaluating the Heterogeneity of Aging Processes Using Vital Statistics Data: Explaining the Black/White Mortality Crossover by a Model of Mortality Selection." *Human Biology* 53 (1981):47–67.

Mare, Robert D., and Christopher Winship "Changes in Race Differentials in Youth Labor Force Status." In *Fifth Annual Report to the President and the Congress of the National Commission for Employment Policy, Expanding Employment Opportunities for Disadvantaged Youth*. Washington, DC: National Commission for Employment Policy, 1979.

_____ "Racial Inequality and Joblessness." *American Sociological Review* 49 (February 1984):39–55.

Marmor, Theodore R., ed. *Poverty Policy: A Compendium of Cash Transfer Proposals.* Chicago: Aldine-Atherton, 1971.

Marshall, Ray, and Virgil L. Christian, Jr. "Economics of Employment Discrimination." In Ray Marshall and Virgil L. Christian, Jr., eds. *Employment of Blacks in the South.* Austin: University of Texas Press, 1978.

Martin, Elmer, and Joanne Martin *The Black Extended Family.* Chicago: University of Chicago Press, 1978.

Masters, Stanley H. *Black-White Income Differentials: Empirical Studies and Policy Implications.* New York: Academic Press, 1975.

McAdoo, Harriette "Factors Related to Stability in Upwardly Mobile Black Families." *Journal of Marriage and the Family* 40 (November 1978):761–78.

McClelland, Peter D., and Richard J. Zeckhauser *Demographic Dimensions of the New Republic.* New York: Cambridge University Press, 1982.

McCormick, M. C.; S. Shapiro; and B. Starfield "High-Risk Young Mothers: Infant Mortality and Morbidity in Four Areas of the United States, 1973–1978." *American Journal of Public Health* 74 (1984):18–25.

McFalls, Joseph A., Jr., and Marguerite Harvey McFalls *Disease and Fertility.* Orlando, FL: Academic Press, 1984.

McFalls, Joseph A., Jr., and George S. Masnick "Birth Control and the Fertility of the U.S. Black Population, 1880 to 1980." *Journal of Family History* 6 (Spring 1981):89–106.

McKenzie, R. D. "The Ecological Approach to the Study of the Human Community." In Robert E. Park, Ernest W. Burgess, and R. D. McKenzie, eds. *The City.* Chicago: University of Chicago Press, 1967 [1925].

McNeil, H. L. "Syphilis in the Southern Negro." *Journal of the American Medical Association* 67 (September 30, 1916):1001–4.

McNeil, John M.; Enrique J. Lamas; and Sheldon E. Haber Lifetime Labor Force Attachment: Retrospective Data from the Survey of Income and Program Participation." Paper presented at the annual meeting of the American Statistical Association, Chicago, August 18–21, 1986.

Mechanic, David *Medical Sociology: A Selective View.* New York: Free Press, 1968.

Meeker, Edward "Mortality Trends of Southern Blacks, 1850–1910: Some Preliminary Findings." *Explorations in Economic History* 13 (January 1976): 13–42.

———— "Freedom, Economic Opportunity and Fertility: Black Americans, 1860–1910." *Economic Inquiry* 15 (July 1977):397–412.

Meier, August, and Elliott Rudwick *Black Detroit and the Rise of the UAW.* New York: Oxford University Press, 1979.

Menken, Jane "The Health and Demographic Consequences of Adolescent Pregnancy and Childbearing." In Catherine S. Chilman, ed. *Adolescent Pregnancy and Childbearing: Findings from Research.* Washington, DC: U.S. Department of Health and Human Services, 1980.

———— "The Health and Social Consequences of Teenage Childbearing." In Frank F. Furstenberg, Jr., Richard Lincoln, and Jane Menken, eds. *Teenage Sexuality, Pregnancy and Childbearing.* Philadelphia: University of Pennsylvania Press, 1981.

Michelson, Stephen *Incomes of Racial Minorities.* Washington, DC: Brookings Institution, 1968.

Miller, Herman P. *Income of the American People.* New York: Wiley, 1955.

———— *Rich Man, Poor Man.* New York: Crowell, 1964.

―――― *Income Distribution in the United States.* Washington, DC: U.S. Government Printing Office, 1966.

Miller, Kelly "Enumeration Errors in Negro Population." *Scientific American* 14 (February 1922):168–87.

Mincer, Jacob *Schooling, Experience and Earnings.* New York: National Bureau of Economic Research, 1974.

―――― "Unemployment Effects of Minimum Wages." *Journal of Political Economy* 84 (August 1976):87–105.

―――― , **and Solomon Polachek** "Family Investment in Human Capital: Earnings of Women." In Theodore W. Schultz, ed. *Economics of the Family.* Chicago: University of Chicago Press, 1973.

―――― "An Exchange: Theory of Human Capital and the Earnings of Women: Women's Earnings Reexamined." *Journal of Human Resources* 13 (Winter 1978):118–34.

Molotch, Harvey Luskin *Managed Integration.* Berkeley: University of California Press, 1972.

Mooney, J. D. "Housing Segregation, Negro Employment and Metropolitan Decentralization: An Alternative Perspective." *Quarterly Journal of Economics* 83 (1969):299–311.

Moore, Joseph H. *The Modern Treatment of Syphilis.* Springfield, IL: Thomas, 1941.

Moore, Kristin A., and Martha R. Burt *Private Crisis, Public Cost: Policy Perspectives on Teenage Childbearing.* Washington, DC: Urban Institute Press, 1982.

Moore, Kristin A., and Linda Waite "Marital Dissolution, Early Motherhood and Early Marriage." *Social Forces* 60 (September 1981):20–40.

Moore, Kristin A.; Margaret Simms; and Charles L. Betsey *Choice and Circumstance: Racial Differences in Adolescent Sexuality and Fertility.* New Brunswick, NJ: Transaction Books, 1986.

Morgan, James; Martin H. David; Wilbur J. Cohen; and Harvey E. Brazier *Income and Welfare in the United States.* New York: McGraw-Hill, 1962.

Moriyama, Iwao M., and Lillian Guralnick "Occupational and Social Class Differences in Mortality." *Milbank Memorial Fund Quarterly* 34 (1956):61–73.

Mosher, William D. *Contraceptive Utilization, United States, 1976.* U.S. National Center for Health Statistics, *Vital and Health Statistics,* series 23, no. 7 (March 1981).

―――― , **and Charles F. Westoff** *Trends in Contraceptive Practice, 1965–76.* U.S. National Center for Health Statistics, *Vital and Health Statistics,* series 23, no. 10 (February 1982).

Moss, A. J., and M. H. Wilder "Use of Selected Medical Procedures Associated with Preventive Care: United States, 1973." U.S. National Center for Health Statistics. *Vital and Health Statistics,* series 10, no. 110 (March 1977).

Moynihan, Daniel Patrick *The Negro Family: The Case for National Action.* Washington, DC: U.S. Department of Labor, 1965.

―――― "The Crisis in Welfare." *Public Interest* 10 (Winter 1968):3–29.

―――― *The Policies of a Guaranteed Income.* New York: Random House, 1973.

Murray, Charles *Losing Ground: American Social Policy: 1950–1980.* New York: Basic Books, 1984.

Muth, Richard F. *Cities and Housing.* Chicago: University of Chicago Press, 1969.

_____ "Residential Segregation and Discrimination." In G. M. von Furstenberg, B. Harrison, and A. H. Horowitz, eds. *Housing Patterns of Racial Discrimination*, vol. 1. Lexington, MA.: Lexington Books, 1974.

Myrdal, Gunnar *An American Dilemma: The Negro Problem and Modern Democracy*. New York: Harper, 1944.

Nam, Charles B., and Susan O. Gustavus *Population: The Dynamic Aspects of Demographic Change*. Boston: Houghton Mifflin, 1976.

Nam, Charles B., and Mary G. Powers "Changes in the Relative Status Level of Workers in the United States, 1950–1960." *Social Forces* 48 (1968):158–77.

National Academy of Sciences *Freedom of Choice in Housing*. Washington, DC: National Academy of Sciences, 1972.

_____ *Counting the People in 1980: An Appraisal of Census Plans*. Washington, DC: National Academy of Sciences, 1978.

National Bureau of Economic Research *An Appraisal of the 1950 Census Income Data*. Princeton, NJ: Princeton University Press, 1958.

Nelson, Nels A., and Gladys L. Crain *Syphilis, Gonorrhea and the Public Health*. New York: Macmillan, 1938.

Nobles, Wade "Toward an Empirical and Theoretical Framework for Defining Black Families." *Journal of Marriage and the Family* 40 (November 1979):679–90.

Oates, Stephen B. *Let the Trumpet Sound*. New York: New American Library, 1983.

O'Connell, Martin, and Carolyn C. Rogers "Out-of-Wedlock Births, Premarital Pregnancies and Their Effect on Family Formation and Dissolution." *Family Planning Perspectives* 16 (July-August 1984):157–62.

O'Connell, Martin, and Maurice J. Moore "The Legitimacy Status of First Births to U.S. Women Aged 15–24, 1939." In Frank F. Furstenberg, Jr., Richard Lincoln and Jane Menken, eds. *Teenage Sexuality, Pregnancy and Childbearing, 1978*, Philadelphia: University of Pennsylvania Press, 1981.

O'Dea, Thomas *American Catholic Dilemma*. New York: Sheed & Ward, 1959.

Ogbu, John U. *Minority Education and Caste: The American System in Cross-Cultural Perspective*. New York: Academic Press, 1978.

Oliver, Melvin L., and Mark A. Glick "An Analysis of the New Orthodoxy on Black Mobility." *Social Problems* 29 (1982):511–23.

Orfield, Gary *Public School Desegregation in the United States, 1968–1980*. Washington, DC: Joint Center for Political Studies, 1983.

Ornstein, Michael D. *Entry into the American Labor Force*. New York: Academic Press, 1976.

Orr, Margaret Terry, and Lynne Brenner "Medical Funding of Family Planning Clinic Services." *Family Planning Perspectives* 13 (November-December 1981):280–86.

Osofsky, Gilbert *Harlem: The Making of a Ghetto: Negro New York, 1890–1930*. New York: Harper & Row, 1963.

Painter, Neil Irvin *Exodusters*. New York: Norton, 1976.

Park, Robert E. "Racial Assimilation in Secondary Groups: With Particular Reference to the Negro." *American Journal of Sociology* 19 (March 1914):606–23.

Parsons, Carole W., ed. *America's Uncounted People*. Washington, DC: National Academy of Sciences, 1972.

Parsons, Donald O. "Racial Trends in Male Labor Force Participation." *American Economic Review* 70 (December 1980):911–20.

Passell, Jeffery S. "Undocumented Immigrants: How Many?" Paper presented at the annual meeting of the American Statistical Association, Las Vegas, August 6, 1985.

―――― "Factors Associated with Variation in Sex Ratios of the Population Across States and MSA's: Findings Based on Regression Analysis of 1980 Census Data." *Proceedings of the Social Statistics Section*, American Statistical Association. Washington, DC: American Statistical Association, 1984.

――――, **and J. Gregory Robinson** "Revised Estimates of the Coverage of the Population in the 1980 Census Based on Demographic Analysis: A Report on Work in Progress." *1984 Proceedings of the Social Statistics Section*, American Statistical Association, Table 3, 1984.

Passell, Jeffrey S.; Jacob S. Siegel; and J. Gregory Robinson "Coverage of the National Population in the 1980 Census, by Age, Sex, and Race: Preliminary Estimates by Demographic Analysis." U.S. Bureau of the Census, *Current Population Reports*, series P-23, no. 115 (February 1982).

Patterson, James T. *America's Struggle Against Poverty: 1900–1985*. Cambridge, MA: Harvard University Press, 1986.

Pearce, Diana "Gatekeepers and Homeseekers: Institutionalized Patterns in Racial Steering." *Social Problems* 26 (February 1979):325–42.

Pearl, Raymond "Preliminary Notes on a Cooperative Investigation of Family Limitation." *Milbank Memorial Fund Quarterly* 11 (January 1933):37–59.

―――― "Second Progress Report on a Study of Family Limitation." *Milbank Memorial Fund Quarterly* 13 (July 1934):258–84.

―――― "Fertility and Contraception in Urban Whites and Negroes." *Science* 83 (May 22, 1936):503–6.

―――― "Third Progress Report on a Study of Family Limitation." *Milbank Memorial Fund Quarterly* 14 (July 1936):363–407.

―――― "Fertility and Contraception in New York and Chicago." *Journal of the American Medical Association* 108 (April 24, 1937):1385–90.

―――― *The Natural History of Population*. Oxford: Oxford University Press, 1939.

Petersen, William *Malthus*. Cambridge, MA: Harvard University Press, 1979.

―――― *Japanese Americans*. New York: Random House, 1971.

Peterson, Robert *Only the Ball Was White*. Englewood Cliffs, NJ: McGraw-Hill, 1984.

Pettigrew, Thomas F. "Attitudes on Race and Housing: A Social-Psychological View." In Amos H. Hawley and Vincent P. Rock, eds. *Segregation in Residential Areas*. Washington, DC: National Academy of Sciences, 1973.

Phillips, Ulrich B. *American Negro Slavery*. Baton Rouge: Louisiana University Press, 1966 [1918].

―――― *The Slave Economy of the Old South: Selected Essays in Economic and Social History*. Edited and with an introduction by Eugene D. Genovese. Baton Rouge: Louisiana State University Press, 1968.

Pilpel, Harriet F. and Peter Ames "Legal Obstacles to Freedom of Choice in the Areas of Contraception, Abortion and Voluntary Sterilization in the United States." In Charles F. Westoff and Robert Parke, Jr., eds. *Aspects of Population Growth and Policy*. Commission on Population Growth and the American Future, vol. 6. Washington, DC: U.S. Government Printing Office, 1972.

Piven, Frances Fox, and Richard A. Cloward *Regulating the Poor: The Functions of Public Welfare*. New York: Vintage Books, 1971.

Pleck, Elizabeth Hafkin *Black Migration and Poverty: Boston, 1865–1900*. New York: Academic Press, 1979.

Plotnick, Robert D., and Felicity Skidmore *Progress Against Poverty: A Review of the 1965–1974 Decade.* New York: Academic Press, 1975.

Pohlmann, Vernon C., and Robert H. Walsh "Black Minority Status and Fertility in the United States, 1970." *Sociological Focus* 8 (April 1975):97–108.

Polochck, Solomon W. "Discontinuous Labor Force Participation and Its Effects on Women's Market Earnings." In Cynthia B. Lloyd, ed. *Sex, Discrimination, and the Division of Labor.* New York: Columbia University Press, 1975.

Potter, J. "The Growth of Population in America: 1700–1860." In D. V. Glass and D. E. C. Eversley, eds. *Population in History: Essays in Historical Demography.* Chicago: Aldine, 1965.

Powell-Griner, Eve "Induced Terminations of Pregnancy: Reporting States, 1982 and 1983." U.S. National Center for Health Statistics. *Monthly Vital Statistics Report*, vol. 35, no. 3, supplement (July 14, 1986).

Prager, Kate "Induced Terminations of Pregnancy: Reporting States, 1981." U.S. National Center for Health Statistics, *Monthly Vital Statistics Report*, vol. 34, no. 4, supplement(2) (July 30, 1985).

Pratt, William F., and Marjorie C. Horn "Wanted and Unwanted Childbearing: United States, 1973–82." U.S. National Center for Health Statistics, *Advancedata*, no. 108 (May 9, 1985).

Preston, Howard L. *Automobile Age Atlanta.* Athens: University of Georgia Press, 1979.

Preston, Samuel H., and Michael R. Haines "New Estimates of Child Mortality in the United States at the Turn of the Century." *Journal of the American Statistical Association* 79 (June 1984):272–81.

Price, Daniel O. *Changing Characteristics of the Negro Population.* Washington, DC: U.S. Government Printing Office, 1969.

Priebe, John "Occupational Classification in the 1980s." Paper presented at the annual meeting of the Southern Sociological Association, March 26–29, 1980.

Rabinowitz, Howard N. *Race Relations in the Urban South: 1865–1890.* New York: Oxford University Press, 1978.

Ragan, James F., Jr. "Minimum Wages and the Youth Labor Market." *Review of Economics and Statistics* (May 1977):129–36.

Rainwater, Lee "Crucible of Identity: The Negro Lower Class Family." *Daedalus* 95 (Winter 1966):172–216.

———, **and William L. Yancey, eds.** *The Moynihan Report and the Politics of Controversy.* Cambridge, MA: MIT Press, 1967.

Ramist, L., and S. Arbeiter *Profiles, College-Bound Seniors, 1983.* New York: College Entrance Examination Board, 1984.

Rees, Albert "The Labor Supply Results of the Experiment: A Summary." In Harold W. Watts and Albert Rees, eds. *Labor-Supply Responses.* The New Jersey Income-Maintenance Experiment, vol. 2. New York: Academic Press, 1977.

———, **and George P. Shultz** *Workers and Wages in an Urban Labor Market.* Chicago: University of Chicago Press, 1970.

Reich, Michael *Racial Inequality: A Political-Economic Analysis.* Princeton, NJ: Princeton University Press, 1981.

Reid, Ira De Augustine *The Negro Immigrant: Characteristics and Social Adjustment, 1899–1937.* New York: Columbia University Press, 1939.

Reid, John "Black America in the 1980's." *Population Bulletin* 37 (December 1982).

Reimers, Cordelia W. "A Comparative Analysis of the Wages of Hispanics,

Blacks, and Non-Hispanic Whites." In George J. Borjas and Marta Tienda, eds. *Hispanics in the U.S. Economy*. New York: Academic Press, 1985.

Rieder, Jonathan *Canarsie: The Jews and Italians of Brooklyn Against Liberalism*. Cambridge, MA: Harvard University Press, 1985.

Rindfuss, Ronald R. "Minority Status and Fertility Revisited Again: A Comment on Johnson." *American Journal of Sociology* 36 (September 1980):372–75.

———, **and James A. Sweet** *Postwar Fertility Trends and Differentials in the United States*. New York: Academic Press, 1977.

Ritchey, P. Neal "The Effects of Minority Group Status on Fertility: A Reexamination of Concepts." *Population Studies* 29 (July 1975):249–57.

Roberts, Jean, and Michael Rowland "Hypertension in Adults 25 to 74 Years of Age, United States, 1971–1975." U.S. National Center for Health Statistics, *Vital and Health Statistics*, series 11, no. 221 (April 1981).

Roberts, Robert E., and Eun Sul Lee "Minority Group Status and Fertility Revisited." *American Journal of Sociology* 80 (September 1974):502–23.

Robins, Philip K. "Labor Supply Response of Family Heads and Implications for a National Program." In Philip K. Robins et al., eds. *A Guaranteed Annual Income: Evidence from a Social Experiment*. New York: Academic Press, 1980.

———, **and Richard W. West** "Labor Supply Response of Family Heads Over Time." In Philip K. Robins et al., eds. *A Guaranteed Annual Income: Evidence from a Social Experiment*. New York: Academic Press, 1980.

Rodgers, Harrell R., Jr. "Fair Employment Laws for Minorities: An Evaluation of Federal Implementation." In Charles S. Bullock III and Charles M. Lamb, eds. *Implementation of Civil Rights Policy*. Monterey, CA: Brooks/Cole, 1984.

Rosenthal, Erich "The Equivalence of United States Census Data for Persons of Russian Stock or Descent with American Jews: An Evaluation." *Demography* 12 (May 1975):275–90.

Ross, Heather, and Isabel Sawhill *Time of Transition: The Growth of Families Headed by Women*. Washington, DC: The Urban Institute, 1975.

Rossiter, W. S. *A Century of Population Growth*. Washington, DC: U.S. Government Printing Office, 1909.

Rudwick, Elliott M. *Race Riot at East St. Louis, July 2, 1917*. Cleveland: World, 1966.

Russell, John H. *The Free Negro in Virginia*. Baltimore: Johns Hopkins University Press, 1913.

Ryder, Norman B., and Charles F. Westoff *Reproduction in the United States: 1965*. Princeton, NJ: Princeton University Press, 1971.

Rytina, Nancy F., and Suzanne M. Bianchi "Occupational Reclassification and Distribution by Gender." *Monthly Labor Review* 107 (March 1984):11–17.

Sandell, S. J., and D. Shapiro "An Exchange: Theory of Human Capital and the Earnings of Women: A Reexamination of the Evidence." *Journal of Human Resources* 13 (Winter 1978):103–17.

Sanderson, Warren "On Two Schools of the Economics of Fertility." *Population and Development Review* 2 (1976):469–77.

Savitt, Todd L. *Medicine and Slavery: The Disease and Health Care of Blacks in Antebellum Virginia*. Urbana: University of Illinois Press, 1978.

Scanzoni, John *The Black Family in Modern Society*. Boston: Allyn and Bacon, 1971.

Schnare, Ann B. *The Persistence of Racial Segregation in Housing*. Washington, DC: Urban Institute, 1978.

Schuman, Howard; Charlotte Steeh; and Lawrence Bobo *Racial Attitudes in*

America: Trends and Interpretations. Cambridge, MA: Harvard University Press, 1985.

Scott, Ann Herbert *Census, U.S.A., Fact Finding for the American People: 1790–1970*. New York: Seabury Press, 1968.

Scott, Emmett J. *Negro Migration During the War*. New York: Oxford University Press, 1920.

Shapiro, Sam; Edward R. Schlesinger; and Robert E. L. Nesbitt, Jr. *Infant, Perinatal, Maternal and Childhood Mortality in the United States*. Cambridge MA: Harvard University Press, 1968.

Shimkin, Demitri; E. Shimkin; and D. Frate *The Extended Family in Black Societies*. Paris: Mouton, 1978.

Shogan, Robert, and Tom Craig *The Detroit Race Riot: A Study in Violence*. Philadelphia: Chilton Books, 1964.

Shryock, Henry S., and Jacob S. Siegel *The Methods and Materials of Demography*. Washington, DC: U.S. Government Printing Office, 1971.

Siegel, Jacob S. "Completeness of Coverage of the Nonwhite Population in the 1960 Census and Current Estimates, and Some Implications." In David M. Heer, ed. *Social Statistics and the City*. Cambridge, MA: Joint Center for Urban Studies of the Massachusetts Institute of Technology and Harvard University, 1968.

_____ "Estimates of Coverage of the Population by Sex, Race, and Age in the 1970 Census." *Demography* 11 (February 1974):1–24.

_____, **and Charles D. Jones** "The Census Bureau Experience and Plans." In *Proceedings of the Conference on Census Undercount*. Washington, DC: U.S. Government Printing Office, 1980.

Seigel, Jacob S.; Jeffrey S. Passell; Norfleet W. Rives, Jr.; and J. Gregory Robinson "Developmental Estimates of the Coverage of the Population of States in the 1970 Census: Demographic Analysis." Washington, DC: U.S. Bureau of the Census, *Current Population Reports*, series P-23, no. 65 (December 1977).

Seigel, Jacob S.; Jeffery S. Passel; and J. Gregory Robinson "Preliminary Review of Existing Studies of the Number of Illegal Residents in the United States." *U.S. Immigration Policy and the National Interest*. Staff Report, Select Commission on Immigration and Refugee Policy, Appendix E: Papers on Illegal Immigration to the U.S., 1980.

Siegel, Paul "On the Cost of Being a Negro." *Sociological Inquiry* 35 (1965):41–57.

Singer, Judith D.; Patricia Granahan; Nancy N. Goodrich; Linda D. Meyers; and Clifford L. Johnson "Diet and Iron Status, A Study of Relationships." U.S. National Center for Health Statistics, *Vital and Health Statistics*, series 11, no. 229 (December 1982).

Sitkoff, Harvard *A New Deal for Blacks: The Emergence of Civil Rights as a National Issue: The Depression Decade*, vol. 1. New York: Oxford University Press, 1978.

Sklar, June, and Beth Berkov "Teenage Family Formation in America." In Frank F. Furstenberg, Jr., Richard Lincoln, and Jane Menken, eds. *Teenage Sexuality, Pregnancy and Childbearing*. Philadelphia: University of Pennsylvania Press, 1981.

Sklare, Marshall *American Jews*. New York: Random House, 1971.

Slater, Courtenay M. "The Impact of Census Undercoverage on Federal Programs." In *Proceedings of the 1980 Conference on Census Undercount*. Washington, DC: U.S. Government Printing Office, 1980.

Sly, David F. "Minority-Group Status and Fertility: An Extension of

Goldscheider and Uhlenberg." *American Journal of Sociology* 76 (November 1970):443–59.

Smeeding, Timothy M. *Alternative Methods for Valuing Selected In-Kind Benefits and Measuring Their Effect on Poverty.* Technical Paper no. 50. Washington, DC: U.S. Bureau of the Census, 1982.

_____ "Is the Safety Net Still Intact?" In D. Lee Bawden, ed. *The Social Contract Revisited.* Washington, DC: Urban Institute Press, 1984.

Smith, Arthur B., Jr.; Charles B. Craver; and Leroy D. Clark *Employment Discrimination Law.* Charlottesville, VA: Michie Bobbs-Merrill, 1982.

Smith, James P., and Finis R. Welch "Black/White Male Earnings and Employment: 1960–70." In F. Thomas Juster, ed. *The Distribution of Economic Well-Being.* Cambridge, MA: Ballinger, 1977.

_____ *Closing the Gap: Forty Years of Economic Progress for Blacks.* Santa Monica, CA.: Rand, 1986.

Smith, O. P. "A Reconsideration of Easterlin Cycles." *Population Studies* 35 (July 1981):247–64.

Smith, T. Lynn *Fundamentals of Population Study.* New York: Lippincott, 1960.

Sørensen, Annemette; Karl E. Taeuber; and Leslie J. Hollingsworth, Jr. "Indexes of Racial Residential Segregation for 109 Cities in the United States, 1940 to 1970." *Sociological Focus* 8 (April 1975):125–42.

Sowell, Thomas *Race and Economics.* New York: Longman, 1975.

_____, ed. *Essays and Data on American Ethnic Groups.* Washington, DC: Urban Institute, 1978.

_____ *Ethnic America: A History.* New York: Basic Books, 1981a.

_____ *Markets and Minorities.* New York: Basic Books, 1981b.

_____ *The Economics and Politics of Race.* New York: Morrow, 1983.

_____ *Civil Rights: Rhetoric or Reality?* New York: Morrow, 1984.

Spain, Daphne, and Larry H. Long *Black Movers to the Suburbs: Are They Moving to Predominantly White Neighborhoods?* Washington, DC: U.S. Bureau of the Census, 1981.

Spanier, Graham B., and Paul C. Glick "Mate Selection Differentials Between Whites and Blacks in the United States." *Social Forces* 58 (March 1980):707–25.

Spear, Allan H. *Black Chicago: The Making of a Negro Ghetto: 1890–1920.* Chicago: University of Chicago Press, 1967.

Spero, Sterling D., and Abram L. Harris *The Black Worker: The Negro and the Labor Movement.* New York: Columbia University Press, 1931.

Stack, Carol B. *All Our Kin.* New York: Harper, 1974.

Stampp, Kenneth M. *The Peculiar Institution.* New York: Vintage Books, 1956.

Stanton, William *The Leopard's Spots: Scientific Attitudes Toward Race in America, 1815–59.* Chicago: University of Chicago Press, 1966 [1960].

Staples, Robert "Towards a Sociology of the Black Family: A Decade of Theory and Research." *Journal of Marriage and the Family* (February 1971):19–38.

_____, and A. Mirande "Racial and Cultural Variations Among American Families: A Decennial Review of the Literature on Minority Families." *Journal of Marriage and the Family* 42 (November 1980):887–903.

Steinberg, Stephen *The Ethnic Myth.* Boston: Beacon Press, 1981.

Stevens, Gillian, and Joo Hyun Cho "Socioeconomic Indexes and the New 1980 Census Occupational Classification Scheme." *Social Science Research* 14 (June 1985):142–68.

462

_____, and David L. Featherman "A Revised Socioeconomic Index of Occupational Status." *Social Science Research* 10 (1981):364–95.

Stiglitz, Joseph E. "Theories of Discrimination and Economic Policy." In George M. von Furstenberg, Ann R. Horowitz, and Bennett Harrison, eds. *Employment and Income Patterns of Racial Discrimination*, vol. 2. Lexington, MA: Heath, 1974.

Stinner, William; Klaus de Albuquerque; and Roy S. Bryce Laporte *Return Migration and Remittances: Developing a Caribbean Perspective.* Washington, DC: Smithsonian Institution, 1982.

Stix, Regine K. "Contraceptive Service in Three Areas, Part I." *Milbank Memorial Fund Quarterly* 19 (April 1941):171–88.

St. John, Craig "Race Differences in Age at First Birth and the Pace of Subsequent Fertility: Implications for the Minority Group Status Hypothesis." *Demography* 19 (August 1982):301–14.

Stolzenberg, Ross M. "Education, Occupation and Wage Differences Between White and Black Men." *American Journal of Sociology* 81 (September 1975):299–323.

Sudarkasa, Niara "Interpreting the African Heritage in Afro-American Family Organization." In Harriette P. McAdoo, ed. *Black Families.* Beverly Hills, CA: Sage, 1981.

Sutch, Richard "The Breeding of Slaves for Sale and the Westward Expansion of Slavery, 1850–1860." In Stanley L. Engerman and Eugene D. Genovese, eds. *Race and Slavery in the Western Hemisphere, Quantitative Studies.* Princeton, NJ: Princeton University Press, 1975.

Swafford, Michael "Three Parametric Techniques for Contingency Table Analysis: A Nontechnical Commentary." *American Sociological Review* 45 (August 1980):644–90.

Sweet, James A. *Women in the Labor Force.* New York: Seminar Press, 1973.

Taeuber, Irene B., and Conrad Taeuber *People of the United States in the Twentieth Century.* Washington, DC: U.S. Government Printing Office, 1971.

Taeuber, Karl E. Residential Segregation." *Scientific American* 213 (August 1965):12–19.

_____ "The Effect of Income Redistribution on Racial Residential Segregation." *Urban Affairs Quarterly* 4 (September 1968):5–14.

_____ *Racial Residential Segregation, 28 Cities, 1970–1980.* Working Paper no. 83-12. Madison: Center for Demography and Ecology, University of Wisconsin, 1983.

_____, and Alma F. Taeuber *Negroes in Cities.* Chicago: Aldine, 1965.

_____ "The Negro Population in the United States." In John P. David, ed. *The American Negro Reference Book.* Englewood Cliffs, NJ: Prentice-Hall, 1966.

Taffel, Selma *Factors Associated with Low Birth Weight.* U.S. National Center for Health Statistics, *National Vital Statistics Systems,* series 21, no. 37 (1980).

Taub, Richard P.; D. Garth Taylor; and Jan D. Dunham *Paths of Neighborhood Change: Race and Crime in Urban America.* Chicago: University of Chicago Press, 1984.

Taylor, D. Garth; Paul B. Sheatsley; and Andrew M. Greeley "Attitudes Toward Racial Integration." *Scientific American* 238 (June 1978):42–49.

Taylor, Howard F. *The I.Q. Game: A Methodological Inquiry Into the Heredity-Environment Controversy.* New Brunswick, NJ: Rutgers University Press, 1980.

Thomas, Gail E. *The Access and Success of Blacks and Hispanics in U.S. Graduate and Professional Education.* Working Paper, Office of Scientific and Engineering Personnel, National Research Council. Washington, DC: National Academy Press, 1986.

Thomas, W. I., and F. Znaniecki *The Polish Peasant in Europe and America,* vols. 1–5. Chicago: University of Chicago Press, 1918.

Thomlinson, Ralph *Population Dynamics.* New York: Random House, 1965.

Thompson, Warren S., and P. K. Whelpton *Population Trends in the United States.* New York: McGraw-Hill, 1933.

Thornton, Arland "Decomposing the Re-Marriage Process." *Population Studies* 31 (July 1977):572–95.

––––– "Marital Instability Differentials and Interaction Insights from Multivariate Contingency Table Analysis." *Sociology and Social Research* 62 (July 1978):572–95.

–––––, **and Willard L. Rodgers** *Changing Patterns of Marriage and Divorce in the United States.* Final report prepared for the National Institute for Child Health and Human Development. Ann Arbor: Institute for Social Research, University of Michigan, 1983.

Thurow, Lester C. *Poverty and Discrimination.* Washington, DC: Brookings Institution, 1969.

Tienda, Marta, and Ronald Angel "Headship and Household Composition Among Blacks, Hispanics and Other Whites." *Social Forces* 61 (December 1982):508–31.

Tobin, James "On Improving the Economic Status of the Negro." *Daedalus* 94 (Fall 1965):878–98.

Toll, William *The Resurgence of Race.* Philadelphia: Temple University Press, 1979.

Tolnay, Stewart E. "Trends in Total and Marital Fertility for Black Americans, 1886–1899." *Demography* 18 (November 1981):443–63.

Triplett, Jack E. "Reconciling the CPI and the PCE Deflator." *Monthly Labor Review* 104 (September 1981):3–15.

Trotter, Joe William, Jr. *Black Milwaukee: The Making of an Industrial Proletariat, 1915–45.* Urbana, IL: University of Illinois Press, 1985.

Turner, Ralph H., ed. *Robert E. Park: On Social Control and Collective Behavior.* Chicago: University of Chicago Press, 1967.

Tygiel, Jules *Baseball's Great Experiment: Jackie Robinson and His Legacy.* New York: Oxford University Press, 1983.

Tyttle, William M., Jr. *Race Riot: Chicago in the Red Summer of 1919.* New York: Atheneum, 1972.

United Nations *Statistical Yearbook: 1979–80.* New York: United Nations, 1981.

––––– *Demographic Yearbook: 1980.* New York: United Nations, 1982.

––––– *Statistical Yearbook: 1981.* New York: United Nations, 1983.

U.S. Department of Health, Education, and Welfare *Equality of Educational Opportunity.* Washington, DC: U.S. Government Printing Office, 1966.

U.S. Department of Labor *The Negro Family: The Case for National Action* (known as the Moynihan Report). Washington, DC: Department of Labor, Office of Policy Planning and Research, 1965.

U.S. National Advisory Commission on Civil Disorders *Report of the National Advisory Commission on Civil Disorders.* New York: Bantam Books, 1968.

Van Valey, Thomas L.; Wade Clark Roof; and Jerome E. Wilcox "Trends in Residential Segregation: 1960–1970." *American Journal of Sociology* 82 (January 1977):826–44.

Vance, Ruppert B. "The Old Cotton Belt." In Carter Goodrich et al., eds. *Migration and Economic Opportunity*. Philadelphia: University of Pennsylvania Press, 1936.

────── "The Development and Status of American Demography." In Philip M. Hauser and Otis Dudley Duncan, eds. *The Study of Population*. Chicago: University of Chicago Press, 1959.

Vaughan, Barbara; James Trussell; Jane Menken; Elise F. Jones; and William Grady *Contraceptive Efficacy Among Married Women Aged 15–44 Years*. U.S. National Center for Health Statistics, *Vital and Health Statistics*, series 23, no. 5 (May 1980).

Vavra, Helen M., and Linda J. Querec *A Study of Infant Mortality from Linked Records by Age of Mother, Total Birth Order and Other Variables, United States, 1960 Live Birth Cohort*. U.S. National Center for Health Statistics, *Vital and Health Statistics*, series 20, no. 14 (September 1973).

Vickery, William *The Economics of the Negro Migration, 1900–1920*. New York: Arno Press, 1977.

Vincent, Clark E. *Unmarried Mothers*. New York: The Free Press, 1961.

Voegeli, A. Jacque *Free But Not Equal*. Chicago: University of Chicago Press, 1967.

Vonderlehr, R. A., and Lida J. Usilton "Syphilis Among Men of Draft Age in the United States." *Journal of the American Medical Association* 120 (December 26, 1942):1369–72.

Vose, Clement E. *Caucasians Only: The Supreme Court, the NAACP, and the Restrictive Covenant Cases*. Berkeley: University of California Press, 1959.

Wade, Richard C. *The Urban Frontier*. Chicago: University of Chicago Press, 1959.

────── *Slavery in the Cities*. New York: Oxford University Press, 1964.

Walker, Francis A. "Statistics of the Colored Race in the United States." *Publication of the American Statistical Association*, vol. 2 (September-December 1890).

Wallace, Phyllis A., ed. *Equal Employment Opportunity and the A.T.&T. Case*. Cambridge, MA: MIT Press, 1976.

Waskow, Arthur I. *From Race Riot to Sit-In*. Garden City, NY: Doubleday, 1966.

Weiss, Nancy J. *Farewell to the Party of Lincoln*. Princeton, NJ: Princeton University Press, 1983.

Weiss, Richard S., and Herbert L. Joseph *Syphilis*. New York: Nelson, 1951.

Welch, Finis "Black-White Differences in Returns to Schooling." *American Economic Review* 63 (December 1973):893–907.

Wells, Robert V. *The Population of the British Colonies in America Before 1776: A Survey of Census Data*. Princeton, NJ: Princeton University Press, 1975.

Westcott, Diane N. "Youth in the Labor Force: An Area Study." *Monthly Labor Review* 99 (July 1976):3–9.

Westoff, Charles F. "The Decline of Unplanned Births in the United States." *Science* 191 (January 1976):38–40.

────── "Some Speculations on the Future of Marriage and the Family." *Family Planning Perspectives* 10 (March-April 1978):79–83.

465

_____ "Marriage and Fertility in the Developed Countries." *Scientific American* 239 (December 1978):51–57.

_____ "The Decline in Unwanted Fertility, 1971–76." *Family Planning Perspectives* 13 (March-April 1981):70–72.

_____, and E. F. Jones "Contraception and Sterilization in the United States, 1965–1975." *Family Planning Perspectives* 9 (July-August 1977):153–57.

_____ "Teenage Fertility in Developed Nations: 1971–1980." *Family Planning Perspectives* 15 (May-June 1983):105–10.

Westoff, Charles F.; Gerard Calot; and Andrew D. Foster "Teenage Fertility in Developed Nations: 1971–1980." *Family Planning Perspectives* (May/June 1983): 106–110.

Westoff, Charles F., and Norman B. Ryder "Contraceptive Practice Among Urban Blacks in the United States, 1965." *Milbank Memorial Fund Quarterly* 48 (April 1970):215–33.

_____ *The Contraceptive Revolution.* Princeton, NJ: Princeton University Press, 1977.

Whelpton, Pascal K.; Arthur A. Campbell; and John E. Patterson *Fertility and Family Planning in the United States.* Princeton, NJ: Princeton University Press, 1966.

Wienk, Ronald E.; Clifford E. Reid; John C. Simonson; and Frederick J. Eggers *Measuring Racial Discrimination in American Housing Markets: The Housing Market Practices Survey.* Washington, DC: Department of Housing and Urban Development, Office of Policy Development and Research, 1979.

Wilcox, R. R. *Textbook of Venereal Disease and Treponenatoses.* Springfield, IL: Thomas, 1964.

Williams, Walter E. *The State Against Blacks.* New York: McGraw-Hill, 1982.

Williamson, Joel *The Crucible of Race.* New York: Oxford University Press, 1984.

Willie, Charles *A New Look at Black Families.* Bayside, NY: General Hall, 1976.

Wilson, William Julius *The Declining Significance of Race: Blacks and Changing American Institutions.* Chicago: University of Chicago Press, 1978.

_____ **"Cycles of Deprivation and the Underclass Debate."** *Social Service Review* 59 (December 1985):541–559.

_____, **and Kathryn M. Neckerman** "Poverty and Family Structure: The Widening Gap between Evidence and Public Policy Issues." In Stanley H. Danziger and Daniel H. Weinberg, eds. *Fighting Poverty: What Works and What Doesn't.* Cambridge, MA: Harvard University Press, 1986.

Wing, Steve; Kenneth G. Manton; Eric Stallard; Curtis G. Hames; and H.A. Tryoler "The Black/White Mortality Crossover: Investigation in a Community-Based Study." *Journal of Gerontology* 40 (1985):78–84.

Wise, Michael B. "Desegregation in Education: A Directory of Reported Federal Decisions." Mimeographed. Notre Dame, IN: Center for Civil Rights, University of Notre Dame Law School, 1977.

Woodson, Carter Goodwin *A Century of Negro Migration.* New York: Russell & Russell, 1918.

_____, *The Rural Negro.* New York: Russell & Russell, 1930.

Woodward, C. Vann *The Strange Career of Jim Crow.* New York: Oxford University Press, 1957.

Woofter, T. J. "What is the Negro Rate of Increase?" *Journal of the American Statistical Association* 26 (December 1931):461–62.

Wright, Carroll D. *The History and Growth of the United States Census.* Washington, DC: U.S. Government Printing Office, 1900.

Wright, Erik Olin *Class Structure and Income Determination.* New York: Academic Press, 1979.

Wright, Gavin *Old South, New South Revolutions in the Southern Economy Since the Civil War.* New York: Basic Books, 1986.

Wright, Paul, and Peter Pirie *A False Fertility Transition: The Case of American Blacks.* Paper no. 90. Honolulu: East-West Population Institute, 1984.

Wright, Richard *Twelve Million Black Voices: A Folk History of the Negro in the United States.* New York: Viking Press, 1941.

Yinger, John "Measuring Racial and Ethnic Discrimination with Fair Housing Audits: A Review of Existing Evidence and Research Methodology." Paper presented at a Department of Housing and Urban Development Conference on Fair Housing Testing, Washington, DC, December 6–7, 1984.

Zabin, Laurie Schwab, and Samuel D. Clark, Jr. "Why They Delay: A Study of Teenage Family Planning Clinic Patients." *Family Planning Perspectives* 13 (September-October 1981):205–17.

Zangrando, Robert L. *The NAACP Crusade Against Lynching, 1909–1950.* Philadelphia: Temple University Press, 1980.

Zelnik, Melvin "Fertility of the American Negro in 1830 and 1950." *Population Studies* 20 (March 1966):77–83.

_____ "Age Patterns of Mortality of American Negroes: 1900–02 to 1959–61." *Journal of the American Statistical Association* 64 (June 1969):433–51.

_____, and John F. Kantner "Reasons for Nonuse of Contraception by Sexually Active Women Aged 15–19." *Family Planning Perspectives* 11 (September-October 1979:189–97.

_____ "Sexual Activity, Contraceptive Use and Pregnancy Among Metropolitan Area Teenagers: 1971–79." *Family Planning Perspectives* 12 (September-October 1980):230–37.

_____, and Kathleen Ford *Sex and Pregnancy in Adolescence.* Beverly Hills, CA: Sage, 1981.

Zoloth, Barbara S. "Alternative Measures of School Segregation." *Land Economics* 52 (1976):278–98.

Zunz, Olivier *The Changing Face of Inequality.* Chicago: University of Chicago Press, 1982.

Name Index

Boldface numbers refer to figures and tables.

A

Abbott, Robert, 115
Abowd, John M., **324–325**
Abraham, Sidney, 56n
Albuquerque, Klaus de, 364n
Allen, Walter R., 161n, 168n, 171n
Alwin, Duane F., 346n
Ames, Peter, 59n
Anderson, Elijah, 249, 358n
Anderson, James, 189n, 190n, 201, 208
Anderson, Jervis, 375n, 396n, 404
Anderson, Kristin, 168n
Anderson, Martin, 241n, 243n, 284n
Angel, Ronald, 168n, 171
Arbeiter, S., 205n
Armstrong, Roger J., 50n, 51n
Arrow, Kenneth, 321n
Athearn, Robert G., 110n

B

Bachrach, Christine A., 85n, 87n
Bancroft, Gertrude, 205n, 228n, 233n, 234n
Bane, Mary Jo, 83n
Banfield, Edward C., 132n, 244
Batchelder, Alan, 295n
Beale, Calvin C., 114n
Bean, Frank D., 74n
Beebe, Gilbert W., 20n, 84n
Beller, Andrea H., 270n
Bennett, Lerone, Jr., 16n
Berkov, Beth, 77n
Bernstein, Blanche, 83n
Betsey, Charles L., 77n
Bianchi, Suzanne M., 151n, 156n, 172n, 233n, 235n, 262n, 263n, 270n, 340n, 342n

Billingsley, Andrew, 161, 228n
Blalock, Hubert M., Jr., 3n
Blassingame, John W., 137n, 257n
Blau, Francine D., 270n
Blau, Peter M., 94n, 132
Blinder, Derrick A., Jr., **322–323**
Bobo, Lawrence, 151n, 231n, 241n, 260n, 300n, 357n, 358n
Bonacich, Edna, 111n, 258
Borchert, James, 31n
Bowen, William G., 204, 205n, 228n, 229n, 233n, 234n
Bridenbaugh, Carl, 8n, 9n
Brookover, Wilbur B., 190n
Bryce, Herrington J., 421n
Burgess, Ernest W., 364n
Burnham, Drusilla, 76n
Burstein, Paul, 261n, 280n, 281n
Burt, Martha P., 83n
Burton, Nancy W., 281n
Butz, William P., 71

C

Cafferty, P. S. J., 248n
Cagan, Phillip, 294n
Cain, Glen, 233n
Calhoun, John C., 420–421
Callahan, David W., 294n
Calot, Gerard, 68n
Campbell, Arthur A., 60n, 63n, 84n, 92n
Campbell, Rex R., 109n, 112n
Carey, Henry, 9
Carlson, Edwood D., 51n
Carlucci, Carl P., 421n
Carmichael, Stokely, 396
Carpenter, Niles, 372n, 378n

469

Cates, W. Jr., 76n
Caudill, Henry, 356
Cebula, Richard J., 123n
Chase, Helen C., 50n, 51n
Chaudacuff, Howard P., 148n
Cherlin, Andrew J., 72n
Chicago Commission on Race Relations, 115n, 257n, 258n, 419n
Chiswick, Barry R., 401n
Cho, Joo Hyun, 262n, 274n, 396n
Chow, Winston, 83n
Christian, Virgil L., 321n
Citro, Constance F., 428n, 429n
Clague, Alice J., 77n
Clark, Leroy D., 280n
Clark, Samuel D. Jr., 75n
Clark, Taliaferro, 22n
Cloward, Richard A., 245n, 301n
Coale, Ansley J., 16n, 18n, 40n, 424n
Cogan, John F., 217n, 241n, 245–246
Cohen, Michael L., 428n, 429n
Cohen, Patricia Cline, 217n, 241n, 245–246
Colasanto, Diane, 151n
Conk, Margo Anderson, 262n
Conot, Robert, 138n
Conrad, Alfred H., 13n
Corcoran, Mary, 93n, **324–325,** 334n, 342n
Courant, Paul N., 155n
Cox, Oliver C., 191n, 424n
Craig, Tom, 116n, 138n, 258n
Crain, Gladys L., 21n
Craver, Charles B., 280n
Crèvecoeur, J. Hector, 363n
Crimmins, Eileen M., 35n, 37n
Curry, Leonard P., 15n, 107n, 108n, 111n, 209, 210, 257n
Curtin, Philip D., 9n
Cutright, Philip, 21n, 63n, 83n

D

Danchik, Kathleen M., 44n
Daniel, Peter, 113n
Darrow, Clarence, 138
Datcher, Linda, **324–325,** 340n
Dauer, Carl C., 37n
Davie, Maurice R., 23n
DeAre, Diane, 136n, 142n, 246n
Demeny, Paul, 24n
Devaney, Barbara, 71n
Dickerson, Dennis C., 257n

Dimond, Paul R., 153n
Drury, Thomas F., 57n
Dryfoos, G., 59n
DuBois, W. E. B., 2–3, 28, 107n, 111n, 137n, 138n, 150, 157, 160, 162n, 209, 210, 257, 258, 408–409, 413, 414
Duncan, Beverly, 228n, 346n
Duncan, Greg J., **324–325,** 334n
Duncan, Otis Dudley, 94n, 132, 228n, 262, 266, **322–323,** 338n, 346n
Dunham, Jan D., 152n

E

Easterlin, Richard A., 70
Eblan, Jack E., 14n, 24
Eckler, A. Ross, 284n
Edwards, G. Franklin, 366n
Edwards, Lydia B., 21n
Eisenhower, Dwight D., 59
Elder, Glen H., Jr., 72, 162n
Eldridge, Hope T., 109n
Ellwood, David T., 83n
Engerman, Stanley L., 9n, 11n, 15, 16n, 18n, 21, 22n, 82n, 106, 164
Espenshade, Thomas J., 248n, 387n
Evans, M. D. R., 99n

F

Farley, Reynolds, 14n, 15n, 24n, 92n, 140n, 151n, 152n, 154n, 156n, 168n, 172n, 205n, 236n, **324–325,** 326n, 331n
Featherman, David L., 132n, 133n, 228n, 262n, **324–325,** 364n
Feldman, Jacob J., 32n, 44n, 50n
Fields, Barbara Jeanne, 107n
Finegan, T. Aldrich, 204, 205n, 228n, 229n, 233n, 234n
Fingerhut, Lois A., 32n, 37n, 40n, 44n
Flieger, Wilhelm, 14n, 24n
Fligstein, Neil, 114, 245n
Fogel, Robert William, 9n, 11n, 15, 16n, 21, 22n, 106, 114
Foner, Philip S., 116n, 257n
Ford, Kathleen, 75n
Forrest, Jacqueline Derroch, 59n
Foster, Andrew, 68n
Franklin, 7n, 16n, 106n, 107n, 115n
Franklin, John Hope, 111n
Frate, D., 179n

Frazier, E. Franklin, 3n, 107n, 161, 162n, 163, 186–187, 364n, 366
Freeman, Richard B., 217n, 244n, 277, 280, 304n, 321n, 345n, 348, 350n
Fugitt, Glenn V., 136n
Fulwood, Robinson, 56n
Furstenberg, Frank J., Jr., 77n, 163–164

G

Garfinkel, Irwin, **322–323**
Garvey, Marcus, 396
Genovese, Eugene D., 15n
Gilder, George, 241–242, 244n
Gingrich, Paul, 24n,
Glass, Evelyn J., 32n
Glazer, Nathan, 364, 365n, 367, 377n, 404
Glick, Paul C., 162n, 424n
Godley, Frank, 49n, 57n
Gold, Rachel Benson, 59n
Goldenberg, Robert L., 51n
Gordon, David M., 321n
Grabill, Wilson H., 63n, 93n
Greeley, Andrew M., 140n
Green, Constance McLaughlin, 136, 137n, 258n
Green, Lorenzo Johnston, 8n
Griliches, Zvi, 346n
Grossman, Michael, 49n
Grove, Robert D., 22n, 23n, 35n, 40n, 58n, 422n
Guralnick, Lillian, 33n,
Gurock, Jeffery S., 137n
Gustavus, Susan O., 59n
Gutman, Herbert G., 16n, 161, 163, 228n, 387–388, 396n, 397, 400, 404
Guttentag, Marcia, 424n
Gwartney, James, **322–323**

H

Haber, Sheldon E., 334n, 342n
Haines, Michael R., 24n, 54n
Halacy, Dan, 284n
Hamill, Peter V., 56n
Handlin, Oscar, 365
Hannerz, Ulf, 82n, 249n
Hanushek, Eric, 320n
Harding, Vincent, 8n
Hare, Bruce, 204n

Harrington, Michael, 356
Harris, Abram L., 111n, 257n
Harris, William H., 111n, 112n, 257n
Hartley, Shirley Foster, 76n
Hauser, Philip M., 33n, 55n, 162–163
Hauser, Robert M., 133n, 228n, 263n, **324–325**
Haveman, Robert H., **322–323**
Hawley, Amos H., 153n
Haworth, C., **322–323**
Haworth, J., **322–323**
Haynes, George Edmund, 136
Hazen, H. H., 22n
Heaton, Tim B., 136n
Heer, David M., 428n
Heiss, Jerold, 161
Helper, Rose, 139n
Heltman, Lynne R., 123n, 133n
Hendricks, Wallace E., 270n
Hendrix, Lewellyn, 18n, 82n
Henshaw, Stanley K., 59n
Hershberg, T., 163n
Hetzel, Alice M., 35n, 40n, 58n
Hiestand, Dale L., 247n
Higgs, Robert, 28n, 29
Hill, Robert B., 161, 204n, 212n, 236n, 418, 424n
Hirsch, Arnold R., 153
Hirschman, Charles, **324–325**
Hofferth, Sandra L., 93n
Hoffman, Frederick L., 55n
Hoffman, Saul D., **324–325**
Hogan, Dennis P., 75n, 83n, 132n, 133n
Hollingsworth, Leslie J., 140n
Horn, Marjorie, 88n
Hough, George C. Jr., 40n, 424n
Hout, Michael, 281n
Hurn, Christopher J., 190n
Hyman, Herbert, 161

J

Jackman, Mary, 156n
Jackman, Robert, 156n
Jackson, Jacquelyne J., 424n
Jacobowitz, Steven, 49n
Jacobson, Paul H., 14n
Janowitz, Morris, 6n
Jarvis, Edward, 420
Jaynes, Gerald David, 111n, 112n
Jencks, Christopher, 294n
Jibou, Robert M., 156n, 246n

Johnson, Charles S., 28, 29, 82n, 138n, 419
Johnson, Clifford L., 56n
Johnson, Daniel M., 109n, 112n
Johnson, James Weldon, 367, 404
Johnson, Nan E., 74n
Johnston, Francis E., 56n
Jones, E. F., 84n
Jones, James H., 21n, 28n, 55n
Jones, Lyle V., 281n
Jordan, Winthrop D., 8n
Joseph, Herbert L., 21n

K

Kain, John F., 155, 156n, 246n
Kampmeier, Rudolph H., 21n
Kantner, John F., 75n, 79
Katzman, David M., 108n, 109n
Keyfitz, Nathan, 14n, 24n
Killingsworth, Charles C., 210n, **324–325**
King, A. Thomas, 155
Kiple, K., 21
Kirby, John B., 114n
Kiser, Clyde Vernon, 63n, 93n, 112n
Kitagawa, Evelyn M., 33n, 55n, 75n, 83n
Klebba, A. Joan, 32n
Kleinman, Joel C., 32n, 44n, 51n
Kluegel, James R., 357n
Korns, Robert F., 37n
Kovar, Mary Grace, 50n
Kramer, D. G., 76n
Kusmer, Kenneth L., 138n, 257n

L

Lake, Robert W., 143n
Lamas, Enrique J., 334n, 342n
Lamb, Charles M., 139n, 152n
Lammamier, Paul J., 164
Land, Kenneth C., 40n, 424n
Lane, Roger, 22, 111n, 136n, 209n, 258n
Lantz, Herman, 18n, 82n
Laporte, Roy S. Bryce, 364n
Lasch, Christopher, 72
Lee, Anne S., 74n
Lee, Eun Sul, 74n
Lee, Everett S., 74n, 119n, 391n
Lee, Ronald, 71n
Leibowitz, A., 83n
Lcmcshaw, Stanley, 56n
Levine, D., 204n

Levitan, Sar A., 59n, 290n
Lewis, Oscar, 356
Lewis, W. Arthur, 396
Lichter, Daniel T., 136n
Lieberson, Stanley, 4n, 122n, 139, 140n, 148n, 247n, 257, 281n
Liebow, Elliot, 82n, 228n, 249, 358n
Lillard, Lee, 285n
Linder, Forrest E., 22n, 23n, 422n
Litwack, Leon F., 82n, 107n, 108n, 228n
Logan, John R., 142n, 246n
Long, Larry, 123n, 132n, 133, 136n, 142n
Lowell, Anthony M., 21n

M

MacMahon, Brian, 50n
Malcolm X, 396
Mandle, Jay R., 114n, 245n
Mannix, Daniel P., 13n
Manton, Kenneth G., 40n
Mare, Robert D., 217n, 241n, 246
Marmor, Theodore R., 243n
Marshall, Harvey H., Jr., 156n, 246n
Marshall, Ray, 321n
Martin, Elmer, 179
Martin, Joanne, 179
Masnick, George S., 18n
Mason, William M., 346n
Masters, Stanley H., 156n, 246n, **322–323**
Mauer, Jeffery D., 32n
McAdoo, Harriette, 179
McClelland, Peter D., 14n, 15n, 104n
McFalls, Joseph A., 18n, 20, 63n
McFalls, Marguerite Harvey, 20n, 63n
McKay, Claude, 396
McKenzie, R. D., 364n
McMillen, Marilyn, 40n, 424n
McNeil, H. L., 22n
McNeil, John M., 334n, 342n
Mechanic, David, 32n
Meeker, Edward, 18n, 24
Meezan, William, 83n
Meier, August, 116n, 259n
Meyer, John R., 13n
Michelson, Stephen, **322–323**
Mieszkowski, P., 155
Miller, Herman P., 284n, 285n, 288n, 295, 303, 304n, 345
Miller, Kelly, 422n
Mincer, Jacob, 244n, **322–323,** 328n, 334n
Mirande, A., 160n

Modell, J., 163n
Molotch, Harvey Luskin, 152n
Mooney, J. D., 156n
Moore, Kristin A., 77n, 83n
Moore, Geoffrey H., 294n
Moore, Joseph H., 21n
Moore, Kristin A., 93n
Moore, Maurice J., 82n, 83n
Morgan, James, 94n
Moriyama, Iwao M., 33n
Mosher, William D., 85n, 87n
Moss, A. J., 57n
Moynihan, Daniel Patrick, 123n, 228, 243n,
 364, 365n, 367, 377n, 404
Murray, Charles, 83n, 242, 281n, 284n, 417
Muth, Richard F., 156
Myrdal, Gunnar, 30, 114, 116, 161, 210,
 257n, 301n, 357

N

Najjar, Matthew F., 56n
Nam, Charles B., 59n, 262n
National Academy of Sciences, 153n, 421n,
 424n, 428n
National Center for Health Statistics
 (NCHS), 56, 82, 87; *Health, United
 States, 1984*, 47n, 56n, 57n, 76n, 87n;
 Health, United States, 1985, 33n, 47n,
 59n, 87n; *Monthly Vital Statistics
 Report*, 14n, 15n, 41n, 49n, 50n, 52n,
 55n, 61n, 65n, 68n, 69n, 71n, 74n, 76n,
 165n; *Vital Statistics of the United
 States: 1973*, 76n; *Vital Statistics of the
 United States: 1979*, 44n, 47n, 52n, 55n,
 61n; *Vital Statistics of the United
 States: 1981*, 74n
Neckerman, Kathryn M., 83n, 424n
Nelson, Nels A., 21n
Nesbitt, Robert E. L., Jr., 40n, 49n
Nestor, Barry, 59n
Nishida, Ryoko, 74n
Nixon, Richard, 153, 243
Nobles, Wade, 171n

O

Oates, Stephen B., 116n
O'Connell, Martin, 77n, 82n, 83n
O'Dea, Thomas, 366n
Ogbu, John U., 191, 208

O'Reilly, Kevin, 68n, 87n
Orfield, Gary, 205n
Osofsky, Gilbert, 137n, 162, 375n, 378n,
 387, 391, 404

P

Painter, Neil Irvin, 110n
Palmer, Carroll E., 21n
Park, Robert E., 363, 419
Parsons, Carole W., 421n, 428n
Parsons, Donald O., 243n
Passell, Jeffrey S., 9n, 110n, 248n, 424n, 428
Patterson, James T., 290n
Patterson, John E., 60n, 63n, 84n, 92n
Pearl, Raymond, 20, 84n
Petersen, William, 366n
Peterson, Robert, 259n
Pettigrew, Thomas F., 151n, 153n
Phillips, Ulrich B., 16n, 22n, 106n
Pilpel, Harriet F., 59n
Pirie, Peter, 23, 99
Piven, Frances Fox, 245n, 301n
Pleck, Elizabeth Hafkin, 31n, 111n, 257,
 378n, 391n, 403–404
Pohlmann, Vernon C., 74n
Poitier, Sidney, 396
Polochek, Solomon W., **322–323,** 328n,
 334n
Poss, Susan S., 40n
Potter, J., 9n
Powell-Griner, Eve, 82n, 87n, 88n
Powers, Mary G., 267n
Prager, Kate, 76n
Pratt, William F., 88n
Preston, Howard L., 138n
Preston, Samuel H., 24n, 54n

Q

Querec, Linda J., 50n, 77n
Quigley, J. M., 155

R

Rabinowitz, Howard N., 21n, 137n
Ragan, James F., Jr., 244n
Rainwater, Lee, 82n, 161, 228n
Ramist, L., 205n
Reed, John, 161
Rees, Albert, 243n, 247n, 338n
Reich, Michael, 321n, 343n

Reid, Ira De Augustine, 365, 366, 372n, 375n, 378n, 382, 396n, 404
Reid, John, 190n
Reimers, Cordelia W., **324–325**
Rieder, Jonathan, 152n
Rindfuss, Ronald R., 74n
Rives, Norfleet W., Jr., 18n
Roberts, Jean, 57n
Roberts, Robert E., 74n
Robins, Philip K., 243n
Robinson, J. Gregory, 9n, 179n, 424n, 428
Rock, Vincent P., 153n
Rodgers, Harrell R., Jr., 241n, 261n, 280n
Rodgers, Willard L., 387n
Rogers, Carolyn C., 77n, 83n
Roof, Wade Clark, 140n
Roosevelt, Franklin D., 113, 116
Rosenberg, Harry M., 37n, 40n
Rosenthal, Erich, 146n
Ross, Heather, 167n
Rossiter, W. S., 7n, 8n, 9n, 11n, 13n
Rowland, Michael, 57n
Rudwick, Elliott M., 111n, 115n, 116n, 258n, 259n
Russell, John H., 107n
Ryder, Norman B., 59n, 60, 84n
Rytina, Nancy F., 262n, 263n, 270n

S

St. John, Craig, 74n
Sandell, S. J., 334n
Sanderson, Warren, 71n
Savitt, Todd L., 15n, 22n
Sawhill, Isabel, 167n
Scanzoni, John, 161
Schlesinger, Edward R., 40n, 49n
Schnare, Ann B., 157n
Schneider, Mark, 142n, 246n
Schuman, Howard, 151n, 231n, 241n, 260n, 300n, 357n, 358n
Schuman, Leonard M., 37n
Scott, Ann Herbert, 284n
Scott, Emmett J., 109n, 112n
Secord, Paul F., 424n
Shapiro, D., 334n
Shapiro, Sam, 40n, 49n
Sheatsley, Paul B., 140n, 151n
Shimkin, Demitri, 179
Shimkin, E., 179n
Shogan, Robert, 116n, 138n, 258n
Shorter, Edward, 21n, 63n

Shryock, Henry S., 119n
Shultz, George P., 247n, 338n
Siegel, Jacob S., 9n, 40n, 52n, 119n, 421n, 424n
Siegel, Paul, **322–323,** 338n
Simms, Margaret, 77n
Singer, Judith D., 56n
Singleton, Pappy, 109–110
Sitkoff, Harvard, 301n
Sklar, June, 77n
Sklare, Marshall, 366n
Slater, Courtenay M., 421n
Smeeding, Timothy M., 284n, 301n
Smith, Arthur B., 280n
Smith, Eliot R., 357n
Smith, James P., 281n, 285n, **322–323, 324–325,** 331n, 355n
Smith, O. P., 71n
Smith, T. Lynn, 74n
Sorensen, Annemette, 140n
Sowell, Thomas, 4n, 244n, 245, 320n, 365n, 366n, 368, 375, 377, 396n, 400, 404–405
Spain, Daphne, 142n, 156n, 340n, 342n
Spanier, Graham B., 424n
Spear, Allan H., 111n, 115n, 136, 137, 138n, 151, 257n
Spero, Sterling D., 111n, 257n
Stack, Carol B., 82n, 161, 177, 228n, 235n
Stallard, Eric, 40n
Stampp, Kenneth M., 15n, 16n, 163n
Stanton, William, 55n, 420n, 421n
Staples, Robert, 160n, 171n
Steeh, Charlotte, 151n, 231n, 241n, 260n, 300n, 357n, 358n
Steinberg, Stephen, 112, 367
Stevens, Gillian, 262n, 274n, 396n
Stiglitz, Joseph E., 321n
Stinner, William, 364n
Stix, Regine K., 84n
Stolzenberg, Ross M., **322–323**
Sudarkasa, Niara, 171n
Sutch, Richard, 106n
Sweet, James A., 94n, 234n
Sweet, Ossian, 138

T

Taeuber, Alma F., 104n
Taeuber, Conrad, 11n, 137n, 140n, 370n, 379n
Taeuber, Irene B., 11n, 137n, 140n, 370n, 379n

Taeuber, Karl E., 104n, 140n
Taffel, Selma, 51n, 77n
Taub, Richard P., 152n
Taylor, D. Garth, 140n, 151n, 152n
Taylor, Howard F., 204n
Terrell, Mary, 137
Thomas, Dorothy Swaine, 109n
Thomas, Gail E., 281n
Thomlinson, Ralph, 74n
Thompson, Warren S., 11n, 30
Thornton, Arland, 387n
Tienda, Marta, 168n, 171
Tietze, C., 76n
Tobin, James, 215n, 246n, 356n
Toll, William, 157n
Tolnay, Stewart E., 18n
Triplett, Jack E., 294n
Trotter, Joe William, 258n
Turner, Ralph H., 363n
Tygiel, Jules, 259n
Tyttle, William M., Jr., 115n

U

United Nations: *Demographic Yearbook:
1980*, 24n, 54n, 76n; *Demographic
Yearbook: 1983*, 68n; *Statistical
Yearbook: 1981*, 35n
U.S. Bureau of the Census, 4, 270, 284;
Annual Housing Survey, 291; *Census of
Housing: 1960*, 156n; *Census of Housing:
1980*, 156n, 291n; *Census of Population:
1950*, 262n, 304n; *Census of Population:
1960*, 19n, 165n, 317n; *Census of
Population: 1970*, 13n, 317n; *Census of
Population: 1980*, 58n, 103n, 134n, 140n,
144n, 180n, 213n, 247n, 248n, 263n,
270n, 351n, 370n, 382n, 384n, 425n,
426n, 429n, 430n, 434n; *Census of
Population and Housing: 1980*, 5n, 89,
122n, 123n, 174n, 210n, 426n, 428n,
432n, 434n; *Current Population Reports*,
7n, 11n, 21n, 33n, 52n, 55n, 59n, 77n,
114n, 126n, 148n, 164n, 165n, 167n,
190n, 241n, 242n, 248n, 260n, 281n,
283n, 285n, 286n, 292n, 293n, 294n,
303n, 304n, 340n, 342n, 351n, 357n,
370n, 387n, 434n; *Current Population
Survey*, 212, 235, 262, 263, 267, 270,
271n, 284–286, 288, 351, 353, 357n, 431;
*Eleventh Census of the United States:
1980*, 18n; *Evaluation and Research

Program of the U.S. Censuses of
Population and Housing, 1960*, 431n;
Fifteenth Census of the United States,
422n; *Historical Statistics of the United
States: Colonial Times to 1970*, 11n,
113n, 114n; *Negro Population of the
United States: 1790–1915*, 12n, 13n, 14n,
28n, 29n, 103n, 104n, 106n, 109n, 110n,
112n, 134n, 378, 421n, 422n; *Negroes in
the United States: 1920–32*, 11n, 29n,
109n, 134n, 190n, 258n, 379n, 422n;
*1970 Census of Population and Housing:
Evaluation and Research Program*, 431n;
*Population and Housing Inquiries in
U.S. Decennial Censuses: 1790–1970*,
104n, 209n; *The Post-Enumeration
Survey: 1950*, 431n; *Proceedings of the
1980 Conference on Census Undercount*,
428n; *Sixteenth Census of the United
States: 1940*, 15n, 21n, 33n, 58n, 241n,
256n; *Social Indicators III*, 281n;
*Statistical Abstract of the United States:
1982–83*, 370n; *Statistical Abstract of
the United States: 1986*, 370n; *Survey of
Income and Program Participation (SIPP)*,
288–289; *Vital Statistics Report*, 74n
U.S. Bureau of Labor Statistics, 30, 210,
213, 294; *Employment and Earnings*,
210n, 212n, 213n, 225n, 247n, 281n,
292n, 412n; *Labor Force Statistics
Derived from the Current Population
Survey: A Databook*, 247n
U.S. Department of Health, Education, and
Welfare, 59
U.S. Department of Labor, 161n, 186n,
228n
U.S. National Advisory Commission on
Civil Disorders, 2, 356n, 410n
U.S. National Commission on
Employment and Unemployment
Statistics, 211n
Usilton, Lida J., 23n

V

Vance, Ruppert B., 113n, 422n
Van Valey, Thomas L., 140n
Vavra, Helen M., 50n, 77n
Ventura, Stephanie J., 77n
Vickery, William, 115n
Vincent, Clark E., 82n
Voegeli, A. Jacque, 108n

Vonderlehr, R. A., 23*n*
Vose, Clement E., 115*n*, 138*n*, 139*n*

W

Wade, Richard C., 107*n*, 108*n*
Walker, Francis A., 421*n*
Wallace, Phyllis A., 261*n*
Walrand, Eric, 396
Walsh, Robert H., 74*n*
Ward, Michael P., 71
Weaver, William, 421
Weiss, Nancy J., 114, 301*n*
Weiss, Richard S., 21*n*
Welch, Finis R., 281*n*, 285*n*, 320*n*,
    322–323, 324–325, 331*n*, 335*n*
Wells, Robert V., 9*n*, 10*n*
West, Richard W., 243*n*
Westcott, Diane N., 156*n*, 246*n*
Westoff, Charles F., 59*n*, 60, 68*n*, 73, 84*n*,
    85*n*
Whelpton, Pascal K., 11*n*, 30, 60*n*, 63*n*,
    84*n*, 92*n*, 93*n*
Wienk, Ronald E., 152*n*
Wilcox, Jerome E., 140*n*
Wilder, M. H., 57*n*
Wilkinson, Christy A., 122*n*
Williams, Walter E., 242*n*, 244*n*, 245*n*
Williamson, Joel, 111*n*, 258*n*

Willie, Charles, 228*n*, 235*n*
Wilson, Ronald, 49*n*, 57*n*
Wilson, William Julius, 83*n*, 111*n*, 257*n*,
    304*n*, 417*n*, 424*n*
Wing, Steve, 40*n*
Winship, Christopher, 217*n*, 241*n*, 246
Wise, David A., 217*n*
Wong, Morrison G., **324–325**
Wood, Charles H., 74*n*
Woodson, Carter Goodwin, 23*n*, 29, 30,
    107*n*, 108*n*
Woodward, C. Vann, 111*n*, 257*n*
Woofter, T. J., 422*n*
Wright, Erik Olin, **324–325**
Wright, Gavin, 113*n*, 114*n*
Wright, Paul, 23, 99
Wright, Richard, 189*n*

Y

Yinger, John, 153*n*

Z

Zabin, Laurie Schwab, 75*n*
Zelnik, Melvin, 14*n*, 15*n*, 18*n*, 75*n*
Zoloth, Barbara S., 140*n*
Zunz, Olivier, 31*n*, 110*n*, 111*n*, 116*n*, 136*n*,
    137, 138*n*, 378*n*

Subject Index

Boldface numbers refer to figures and tables.

A

abolitionists, 107
abortion: and fertility rates, 18, 20, 49, 58; frequency of use of, 87–88; and teenage fertility, 68, 75–76; and unmarried women, 82
accidents, 41, **42, 43,** 45, **46**
accommodation stage of assimilation model, 363
administrative occupations: and census allocation, **433;** in census vs. monograph samples, **437;** distribution of workers in, **264, 265,** 267, **268, 272, 273,** 275–278, **276, 279, 282;** earnings from, 270, **272, 273;** and educational attainment, 198, **200;** foreign-born population in, **398, 399**
administrative support occupations: and census allocation, **433;** in census vs. monograph samples, **437;** distribution of workers in, **272, 273;** earnings from, 271, **272, 273, 402;** and educational attainment, 198, **200;** and fertility rates, **91;** foreign-born population in, **398, 399;** and migration, **124, 125**
affirmative action, 261, 281
African immigrants, **374,** 375
African language-speaking people, **383**
Africans: as indentured servants, 7–8; *see also* slavery; slaves
age: of allocated population, **427;** in census vs. monograph samples, **435;** of children, and family income, 174, **176;** of children, in single-parent families, 170; and earnings, 317, **318–319,** 321, 348–351, **349, 361;** and educational attainment, 193–195, **194;** and employment-population ratios, 222–224, **223, 226,**

227, 238, 239; of enumerated and reported population, **427;** and fertility rates, 65–70, **66–67;** at first marriage, 18–19, 170; of foreign-born population, 384, **385;** and hours of employment, **226, 227,** 231–233, **232, 238, 239, 254;** and income, **183,** 184, 306–312, **307, 310–311, 312;** and labor force participation, 218–220, *219;* and migration, 118–119, **120–121,** 126–128, **127, 128;** and mortality rates, **34,** 35–41, **38–39, 42–43;** of mother, and infant mortality rates, 50; and net census undercount, 423, **423;** and occupational achievement, 275, **276,** 278, **279, 282;** of substituted population, **427;** and unemployment prevalence and duration, **226, 227;** and unemployment rates, 213–218, **214, 216, 226, 227, 238, 239, 251, 252, 255**
Agricultural Adjustment Act (AAA), 113
agricultural labor, 11–13
Aid to Families with Dependent Children, 83, 123, 292, 357, 418
Aleutian immigrants, 371, 372
allocation, census, **427,** 428–430, **432, 433**
American Dilemma, An (Myrdal), 210
American Indians, 371, 372, **383**
American Telephone and Telegraph Company, 261
Annual Housing Survey (1977), 156
annuity income, **287,** 291
Arkansas, 109
Asian language-speaking people, **383**
Asians: employment of, 248; families of, 168, 172; immigration of, *see* foreign-

born population; urban populations of, 144, **145, 146,** 150
asset holdings, 288–289, **289, 290**
assimilation stage of assimilation model, 363
assimilation theories, 363–368
Atlanta, Georgia, 134; public school enrollment in, **207,** 208; residential segregation in, 137, **141, 143, 145, 147**
Atlantic Islands, 378
Australian immigrants, **369**
Austria, 35

B

baby boom–birth dearth theories, 70–73
Baltimore, Maryland, 134; public school enrollment in, **207;** residential segregation in, 138, **141, 143, 145, 147**
Barbados, 76
baseball, exclusion of blacks from, 259
Beyond the Melting Pot (Glazer and Moynihan), 364
biological differences, 55
Birmingham, Alabama: residential segregation in, **141**
birth control, 18, 20, 49, 58–60, 75, 83–88, **86, 87, 88**
birth order: and infant mortality rates, 50
birthplace, *see* region of birth
birth rates, *see* fertility rates
birth weight: and infant mortality rates, 51
Black Belt, 134
black men: crude and adjusted mortality rates for, 35, **36;** earnings of, 317, **319,** 321, **322–325,** 326, 329, **330,** 331–339, **332, 333, 336, 338, 344,** 346, **347,** 348, **349,** 350, 351, **352,** 353, **359, 360, 361,** 412; educational attainment of, 33, 193, **193, 197, 226, 230, 232, 238, 240, 252, 254, 255, 276, 318, 330, 333, 337, 347,** 350, 351, **352, 359, 361;** employment-population ratios for, 222, **223,** 224, 225, **226,** 228, 229, **238;** and fertility rates, 98, **100–101;** foreign-born, 384, **386,** 388, **389,** 390, 391, **392,** 393–394, **395,** 397, 398, 401, **402, 404;** as heads of household, **197;** hourly wages of, 317, **318,** 331–334, **332, 333, 352;** hours of employment of, **226,** 229, 231–233, **232, 238, 254, 330;** income amounts of, 283, **287,** 290–291, **315;** income sources of,

286–288, **287;** income trends (general) for, 295–299, **296, 298, 300;** income trends by age for, 306–312, **307, 310–311, 312;** income trends by educational attainment for, 304–306, **305;** income trends by region for, 301–303, **302;** labor force participation by, 218, **219, 221,** 222, **252,** 411–412; life span of, **26,** 45, **46,** 52–53, *53;* migrant, 122–124, **124,** 126–132, **127, 130–131;** motality rates by age for, 37–40, **38–39;** mortality rates by cause for, 41–44, **42;** net census undercount of, **423,** 423–425; occupational achievement of, 29, 263, **264,** 267, **268, 269,** 270, 271, **272, 274,** 275, **275, 276,** 278, **279,** 280, **282,** 412; overweight, 56; ratio to black women, 170; and syphilis, 22–23; unemployment determinants for, **252–253, 255;** unemployment prevalence and duration for, **226;** unemployment rates for, 213, **214,** 215, **216,** 217, 218, **226,** 229–231, **230, 238, 251,** 412
blacks: and comparative status of whites, 410–413; *see also* black men; black women
"black underclass," 416
black women: and abortion, 82; and birth control, 49, 59, 60, 75, 83–88, **86, 87, 88;** and birth weight, 51; childless, 21, 68–69, **69;** crude and adjusted mortality rates for, 35, **36;** earnings of, 317, **319, 322–325,** 326, **330, 332, 333,** 334, **336,** 337, **338,** 339–342, **341, 344, 347,** 348, **349,** 350–355, **352, 354, 359, 360, 361,** 412; educational attainment of, 50, **91,** 92, 94–97, **95, 97, 100,** 193, **193,** 195, **197, 227, 230, 239, 240, 252, 254, 255, 276, 319, 330, 333,** 334, 340, **347,** 351, **352, 359, 361;** employment-population ratios for, **223,** 224, **227, 239;** and family size, 63–65, **64;** as heads of household, 195, **197;** fertility rates and educational attainment of, **91,** 92, 94–97, **95, 97, 100;** fertility rates and employment of, **91,** 92–96, **95, 100;** fertility rates and marital status of, 76–83, **78, 79, 81,** 85–89, **86, 88, 90, 95, 97;** fertility rates by age of, 65–70, **66–67;** foreign-born, **386,** 387, 388, **389,** 390, 391, **392,** 393, 394, **395,** 397, **399, 404;** hourly wages of, 317, **319, 332, 333,** 334, **352;** hours of employment of, **227,**

234, 235, **239, 254, 330**; income amounts of, 283, **287,** 290–291, **315**; income sources of, 286–288, **287**; income trends (general) for, 295, **296, 298,** 299–300, **300**; income trends by age for, 306–309, **307, 310–311, 312**; income trends by educational attainment for, **305,** 306; income trends by region for, **302,** 303; labor force participation by, 218–220, **219, 221,** 222, **252,** 412; life span of, 14, 24, **26,** 45, **46,** 47, 53, 53–54; mortality rates by age for, **38–39,** 40–41; mortality rates by cause for, **43,** 44–45; net census undercount of, **423,** 423–425; obesity in, 56; occupational achievement of, 263, **265,** 266, **268, 269,** 271, **273, 274,** 275, **275, 276,** 278, **279,** 280, **282,** 412; and prenatal care, 51–52; ratio to black men, 170; rural, 60; teenage fertility, 65, **66,** 67, 74–76, **79,** 82–83; total fertility rates of: 1940–1984, 61, **62**; unemployment determinants for, **252–253, 255**; unemployment prevalence and duration for, **227**; unemployment rates for, 213, **214, 216,** 217, 218, **227,** 233–235, **239, 251,** 412; unmarried mothers, 76–83, **78, 79, 81,** 89, **90**

Boston, Massachusetts, 8, 9, 257; foreign-born blacks in, 378; residential segregation in, **141,** 152–153

business ownership, **289**

C

California: foreign-born blacks in, 379, **380**

California Achievement Test, 204

Canadian immigrants, 369, 370, **374,** 378

cancer, 41, **42, 43,** 44, 45, **46,** 56

cash transfer payments, 284

caste membership, and educational attainment, 191

census data, 4–5, 420–438

census procedures, 426–430, **427**

census undercount, net, 421–425, **423**

Central American immigrants, **374,** 375, 382

cerebrovascular diseases, 41, **42, 43,** 44, 45, **46**

certification, 244–245

Charleston, South Carolina, 8, 9, 13, 137

checking account ownership, **289**

chemical industry, 354

Chicago, Illinois: foreign-born blacks in, **381,** 382; public school enrollment in, **207**; residential segregation in, 136–138, **141,** 142, **143, 145, 147,** 150, 151

Chicago Defender, 112, 247

childbearing, *see* fertility rates

childlessness, 21, 61, 63, **64,** 68–70, **69**

children: age distribution in single-parent families, 170; distribution by family income, 175, 177; family income and age of, 174, **176**; living with both parents, 166; mortality rates by sex for, 37–40, **38–39**; and narcissistic personality style, 72–73; and preventive health care, 57; *see also* infant mortality; school enrollment

church-affiliated schools: enrollment in, 203, **203**

Cincinnati, Ohio, 108, 109, **141**

cirrhosis, 41, **42, 43,** 44, 45, **46**

citizenship of foreign-born population, 375, 377–378

Civil Rights Act of 1964, 116, 241, 245, 280

Civil War period: population growth during, 11–24

cleaners: and census allocation, **433**; in census vs. monograph samples, **437**; foreign-born population as, 396, **398, 399**

clerical occupations: distribution of workers in, 263, **264, 265**

Cleveland, Ohio, 257; public school enrollment in, **207**; residential segregation in, 138, **141,** 142, **143, 145, 147**

cohabitation patterns: and fertility rate reduction, 18, 20

cohort fertility rates, 63, **64**

college attendance, 192, **192**; by age, **194**; and census allocation, **432**; in census vs. monograph samples, **436**; and earnings, 333, 334, **352, 361**; and employment-population ratios, **226, 227, 238, 239**; and employment status, **199**; and family income, **202**; and fertility rates, **91, 95, 97, 100, 101**; by foreign-born population, **392**; and hours of employment, **226, 227**; and household type, **197**; and income, **305**; and migration, **127, 128**; and occupational achievement, 198, **200–201, 276, 279, 282**; by region, **196**; by sex, 193, **193**; and unemployment prevalence and

duration, **226, 227, 238, 239;** and
unemployment rates, **226, 227, 230, 238, 239, 240**
college graduates, 190, 192, **192;** by age, 193–195, **194;** and census allocation, **432;** in census vs. monograph samples, **436;** earnings of, **318, 319,** 346–348, **347, 352, 361;** employment-population ratios for, **226, 227, 238, 239;** employment status of, 198, **199;** family income of, **202;** fertility rates of, **91, 95, 97, 100, 101;** foreign-born population as, 391–393, **392;** as heads of household, 172, 177, 179, 181, **183;** hours of employment of, **226, 227, 232;** and household type, 195, **197;** income of, 304–306, **305;** and migration, **124, 125,** 126, **127, 128;** occupational achievement of, 198, **200–201,** 275, **276, 279, 282;** by region, 195, **196;** Scholastic Aptitude Test scores of, 206; by sex, 193, **193;** unemployment prevalence and duration for, **226, 227, 238, 239;** unemployment rates for, **226, 227, 230, 238, 239**
Colonial period, population growth in, 7–10
Columbus, Ohio: residential segregation in, **141,** 142
Committee on Social Trends, 30
community institutions: black, 416
competition stage of assimilation model, 363
Comstock Law of 1873, 58–59
conflict stage of assimilation model, 363
Congress of Industrial Organizations (CIO), 116
Connecticut, 9, 107
conservative ideology, 417
Constitutional Convention, 13
Consumer Price Index (CPI), 294
contraception, *see* birth control
Costa Rica, 35
cotton farming, 11, 13, 106, 112–114
craft occupations: and census allocation, **433;** in census vs. monograph samples, **437;** distribution of workers in, 263, **264, 265,** 270, **272, 273;** earnings from, 271, **272, 273**
credentialism, 244–245
crime rates, 151
crude mortality rates, 14, 33–35, **34, 36**
Cuba, 47

Cuban immigrants, 372, **374,** 382
cultural determinism vs. racial discrimination, 365–368, 403–405
cultural differences in family organization, 171, 179–180, 186
cultural values: black, 417
Czechoslovakia, 68

D

Dallas, Texas: public school enrollment in, **207;** residential segregation in, **141,** 142, **143, 145, 147**
Dearborn, Michigan, 154
death rates, *see* mortality rates
Death Registration Areas (DRAs), 25, **26, 27,** 30, 31
Denmark, 68, 76
Denver, Colorado, 152, 243–244
Depression, 72
Detroit, Michigan, 109; foreign-born blacks in, **381;** public school enrollment in, **207,** 208; residential segregation in, 136–138, **141, 143, 145, 147,** 150, 153–154
diabetes, 41, **42, 43,** 56
diet, *see* nutrition
disability, **226, 227,** 229
discouraged workers, 211–212
discrimination, 410–413; vs. cultural determinism theories, 365–368, 403–405; and earnings, 317, 320–321, 334–335, 340, 355; and educational attainment, 189–191, 195, 205, **207,** 208; and foreign-born immigrants, 365–368, 403–405; and income, 283; and occupational achievement, 257–261, 270, 278, 280
diseases: fertility inhibiting, 18, 20–23; genetic factors, 55; and mortality rates, 27–28, 32, 40–45, **42–43;** venereal, 18, 21–23
District of Columbia, *see* Washington, D.C.
divorced men: foreign-born, **389;** occupational achievement of, **276**
divorced women: earnings of, **341;** fertility rates for, 89, **90, 95;** foreign-born, **389;** as heads of household, 165; occupational achievement of, **276**
domestic service occupations: distribution of workers in, 263–267, **264, 265;** and fertility rates, **91;** *see also* private household occupations

Dominican Republic immigrants, 382
Dred Scott decision, 410

E

earnings, 284, 316–361, 412; and age, 317,
318–319, 321, 348–351, **349, 361;** annual,
323, 325, 329, 331, **332,** 335–342, **336,**
338, 341, 352, 353–355, **354, 359,**
400–401, *402;* and the baby boom, 70–71;
and discrimination, 317, 320–321,
334–335, 340, 355; and educational
attainment, 317, **318–319,** 320, 321, 327,
330, 331, **333,** 334–340, **338,** 345–348,
347, 350–351, **352, 359, 361;** and family
organization, 340–342, **341;** of foreign-
born population, 400–403, **402, 406–407;**
and government spending, 356–357;
hourly wage rates, *see* hourly wage rates;
and hours of employment, 235–236, 329,
330, 337, **338, 352;** and labor market
experience, 321, 327, 328, 331, **333,** 334,
337, **338, 359;** and macroeconomic
trends, 356; and marital status, 340–342,
341; measurement of differences in,
320–329; and migration, 122–123, **124,**
125, 128–132, **130–131;** national studies
of differences in, **322–325;** in 1960
census vs. 1980 census, 329, 331–335,
332, 333; and occupational achievement,
270–271, **272–273,** 317, **318–319;** as
percent of total income, **287;** and racial
attitudes of blacks, 358; and racial
attitudes of whites, 357–358; and rates of
return associated with labor market
characteristics, 331, **333,** 334–342, **336;**
and region of residence, 321, 328, 331,
333, 334–335, 337, 343–345, **344, 359,**
361; sexual differences in, **287,** 331, **332,**
340; and welfare, 357
Eastern European countries, 68
economic differences view of residential
segregation, 148–150, **149**
economic growth: and unemployment, 213,
215–217, **216**
economic status, 2–4; and family
organization, 166–167, 171–179, **174,**
175, 176, 178, 185–187; *see also*
earnings; educational attainment;
employment; income; occupational
achievement

economic structure: and employment,
245–246
education, government-backed loans for,
284
educational attainment, 188–208, 411; by
age, 193–195, **194;** and caste
membership, 191; and census allocation,
430, **432;** in census vs. monograph
samples, **436;** and discrimination,
189–191, 195, 205, **207,** 208; and
earnings, 317, **318–319,** 320, 321, 327,
330, 331, **333,** 334–340, **338,** 345–348,
347, 350–351, **352, 359, 361;** and
employment, 189, 198, **199,** 228, 229,
230, 231, **232,** 233, 236–237, **238, 239,**
240, 254; and employment-population
ratios, **226, 227, 238, 239;** and
extendedness of household, 177, 179; and
family income, 181, **183,** 198, 200, **202;**
and fertility rates, 60, 76, **91,** 92, 94–98,
95, 97, 100, 101; of foreign-born
population, 391–393, **392;** of head of
household, 172, 173, 177, 179, 181, **183,**
195, **197,** 198, 200, **202;** historical
perspective on, 189–191; and hours of
employment, **226, 227,** 231, **232,** 233,
235, **238, 239, 254;** and household type,
172, 173, 177, 179, 181, **183,** 195, **197,**
198; and income, 181, **183,** 198, 200, **202,**
303–306, **305;** and migration, 122, 123,
125, 126–129, **127, 128,** 133; and
mortality rates, 33, 50, 55; and
occupational achievement, 198, **200–201,**
260, 270–271, 275, **276,** 278, **279,**
280–281, **282;** for the population aged 16
and over, **192,** 192–193; and quality of
education, 204; by region, 195, **196;**
research on, 415; and residential
segregation, 148–150, **149,** 157, 205, **207,**
208; and school enrollment rates, **203,**
203–205, **207,** 208; sexual differences in,
193, **193;** and slavery, 189, 195; and
standardized tests, 204–205, **205, 206,**
and unemployment rates, **226, 227,** 229,
230, 233, 235, 237, **238, 239, 240**
elementary school education: and census
allocation, **432;** in census vs. monograph
samples, **436;** and earnings, 331, **333,**
346, **347, 352, 359, 361;** and
employment-population ratios, **226, 227;**
and fertility rates, **91, 95, 100, 101;** and
foreign-born population, **392;** and hours

of employment, **226, 227**; and income, 304, **305**; and migration, 126, **127, 128**; and occupational achievement, **276, 279, 282**; and unemployment prevalence and duration, **226, 227**; and unemployment rates, **226, 227**; *see also* primary school education

El Salvador, 26

employed, classification as, 210–211

employment, 209–255, 411–412; of adult men, 225, **226**, 228–233, **230, 232**; of adult women, **227, 230**, 233–236; and age, 218–220, **219**, 222–224, **223, 226, 227**, 231–233, **232**, 236, **254**; definitions of, 210–211; and demographic structure, 245–246; and economic structure, 245–246; and educational attainment, 189, 198, **199**, 225, **226, 227**, 228, 229, **230**, 231, **232**, 233, 236–237, **238, 239, 240, 254**; and fertility rates, 70–71, **91**, 92–96, **95, 100, 101**; of foreign-born population, 393–400, **395, 398–399**; and government regulations, 244–245; and health, **226, 227**, 229; hours of, *see* hours of employment; and household type, 173–174, **175**, 228; and increased competition hypothesis, 246–249; labor force participation, *see* labor force participation; and marital status, **226, 227**, 228–233, **230, 232**, 235, 237, **238, 239, 254**; and migration, 111–116, 122–123, **124, 125**, 133, **226, 227**; and minimum wages, 244, 249; occupational achievement, *see* occupational achievement; -population ratios, *see* employment-population ratios; and region of residence, **226, 227**, 229, 237, **238, 239**; research on, 415; and residential segregation, 246; sexual differences in, 394; social-psychological explanations of differences in, 249–250; and welfare, 210, 241–244; of youth, 236–240, **238–239, 240**; *see also* unemployment; unemployment rates

employment-population ratios, 222–224, **223**; for adult men, 225, **226**, 228; for adult women, **227**, 233; and age, 222–224, **223, 226, 227, 238, 239**; defined, 212; and disability, **226, 227**; and educational attainment, **226, 227, 238, 239**; and health, 229; and marital status,

226, 227, 229, 233, **238, 239**; and migration, **226, 227**; and place of residence, **226, 227, 238, 239**; and region of residence, **226, 227, 238, 239**; for youth, **238, 239**

England, 68

Equal Employment Opportunity Commission (EEOC), 260–261, 281

Eskimo immigrants, 371, 372

ethnic homogeneity view of residential segregation, 145–148, **147**

European immigrants, 368, **369, 374**

executive occupations: and census allocation, **433**; in census vs. monograph samples, **437**; distribution of workers in, 274–280, **276, 279, 282**; and earnings, 270, **272, 273**; and educational attainment, 198, **200**; and fertility rates, **91**, 93; of foreign-born population, 396, **398, 399**; and household type, 172, 174, 179; and migration, **124, 125**

Exoduster Movement, 110

extendedness of families, 168, 177–180; and educational attainment, 177, 179; and family income, 177, **178**, 181, **182**; and hours of work, 235; and occupation of household head, 179; social and cultural factors in, 179–180

F

Fair Employment Practices Committee (FEPC), 116, 260, 316

Fair Housing Act, 139, 130

family: defined, 167

family households, *see* households

family income, 166–167; and age of children, 174, **176**; by age of household head, **183**, 184; by area type, **183**, 184; in census vs. monograph samples, **437**; and educational attainment, 181, **183**, 198, 200, **202**; of foreign-born population, 384, **385**, 400, 403, 404; and household composition, 177, **178**, 181, **182**; and household type, 166, 167, 172–173, **174–175**, 181, **182**; per capita, 172, 180–184, **182–183**; by region, 166, **182**, 183

family organization, 160–187, 410–411; and cultural differences, 171, 179–180, 186; and earnings, 340–342, **341**; and economic differences, 166–167, 171–179,

174, 175, 176, 178, 185–187; of foreign-born immigrants, 387–391, 403, 404; historical perspective on, 162–167; household composition, *see* household composition; household type, *see* household type; multivariate analysis of, 180–184, 182–183; and 1980 census, 167–171; research on, 161–167, 415

family size, 63–65, 64, 83–84, 164–165

farming occupations: and census allocation, 433; in census vs. monograph samples, 437; distribution of workers in, 263, 264, 265, 267, 272, 273; earnings from, 271, 272, 273; and educational attainment, 198, 201; and fertility rates, 91, 93; of foreign-born population, 398, 399

female-headed households, 163–165; age distribution of children in, 170; distribution by race, 168, 169; and educational attainment of head, 195, 197, 198; and employment status, 173, 175; and family income, 167, 172–173, 174–175; of foreign-born population, 390, 403, 404; and net worth, 290; and welfare, 242

females, *see* black women; white women

female sterilization, 85, 86

fertility rates, 165, 410; by age, 65–70, 66–67; and baby boom–birth dearth theories, 70–73; and birth control, 18, 20, 49, 58–60, 75, 83–88, 86, 87, 88; cohort, 63, 64; in Colonial period, 9–10, and demographic factors, 60, 89–98, 90–91, 95, 97, 100; and disease, 18, 20–23; and educational attainment, 60, 76, 91, 92, 94–98, 95, 97, 100, 101; and employment, 70–71, 91, 92–96, 95, 100, 101; and family size, 63–65, 64, 83–84; and husbands, 98–99, 100–101; and marital status of mothers, 18–20, 76–83, 78, 79, 81, 85–89, 86, 88, 90, 95, 97; net effect of factors related to the fertility of married women, 94–98, 95, 97; in post-Civil War period, 16–23; in pre-Civil War period, 14–16; reductions in chilbearing: 1850–1985, 16–23, 17; of teenagers, 65, 66, 67, 74–76, 79, 82–83; total: 1860–1984, 18, 19; total: 1940–1984, 58, 61, 62

fishing occupations: and census allocation, 433; in census vs. monograph samples,

437; distribution of workers in, 272, 273; earnings from, 271, 272, 273; and educational attainment, 198, 201; of foreign-born population, 398, 399

Florida: foreign-born population in, 379, 380

food stamps, 242, 284, 292, 293, 293

foreign-born population, 362–407; age structure of, 384, 385; arrival dates of, 375, 376, 377; assimilation of, 363–368; blacks as a special case, 365–368; citizenship of, 375, 377–378; classification of, 371–372; countries or areas of birth of, 372, 374, 375; cultural determinism vs. racial discrimination theories, 365–368, 403–405; earnings of, 400–403, 402, 406–407; educational attainment of, 391–393, 392; employment of, 393–400, 395, 398–399; family income of, 384, 385, 400, 403, 404; family organization of, 387–391, 403, 404; geographic distribution of, 378–382, 380–381; languages spoken by, 382, 383; marital status of, 387–391, 389; and migration flow changes, 368–370, 369; numbers of, 371, 371; occupational achievement of, 394–400, 398–399; racial composition of, 372, 373; sex structure of, 384, 386, 387

forestry occupations: and census allocation, 433; in census vs. monograph samples, 437; distribution of workers in, 272, 273; earnings from, 271, 272, 273

France, 68

Freedman's Bureau, 112

French-speaking people, 382, 383

fringe benefits, 284, 291–292

G

gang labor, 8

Gary, Indiana, 243

genetically linked diseases, 55

German-speaking people, 383

ghettoes, 136–138, 156, 228

gonorrhea, 21, 23

governmental transfer payments, 284, 286, 287, 292–293, 293, 356–357

Great Britain, 191

Greece, 47

Griggs v. *Duke Power Company* (1971), 261

Gross National Product (GNP): and
unemployment rates, 213, 215–217, **216,**
220
guaranteed income, 243–244
Guatemala, 68
Guinea, 24

H

Haitian immigrants, 372, **374,** 382
handlers: and census allocation, **433;** in
census vs. monograph samples, **437;**
distribution of, 271, **272, 273;** foreign-
born population as, 366, **368, 399;** hourly
wages of, **318, 319**
Harlem, New York, 137
Hart-Celler Immigration Act of 1965, 368
heads of household, *see* female-headed
households; husband-wife families;
male-headed households
health, and employment, **226, 227,** 229
health care: and mortality rates, 30–33,
56–57; research on, 414–415
health practices: and mortality rates, 55–56
heart disease, 32, 41, **42, 43,** 44, 45, **46,** 47,
56
helpers, 271, **272, 273**
high school attendance, 192, **192;** by age,
194; and census allocation, **432;** in
census vs. monograph samples, **436;** and
earnings, **318, 319, 352, 359,** 361; and
employment-population ratios, **226, 227,**
238, 239; and employment status, **199;**
and family income, **202;** and fertility
rates, **91, 95, 97, 100, 101;** and foreign-
born population, **392;** and hours of
employment, **226, 227, 232;** and
household type, **197;** and income, **202,**
305, and migration, **127, 128;** and
occupational achievement, 198, **200–201,**
276, 279, 282; by region, **196;** sexual
differences in, 193, **193;** and
unemployment prevalence and duration,
226, 227, 238, 239; and unemployment
rates, **226, 227, 230, 238, 239, 240**
high school graduates, 190, 192, **192;** by
age, 193, **193;** in census vs. monograph
samples, **436;** earnings of, **318, 319, 347,**
352, 361; employment-population ratios
for, **226, 227, 238, 239;** employment
status of, 198, **199;** family income of,
202; fertility rates of, **91, 95, 97, 100,**

101; foreign-born population as, **392;**
heads of household as, 177, **183,** hours of
employment of, **226, 227, 232;** income of,
202, 305, 306; and infant mortality rates,
50; and migration, **124, 125, 127, 128;**
occupational achievement of, 198,
200–201, 276, 279, 282; by region, **196;**
Scholastic Aptitude Test scores of, **206;**
by sex, 193, **193;** unemployment
prevalence and duration for, **226, 227,**
238, 239; unemployment rates for, **226,**
227, 230, 238, 239, 240
Hispanics, 248; families of, 172;
immigration of, *see* foreign-born
population; urban populations of, 144,
145, 146
home ownership, **289,** 291, **292**
homicide, 41, **42, 43,** 44, 45, **46,** 56
Honduras, 68
hospitalization, 57
hourly wage rate, 321, **323, 325,** 326–327,
332, 359; and educational attainment,
317, **318–319,** 331, **333, 334, 352;** of
foreign-born population, 400, **406–407;**
and labor market experience, 331, **333,**
334; minimum, 244, 249; and region of
residence, 331, **333,** 334–335
hours of employment: of adult men, **226,**
231–233, **232, 254;** of adult women, **227,**
234, 235, **254;** and age, **226, 227,**
231–233, **232, 238, 239, 254;** and
disability, **226, 227,** 229; and earnings,
235–236, 329, **330,** 337, **338,** 352; and
educational attainment, **226, 227,** 231,
232, 233, 235, **239, 239, 254;** and fertility
rates, **91,** 92; and guaranteed income,
243–244; and household composition,
235; and marital status, **226, 227,** 229,
231–233, **232,** 235, **238, 239, 254;** and
migration, **124, 125, 226, 227;** and place
of residence, **226, 227, 238, 239;** and
region of residence, **226, 227, 238, 239;** of
youth, **238, 239**
household composition: and cultural
differences, 179–180; defined, 168;
distribution by race, 168; and educational
attainment, 177, 179; and family income,
177–179, **178,** 181, **182, 183;** and
occupational achievement of household
head, 179; social and cultural factors in,
179–180
household income, *see* family income

household occupations, *see* private household occupations

households: asset holdings of, 288–289, **289, 290;** defined, 167; increase in: 1890–1984, 164; size of, 63–65, **64,** 83–84, 164–165; *see also* household composition; household type

household type, 163–166; and age distribution of children, 179; defined, 168; distribution by race, 168, **169;** and educational attainment, 172, 173, 177, 179, 181, **183,** 195, **197,** 198; and employment status, 173–174, **175,** 228; and family income, 166, 167, 172–173, **174–175,** 181, **182;** and net worth, **290;** and occupational achievement, 174

housing: public, 284, 292, 293, **293;** *see also* residential segregation

Houston, Texas: public school enrollment in, **207;** residential segregation in, **141,** 142, **143, 145, 147**

Hungary, 68

husbands, *see* married men

husband-wife families, 163, 165; and educational attainment, 195, **197,** 198; distribution by race, 168, **169;** and employment status, 173, **175;** and family income, 172, 173, **174–175;** of foreign-born immigrants, 390, 403, **404;** and occupational achievement, 174

hypertension, 57

I

Iceland, 76

illegitimacy, 76–80, **78, 79, 81**

Illinois, 107; foreign-born blacks in, **380**

illiteracy rates, 190

income, 283–315; and age, **183,** 184, 306–312, **307, 310–311, 312;** and census allocation, 430, **433;** and Consumer Price Index, 294; defined, 284; differences in sources of, 286–293, **287;** and discrimination, 283; earnings, *see* earnings; and educational attainment, 181, **183,** 198, 200, **202,** 303–306, **305;** general trends in, 293–300, **296, 298, 300;** guaranteed, 243–244; mean, 294, **315;** median, *see* median income; and mortality rates, 32–33; per capita, 299, **300, 315;** and region of residence, 300–303, **302;** reporting of, 285–286; and

residential segregation, 148–150, **149;** and Scholastic Aptitude Test scores, **206;** sexual differences in, 294, 299, 309

increased competition hypothesis, 246–249

indentured servants, 7–8

Indiana, 107

Indianapolis, Indiana: public school enrollment in, **207;** residential segregation in, **141**

infant mortality, 47–52; decline in, **48,** 48–49; in early twentieth century, 25; in pre-Civil War period, 14; and syphilis, 22

influenza, 40, 41, **42, 43**

integration of immigrant population: theories of, 363–368

interest-bearing deposit ownership, **289**

Internal Revenue Service (IRS), 285, 286

International List of Causes of Death, 41

IRA plan ownership, **289**

Ireland, 68

Irish immigrants, 107, 367

iron consumption, 56

Israel, 191

Italian-speaking people, **383**

J

Jacksonville, Florida: public school enrollment in, **207;** residential segregation in, 137, **141,** 142

Jamaica, 47

Jamaican immigrants, **374**

Japan, 47, 51, 191

Japanese immigrants, 367

Jewish immigrants, 366, 367

Jim Crow system, 111, 137–139

job-seeking activities, 211

K

Kansas, 109–110, 112

Kansas City, Missouri, **141**

Kansas Exodus, 109

Kansas Fever, 110

Keogh plan ownership, **289**

Kerner Commission, 410

Ku Klux Klan, 109, 116

L

labor, racial division of, 413–414

laborer occupations: and census allocation, **433;** in census vs. monograph samples,

437; distribution of workers in, 263, **264,**
265, 272, 273; earnings from, 271, **272,**
273; foreign-born population as, 396, **398,**
399; hourly wages of, **318, 319**
labor force: defined, 210
labor force participation, 218–222, **219,**
221, 236, 411–412; and educational
attainment, 198, **199;** and fertility rates,
70–71; and migration, **124, 125;** *see also*
employment-population ratios
labor market characteristics: rates of return
associated with, 331, **333,** 334–342, **336**
labor market experience: and earnings, 321,
327, 328, 331, **333,** 334, 337, **338, 359**
labor supply shortages: and employment of
blacks, 246–249; and occupational
achievement, 258, 281
Law School Aptitude Test, 204
liberal ideology, 417, 418
Liberty City, Miami, Florida, 1–2
licensing, 244–245
life span: of black women, 14, 24, **26,** 45,
46, 47, **53,** 53–54; in early twentieth
century, 25, **26,** 31; estimates of: with
the elimination of causes of death,
45–47, **46;** expectations at birth by sex
and race, 25, **26;** in post-Civil War period,
24; in pre-Civil War period, 14; in
Sweden, 24
literacy, 28, 190
Los Angeles, California: foreign-born
blacks in, 379, 381, 382; public school
enrollment in, **207;** residential
segregation in, 140, **141, 143, 145, 147,**
150
Louisiana, 134
Louisiana Territory, 104
Louisville, Kentucky, 138–139

M

machine operator occupations, *see* operator
occupations
Macon, Georgia, 112
Making the Second Ghetto (Hirsch), 153
male-headed households, 388; age
distribution of children in, 170;
distribution by race, **169;** and educational
attainment, **197;** and employment status,
173, **175;** and family income, 167;
foreign-born, 390
males, *see* black men; white men

Mali, 24
malignant neoplasms, 41, **42, 43,** 56
managerial occupations: distribution of
workers in, 263, **264, 265,** 267, **268,** 270,
272, 273, 275–278, **276, 279, 282;**
earnings from, 270, **272, 273;** of foreign-
born population, **398, 399;** and household
type, 172, 174, 179
marital status, 410–411; of allocated
population, **427,** 429; in census vs.
monograph samples, **435;** distribution by
race, 170; and earnings, 340–342, **341;**
and employment, **226, 227,** 228–233, **230,**
232, 235, 237, **238, 239, 254;** and
employment-population ratios, **226, 227,**
229, 233, **238, 239;** of enumerated
population, **427;** and fertility rates,
18–20, 76–83, **78, 79, 81,** 85–89, **86, 88,**
89, **90;** of foreign-born population,
387–391, **389;** and hours of employment,
226, 227, 229, 231–233, **232,** 235, **238,**
239, 254; and migration, 122, **124, 125,**
126; and occupational achievement, 228,
276, 276–277, **282;** and unemployment
rates, **226, 227,** 229–231, **230,** 233–234,
238, 239
marriage: age at first, 18–19, 170; and sex
ratios, 170
married couples: and net worth, **290;** *see*
also husband-wife families; married
men; married women
married men: educational attainment of,
98, **101;** and fertility rates, 98–99;
foreign-born, 387, **389;** hours of
employment of, **330;** income of, 98, **101;**
occupational achievement of, **276,**
276–277, **282**
married women: and birth control, 85, **86,**
88; earnings of, **341,** 341–342; fertility
rates for, 18–20, **79,** 79–80, 89, **90,** 94–98,
95, 97; foreign-born, 388, **389;** hours of
employment of, **330;** and migration, **124,**
125, 126; occupational achievement of,
276, 282
Maryland, 8, 9, 14
Massachusetts, 14; foreign-born blacks in,
378–379
mean income, 294, **315**
median income: by age, 307, **310–311, 312;**
defined, 294; by educational attainment,
305; general trends in, **296, 298;** by
region, **302**

Subject Index

Medicaid, 288, 292, 293, **293**
Medicare, 284
Memphis, Tennessee, **207**
men, *see* black men; white men
Mexican immigrants, 370, **374**, 393
Miami, Florida: foreign-born blacks in, 379,
   381, 382; residential segregation in, **142,
   143, 145, 147**
Michigan, 108
Michigan, University of: Panel Study of
   Income Dynamics, 321
Mid-Atlantic states: foreign-born blacks in,
   380
migration, 103–136, 411; and age, 118–119,
   120–121, 126–128, **127, 128**; causes and
   consequences of, 132–133, between Civil
   War and World War I, 109–112; and
   earnings, 122–123, **124, 125,** 128–132,
   130–131; and educational attainment,
   122, 123, **125,** 126–129, **127, 128,** 133;
   and employment, 111–116, 122–123, **124,
   125,** 133, **226, 227;** in-migrants and out-
   migrants by region: 1965–1980, **118;**
   international, 368–370, **369;** and marital
   status, 122, **124, 125,** 126; since 1970,
   117–119; recent, selectivity of, 126–128,
   127, 128; during slavery period, 13, 14,
   104, 106–109; social and economic
   characteristics of migrants, 119, 122–126,
   124–125; and unemployment rates, **226,
   227,** 248; and urbanization, 134–136,
   135; and welfare recipiency, 123, **124,
   125,** 132, 133; from World War I to 1970,
   112–117; *see also* foreign-born
   population
military service, 246
Milwaukee, Wisconsin: public school
   enrollment in, **207;** residential
   segregation in, **141**
minimum wages, 244, 249
minority group status hypothesis, 73, 74
Mississippi, 134
Montgomery, Alabama, 137
mortality rates, 410; and age, **34,** 35–41,
   38–39, 42–43; and biological differences,
   55; by cause, 41–45, **42–43;** in Colonial
   period, 9, 10; crude, 33–35, **34, 36;** and
   disease, 27–28, 32; in early twentieth
   century, 25–31, **26, 27;** and educational
   attainment, 33, 50, 55; and health care,
   30–33, 56–57; and health practices,
   55–56; and income, 32–33; for infants,

see infant mortality; from 1940 to the
   present, 32–57; and occupational
   attainment, 29; in post-Civil War period,
   23–24; in pre-Civil War period, 14–15;
   and standard of living, 28–33, 55; and
   syphilis, 22; *see also* life span
motivation, 417
motor vehicle accidents, 41, **42, 43, 46**
motor vehicle ownership, **289**
Mountain states: foreign-born blacks in,
   380
mutual funds ownership, **289**

N

narcissistic personality style, 72–73
Nashville, Tennessee: public school
   enrollment in, **207;** residential
   segregation in, 137
National Association for the Advancement
   of Colored People (NAACP), 139
National Association of Real Estate Boards,
   139
National Labor Relations Act of 1935, 116
National Origins Act of 1924, 383
National Survey of Family Growth (1982),
   85
National Urban League, 212
Negro Immigrant, The (Reid), 365
Netherlands, 47, 68
net reproduction rate, 61, **62**
New Amsterdam, 8
Newark, New Jersey: residential
   segregation in, **141, 143, 147**
New Bedford, Massachusetts, 378
*Newcomers, The: Negroes and Puerto
   Ricans in a Changing Metropolis*
   (Handlin), 365
New England: foreign-born blacks in, 379,
   380
New Jersey, 9, 107, 243; foreign-born
   blacks in, 378; residential segregation in,
   142–143
New Orleans, Louisiana, 257; public school
   enrollment in, **207,** 208; residential
   segregation in, 134, **141, 143, 145, 147**
Newport, Rhode Island, 8, 9
New York City, New York: foreign-born
   blacks in, 365, 367, 378, 379, **381,** 382;
   public school enrollment in, **207;**
   residential segregation in, **141, 143,** 144,
   145, 147, 150

487

New York State, 9, 25, 107; foreign-born blacks in, 379
New Zealand, 191
non-family households: distribution of, **169**; and educational attainment, 195, **197**; and employment status, **175**; and family income, **174–175**
North Central states: foreign-born blacks in, **380**
Northwest Ordinance, 107
nuclear family, 168; *see also* household composition
nutrition: of black sharecroppers, 21; racial differences in, 56; of rural blacks, 29–30; of slaves, 15

O

Oakland, California: foreign-born blacks in, **381**; residential segregation in, **141**, 142, **143**, 150
obesity, 56
occupational achievement, 256–282, 412; and affirmative action, 281; and age, 275, **276, 278, 279, 282**; assessment of changes in, 261–263; and census allocation, 430, **433**; in census vs. monograph samples, **437**; and civil rights laws, 260–261, 270; and discrimination, 257–261, 270, 278, 280; and earnings, 270–271, **272–273**, 317, **318–319**; and educational attainment, 198, **200–201**, 260, 270–271, 275, **276**, 278, **279**, 280–281, **282**; and fertility rates, **91**, 92–94; of foreign-born population, 394–400, **398–399**; of heads of household, 172; historical perspectives on, 256–261; and household extendedness, 179; and household type, 172, 174, 179; and indexes of occupational dissimilarity, 266–270, **268, 269, 274, 275, 276**; and labor supply shortages, 258, 281; and marital status, 228, **276**, 276–277, **282**; and migration, **124, 125**; and mortality rates, 29; occupational change: 1940–1980, 263–270, **264–265**; occupational change: 1970–1986, 270–274, **272–273**; and region of birth, 275–276, **276**; and region of residence, 275, **276, 282**; sexual differences in, **264, 265**, 267, **268, 269**, 270, **272, 273**, 278, **279, 282**

occupational classification: 1980 change in, 262–263, 270–274, **272–273**
occupational dissimilarity indexes, 266–270, **268, 269, 274, 275, 276**
Office of Federal Contract Compliance Programs, 261
Ohio, 108
Oklahoma, 109
operator occupations: and census allocation, **433**; in census vs. monograph samples, **437**; distribution of workers in, 263, **264, 265, 272, 273**; earnings from, 271, **272, 273, 402**; and educational attainment, 198, **201**; and fertility rates, **91**; of foreign-born population, **398, 399**; and hourly wages, **318, 319**; and household type, 172, 174, 179; and migration, **124, 125**

P

Pacific Islander immigrants, 371, 372
Pacific states: foreign-born blacks in, **380**
Panama, 68, 76
Pennsylvania, 107
per capita income, 299, **300, 315**; family, 172, 180–184, **182–183**
personal income, *see* income
Philadelphia, Pennsylvania, 8, 21–22, 107, 163–164, 257; public school enrollment in, **207**; residential segregation in, 136–138, **141**, 142, **143, 145, 147**, 150
Plessy vs. Ferguson, 189
pneumonia, 40, **42, 43**
Poland, 68
population: distribution by region: 1790–1985, 104, **105**; foreign-born, *see* foreign-born population; percent black for U.S. and regions: 1790–1985, **108**; *see also* population growth
population growth: average annual rates of, **12**; following the Civil War, 16–31, **17, 19, 26, 27**; from Colonial period to World War II, 7–31; family size, 63–65, **64**, 83–84, 164–165; before the Revolutionary War, 7–10; from the Revolutionary War to the Civil War, 11–16; 1790–1986, **10**, 10–11, **12**; *see also* fertility rates; migration; mortality rates
Portsmouth, Ohio, 108
Portuguese-speaking people, 382

post-graduate education, **192;** by age, **194,** and census allocation, **432;** in census vs. monograph samples, **436;** and employment-population ratios, **226, 227;** and employment status, 198, **199;** and family income, **202;** and fertility rates, **91, 95, 97, 100, 101;** and hourly wages, **318, 319, 352, 361;** and hours of employment, **226, 227;** and household type, **197;** and migration, **127;** and occupational achievement, **200–201, 276, 279, 282;** by region, 195, **196;** sexual differences in, 193, **193;** and unemployment prevalence and duration, **226, 227;** and unemployment rates, **226, 227**

poverty, 55, 172; "culture" of, 418; "feminization" of, 167; "War" on, 356–357

precision production occupations: and census allocation, **433;** in census vs. monograph samples, **437;** distribution of workers in, **272, 273;** earnings from, 271, **272, 273;** and fertility rates, **91;** of foreign-born population, **398, 399;** and migration, **124, 125**

pregnancy: prenatal care, 51–52; and syphilis, 21; *see also* fertility rates

prenatal care, 51–52

preventive health care, 57

primary school education, **192;** by age, **194;** and employment status, 198, **199;** and family income, 200, **202;** and household type, 195, **197;** and occupational achievement, **200–201;** by region, 195, **196;** and Scholastic Aptitude Test scores, **206;** by sex, 193, **193;** *see also* elementary school education

private household occupations: and census allocation, **433;** in census vs. monograph samples, **437;** earnings from, **272, 273;** and educational attainment, 198, **201;** of foreign-born population, **398, 399**

private schools, enrollment in, **203,** 203–204

professional specialty occupations: and census allocation, **433;** in census vs. monograph samples, **437;** distribution of workers in, 263, **264, 265,** 267, **268,** 270, **272, 273,** 274–280, **276, 279, 282;** earnings from, 270, **272, 273, 402;** and educational attainment, 198, **200;** and

fertility rates, **91,** 93; of foreign-born population, 396, **398, 399;** and hourly wages, **318, 319;** and migration, **124, 125**

property income, 284, **287,** 288

property ownership, 28

protective service occupations: and census allocation, **433;** in census vs. monograph samples, **437;** distribution of workers in, **272, 273;** and educational attainment, 198, **201;** of foreign-born population, **398, 399**

public assistance, *see* welfare

Public Health Service, 23

public housing, 284, 292, 293, **293**

public schools: enrollment in, 203, **203,** 205, **207,** 208; racial segregation in, 157, 205, **207,** 208

Public Use Microdata Samples (PUMS), 5

Puerto Rican immigrants, 372, 382

R

race: as attributed quality, 6

race-differences paradigm, 6

racial attitudes, 416–417; and earnings, 357–358; and residential segregation, 150–155; *see also* discrimination

racial conflict, 1–3, 409, 419; and migration, 107–109, 115; and residential segregation, 138

racial steering, 152

racial stratification, 414–415

Raleigh, North Carolina, 137

real estate ownership, **289**

Reconstruction, 110–111

redistribution of black population, *see* migration

regional migration, *see* migration

region of birth: and fertility rates, 60, 89–98, **90–91, 95, 96–97, 97, 100;** and migration, *see* migration; and occupational achievement, 275–276, **276**

region of residence: in census vs. monograph samples, **435;** and earnings, 321, 328, 331, **333,** 334–335, 337, **338,** 343–345, **344,** 359, 361; and educational attainment, 195, **196;** and employment, **226, 227,** 229, 237, **238, 239;** and employment-population ratios, **226, 227, 238, 239;** and family income, 166, **182,** 184; and fertility rates, 60, 89–98, **90–91, 95, 97, 100;** of foreign-born population,

378–382, **380–381;** and hours of employment, **226, 227, 238, 239;** and income, 166, **182,** 184, 300–303, **302;** and occupational achievement, 275, **276, 282;** population distribution by: 1790–1985, 104, **105;** and unemployment rates, **226, 227,** 229, 231, 233, 234, **238, 239**

rental property ownership, **289**

reported information: accuracy of, 431

research: on family organization, 161–167, 415; future agenda for, 413; macro-level issues, 413–414; micro-level issues, 416–417; middle-range issues, 414–416

residence, place of: and employment-population ratios, **226, 227, 238, 239;** and hours of employment, **226, 227, 238, 239;** and unemployment rates, **226, 227, 238, 239**

residence, region of, *see* region of residence

residential segregation, 136–157; 411, 418; consequences of, 155–157; economic differences view of, 148–150, **149;** and educational attainment, 148–150, **149,** 157, 205, **207,** 208; emergence of, 136–139; and employment, 246; ethnic homogeneity view of, 145–148, **147;** and income, 148–150, **149;** racial attitude and discrimination practices view of, 150–155; research on, 416; trends in, 139–145, **141, 143, 145**

resilient-adaptive model, 162, 170

respiratory cancer, 41, **42, 43,** 44

retirement income, **287,** 291

Rhode Island, 9

Richmond, Virginia: residential segregation in, 137, **141,** 142

Romania, 68

Rosenwald, Julius, Fund, 22, 31

rural blacks: and migration, *see* migration; and mortality rates, 28–31

Rural Negro, The (Woodson), 29

S

St. Louis, Missouri, 134; residential segregation in, **141, 143, 145, 147**

St. Lucia, 76

salary, 284, **287**

sales occupations: and census allocation, **433;** in census vs. monograph samples, **437;** distribution of workers in, **264, 265,** 270, **272, 273;** earnings from, 271, **272,**

273; and educational attainment, 198, **200;** and fertility rates, **91;** of foreign-born population, **398, 399;** and hourly wages, **318, 319;** and migration, **124, 125**

salpingitis, 21

San Antonio, Texas: public school enrollment in, **207,** 208

San Diego, California: public school enrollment in, **207**

San Francisco, California: foreign-born blacks in, 379, **381;** public school enrollment in, **207,** 208; residential segregation in, **143, 145,** 146, **147,** 150

Scholastic Aptitude Test, 204–205, **205,** 206

school attendance, *see* college attendance; elementary school education; high school attendance; post-graduate education; primary school education

school enrollment, **203,** 203–205, **207,** 208, **255,** 411; and employment, 236, 237, **238, 239, 240;** and hours of employment, **330**

school lunches, 284, 292, 293, **293**

Seattle, Washington, 243–244

segregation: occupational, 116, 257–261, 270, 278, 280; in public schools, 157, 205, **207,** 208; residential, *see* residential segregation

self-concept, black, 417

self-employed: earnings of, 284; foreign-born population as, **396, 397;** income of, **287,** 288

separated men: foreign-born, **384;** occupational achievement of, **276**

separated women: earnings of, **341;** fertility rates for, 89, **90, 95;** foreign-born, **389;** occupational achievement of, **276**

service occupations: and census allocation, **433;** in census vs. monograph samples, **437;** distribution of workers in, **264, 265,** 272, **273;** and educational attainment, 198, **201;** and fertility rates, **91,** 93; of foreign-born population, **398, 399;** and household type, 172, 179; and migration, **124, 125**

sex ratios, 170

sex structure of foreign-born population, 384, **386,** 387

sexual differences: in earnings, **287,** 331, **332,** 340; in educational attainment, 193, **193;** in expectations of life at birth, **26;** in

income, 294, 299, 309; in mortality rates, 35, **36;** in occupational achievement, **264, 265,** 267, **268, 269,** 270, **272, 273,** 278, **279, 282;** in unemployment rates, 213–218, **214, 216, 240, 251**
sexual discrimination, 317, 340
Shadow of the Plantation (Johnson), 29
sickle cell anemia, 55
single men: foreign-born, 388, **389;** occupational achievement of, **276;** *see also* male-headed households; single-parent families
single-parent families, 163–165; age distribution of children in, 170; distribution by race, 168, **169;** income of, 167
single women: earnings of, 341, **341,** 342; foreign-born, 388, **389;** as mothers, 76–80, **78, 79, 81,** 89; occupational achievement of, **276;** *see also* female-headed households; single-parent families
slave breeding, 16
slavery, 8, 10; and contemporary black family life, 162–163; and educational attainment, 189, 195; and family organization, 387; migration during period of, 13, 14, 104, 106–109
slaves: in Colonial period, 8–10; migration of, 104, 106; nutrition of, 15; from the Revolutionary War to the Civil War, 11–16
smoking, 55, 56
Social Darwinists, 55, 111, 257, 414
social mobility, 363–368
social policy, 417–419
Social Security Act of 1935, 23, 31
Social Security benefits, 123, 284, **287,** 290–291
South American immigrants, 368–370, **369, 374,** 375
South Atlantic states: foreign-born blacks in, **380**
South Carolina, 8, 9, 13, 134
South Central states: foreign-born blacks in, **380**
Spanish-speaking people, 382, **383**
standardized tests, 204–205, **205, 206,** 317
standard of living: and mortality rates, 28–33, 55
sterilization, 85, **86**
stock ownership, **289**

stroke, 32, 41, **42, 43,** 44, 45, **46,** 47
study methods, 5
substitutions, census, 426, **427,** 428
suburbanization, 142–143
suicide, 41, **42, 43,** 44, 45, 56
Sweden, 14, 24, 76
syphilis, 21–22

T

technical occupations: and census allocation, **433;** in census vs. monograph samples, **437;** distribution of workers in, 267, **268, 272, 273;** earnings from, 271, **272, 273;** and educational achievement, 198, **200;** and fertility rates, **91;** of foreign-born population, **398, 399;** and migration, **124, 125**
teenagers: fertility rates of, 65, 66, 67, 74–76, **79,** 82–83; out-of-wedlock births to, 76, 79, **79**
tenant farmers, 28, 29, 114
Texas, 109, 113: foreign-born blacks in, **380**
tobacco farming, 13
Tobagan immigrants, **374**
total fertility rates: 1860–1984, 18, **19;** 1940–1984, 58, 61, **62**
transfer payments, 284, 286, **287,** 292–293, **293,** 356–357
transportation occupations, 71; and census allocation, **433;** in census vs. monograph samples, **437;** distribution of workers in, **272, 273;** of foreign-born population, **398, 399**
Trinidadian immigrants, **374**
tuberculosis, 18, 20–21, 28, 40, **41, 42, 43**

U

underemployment, 211
unemployed: classification as, 211
unemployment: and adult men, **226,** 228; and adult women, **227,** 233–234; concept of, 211; determinants of, **226, 227, 238, 239,** 252–253, **255;** duration of, 225, **226, 227,** 229, **238, 239;** and educational attainment, 198, **199;** and foreign-born population, 394, **395;** and household type, **175;** prevalence of, 225, **226, 227, 238, 239;** rates of, *see* unemployment rates
unemployment compensation, 284

unemployment rates, 412; for adult men, 225, **226,** 229–231, **230;** for adult women, **227,** 233–235; and age, 213–218, **214, 216, 226, 227, 238, 239, 252, 255;** and census allocation, **432;** definition of, 211–212; and disability, **226, 227;** and discouraged workers, 211–212; and educational attainment, **226, 227,** 229, **230,** 233, 235, 237, **238, 239, 240;** for foreign-born population, 394, 405; and Gross National Product, 215–217, **216,** 220; and marital status, **226, 227,** 229–231, **230,** 233–234, **238, 239;** and migration, **226, 227,** 248; and minimum wages, 244; and place of residence, **226, 227, 238, 239;** and region of residence, **226, 227,** 229, 231, 233, 234, **238, 239;** sexual differences in, 213–218, **214, 216, 240, 251;** trends in, 213–218; for youth, 236, 237, **238–239, 240**
unions, 116, 257, 259
U.S. Department of Agriculture, 29–30
U.S. Department of Housing and Urban Development (HUD), 152
United States v. *Local 189, United Paperworkers* (1970), 261
U.S. Savings Bond ownership, **289**
U.S. Supreme Court decisions, 410; and education, 189, 191; and residential segregation, 115, 138–139; and states' rights, 111
United Steel Workers v. *Weber* (1979), 261
unmarried mothers, 71, 76–83, **78, 79, 81;** fertility rates for, 89, **90;** birth control use by, 85–88, **86, 88**
urbanization, 134–136, **135**

V

venereal diseases, 18, 21–23
veteran's benefits, 284
Virginia, 8, 138
vitamin consumption, 56

W

wages, *see* hourly wage rates
Wagner Act of 1935, 259
War on Poverty, 356–357
Warren, Michigan, 153

Washington, D.C., 25; foreign-born blacks in, **381,** 382; public school enrollment in, **207,** 208; residential segregation in, 136, 137, 140, **141, 143, 145, 147,** 148–149
Watts, Los Angeles, 1
welfare, 357; and employment, 210, 241–244; and fertility rates, 83; income from, 290, 292; and migration, 123, **124; 125,** 132, 133; and work ethic, 241–244
West Indian immigrants, 362, 372, **374,** 375, 378, 382, 390–391, 393, 403–405, 413; income of, 367–368, 400, 401, 403; in New York City politics, 377; occupational achievement of, 394, 396, 397
white men: crude and age-adjusted mortality rates for, 35, **36;** earnings of, 317, **319,** 321, **322–323,** 329, **330,** 331–334, **332, 333, 336,** 336–337, **338, 344,** 346, **347,** 348, **349, 359, 360, 361;** educational attainment of, 193, **193, 226, 230, 232, 238, 240, 252, 254, 255,** 304–306, **305, 318, 330, 333,** 336, 351, **352, 359, 361;** employment-population ratios for, 222, **223,** 224, 225, **226,** 228, 229, **238;** and fertility rates, 98, **100–101;** foreign-born, **386, 389,** 390, 391, **392,** 394, **395,** 397, **398,** 401, **402, 404;** as heads of household, 197; hourly wages of, 317, **318,** 331–334, **332, 333, 352;** hours of employment of, **226,** 229, **232,** 233, **238, 254, 330;** income amounts of, 283, **287,** 290–291, **315;** income sources of, 286–288, **287;** income trends (general) for, 295–299, **296, 298, 300;** income trends by age for, 306–312, **307, 310–311, 312;** income trends by educational attainment for, 304–306, **305;** income trends by region for, **302;** labor force participation by, 218, **219,** 220, **221,** 222, **252;** life span of, **26,** 46, 53, **53;** migrant, 124–126, **125, 127, 130–131,** 132; mortality rates by age for, 37–40, **38–39;** mortality rates by cause for, 41–44, **42;** and net census undercounts, **423,** 425; occupational achievement of, 263, **264,** 267, **268, 269,** 271, **272, 276,** 278, **279,** 280, **282;** ratio to white women, 170; and syphilis, 23; unemployment determinants for, **252–253, 255;** unemployment rates for, 213, **214, 216,** 217, **226,** 229–231, **230, 238, 251**

whitcs: and comparative status of blacks, 410–413; *see also* white men; white women

white women: and abortion, 82; childless, 68–70, **69**; crude and age-adjusted mortality rates for, 35, **36**; earnings of, 317, **319, 322–325, 330, 332, 333,** 334, **336,** 337, **338,** 339–342, **341, 344,** 347, 348, **349, 352,** 353, **354,** 355, **359, 360, 361**; educational attainment of, 193, **193,** 195, **197, 227, 230, 239, 240, 252, 254, 255, 276, 319, 330,** 333, 340, 351, **352, 359, 361**; and family sizc, 63, **64,** 65; fertility rates and birth control use by, 20, 59, 83–88, **86, 87, 88**; fertility rates and educational attainment of, **91,** 92, 94–97, **95, 97, 100**; fertility rates and employment of, **91,** 92–96, **95, 100**; fertility rates and marital status of, 77, **78,** 79–82, 85–89, **86, 88, 90, 95, 97**; foreign-born, 386, 387, **389,** 390, 391, **392, 394, 395, 397, 399, 404**; as heads of household, 195, **197,** 198; hourly wages of, 317, **319, 332, 333, 334, 352**; hours of employment of, **330**; income amounts of, 283, **287,** 290–291, **315**; income sources of, 286–288, **287**; income trends (general) for, 295, **296, 298,** 299–300, **300**; income trends by age for, 306–309, **307,** 310–311, **312**; income trends by educational attainment for, **305,** 306; income trends by region for, **302,** 303; labor force participation by, 218–220, **219, 221,** 222, 237, **252**; life span of, 14, **26,** 45, **46,** 47, **53, 53**–54; mortality rates by age for, **38–39,** 40–41; mortality rates by cause for, **43,** 44–45; and net census

undercount, **423,** 425; occupational achievement of, **265,** 266, 267, **268, 269,** 271, **273, 276,** 278, **279,** 280, **282**; and prenatal care, 51–52; ratio to white mcn, 170; reduction in childbearing: 1850–1985, **17,** 18; teenage fertility, 74–76; total fertility rates for: 1860–1984, **19**; total fertility rates for: 1940–1984, 61, **62**; unemployment determinants for, **252–253, 255**; unemployment prevalence and duration for, **227**; as unmarried mothers, 77, **78, 79,** 79–82, 89, **90**

widowed men: foreign-born, **389**; occupational achievement of, **276**

widowed women: earnings of, **341**; fertility rates for, 89, **90, 95**; foreign-born, **389**; occupational achievement of, **276**

wife-husband families, *see* husband-wife families

women, *see* black women; white women

work ethic, 241–244, 393, 418

World War I: black occupational achievement during, 258

World War II: black occupational achievement during, 258, 260

Y

youth: employment of, 236–240, **238–239, 240**; unemployment rates for, 236, 237, **238–239, 240**; *see also* teenagers

Z

Zambia, 24